Principles of
Geomorphology

WILLIAM D. THORNBURY
Department of Geology
Indiana University

Principles of

Geomorphology

JOHN WILEY & SONS, INC.

New York · London · Sydney

TENTH PRINTING, NOVEMBER, 1966

Library of Congress Catalog Card Number: 54—7553

Printed in the United States of America

To the memory of

CLYDE A. MALOTT

My former teacher, colleague
and friend
To whom the writer is chiefly
indebted for his interest in
geomorphology

Preface

In writing a textbook on *Principles of Geomorphology,* I have been guided by two firm convictions. The first of these is that world landscapes cannot be properly interpreted unless the many complex factors which have influenced their evolution are fully appreciated. With this in mind, I have included early in the book a chapter on fundamental concepts, despite the fact that some of the principles discussed in this chapter cannot be fully appreciated until much of the book has been read. Secondly, I believe that the practical aspects of geomorphology have not been given the attention which they merit, and with this in mind I have included a chapter on applied geomorphology, which I trust will have some appeal to the "practical geologists." Certain other innovations will be found in the book which I hope will add to its value.

The ever-present problem of publication costs has made it impractical to treat all aspects of the subject as fully as might be desired. References at the ends of chapters will enable students to pursue more fully particular phases of the subject which may appeal to them.

To a large degree the book is patterned upon a course in Principles of Geomorphology which I have taught for several years at Indiana University. Students in this course presumably have had, in addition to beginning geology, courses in mineralogy and petrology.

I wish to extend my sincere thanks to the many individuals and organizations who so willingly supplied illustrative materials. Without their assistance adequate illustration of the book would have been difficult, if not impossible. An effort was made to avoid photographs which have become stereotyped through repeated use, but I was unable to resist a few particularly striking ones. My thanks also go to Warren C. Heisterkamp, John Minton, John Peace, and William J. Wayne for their help in the preparation of illustrations.

Much of my philosophy of what geomorphology should be came from the late C. A. Malott, and many of his ideas appear in the book. To Professor A. O. Woodford of Pomona College, where the first draft was written, I am deeply grateful for a congenial environment in which to work as well as for innumerable suggestions as to how to improve the manuscript, over half of which he read and constructively criticized. Professor George W. White contributed many helpful ideas as to what should be included in the first two and last two chapters. I am particularly indebted to Professor Leland Horberg, who read the entire manuscript and gave invaluable advice as to how to improve many sections. Finally, to my wife, Doris, I express my deep gratitude. Without her encouragement, I would never have undertaken the writing of the book. She also aided greatly in typing and proofreading.

To each of these persons, I extend my deepest thanks with the hope that they will find some slight reward for their efforts in the final results without feeling in any way responsible for shortcomings of the book attributable to the author's failure to make the most of their suggestions.

WILLIAM D. THORNBURY

Bloomington, Indiana
February, 1954

Contents

1 · Backgrounds of Geomorphology

THE SCOPE OF GEOMORPHOLOGY

Definitions. If we defined geomorphology in terms of the three Greek roots from which the word was derived, it would mean "a discourse on earth forms." Generally, it is thought of as "the science of land forms" and it will be so used, although we shall extend it to include submarine forms. As Worcester (1939) defined geomorphology, it is a description and interpretation of the earth's relief features. So defined, it is considerably more comprehensive than the science of land forms, for we may include within its scope a discussion of the origin of major earth forms such as the ocean basins and continental platforms, as well as lesser structural forms such as mountains, plains, and plateaus. Ocean basins and continental platforms are relief features, but it seems that their interpretation belongs to dynamic and structural geology. We shall concern ourselves primarily with the lesser forms developed upon these major relief features.

Designation of the study of land forms as geomorphology has come about as a result of dissatisfaction with the term physiography, which was formerly applied to this subject. Physiography, particularly as used in Europe, includes considerable climatology, meteorology, oceanography, and mathematical geography. Rather than continue the practice, previously common in this country, of restricting physiography to the discussion of land forms, it seemed preferable to have a name for this branch of geology which at least reduces confusion as to its scope.

Geomorphology is primarily geology, despite the fact that some geomorphology is taught both in Europe and in this country as a part of physical geography. In most geography courses land forms are treated rather incidentally as a part of the discussion of the physical environment of man, but emphasis usually is placed upon man's adjustments to and uses of land forms rather than upon land forms per se.

HISTORY OF DEVELOPMENT OF GEOMORPHIC IDEAS

Desirability of an historical background. Serious students should be interested in the growth of scientific thought and in the men who have con-

1

tributed to its advancement. To consider this ancient history which in no way contributes to an appreciation of present thought is actually a short-sighted viewpoint. The historical approach is unequaled in giving the student an insight into the scientific method (inductive logic). At least three distinct benefits result from familiarity with the growth of geomorphic thought. The student gains a better perspective from which to view present-day thinking. He is impressed with the fact that the subject is not static and will more likely keep his mind open to new ideas. Furthermore, he realizes that most ideas which we today accept as self-evident met with resistance when first proposed and were slowly and reluctantly accepted as correct and that perhaps some of the new ideas which we scorn today may ultimately stand the test of time.

Views of the Ancients

As land forms are the most widespread geologic phenomena, speculation as to their origin has gone on since the days of the ancient philosophers. It was not until "the present became the key to the past," however, that the study of geomorphic processes came to have great significance. This was not until near the beginning of the nineteenth century, and it was not until the later decades of that century that the systematic evolution of land forms was comprehended. Although separation of geomorphology as a distinct sub-science came late, many of its basic ideas had an early origin.

A discussion of the development of geomorphic thought may well begin with the Greek and Roman philosophers. Space will not permit a detailed discussion of their ideas, but we shall get a fairly good picture of them if we briefly consider the views of four men: Herodotus, Aristotle, Strabo, and Seneca.

Herodotus (485?–425 B.C.), who is known as the "father of history," is also remembered for some of his geological observations. He recognized the importance of the yearly increments of silt and clay deposited by the Nile, and to him is attributed the statement that "Egypt is the gift of the river." He thought it more likely that earthquakes were responsible for the rending asunder of mountains than that they were the expression of the wrath of the Gods. He noted shells in the hills of Egypt and concluded from their presence there that the sea at some time must have extended over lower Egypt, thus anticipating to some degree the idea of changing sea levels, a matter of great geomorphic significance.

Aristotle (384–322 B.C.) in his writings reflected the thinking of his day. His views on the origin of springs are of interest. He believed that the waters which flowed out of springs consisted of: (*a*) some rainwater which had percolated downward; (*b*) water that had formed within the earth by condensation from air which had entered the earth; and (*c*) water that had condensed within the earth from vapors of uncertain origin. All these waters were held by the mountains as if they were great sponges. The term river was then applied only to running waters fed by springs. Aristotle thought that rainfall might produce a temporary torrent, but doubted that it could maintain the flow of a river. Incidentally, the true explanation of spring waters and the discharge of streams long after periods of rainfall was not understood until Bernard Palissy deduced in 1563 and 1580 and Pierre Perrault in 1674 demonstrated the sufficiency of rainfall to maintain them. Aristotle believed that earthquakes and volcanoes were closely related in origin, and he attributed earthquakes to the effects of the mingling of moist and dry air within the earth. Along with others, Aristotle recognized that the sea covered tracts that were formerly dry land and also the probability that land would reappear where now the sea exists. He alluded to the disappearance of rivers underground. (We would call them sinking creeks today. Greece is a country with much limestone and marble in which features produced by solution by groundwater are common.) He recognized that streams removed material from the land and deposited it as alluvium and cited examples from the Black Sea region where river alluvium had accumulated so rapidly in a matter of 60 years as to necessitate the use of smaller boats.

Strabo (54 B.C.–A.D. 25), who traveled widely and observed carefully, noted examples of local sinking and rise of the land. He considered the Vale of Tempe a result of earthquakes which, along with volcanic activity, were still attributed to the force of winds within the earth's interior. He rightly inferred from the nature of its summit that Mt. Vesuvius was of volcanic origin, although it was never active during his lifetime. He, too, recognized the importance of river alluvium and thought that the delta of a river varied in size according to the nature of the region drained by the river and that the largest deltas are found where the regions drained are extensive and the surface rocks are weak. He observed that some deltas are retarded in their seaward growth by the ebb and flow of the tides.

Seneca (? B.C.–A.D. 65) recognized the local nature of earthquakes, but he still believed that they were an effect of the internal struggle of subterranean winds. He likewise held the idea that rainfall was insufficient to account for rivers, although he recognized the power of streams to abrade their valleys. Thus, though the concept that streams make their valleys was in a sense born, the manifold implications of this fact were not sensed until many centuries later. The ancients seem to have realized that there is a genetic relationship between earthquakes and deformation of the earth's crust, although they confused cause and effect and thought that earthquakes caused the deformation.

The Dawn of Modern Geomorphic Ideas

During the many centuries which followed the decline of the Roman Empire there was little or no scientific thinking in Europe. Such knowledge as survived was largely in monasteries, but it was not natural science. Some survival of learning persisted in Arabia, and we find certain ideas expressed there that have a modern flavor. Avicenna (Ibn-Sina, 980–1037) held views upon the origin of mountains which divided them into two classes, those produced by "uplifting of the ground, such as takes place in earthquakes," and those which result "from the effects of running water and wind in hollowing out valleys in soft rocks." Thus the concept of mountains resulting from differential erosion was expressed. The idea of slow erosion over long periods of time was also held by him. Such views have a decidedly modern stamp, but they made no imprint on the thinking in western Europe, if they were even known there. As Fenneman (1939) has pointed out, so little progress was made in Europe from the days of the first century A.D. until the opening of the sixteenth century that little need be said about it. Indeed it may be questioned whether the slight progress made by the ancients in the explanation of the surface features of the earth had any influence upon the eventual emergence of a science of land forms. What the ancients had thought was largely lost and geologic ideas had to evolve anew.

During the fifteenth, sixteenth, and seventeenth centuries land forms were largely explained in terms of the then-prevailing philosophy of *catastrophism,* according to which the features of the earth were either specially created or were the result of violent cataclysms which produced sudden and marked changes in the surface of the earth. As long as the earth's age was measured in a few thousand years there was not much chance for importance to be attached to geologic processes so slow that little change could be noted in a lifetime.

Predecessors of Hutton. The concept of a wasting land, in contrast to the idea of permanence of the landscape as envisaged by early thinkers, is fundamental in modern geomorphic thought. We have seen that some of the ancient philosophers had the idea of land destruction by erosional processes but the time was not ripe for carrying the idea to its logical conclusion. Space will not permit a detailed discussion of the long and slow development of geologic thinking which finally laid the groundwork for the father of modern geomorphic thought, James Hutton, but a few of the men who blazed the trail may be mentioned.

Leonardo da Vinci (1452–1519) can be taken as one of the first representatives of the formative period in modern geologic thinking. He recognized that valleys were cut by streams and that streams carried materials from one part of the earth and deposited them elsewhere. Nicolaus Steno (1638–1687), a Dane who spent much of his life in Italy, also recognized that running water was probably the chief agent in the sculpturing of the earth's surface.

The Frenchman Buffon (1707–1788) recognized the powerful erosive ability of streams to destroy the land and thought that the land would eventually be reduced to sea level. He was one of the first to suggest that the age of the earth was not to be measured in terms of a few thousand years and he suggested that 6 days of creation in the Biblical narrative were not days in the ordinary sense of the word. He was forced, however, to recant these heretical views.

The Italian Targioni-Tozetti (1712–1784) was another who recognized the evidence of stream erosion. He also had the idea that the irregular courses of streams were related to the difference in the rocks in which they were being cut and thus recognized the principle of differential erosion as related to varying geologic materials and structures.

The Frenchman Guetthard (1715–1786) was a geologist in the fullest sense of the word, although the terms geology and geologist were as yet unused. He discussed the degradation of mountains by streams and recognized that not all of the material removed by streams was immediately carried to the sea but that a considerable part of it went into the building of floodplains. He believed the sea to be an even more powerful destroyer of the land than streams and cited the rapid destruction by the sea of the chalk cliffs of northern France in support of his contention. He grasped the fundamental principles of denudation, but unfortunately his ideas were largely buried in many volumes of cumbrous writing. Guetthard is most remembered for his recognition of the volcanic origin of the numerous hills or *puys* in the Auvergne district of central France.

Desmarest (1725–1815), another Frenchman, merits more recognition than is generally given him. By sound reasoning and citation of specific examples, he propounded the idea that the valleys of central France were the products of the streams occupying them. He was apparently the first to attempt to trace the development of a landscape through successive stages of evolution.

The Swiss De Saussure (1740–1799), the first to use the terms geology and geologist in their present sense, was the first great student of the Alps. He was much impressed with the power of streams to sculpture mountains and contended that valleys were produced by the streams which flow in them. He also recognized the ability of glaciers to carry on erosional work. Although he did not always interpret correctly the things that he saw, he amassed a great stock of information, upon which Hutton later drew heavily in developing the doctrine of uniformitarianism.

These three men, Guetthard, Desmarest, and De Saussure, sometimes alluded to as the French school, more, perhaps, than any others paved the way for Hutton, who gratefully acknowledged their help. In America, these men have failed to get the recognition that they deserve, largely because we have not bothered to read French geologic literature.

Hutton and Playfair. James Hutton (1726–1797) was born in Edinburgh, Scotland, and was educated as a physician, but his interests were in science, especially chemistry and geology. He is most famous, perhaps, for the role he played as a leader of a group known as the Plutonists, which maintained that granite was of igneous origin in opposition to the Wernerian school, known as the Neptunists, which contended that granite was a chemical precipitate. He also recognized the evidence for metamorphism of rocks, but his greatest contribution came in expounding the concept that "the present is the key to the past," thus establishing the doctrine of *uniformitarianism* in opposition to that of catastrophism. Hutton's views were first presented in a paper, read by him before the Royal Society of Edinburgh in 1785, which three years later appeared in print in Volume 1 of the *Transactions* of that society under the title, "Theory of the Earth; or an Investigation of the Laws Observable in the Composition, Dissolution and Restoration of Land upon the Globe." In 1795, his views appeared in an expanded form in a two-volume book entitled *Theory of the Earth, with Proofs and Illustrations*. This edition was limited in number and was expensive. It is likely that Hutton's ideas might have been lost or much delayed in acceptance had it not been that his friend, John Playfair (1748–1819), a professor of mathematics and philosophy at Edinburgh, after the death of Hutton, published in 1802 his *Illustrations of the Huttonian Theory of the Earth,* which elaborated and expanded Hutton's principles in a form of scientific prose which has rarely been equaled for clarity and beauty of expression. This book was also smaller and cheaper than Hutton's and hence was more widely read. In it Playfair presented Hutton's ideas and conclusions so clearly that their impact was enormous, particularly upon Sir Charles Lyell, who was later to become the great exponent of uniformitarianism.

Hutton projected the results of the processes, which he observed in operation, both into the past and future. He was impressed with the evidence of land wastage by both mechanical and chemical processes. Others, before Hutton, had observed this, but they failed, with the possible exception of Desmarest, to see the implications that were envisioned by Hutton. The concept of a river system and its geomorphic significance has never been more beautifully expressed than by Playfair,* when he stated:

* Every student of land forms ought to read parts of Playfair's book. It is available in most large geology libraries, and, if not, microfilm copies can be procured at reason-

Every river appears to consist of a main trunk, fed from a variety of branches, each running in a valley proportioned to its size, and all of them together forming a system of valleys, communicating with one another, and having such a nice adjustment of their declivities, that none of them join the principal valley, either on too high or too low a level, a circumstance which would be infinitely improbable if each of these valleys were not the work of the stream which flows in it.

If, indeed, a river consisted of a single stream without branches, running in a straight valley, it might be supposed that some great concussion, or some powerful torrent, had opened at once the channel by which its waters are conducted to the ocean; but, when the usual form of a river is considered, the trunk divided into many branches, which rise at great distance from one another, and these again subdivided into an infinity of smaller ramifications, it becomes strongly impressed upon the mind that all these channels have been cut by the waters themselves; that they have been slowly dug out by the washing and erosion of the land; and that it is by the repeated touches of the same instrument that this curious assemblage of lines has been engraved so deeply on the surface of the globe.

The basic concepts of our modern ideas on earth sculpture are to be found in Hutton's theory. Hutton recognized marine as well as fluvial erosion, but he gave most attention to the development of valleys by streams. Like most great prophets, Hutton was ahead of his times. Three-quarters of a century elapsed before a group of American geologists working in western United States provided the clinching arguments for it. Playfair, from his familiarity with De Saussure's writings, proclaimed the ability of glaciers to erode their valleys deeply, and he seems to have been the first to have suggested a former, much greater extension of the Alpine glaciers in Switzerland, although he did not recognize the effects of glaciation in Scotland.

After Hutton

Developments in Europe. Sir Charles Lyell (1797–1875) became the great exponent of uniformitarianism, and through his series of textbooks he probably did more to advance this principle and geologic knowledge in general than any other man. Even he could not accept the full implication of stream erosion as conceived by Hutton, for, in the eleventh edition of his *Principles of Geology* (1872), we find the following statement: "It is probable that few great valleys have been excavated in any part of the world by rain and running water alone. During some part of their formation, subterranean movements have lent their aid in accelerating the process of erosion."

One of the most significant developments of the nineteenth century, in Europe, was the recognition of the evidence for an ice age during which ice

able costs. Hutton's book is exceedingly rare, but microfilm copies of it are available to libraries.

sheets covered much of northern Europe. The man who is usually given the credit for establishing this fact is Louis Agassiz (1807–1873). Perhaps he does deserve the major credit for establishing its validity, but several men to whom due credit is seldom given anticipated his conclusions. The idea that the Alpine glaciers had once been more extensive was common enough with the Swiss peasants, who observed moraines in the valleys below the termini of present-day glaciers. Playfair on a trip to the Jura Mountains in 1815 recognized the possibility that large boulders present there may have been transported by Alpine glaciers. In 1821, a Swiss engineer named Venetz stated his belief that Alpine glaciers had once been much more extensive, and, in 1824, Esmark, in Norway, expressed the same idea regarding the glaciers of Norway. In 1829, Venetz put forth the view that the glaciers probably extended down onto the plains below. Bernhardi published in Germany, in 1832, a paper in which he proposed the former existence of continental ice sheets. A colleague of Venetz, Jean de Charpentier, wrote a paper in 1834 in which he agreed with Venetz's conclusions. It was largely through Charpentier that Agassiz became interested in glaciation. He was skeptical about the conclusions that had been drawn but agreed to visit the area in which Venetz and Charpentier had been working. In 1836, he visited the area, and, although he arrived a skeptic, he left a convert to Charpentier's ideas. He became convinced that they had not gone far enough in their conclusions. In 1837, he read a paper before the Helvetic Society, the substance of which was published in 1840 as *Études sur les glaciers,* in which his concept of a great ice age was proclaimed. Some coolness developed between the two men because Charpentier thought that Agassiz should have delayed publication of his paper until Charpentier's paper had appeared, which was not until 1841. Although some question may well exist as to whether Agassiz should be proclaimed the father of the concept of continental glaciation, there is no doubt that it was to a large degree through his untiring efforts that the fact of the great ice age was finally acknowledged. By 1870 its validity was fairly widely accepted and multiplicity of glaciation was coming to be recognized, although opposition to the idea continued until as late as 1905.

The importance of marine abrasion was stressed by such geologists as Sir Andrew Ramsay (1814–1891) and Baron Ferdinand von Richthofen (1833–1905). Ramsay described what he believed to be a plain of marine abrasion in the highlands of Wales and southwest England, and Richthofen supported the idea of the importance of marine erosion from evidence observed during his travels in China.

Jukes, in 1862, presented what has become a classical paper on the rivers of southern Ireland in which he recognized the existence of two main types of streams, transverse streams, which flow across geologic structure, and longitudinal streams, which develop along belts of weaker rock parallel to the structure or strike of rock strata. He believed that longitudinal streams developed later than transverse streams and, hence, should be recognized as subsequent streams. These ideas, as we shall see later, have become basic in the philosophy of stream development. He also recognized the fact that one stream could apparently deprive another stream of part of its drainage by a process that we now refer to as stream piracy.

Textbooks which might be considered geomorphic in scope began to appear in the latter third of the nineteenth century. In 1869, Peschel attempted to bring together the principles of land form development into systematic form and attempted what was essentially a genetic discussion of them. Richthofen, in 1886, succeeded to an even greater degree in doing this. A. Penck, in 1894, published his *Morphologie der Erdoberfläche,* which was largely a genetic treatment of land forms. Thus, in Europe, by the close of the last century a sufficient amount of knowledge had accumulated to give rise to a special branch of geology called physiographic geology.

Developments in North America. Zittel (1901) has referred to the period between 1790 and 1820 as "the heroic age of geology." We might well refer to the period between 1875 and 1900 as "the heroic age in American geomorphology," for during this quarter century there evolved most of the grand concepts in this branch of geology. To a large degree these were the outgrowths directly or indirectly of the work of a group of geologists who were connected with the series of geological surveys of the western United States initiated after the Civil War. Three men in particular may be mentioned who did pioneer thinking in the field of geomorphology. They were Major J. W. Powell (1834–1902), G. K. Gilbert (1843–1918), and C. E. Dutton (1841–1912). These men, along with others, collectively laid the groundwork upon which W. M. Davis later was to build the concept of a geomorphic cycle.

It is not too much to say that Major Powell, the one-armed Civil War veteran and first conqueror of the treacherous rapids of the Colorado Canyon, from his work in the Colorado Plateaus and Uinta Mountains laid the foundation for the American school of geomorphology. Davis (1915) in describing Powell's contributions stated that he "was in more senses than one a scientific frontiersman. His life reveals the energetic working of a vigorous and independent personality, not trammeled by traditional methods and not so deeply versed in the history, the content, and the technique of the sciences as to be guided by them, but impelled to the rapid discovery of new principles by the inspiration of previously unexplored surroundings."

Powell from his studies in the Uinta Mountains was impressed by the importance of geologic structure as a basis for classification of land forms. He devoted much attention to the results of stream erosion and proposed two classifications of stream valleys: (1) based upon the relationships between valleys and the strata which they cross and (2) a classification of valleys according to their origin.   In this latter classification he recognized antecedent, consequent, and superimposed valleys, terms which are still widely used in describing valleys.   Probably the most widely applied of Powell's generalizations is his concept of a limiting level of land reduction, which he called base level.   Previous to Powell no one had ventured to carry the carving of the land by rain and streams beyond what today we would call late maturity in the geomorphic cycle, but Powell recognized that the processes of erosion operating undisturbed upon the land would eventually reduce it to a lowland little above sea level.   It remained for Davis to suggest later the name peneplain for such an area, but the idea of the peneplain was to a large degree anticipated by Powell.   He further recognized that the great unconformities in the rocks in the Grand Canyon of the Colorado recorded geologically ancient periods of land erosion.   Powell also noted the geomorphic differences between scarps resulting from erosion and those produced by displacement of rocks (fault scarps, as we call them today).   He recognized that stream divides migrate, but it remained for Gilbert to realize the full implications of this fact.

G. K. Gilbert merits recognition as the first true geomorphologist produced in this country.   Although his interests were catholic, he was above all interested in physiographic geology.   To him we are indebted for a keen analysis of the processes of subaerial erosion and the many modifications which valleys undergo as streams erode the land.   He particularly recognized the importance of lateral planation by streams in the development of valleys.   His study of hydraulic mining debris in California represented one of the first attempts at a quantitative approach to the relations between stream load and such factors as river volume, velocity, and gradient.   His interpretation of the Pleistocene history of Lake Bonneville, the predecessor of Great Salt Lake, from a study of its shore lines and outlets remains one of the classics in American geology.   Equally famous is his explanation of the Henry Mountains of Utah as the result of the erosion of intrusive bodies which he called laccoliths.   He was the first to cite geomorphic evidence for a fault-block origin of the topography of the Great Basin region.

Dutton is remembered most for his penetrating analysis of individual land forms and his recognition of evidence in the Colorado Plateau area of a period of land reduction preceding the cutting of the present canyons when the landscape had been reduced to one of low relief.   This erosional epoch he designated as "the Great Denudation."

Davis—"the great definer and analyst."   The impact on geomorphology of W. M. Davis (1850–1934) was greater than that of any other one man. The Davisian school of geomorphology and the American School are practically synonymous terms.   Davis was basically a great definer, analyst, and systematizer. Before his time, geomorphic descriptions were largely in empirical terms. Even more important than the many new concepts which he introduced was the fact that he breathed new life into geomorphology through introduction of his genetic method of land form description. He will probably be remembered longest for his concept of the geomorphic cycle, an idea perhaps vaguely glimpsed by Desmarest, which in its simplest analysis is the idea that in the evolution of landscapes there is a systematic sequence of land forms which makes possible the recognition of stages of development, a sequence that he designated as youth, maturity, and old age. His idea that differences in land forms are largely explainable in terms of differences in geologic structure, geomorphic processes, and stage of development has become firmly rooted in the thinking of most students of land forms.

During the 1920's and 1930's Walther Penck and his followers staged a minor revolt against some of Davis's ideas and found some adherents. Davis in much of his development of the idea of the geomorphic cycle as effected by running water assumed that relatively rapid initial uplift of the land might be followed by a period of stillstand which permitted a cycle to run its course, culminating in the reduction of the land to a nearly featureless condition. Penck and others have maintained that this is not the normal sequence but that more commonly rise of the land at the beginning of a period of uplift is extremely slow and is followed by an accelerated rate of uplift which would prevent the landscape from passing through stages of development that would terminate in a region of low relief. Several American geologists, particularly from the Pacific coast "mobile belt," have been skeptical of the postulation that the earth's surface is ever stable long enough to allow a cycle to proceed to completion.

Despite objections which have arisen to some of Davis's ideas, it can hardly be denied that geomorphology will probably retain his stamp longer than that of any other single person.

Fenneman (1939) briefly described the various stages through which geomorphic thinking has passed in these words:

The understanding of the work of rain and running water may be considered in three stages or in four if we count the primitive conception of "everlasting hills" which even yet dominates the thinking of many people · · ·. Three advances followed. There appeared first the mere universal fact of degradation as known to some men of the ancient world and others down to the late eighteenth century. Then came the daring proposition that streams make their own valleys. It sounds very simple, but it means that topography is, in the main, carved out and not built up. Some Greeks, Romans, and Arabs saw this, and James Hutton, who died in 1797, saw it clearly. His friend and interpreter, John Playfair, expressed it in language which has never been excelled, but which was a little too sweeping for Lyell, or for us. Even in the geological world (so far as there was one) this principle did not cease to be debated before the time of the Civil War. It barely had time to take its place among the fundamental data of the science when the third stage arrived, in which moving water does not act aimlessly, carving valleys at haphazard, and leaving hills distributed fortuitously, but works to a pattern whose specifications are as distinctive as the sutures of a nautilus or the venation of a leaf. This is the stage of modern physiography or geomorphology.

Certainly Davis played the major role in the attainment of this latest stage.

From this sketch of the development of geomorphic ideas, two things stand out. The first is what we may call the time element. Many ideas failed of acceptance at the time of their proposal because they were "ahead of their time." As an example of this, we may take the several centuries preceding Hutton. Several men (Palissy, Perrault, Buffon, Desmarest, and others) had the germs of modern ideas, but their ideas could not be carried to a logical conclusion or expect a receptive hearing as long as the "intellectual climate" of the time was unfavorable. Acceptance of the ability of erosional processes to shape the earth's surface was impossible so long as the age of the earth was considered to be about 6000 years. Not until after the birth of the inductive method with its dependence upon observation and experimentation rather than upon accepted authority was there much chance for such ideas as were held by Hutton to receive serious consideration.

Secondly, we may note that, although we commonly attribute some new concept to one man, we find that almost always the groundwork for it had been laid by others, who too often do not receive the credit that they deserve. It may be convenient to think of the development of geologic thinking in terms of a few great men, but we should not lose sight of the fact that it has come about as the result of contributions from innumerable individuals.

RECENT TRENDS IN GEOMORPHOLOGY

Some of the significant trends of the past few decades are: (1) a tendency for geomorphology, at least in America, to become more strictly geological than geographical as a result of (*a*) an increasing application to geomorphic studies of other phases of geology, e.g., mineralogy in the study of weathering, stratigraphic methods in paleogeomorphology, and paleontologic techniques in the study of glacial deposits, and (*b*) a decline of interest among geographers in physical geography as they give increasing emphasis to human geography; (2) the development of regional geomorphology, which attempts to divide the continents into areas of similar geomorphic features and history; (3) an increasing recognition of the practical applications of geomorphic principles to such fields as groundwater geology, soil science, and engineering geology; and (4) the dawn of the quantitative and experimental stage with attempts to apply the laws of hydrodynamics to a better understanding of

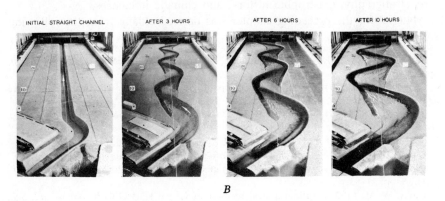

FIG. 1.1.   *A*. Model of a typical bend in the Mississippi River in which simulated articulated concrete mattress-type revetment is being tested as part of a channel stabilization program.   (Corps of Engineers photo.)   *B*. Laboratory study of stream meandering. (Corps of Engineers photo.)

the geomorphic processes. Experimental laboratories have been set up in this country and abroad at which attempts have been made to determine more accurately the application of hydraulic laws to wave, current, and stream action. More of the work done at these laboratories has been carried on by engineers than by geologists with the result that the impact of some of the work has not yet been fully appreciated by geomorphologists.

No one questions the need for more quantitative work in the field of geomorphology. Too many conclusions have been drawn upon inadequate quantitative data. A certain danger, which should be recognized and guarded against, does exist in the cry from some for a quantitative geomorphology. We should be careful that geomorphology does not become so immersed in mathematics, physics, and chemistry that it ceases to be the study of land forms. Sometimes elaborate formulae and equations may appear to add erudition to a discussion but contribute little to the real interpretation of the features described. We should certainly avoid making a fetish of mathematical discussion. It is indeed questionable whether the many variable factors which are involved in the origin of complex landscapes can ever be reduced to mathematical equations. As Baulig (1950) has stated, "the laws of geomorphology are complex, relative and rarely susceptible of numerical expression." It is to be hoped that a knowledge of mathematics, physics, and chemistry never becomes more essential to an understanding of geomorphic discussions than a sound appreciation of lithology, geologic structure, stratigraphy, diastrophic history, and climatic influences.

Material for the preceding chapter was derived chiefly from the following sources.

Bailey, E. B. (1934). The interpretation of Scottish scenery, *Scot. Geog. Mag., 50,* pp. 308–330.

Baulig, Henri (1950). Essais de géomorphologie, *Publs. de la faculté de l'université de Strasbourg, Paris,* p. 35.

Baulig, Henri (1950). William Morris Davis: Master of method, *Assoc. Am. Geog., Ann., 40,* pp. 188–195.

Bryan, Kirk (1941). Physiography, *Geol. Soc. Am., 50th Ann. Vol.,* pp. 3–15.

Darrah, W. C. (1951). *Powell of the Colorado,* Princeton University Press.

Davis, W. M. (1915). Biographical memoir of John Wesley Powell, *Nat. Acad. Sci., Mem. 8,* pp. 11–83.

Davis, W. M. (1922). Biographical memoir of Grove Karl Gilbert, *Nat. Acad. Sci., Mem. 21,* pp. 1–303.

Fenneman, N. M. (1939). The rise of physiography, *Geol. Soc. Am., Bull. 50,* pp. 349–360.

Fenton, C. L., and M. A. Fenton (1945). *The Story of the Great Geologists,* Doubleday, Doran and Co., New York.

Flint, R. F. (1947). *Glacial Geology and the Pleistocene Epoch,* John Wiley and Sons, New York.

Geikie, Sir Archibald (1905). *The Founders of Geology,* Macmillan and Co. Ltd., London.

Gregory, H. E. (1918). A century of geology.—Steps of progress in the interpretation of land forms, *Am. J. Sci., 4th ser., 46*, pp. 104–132. Also in *A Century of Science in America,* pp. 122–152, Yale University Press, New Haven.

Longwell, C. R. (1941). The development of the sciences, Chapter 5, in *Geology,* pp. 147–196, Yale University Press, New Haven.

Lyell, Sir Charles (1872). *Principles of Geology,* 11th ed., Chapters 2, 3, and 4, D. Appleton and Co., New York.

Mather, K. F., and S. L. Mason (1939). *A Source Book of Geology,* McGraw-Hill Book Co., New York.

Merrill, G. P. (1924). *The First One Hundred Years of American Geology,* Yale University Press, New Haven.

Playfair, John (1802). *Illustrations of the Huttonian Theory of the Earth,* William Creech, Edinburgh.

Woodward, H. B. (1911). *History of Geology,* Watts and Co., London.

Zittel, Karl von (1901). *History of Geology and Paleontology,* Walter Scott, London.

2 · Some Fundamental Concepts

We shall attempt in this chapter to establish a few basic concepts which should be grasped by students if, in their study of subsequent chapters, they hope to go beyond rote memorization of facts. The concepts discussed in this chapter are not the only ones used in the interpretation of landscapes, but if these few are even partially grasped they will aid enormously in the evaluation of much that follows.

Concept 1. *The same physical processes and laws that operate today operated throughout geologic time, although not necessarily always with the same intensity as now.*

This is the great underlying principle of modern geology and is known as the *principle of uniformitarianism.* It was first enunciated by Hutton in 1785, beautifully restated by Playfair in 1802, and popularized by Lyell in the numerous editions of his *Principles of Geology.* Hutton taught that "the present is the key to the past," but he applied this principle somewhat too rigidly and argued that geologic processes operated throughout geologic time with the same intensity as now. We know now that this is not true. Glaciers were much more significant during the Pleistocene and during other periods of geologic time than now; world climates have not always been distributed as they now are, and, thus, regions that are now humid have been desert and areas now desert have been humid; periods of crustal instability seem to have separated periods of relative crustal stability, although there are some who doubt this; and there were times when vulcanism was more important than now. Numerous other examples could be cited to show that the intensity of various geologic processes has varied through geologic time, but there is no reason to believe that streams did not cut valleys in the past as they do now; that the more numerous and more extensive valley glaciers of the Pleistocene behaved any differently from existing glaciers; that the winds which deposited the Navajo sandstone during Jurassic times obeyed any different laws from those which control wind movements today. Groundwater opened up solutional passageways in limestones and other soluble rocks and formed surface depressions which we now call sinkholes during the Permian and Pennsylvanian periods as it does today in many parts of the world. Without the principle of uniformitarianism there could hardly be a science of geology that was more than pure description.

Concept 2. *Geologic structure is a dominant control factor in the evolution of land forms and is reflected in them.*

Most students who have had an elementary course in physical geology or geography are likely to have been exposed to the teaching of W. M. Davis

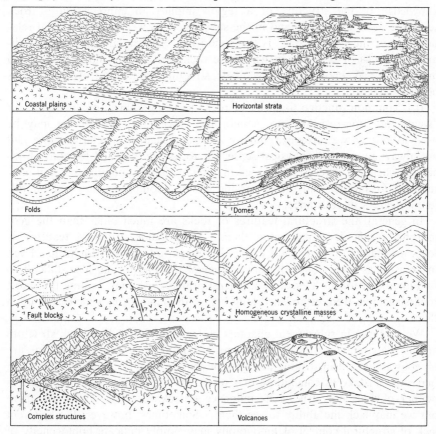

FIG. 2.1. Diagram showing how varying geologic structure and lithology affect the morphology of landscapes. (After A. N. Strahler, *Physical Geography,* John Wiley & Sons.)

that structure, process, and stage are the major control factors in the evolution of land forms. It is with the first of this trilogy that we now concern ourselves. The term *structure* as it is used in geomorphology is not applied in the narrow sense of such rock features as folds, faults, and unconformities but it includes all those ways in which the earth materials out of which land forms are carved differ from one another in their physical and chemical attributes. It includes such phenomena as rock attitudes; the presence or absence of joints, bedding planes, faults, and folds; rock massiveness; the physical hardness of the constituent minerals; the susceptibility of the mineral constituents to chemical alteration; the permeability or impermeability of rocks; and various other ways by which the rocks of the earth's crust differ from one another. The term structure also has stratigraphic implications, and knowledge of the structure of a region implies an appreciation of rock sequence, both in outcrop and in the subsurface, as well as the regional relationships of the rock strata. Is the region one of essentially horizontal sedimentary rocks or is it one in which the rocks are steeply dipping or folded or faulted? A knowledge of geologic structure in the narrow sense thus becomes essential.

It is common practice to speak of rocks as being "hard" or "resistant" or "weak" or "non-resistant" to geomorphic processes. Such terms may be used so long as we recognize that we are using them in a relative sense and not always in a strictly physical sense, for rocks are attacked by both physical and chemical processes. A rock may be resistant to one process and non-resistant to another and under varying climatic conditions may exhibit different degrees of resistance.

In general, the structural features of rocks are much older than the geomorphic forms developed upon them. Such major structural features as folds and faults may go back to far distant periods of diastrophism. Even in areas of as recent diastrophism as that of the Pleistocene it is difficult to find uneroded folds. Hence as a general principle we may assume that most rock structures were established long before the land forms which exist upon them.

We shall see in subsequent chapters many examples of how rock structure affects the characteristics of land forms. Commonly these relationships are obvious and result in striking topographic features. We should not, however, make the mistake of concluding that where the effect of geologic structure is neither obvious nor striking its influence is lacking. The effects are there but we may lack the ability to see them. Sometimes, however, the apparent lack of structural control of topography may simply indicate homogeneity of structure with resultant homogeneity of topography or may result from the large scale of the structural units. It is perhaps not going too far to say that no variation in rock structure is too slight to have significance over a sufficient span of geologic time. The increasing application of geomorphic interpretation to aerial photographs is likely to bring us to a fuller appreciation of the broad applicability of this principle.

Concept 3. *Geomorphic processes leave their distinctive imprint upon land forms, and each geomorphic process develops its own characteristic assemblage of land forms.*

The term *process* applies to the many physical and chemical ways by which the earth's surface undergoes modification. Some processes, such as diastrophism and vulcanism, originate from forces within the earth's crust and have been designated by Penck as *endogenic,* whereas others, such as weathering, mass-wasting, and erosion, result from external forces and have been called *exogenic* in nature. In general, the endogenic processes tend to build up or restore areas which have been worn down by the exogenic processes; otherwise, the earth's surface would eventually become featureless. The concept of geomorphic processes operating upon the earth's crust is not a new one. Even the ancients recognized it to some degree, but the idea that the individual processes leave their stamp upon the earth's surface is rather recent. Just as species of plants and animals have their diagnostic characteristics, so land forms have their individual distinguishing features dependent upon the geomorphic process responsible for their development. Floodplains, alluvial fans, and deltas are products of stream action; sinkholes and caverns are produced by groundwater; and end moraines and drumlins in a region attest to the former existence of glaciers in that area.

The simple fact that individual geomorphic processes do produce distinctive land features makes possible a *genetic classification* of land forms. The recognition of this fact and his insistence upon its superiority to other types of landscape description was one of Davis's important contributions to geomorphology; for it changed the subject from one in which land forms were classified upon a purely morphological basis without regard to interpretations which could be made from them as to their geomorphic history. It doubtless would be possible to describe the multitudinous land forms of the world in terms of perhaps a dozen or fifteen primary or elemental forms such as plain, slope, scarp, mount, ridge, table, column, depression, valley, trough, col, niche, arch, hole, and cavern; but in general such terms tell little or nothing about the origins of the forms or the geologic history of the regions in which they exist. How much more illuminating are such terms as floodplain, fault scarp, sinkhole, sand dune, and wave-cut bench. Even though they are in part descriptive they have genetic implications.

A proper appreciation of the significance of process in land form evolution not only gives a better picture of how individual land forms develop but also emphasizes the genetic relationships of land form assemblages. Land forms are not haphazardly developed with respect to one another but certain forms may be expected to be associated with each other. Thus, the concept of certain types of terrain becomes basic in the thinking of a geomorphologist.

Knowing that certain forms are present, he should be able to anticipate to a considerable degree other forms which may be expected to be present because of their genetic relationships to one another. Even though a hill or mountain may obstruct his vision, in his mind's eye he may be able to see through it and visualize the topography on the other side. Terrain analysis, which has become so important in modern warfare and other fields, is more dependent upon this keen appreciation of land form assemblages than upon any other one thing.

Concept 4. *As the different erosional agencies act upon the earth's surface there is produced a sequence of land forms having distinctive characteristics at the successive stages of their development.*

That land forms possess distinctive characteristics depending upon the stage of their development is the idea that Davis stressed most. A logical outgrowth of this principle is the concept of a *geomorphic cycle,* which we may define as the various changes in surface configuration which a land mass undergoes as the processes of land sculpture act upon it. This concept, if properly conceived and applied, is one of the most useful tools in geomorphic interpretation. The basic idea is that, starting with a given type of initial surface underlain by a certain type of geologic structure, the operation of geomorphic processes upon this mass results in a sequential rather than haphazard development of land forms. The metaphorical terms, *youth, maturity,* and *old age,* are commonly used to designate the stages of development, and it is customary to add such qualifying adjectives as early and late to designate substages.

Although sequence of time is also implied in a geomorphic cycle it is in a relative rather than in an absolute sense. It should be emphasized that there is no implication that these stages are of equal duration nor that separate regions which have attained the same stage of development have necessarily required the same length of time for attainment of that stage, for there are numerous factors which may cause the rate of degradation to vary from place to place. Neither is it implied that two areas in the same stage would be similar in the details of their topography. This could be expected only if the initial surface, the geologic structure, and the climatic and diastrophic conditions in the two areas were the same. Infinite variety may be expected in the details of topographic forms, but this does not negate the idea that there is a systematic evolution of land forms which makes possible a recognition of their stage of development.

Much more progress has been made in working out the details of the geomorphic cycle as carried out by running water than by any other geomorphic agency, but the basic idea seems to apply to the other erosional agencies with the possible exception of glaciers. It seems to apply to a limited degree to erosion by mountain glaciers but as yet no one has seen its application to the work of ice sheets.

It should be noted that to explain land forms in terms of the three control factors, structure, process, and stage, as these terms were used by Davis, the diastrophic history of a region while undergoing degradation must be encompassed either under structure or process. Usually it is considered under process, but much might be said for considering this factor equally important as the other three. Particularly in such tectonically unstable regions as California and New Zealand this factor may be a critical one in the evolution of land forms and may actually seem to obscure any tendency toward sequential development. Under conditions of continuing uplift or rapidly repeated intermittent uplift a landscape may be kept perpetually youthful, mature, or old without running the course of a normal cycle. It is significant that it has been to a large degree from the geologists in these instable belts that skepticism or doubt as to the validity of the geomorphic cycle has arisen. Although it must be admitted that the proponents of the idea of cyclical development of land forms have in many instances been too enthusiastic about its ability to explain landscapes, to deny its reality is to deprive us of one of the most useful and fundamental concepts in geomorphology.

A necessary corollary to the concept of a completed geomorphic cycle is that of a partial cycle. In fact, a partial cycle is far more likely to exist than a completed cycle. We should recognize that much of the earth's crust is restive and is subject to intermittent and differential movement, but in spite of this there does appear to be considerable evidence to indicate that portions of the earth's crust do at times remain sufficiently stable for a geomorphic cycle nearly to run its course. Continued uplift may suspend cyclical development for a long period of geologic time, but there is no reason to assume that such conditions will be permanent. There is good reason to believe that there are regions that were once areas of intense diastrophism that are now relatively stable. Even a partial cycle may leave its imprint upon a landscape, and a geomorphologist needs to be able to recognize the effects of partial cycles because they are probably more the rule than the exception.

Concept 5. *Complexity of geomorphic evolution is more common than simplicity.*

It is probably a fundamental human trait to prefer a simple explanation to a more involved one, but a simple explanation is probably rarely the correct one. Many of the great controversies in science have arisen from this preference for a simple explanation. The serious student of land forms does not progress far in his study of them before he comes to realize that little of the earth's topography can be explained as the result of the operation of a single geomorphic process or a single geomorphic cycle of development. Usually, most of the topographic details have been produced during the current cycle of erosion, but there may exist within an area remnants of features produced during prior cycles, and, although there are many individual land forms which can be said to be the product of some single geomorphic process, it is a rare thing to find landscape assemblages which can be attributed solely to one geomorphic process, even though commonly we are able to recognize the dominance of one. It may facilitate our interpretation of landscapes to group them, as Horberg (1952) did, in five major categories: (1) simple, (2) compound, (3) monocyclic, (4) multicyclic, and (5) exhumed or resurrected landscapes.

Fig. 2.2. Diagram of a simple landscape upon moderately dipping sedimentary rocks. (Drawing by William J. Wayne.)

Simple landscapes are those which are the product of a single dominant geomorphic process; *compound landscapes* are those in which two or more geomorphic processes have played major roles in the development of the existing topography. It might, of course, be argued that nearly all landscapes are compound in nature, and in a strict sense this is true, for rarely do we find any extensive area in which the land forms can be attributed solely to the action of one process. But to press for this narrow usage of the word would to a large degree leave us in the hopeless dilemma of having to discuss all types of landscapes together without recognizing the dominance of certain processes in their development. It is perfectly logical to designate a certain landscape as being primarily the work of running water even though we may realize full well that weathering, movement of material under the direct influence of gravity, and removal of loose materials by the wind may have contributed to its development. It is equally logical to designate another type of landscape as being largely the product of solution by groundwater even though erosion by surface waters, weathering, and other processes may have contributed to its development. The type of landscape which we may logically designate as compound in nature is well illustrated in areas that were subjected to Pleistocene glaciation or felt the indirect effects of it. This is particularly true of the areas glaciated before the Wisconsin age. In such areas, there may be found upland tracts which still retain mainly their original glacial characteristics, yet along stream courses running water has formed the topography. Locally features may also exist resulting from wind deposition of materials derived from the streams which carried glacial meltwaters. Even beyond the limits of actual glaciation the topography may be compound ·in

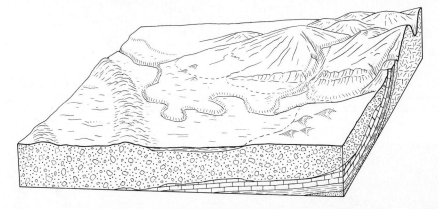

FIG. 2.3. A compound landscape consisting of glacial, glacio-lacustrine, and glacio-eolian features bordering an ancient domal structure. (Drawing by William J. Wayne.)

nature. The main topographic features may be the result of stream action, but along some stream courses there may be found features resulting from glacial outwash into the area and from wind action upon this outwash. In portions of such states as Arizona, New Mexico, Nevada, and Utah we find examples of compound landscapes in which stream-cut topography has within it volcanic cones and lava flows and locally prominent scarps produced by faulting of blocks of the earth's crust.

Monocyclic landscapes are those that bear the imprint of only one cycle of erosion; *multicyclic landscapes* have been produced during more than one erosion cycle. Monocyclic landscapes are less common than multicyclic and are in general restricted to such newly created land surfaces as a recently uplifted portion of the ocean floor, the surface of a volcanic cone or lava plain or plateau, or areas buried beneath a cover of Pleistocene glacial materials. Much, if not most, of the world's topography bears the imprint of more than one erosion cycle. The older cycles may be represented only by limited upland remnants of ancient erosion surfaces or deposits, or they may be marked by benches along valley sides above present valley floors. They may be

FIG. 2.4. A multicyclic landscape. Three cycles are represented. The first cycle is represented by peneplain remnants on the hilltops; the second by bedrock terraces along the valleys; and the third by the entrenched stream courses. (Drawing by William J. Wayne.)

FIG. 2.5. Hanging Rock, near Lagro, Indiana, an exhumed Silurian bioherm (reef). Half of the bioherm has been cut away by the Wabash River. (Photo by C. A. Malott.)

marked by wave-cut terraces above the present shore line or be represented along a glaciated valley by morainal deposits one above another. There is evidence to suggest that the present-day topography of much of eastern North America contains features that are the products of at least three major cycles of erosion. Features of multicyclic origin have been described from all the continents except Antarctica. They are more likely to be easily recognizable in the so-called stable areas, where periods of marked crustal movement are likely to be farther apart, than in the tectonically active belts, but even in such unstable areas as the Pacific coast of the United States old erosion surfaces have been recognized as comprising a part of the present topography. It will be recognized that a monocyclic landscape may be either simple or compound in nature as may a multicyclic landscape.

Exhumed or *resurrected landscapes* are those that were formed during some past period of geologic time, then buried beneath some sort of cover, and then within more recent geologic time exposed through removal of the cover. Topographic features now being exhumed may date back as far as the Pre-Cambrian or they may be as recent as the Pleistocene. Around Hudson Bay, we find an erosion surface which was developed across Pre-Cambrian igneous and metamorphic rocks and then buried beneath Paleozoic sediments now in the process of exposure through the removal of the Paleozoic sediments. Throughout the area covered by Pleistocene glacial materials there are hundreds of streams in the process of exhuming buried ridges and valleys. Most resurrected features are of local extent and constitute a small portion of the present-day topography, but they may be striking features. For example, along one stretch of the Wabash River in northern Indiana, where it flows across a buried preglacial topography, its valley is less than one-half mile wide

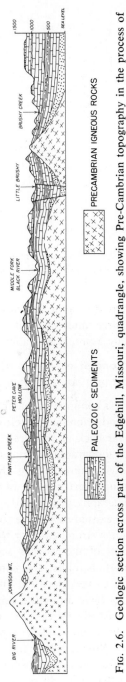

Fig. 2.6. Geologic section across part of the Edgehill, Missouri, quadrangle, showing Pre-Cambrian topography in the process of exhumation. (After C. L. Dake, *Missouri Geol. Survey, Bull. 23.*)

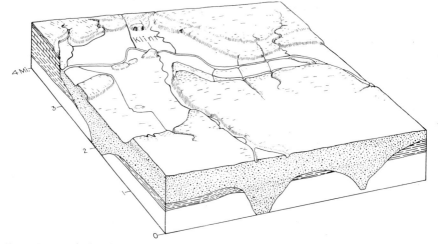

Fig. 2.7. Marked widening of the Wabash Valley where it intersects a preglacial valley filled with easily erosible glacial drift. A klint or exhumed bioherm similar to that in Fig. 2.5 is present in the valley. (Drawing by William J. Wayne.)

where it is being cut into the bedrock of a buried ridge, but it suddenly widens to three miles where it is being cut in the more easily eroded glacial materials filling a preglacial valley.

Concept 6. *Little of the earth's topography is older than Tertiary and most of it no older than Pleistocene.*

It is common in reading older discussions on the age of topographic features to find references to erosion surfaces dating back to the Cretaceous or even as far back as the Pre-Cambrian. We have gradually come to a realization that topographic features so ancient are rare, and, if they do exist, are more likely exhumed forms than those which have been exposed to degradation through vast periods of geologic time. Most of the details of our present topography probably do not date back of the Pleistocene, and certainly little of it existed as surface topography back of the Tertiary. Ashley (1931) has made a strong case for the youthfulness of our topography. He believed that "most of the world's scenery, its mountains, valleys, shores, lakes, rivers, waterfalls, cliffs, and canyons are post-Miocene, that nearly all details have been carved since the emergence of man, and that few if any land surfaces today have any close relation to pre-Miocene surfaces." He estimated that at least 90 per cent of our present land surface has been developed in post-Tertiary time and perhaps as much as 99 per cent is post-middle Miocene in age. Whether these figures are correct is unimportant, but they certainly point the way to a conclusion which geomorphologists should accept, despite the fact that it is still possible to find geologists who believe otherwise.

It is, of course, true that many geologic structures are very old. It has been previously stated that geologic structures are in general much older than the topographic features developed upon them. The only notable exceptions are to be found in areas of late-Pleistocene and Recent diastrophism. The Cincinnati arch and the Nashville dome began to form as far back as the Ordovician but none of the topography developed on them today goes back of the Tertiary; the Himalayas were probably first folded in the Cretaceous and later in the Eocene and Miocene but their present elevation was not attained until the Pliocene and most of the topographic detail is Pleistocene or later in age; the structural features which characterize the Rocky Mountains were produced largely by the Laramide revolution, which probably culminated at the close of the Cretaceous, but little of the topography in this area dates back of the Pliocene and the present canyons and details of relief are of Pleistocene or Recent age.

The Colorado River with its magnificent canyon might seem to some an example of a stream of ancient geologic age yet it seems probable that the river as a through-flowing stream dates back no farther than Miocene and perhaps not that far back. Powell (1875) and Dutton (1882) believed that the river was present far back in the Tertiary (probably in the Eocene) and was older than the structural features which it crosses. Davis (1901) later pointed out the likely error of this conclusion and suggested that a mid-Tertiary date for its beginning was more likely correct. Blackwelder (1934) suggested the possibility that the river might date back no farther than late Pliocene or possibly early Pleistocene. Longwell (1946) later cited some stratigraphic reasons for an assumption that the present river acquired its course after late Miocene time, presumably in the early Pliocene. Thus, the date of its origin is still uncertain, but the evidence seems to suggest that it is no older than the middle Tertiary, and it may possibly be younger.

Concept 7. *Proper interpretation of present-day landscapes is impossible without a full appreciation of the manifold influences of the geologic and climatic changes during the Pleistocene.*

Correlative with the realization of the geologic recency of most of the world's topography is the recognition that the geologic and climatic changes during the Pleistocene have had far-reaching effects upon present-day topography. Glaciation directly affected many million square miles, perhaps as much as 10,000,000 square miles, but its effects extended far beyond the areas actually glaciated. Glacial outwash and wind-blown materials of glacial origin extended into areas not glaciated, and the climatic effects were probably world-wide in extent. Certainly in the middle latitudes the climatic effects were profound. There is indisputable evidence that many regions that are today arid or semiarid had humid climates during the glacial ages.

Freshwater lakes existed in many areas which today have interior drainage. At least 98 of the 126 closed basins in western United States had Pleistocene lakes in them. Similar evidence of pluvial conditions in regions now arid or semiarid has been found in Asia, Africa, South America, and Australia, so that there can be no doubt about the world-wide influence of glacial conditions upon climates.

We know also that many regions now temperate experienced during the glacial ages temperatures such as are found now in the subarctic portions of North America and Eurasia, where there exists permanently frozen ground or what has come to be called permafrost conditions. Stream regimens were affected by the climatic changes, and we find evidence of alternation of periods of aggradation and downcutting of valleys. Many stream courses were profoundly altered as a result of ice invasions. Such streams as the Ohio and Missouri Rivers, and to a considerable extent the Mississippi, have the courses which they do today largely as a result of glacial modifications of their preglacial courses. World sea levels were also affected. Withdrawal of large quantities of water from the oceans to form great ice sheets produced a lowering of sea level of at least 300 feet and perhaps more. Return of this water to the oceans during interglacial ages caused a return of high sea levels such as characterize present geologic time. The discharge of cold glacial meltwaters to the oceans may have had significant effects upon certain marine organisms such as the reef-building corals. Winds blowing across glacial outwash or fresh glacial deposits in many areas built up dunal accumulations of sand or deposited over areas a mantle of silt and clay called loess. Glaciation has been responsible for the formation of more lakes than all other causes combined. The Great Lakes, the world's greatest internal waterway system, are the result of glacial modifications of preglacial lowlands and valleys.

Although glaciation was probably the most significant event of the Pleistocene, we should not lose sight of the fact that in many areas the diastrophism which started during the Pliocene continued into the Pleistocene and even into the Recent. Around the margins of the Pacific Ocean, Pleistocene diastrophism has played a most significant role in the production of present-day landscapes. It is further evidenced in the Rocky Mountains by deep canyon cutting, in many places over 1000 feet, which took place between the earlier glacial ages and the later Wisconsin glaciations.

Concept 8. *An appreciation of world climates is necessary to a proper understanding of the varying importance of the different geomorphic processes.*

That climatic factors, particularly those of temperature and precipitation, should influence the operation of the geomorphic processes seems self-evident, yet there have been surprisingly few detailed studies made which attempted to show to just what degree climatic variations influence topographic details. The reason for this somewhat paradoxical situation is not apparent. It may possibly be in part a result of the fact that geologists in general are not climate-minded and that geographers, who are better acquainted with the

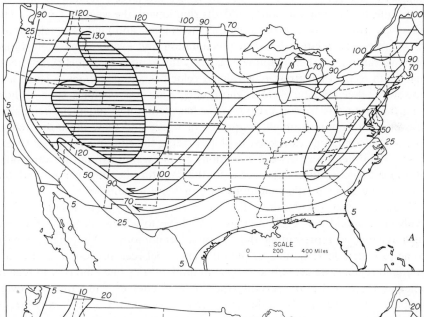

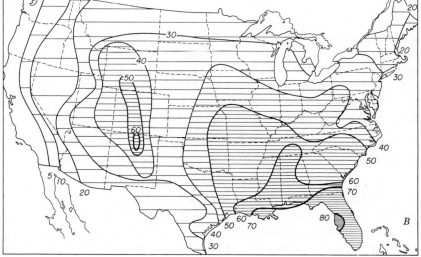

FIG. 2.8. Two climatic maps which have geomorphic significance. *A*. Average annual number of times of freeze and thaw. *B*. Number of days per year with thunderstorms. (After S. S. Visher.)

details of climate, in recent years have been more concerned with the adjustments of man's activities to varying landscapes than with the origin of the
landscapes themselves.

Climate variations may affect the operation of geomorphic processes either
indirectly or directly. The indirect influences are largely related to how climate affects the amount, kind, and distribution of the vegetal cover. The
direct controls are such obvious ones as the amount and kind of precipitation,
its intensity, the relation between precipitation and evaporation, daily range
of temperature, whether and how frequently the temperature falls below freezing, depth of frost penetration, and wind velocities and directions. There are,
however, other climatic factors whose effects are less obvious, such as how
long the ground is frozen, exceptionally heavy rainfalls and their frequency,
seasons of maximum rainfall, frequency of freeze and thaw days, differences
in climatic conditions as related to slopes facing the sun and those not so exposed, the differences between conditions on the windward and leeward sides
of topographic features transverse to the moisture-bearing winds, and the
rapid changes in climatic conditions with increase in altitude.

Most of what we consider the basic concepts in geomorphology have
evolved in humid temperate regions. We have to a large degree considered
such regions as representing the "normal" conditions. Although we have
come to realize to some degree the differences between processes in arid and
in humid regions, we are still too prone to think of only one kind of arid region,
whereas there really are several types of arid climates, and thus we oversimplify when we speak of the arid cycle. As yet we have hardly begun
formulation of basic geomorphic concepts with respect to the humid tropical,
arctic, and subarctic regions. Of this much we may feel certain, that the
processes which are dominant in the humid middle latitudes are not necessarily important to the same degree in the lower and higher latitudes and
that their significant differences will not be fully comprehended until we take
into full account climatic variations. High altitudes within any climatic
realm impose modifications which should also be recognized.

Concept 9. *Geomorphology, although concerned primarily with present-
day landscapes, attains its maximum usefulness by historical extension.*

Geomorphology concerns itself primarily with the origins of the present
landscape but in most landscapes there are present forms that date back to
previous geologic epochs or periods. A geomorphologist is thus forced to
adopt an historical approach if he is to interpret properly the geomorphic
history of a region. Application of the principle of uniformitarianism makes
this approach possible.

At first thought it might seem that the recognition of ancient erosion surfaces and the study of ancient topographies does not belong in the field of

geomorphology, but the approach of the geomorphologist may be the most logical one. This aspect of geomorphology may well be called *paleogeomorphology*. Two examples of how paleogeomorphology has practical application may be cited.

10 0 10 20 30 40 MILES

FIG. 2.9. A paleophysiographic diagram showing the major features of the preglacial topography of Illinois. (After Leland Horberg, *Illinois Geol. Survey, Bull. 73.*)

Certain oil and gas fields are in long narrow sandstone bodies which have come to be called shoestring sands. Numerous examples are found in southeastern Kansas and northeastern Oklahoma. Shoestring sands have several origins, the more common of which are as beach sand deposits, offshore bars, spits and hooks, river channel fillings, and delta distributary channel fillings. It is important that the type of sand deposit be recognized for proper planning of the development of an oil or gas field. What is being dealt with is a buried land form, and a proper realization of its likely shape, pattern, and extent is dependent upon a recognition of the particular type of geomorphic feature that exists.

It has been found that preglacial valleys now buried beneath glacial materials are often important sources of groundwater, and a reconstruction of the preglacial topography becomes significant. The problem is fundamentally one of reconstructing the late Tertiary topography of glaciated areas. This topography was intimately related to that of the adjacent unglaciated areas. Thus, in the search for buried valleys it becomes necessary to know what and where the major preglacial drainage lines were, and this is most easily conceived if the preglacial geomorphic history is fully appreciated. In fact, one of the most fruitful future uses of geomorphology will involve the study of ancient topographies; in Chapter 22 we cite several examples of how this technique has already proved helpful in the search for economic resources.

The historical nature of geomorphology was recognized by Bryan (1941) when he stated:

> If land forms were solely the result of processes now current, there would be no excuse for the separation of the study of land forms as a field of effort distinct from Dynamic Geology. The essential and critical difference is the recognition of land forms or the remnants of land forms produced by processes no longer in action. Thus, in its essence and in its methodology, physiography [geomorphology] is historical. Thereby, it is a part of Historical Geology, although the approach is by a method quite different from that commonly used.

Horberg (1952) has strongly espoused the idea that geomorphology to attain its maximum usefulness must become more historical in its approach. In support of this viewpoint, he stated: *

> The historical approach broadens the field of geomorphology to include the application of the geological methods of sedimentology and stratigraphy and attention to such related fields as soil science and climatology.
>
> It also imposes the responsibility of developing a body of principles by which geomorphic history can be interpreted. These principles as such are seldom considered and are yet to be defined and organized. ··· It also may be desirable, in search of new approaches, to apply familiar stratigraphic concepts leading to such topics as: (1) the nature and significance of geomorphic unconformities [see p. 144]; (2) criteria for determining relative age of cyclical surfaces; (3) bases for correlation of erosional and depositional surfaces; and (4) the concept of facies applied to landscapes ···

---

* Quoted by permission of University of Chicago Press from Leland Horberg, Interrelations of geomorphology, glacial geology, and Pleistocene geology, *J. Geol., 60,* p. 188.

The historical approach emphasizes the need for recognition of relict or "fossil" features inherited from past conditions and especially those relict features which are subordinate elements in the over-all landscape ⋯

When geomorphologists themselves fully realize and in turn convince other geologists of this use which can be made of geomorphic principles and knowledge, the subject will become a true working tool in the practical applications of geology.

REFERENCES CITED IN TEXT

Ashley, G. H. (1931). Our youthful scenery, *Geol. Soc. Am., Bull. 42*, pp. 537–546.

Blackwelder, Eliot (1934). Origin of the Colorado River, *Geol. Soc. Am., Bull. 45*, pp. 551–566.

Bryan, Kirk (1941). Physiography, *Geol. Soc. Am., 50th Ann. Vol.*, pp. 3–15.

Davis, W. M. (1901). An excursion to the Grand Canyon of the Colorado, *Bull. Mus. Comp. Zool., Harvard, 38*, pp. 108–200.

Dutton, C. E. (1882). Tertiary history of the Grand Canyon district, *U. S. Geol. Survey, Mon. 2*, pp. 206–229.

Horberg, Leland (1952). Interrelations of geomorphology, glacial geology and Pleistocene geology, *J. Geol., 60*, pp. 187–190.

Longwell, C. R. (1946). How old is the Colorado River?, *Am. J. Sci., 244*, pp. 817–835.

Powell, J. W. (1875). *Exploration of the Colorado River of the West and Its Tributaries*, Smithsonian Institution, Washington, pp. 149–214.

Additional References

Baulig, Henri (1950). Les concepts fondamentaux de la géomorphologie, Essais de géomorphologie, *Publs. de la faculté des lettres de l'université de Strasbourg, Paris*, pp. 31–42.

Davis, W. M. (1909). The geographical cycle, Chapter 13 in *Geographical Essays*, Ginn and Co., New York.

Visher, S. S. (1941). Climate and geomorphology; some comparisons between regions, *J. Geomorph., 5*, pp. 54–64.

Visher, S. S. (1945). Climatic maps of geologic interest, *Geol. Soc. Am., Bull. 56*, pp. 713–736.

3 · An Analysis of the Geomorphic Processes

GEOMORPHIC AGENTS AND PROCESSES

The *geomorphic processes* are all those physical and chemical changes which effect a modification of the earth's surficial form. A *geomorphic agent* or *agency* is any natural medium which is capable of securing and transporting earth material. Thus running water, including both concentrated and unconcentrated runoff, groundwater, glaciers, wind, and movements within bodies of standing waters including waves, currents, tides, and tsunami are the great geomorphic agencies. They may further be designated as mobile agents because they remove materials from one part of the earth's crust and transport and deposit them elsewhere. Most of the geomorphic agencies originate within the earth's atmosphere and are directed by the force of gravity. Gravity is not a geomorphic agent because it can not secure and carry away materials. It is better thought of as a directional force.

The agencies thus far mentioned and the processes performed by them originate outside the earth's crust, and for that reason have been designated by Lawson as *epigene* and by Penck as *exogenous*. To the agents already mentioned should be added such minor ones as man and other organisms, although in some regions there may be a question whether man is a minor agent. Other geomorphic processes have their origin within the earth's crust and were classed by Lawson as *hypogene* and by Penck as *endogenous*. Vulcanism and diastrophism belong in this class. A geomorphic process that may have local significance and that does not fall within either of the above categories is the impact of meteorites. No good single name has yet been coined for this process.

An outline of the geomorphic processes. The following is an outline of the processes which shape the earth's surface.

Geomorphic processes.
   Epigene or exogenous processes.
      Gradation.
         Degradation.
            Weathering.

34

Mass wasting or gravitative transfer.
Erosion (including transportation) by:
    Running water.
    Groundwater.
    Waves, currents, tides, and tsunami.
    Wind.
    Glaciers.
Aggradation by:
    Running water.
    Groundwater.
    Waves, currents, tides, and tsunami.
    Wind.
    Glaciers.
Work of organisms, including man.
Hypogene or endogenous processes.
  Diastrophism.
  Vulcanism.
Extraterrestrial processes.
  Infall of meteorites.

Unfortunately, there is a confusing variation in usage of the terms that designate the common geomorphic processes. To a certain extent this confusion results from differences in opinion as to what is included under certain processes, but to a considerable degree it is the result of carelessness in thinking and writing. Writers do not agree on what the seemingly simple process of erosion includes. Some include weathering, although there has been an increasing tendency to recognize that weathering is not a part of erosion. There also may be a difference of opinion as to whether transportation is a part of erosion, but it is usually so considered. Certainly, it is extending the meaning of erosion too far to consider aggradation as a part of it.

The author uses the term *gradation* in the sense originally used by Chamberlin and Salisbury (1904) to include "all those processes which tend to bring the surface of the lithosphere to a common level." They recognized that gradational processes belong to two categories—those which level down, *degradation,* and those which level up, *aggradation.*

Some geologists have used the term denudation as if it were synonymous with gradation, but, as this term implies removal of material, it is hardly logical to include deposition under it. Planation has also been used in the sense in which gradation is being used but it implies erosion and not deposition. There may be objection to the use of gradation because of its similarity to such terms as grade and grading, which have restricted meanings as applied to stream erosion. Despite this objection, it seems to the writer that gradation comes nearer to implying the leveling off of the earth's surface by both destructional and constructional processes than any other available term. It does not, however, encompass such activity as building up of a lava plateau through igneous extrusion.

Gradational Processes

Degradation. The three distinct degradational processes are weathering, mass-wasting, and erosion. *Weathering* may be defined as the disintegration or decomposition of rock in place. It is really a name for a group of processes which act collectively at and near the earth's surface and reduce solid rock masses to the clastic state. It is a static process and does not involve the seizure and removal of material by a transporting agency.

Mass-wasting involves the bulk transfer of masses of rock debris down slopes under the direct influence of gravity. Mass-wasting is usually aided by the presence of water, but the water is not in such an amount as to be considered a transporting medium. The distinction between mass-wasting and stream erosion may seem simple, but actually it may be difficult at times to say when the amount of water present is sufficient to be considered a transporting agency and when not. Thus it may be hard to draw the line between a mudflow and an extremely muddy stream, but this merely emphasizes the point, made by Sharpe (1938), that there is a continuous series of types of mass movements grading from those that are imperceptibly slow and with little associated water to those that are rapid and involve large amounts of water. This latter group in turn grades imperceptibly into streams in which water is predominant over waste material.

The gradational series is as follows:

Landslides: characterized by little water and large load moving over moderate to high slopes.
Debris avalanches
Earthflows
Mudflows
Sheetfloods
Slope wash
Streams: characterized by much water and relatively small load moving over low-angle slopes.

The difficulty encountered in drawing sharp lines between these different types of transport merely emphasizes that in setting up definitions we often suggest fine distinctions which do not exist in nature. Consequently most definitions have a certain degree of artificiality.

Erosion (Latin, *erodere,* to gnaw away) is a comprehensive term applied to the various ways by which the mobile agencies obtain and remove rock debris. If we wished to be overly technical we might restrict erosion to the acquisition of material by a mobile agent and thereby not consider transportation as a part of it, but most geologists are likely to consider transportation as an integral part of erosion. It is, however, certainly extending the term weathering too far to consider it a part of erosion, although this is often carelessly done. The two processes are entirely distinct. Weathering can take place without subsequent erosion, and erosion is possible without previous weathering. It is true, of course, that weathering is a preparatory process and may make erosion easier, but it is not prerequisite to nor necessarily followed by erosion.

Rock Weathering

Conditioning factors. At least four variable factors influence the type and rate of rock weathering. These are rock structure, climate, topography, and vegetation. Rock structure, as indicated in Chapter 2, is used in the broad sense to include the many physical and chemical characteristics of rocks. It includes mineralogic composition as well as such physical features as joints, bedding planes, faults, and minute intergrain fractures and voids. The minerals forming the rock in part determine whether it is more susceptible to chemical or physical weathering. Physical features such as joints, lesser fractures, bedding planes, and faults to a large degree determine the ease with which moisture enters the rock. The major climatic factors of temperature and humidity determine not only the rate at which weathering proceeds but also whether chemical or physical processes predominate. Topography affects the amount of rock exposure and also has important effects upon such factors as the amount and kind of precipitation, temperatures, and indirectly the kind and amount of vegetation. The abundance and type of vegetation influence the rate and type of weathering by determining the extent of rock outcrops and the amount of decaying organic matter from which carbon dioxide and humic acids may be derived. Because of the several factors which influence the rate of weathering, we usually find that even within relatively small areas rock variations not detectable by the eye become evident through differential weathering. This is particularly true where diversity of rock types exists.

Physical weathering processes. Four, or possibly five, of the weathering processes are physical in nature and lead to fragmentation of rock. These, following Reiche (1950), may be designated: *expansion resulting from unloading, crystal growth, thermal expansion, organic activity,* and *colloid plucking.* Expansion or dilation accompanying the unloading of rock masses, particularly igneous rocks formed at great depth, leads to the development of large-scale fractures which are roughly concentric with the surface topography (Farmin, 1937). Sheety structure in granitoid rocks is believed to be produced in this way (Jahns, 1943). The individual sheets usually become progressively more closely spaced as the earth's surface is approached.

Unloading of rocks formed at considerable depth may have contributed significantly to the formation of such large monoliths as Stone Mountain, Georgia, and Half Dome in Yosemite, which are often called *exfoliation domes*. Matthes (1937) concluded that the concentric exfoliation which is particularly characteristic of the many domes in Yosemite and Sequoia National Parks resulted from expansion accompanying relief of load. Microscopic examination of the rock indicated that the expansion was mechanical

FIG. 3.1.   Looking Glass Rock, North Carolina, an exfoliation dome.   (Photo by Elliot Lyman Fisher.)

in nature and not the result of hydration or other chemical changes. Exfoliation shells resulting from dilation accompanying release of load may measure hundreds or even thousands of feet in horizontal extent.

Expansion attendant upon crystal growth may result in rock fracturing. This includes not only the formation of ice crystals in rocks but also the growth of other crystals, particularly the salines, which form in dry climates as a result of the capillary action of water containing salts in solution. Great efficacy to fracture rocks has been attributed to freezing water, and rightly so, although it probably seldom exerts the enormous pressures sometimes attributed to it, for such pressures are attained only when water is completely confined. Taber (1930) has shown that frost heaving to a large degree is dependent upon the formation of ice masses rather than upon the freezing of interstitial water; hence alternate freezing and thawing will be most effective in rocks containing numerous fractures or bedding planes. Formation of ice crystals is most effective as a weathering process when there is repeated freeze and thaw. This happens in the middle latitudes and high altitudes during late fall and early winter and again in late winter and early spring. Large daily temperature ranges and frequent alternations of cyclonic and anti-cyclonic weather with their attendant warm and cold air masses are conducive to maximum freeze and thaw. The scaling off of rock surfaces through

FIG. 3.2. Exfoliation of massive siltstone. Alternate freezing and thawing are mainly responsible for the spalling off of the rock surface. (Photo by P. B. Stockdale.)

growth of salines by capillary action has been called *exsudation*. It probably has only local importance as a weathering process.

Formerly, much significance was attributed to the role played by repeated thermal expansion and contraction of rock surfaces in the formation of such features as spheroidal boulders. Blackwelder (1925, 1933) questioned its effectiveness as a process of rock disintegration, but belief in it dies slowly because the process as usually described seems so logical. Rock is admittedly a poor conductor of heat, and most rocks are composed of several minerals with different coefficients of expansion. It would seem that repeated heating and cooling of a rock surface should, because of the slow rate at which the absorbed heat is conducted into the rock mass, produce strains in a thin surface layer which would ultimately cause it to spall.   This has been called *mass ex-foliation*.   Rocks composed of a variety of minerals having different coefficients of expansion would as a result of repeated differential expansion and contraction of the individual minerals undergo *granular exfoliation* or *disinte-*

Fig. 3.3.   Granular exfoliation of quartz monzonite near Butte, Montana.

gration.   There seems to be little field or experimental evidence to support the idea that mass exfoliation results largely from repeated heating and cooling of a rock surface, but Blackwelder concluded that temperature changes may be significant in granular exfoliation.   Chapman and Greenfield's (1949) studies of the processes involved in the formation of spheroidal boulders indicated that the shells which spalled off of them contained many secondary minerals, such as kaolinite, sericite, serpentine, montmorillonite, and chlorite.

From this evidence they concluded that spheroidal scaling probably resulted largely from the effects of oxidation and hydration of silicate minerals.

Organisms are of minor importance in physical weathering. Growth of plant roots may aid in the widening of joints and other fractures. On the other hand, decay of plant and animal matter may contribute to chemical weathering through the formation of carbon dioxide and organic acids. Earthworms, ants, prairie dogs, and other animals may, however, bring fresh rock material to or near the surface where it is more readily attacked by the chemical weathering processes.

A weathering process of uncertain importance has been called colloid plucking by Reiche. It seems probable that soil colloids may have the power to loosen or pull off small bits of rock from the surfaces with which they come in contact. Gelatin drying in a glass has been observed to pull off flakes of glass, and it has been inferred that similar plucking might take place after each wetting and drying of soil colloids. As yet, the importance of this process is unknown.

Chemical weathering processes. The chief chemical weathering processes are hydration, hydrolysis, oxidation, carbonation, and solution. In general, it is probably true that chemical weathering is more important than physical weathering. This may be true even in arid regions, although the more advanced chemical weathering processes are not significant there. The fact that arkoses are common in arid regions is often cited as evidence of the predominance of physical over chemical weathering. It is probably more nearly correct to say that they merely indicate that advanced chemical weathering is not typical of such regions.

Most chemical weathering results in: (*a*) an increase in bulk with resulting strains and stresses within the rocks, (*b*) lower density materials, (*c*) smaller particle size and hence an increased surface area per unit of volume, and (*d*) more stable minerals. The principle of mineral stability must be understood to appreciate fully the persistence of certain minerals in nature. The minerals in igneous and metamorphic rocks were at equilibrium under the conditions of temperature and pressure at which the rocks formed, but under the temperature and pressure conditions at the earth's surface some of the minerals are not the most stable minerals. In general, chemical weathering progresses toward the formation and retention of those minerals which are at equilibrium at the earth's surface. As a result, some of the minerals of igneous and metamorphic rocks are more susceptible to chemical alteration than others. Although there is no complete agreement upon the exact order of mineral stability, the general order is known. Goldich (1938) in discussing this principle tabulated mineral stabilities as shown in the accompanying list. The least stable minerals are at the top and the most stable at the bottom.

Mineral-stability series in weathering

Olivine	Calcic plagioclase
Augite	Calcic-alkalic plagioclase
Hornblende	Alkalic-calcic plagioclase
Biotite	Alkalic plagioclase
	Potash feldspar
	Muscovite
	Quartz

It will be evident to a student versed in petrology that the above arrangement of minerals according to their relative stabilities is essentially that of Bowen's (1922) reaction series. It is not to be inferred, however, that olivine weathers to augite or augite to hornblende but rather that in a rock containing both olivine and augite the olivine will weather more rapidly than the augite. It should be evident from this table why quartz and muscovite are common constituents of rocks composed of weathering residues. Of the minerals which are common in igneous and metamorphic rocks, they are the most nearly in equilibrium with the conditions of temperature and pressure which exist at the earth's surface.

Hydration as commonly described actually involves two processes, hydration and hydrolysis. The process of *hydration* involves adsorption of water. It is illustrated in the conversion of anhydrite to gypsum according to the following equation:

$$CaSO_4 + 2H_2O \rightarrow CaSO_4 \cdot 2H_2O$$

The conversion of hematite to limonite also involves this process:

$$2Fe_2O_3 + 3H_2O \rightarrow 2Fe_2O_3 \cdot 3H_2O$$

Both of the above reactions are exothermic and are rather easily reversible upon application of heat, which indicates that there has been no fundamental chemical change. *Hydrolysis* involves the formation of hydroxyl and does represent a chemical change. It is common in the weathering of the feldspars and micas and is illustrated by the following reaction involving orthoclase and hydroxyl:

(1) $$KAlSi_3O_8 + HOH \rightarrow HAlSi_3O_8 + KOH$$

The aluminosilicic acid formed is unstable and undergoes further change to produce colloidal silica and a colloidal complex which under the proper conditions forms clay minerals. The hydroxide (KOH) that is formed will react in the presence of carbon dioxide to form potassium carbonate according to the following equation:

(2) $$2KOH + H_2CO_3 \rightarrow K_2CO_3 + 2HOH$$

This reaction is commonly referred to as *carbonation*. The potassium carbonate so formed is soluble in water and will be carried away in solution.

It should be stressed that the various weathering processes are interrelated. If it were otherwise, weathering would be an extremely slow process. The weathering of feldspar as it is commonly written illustrates this point:

$$(3) \quad 2KAlSi_3O_8 + 2H_2O + CO_2 \rightarrow H_4Al_2Si_2O_9 + K_2CO_3 + 4SiO_2$$

Hydrolysis and carbonation have produced a water-soluble carbonate. The aluminosilicic acid formed (equation 1) has taken on water of combination and from it colloidal clay and silica have formed.

The effects of oxidation are usually the first to be noticed, and it is commonly assumed that it is the first to take place. Actually, hydrolysis usually precedes it. The effects of oxidation are most apparent in rocks containing iron in the sulfide, carbonate, and silicate forms. Discoloration by the oxides is readily noticeable. The amphiboles and pyroxenes are also notably affected by this process. The oxidation of olivine well illustrates the course which oxidation may run:

$$(4) \quad MgFeSiO_4 + 2HOH \rightarrow Mg(OH)_2 + H_2SiO_3 + FeO$$

Olivine	Hydroxyl	Magnesium hydroxide	Silicic acid	Ferrous oxide

Hydrolysis has resulted in the formation of magnesium hydrate, silicic acid, and ferrous iron from the olivine. The ferrous iron then reacts as follows to form limonite:

$$(5) \quad 4FeO + 3H_2O + O_2 \rightarrow 2Fe_2O_3 \cdot 3H_2O$$

The effects of hydration, hydrolysis, and oxidation upon the feldspars, micas, olivine, and other minerals are such as to make them soft and cause them to lose their luster and elasticity. Usually an increase in bulk takes place as a result of the addition of water. The significant result is that the new minerals formed are more readily attacked by the chemical and physical weathering processes.

Perhaps the most common *solution* reaction is that in which calcium carbonate is dissolved to form the bicarbonate as follows:

$$CaCO_3 + H_2O + CO_2 \rightarrow Ca(HCO_3)_2$$

The carbonate and bicarbonate minerals are most commonly taken into solution, and carbon dioxide derived particularly from decaying organic matter aids greatly in this reaction. Even minerals commonly considered insoluble may go slowly into colloidal solution, as when the feldspars break down into colloids and ultimately form clay minerals.

The complex interrelationships of the weathering processes are well summarized as follows by Lyon and Buckman (1943): *

An understanding, even though it be limited, of the influences of the separate forces of weathering makes it possible to picture in a simple way the development of soil material from bedrock. A physical weakening, due usually to temperature changes, initiates the process, but it is accompanied and supplemented by certain chemical transformations. Such minerals as the feldspars, mica, hornblende, and the like suffer hydrolysis and hydration, while part of the combined iron is oxidized and hydrated. The minerals soften, lose their luster, and increase in volume. If hematite or limonite is formed, the decomposing mass becomes definitely yellow or red. Otherwise the colors are subdued.

Coincident with these changes such active cations as calcium, magnesium, sodium, and potassium suffer carbonation, and soluble mineral products appear in the water present in the mass. When this water drains away, these soluble constituents are removed, leaving a residue more or less bereft of its easily soluble bases. As the process goes on, all but the most resistant of the original minerals disappear, and their places are occupied by secondary hydrated silicates and often recrystallize into highly colloidal clay. When this clay is small in amount, a sandy soil material is developed, but when dominant, the mass is heavy and plastic.

Such a brief statement of rock weathering demands certain supplementary explanations. In the first place it must be recognized that the intensity of the various agencies will fluctuate with climate. Under arid conditions, the physical forces will dominate, and the resultant soil material will be coarse. Temperature changes, wind action, and water erosion will be accompanied by a minimum of chemical action.

In a humid region, however, the forces are more varied, and practically the full quota will be at work. Vigorous chemical changes will accompany disintegration, and the result will be shown in greater fineness of the product. Clayey minerals will be more common, and a higher colloidicity can be expected.

Again, it must be remembered that the forces of weathering not only lose their intensity in the lower layers of the mass, but also the transformations are somewhat different. At the surface, the full effects of climatic agencies are apparent, the influence of the decaying organic matter usually greatly augmenting the chemical and physical changes. Below, the action is much less vigorous. This is due to the presence of larger amounts of water and a decrease both in porosity and aeration. This differentiation is a forerunner of a definite profile development and the genesis of a soil from the decomposing mass of rock materials.

Mass-Wasting or the Gravitative Transfer of Material

Only within recent years has mass-wasting received the attention which it merits. Neglect of this geomorphic process resulted partly from the fact that only the more precipitous types of gravitative transfer produce immediate and

* Quoted with the permission of The Macmillan Company from Lyon and Buckman's *The Nature and Properties of Soils,* 4th edition, pp. 238–239, copyright 1943.

perceptible modification of the earth's surface and also because few systematic studies of the process have been made. We are particularly indebted to Sharpe (1938) for an analysis of this aspect of gradation. Although there may be some objections to Sharpe's classification of types of mass-wasting, it seems to be the best available and has come into rather general usage.

Fig. 3.4. Turtle Mountain rockslide near Frank, Alberta. (Royal Canadian Air Force photo.)

Sharpe recognized four classes of mass-wasting, which he designated as slow flowage, rapid flowage, landslides, and subsidence. Various types and subtypes were recognized under each class and defined as follows:

Slow flowage types.
  Creep: the slow movement downslope of soil and rock debris which is usually not perceptible except through extended observation.
  Soil creep: downslope movement of soil.
  Talus creep: downslope movement of talus or scree.
  Rock creep: downslope movement of individual rock blocks.
  Rock-glacier creep: downslope movement of tongues of rock waste.

Solifluction: the slow-flowing downslope of masses of rock debris which are
saturated with water and not confined to definite channels.
Rapid flowage types.
Earthflow: the movement of water-saturated clayey or silty earth material
down low-angle terraces or hillsides.
Mudflow: slow to very rapid movement of water-saturated rock debris down
definite channels.
Debris avalanche: a flowing slide of rock debris in narrow tracks down steep
slopes.
Landslides: those types of movement that are perceptible and involve relatively
dry masses of earth debris.
Slump: the downward slipping of one or several units of rock debris usually
with a backward rotation with respect to the slope over which movement
takes place.
Debris slide: the rapid rolling or sliding of unconsolidated earth debris without backward rotation of the mass.
Debris fall: the nearly free fall of earth debris from a vertical or overhanging
face.
Rockslide: the sliding or falling of individual rock masses down bedding,
joint or fault surfaces.
Rockfall: the free falling of rock blocks over any steep slope.
Subsidence: downward displacement of surficial earth material without a free
surface and horizontal displacement.

The conditions which favor rapid mass-wasting were divided by Sharpe
(1938) into passive and activating or initiating causes. Passive causes include: (*a*) *Lithologic factors,* unconsolidated or weak materials or those which
become slippery and act as lubricants when wet; (*b*) *stratigraphic factors,*
laminated or thinly bedded rock and alternating weak and strong or permeable and impermeable beds; (*c*) *structural factors,* closely spaced joints,
faults, crush zones, shear and foliation planes, and steeply dipping beds; (*d*)
topographic factors, steep slopes or vertical cliffs; (*e*) *climatic factors,* large
diurnal and annual range of temperature with high frequency of freeze and
thaw, abundant precipitation, and torrential rains; and (*f*) *organic factors,*
scarcity of vegetation.

Activating causes are: removal of support through natural or artificial
means, oversteepening of slopes by running water, and overloading through
water saturation or by artificial fills.

Erosion and Transportation Agencies

Each of the erosional agencies accomplishes erosion in one or more ways.
For some agents the processes involved are essentially the same; for others
they are distinctly different because of inherent physical differences between

the erosional agents. Specific names have been given to the various erosional processes but there is considerable confusion and loose usage. An attempt is made in the accompanying outline to systematize this terminology, as well as to indicate the interrelationships of the processes.

EROSIONAL PROCESSES

Agency Involved	Processes by Which Loosened Material Is Acquired	Processes by Which Earth Surfaces Are Eroded by Materials in Transit	Processes of Wear of Materials While in Transit	Methods of Transport
Running water	Hydraulic action or fluviraption	Corrasion or abrasion Corrosion	Attrition	Traction Saltation Suspension Solution Flotation
Groundwater *		Corrosion		Solution
Waves and currents	Hydraulic action	Corrasion or abrasion Corrosion	Attrition	Traction Saltation Suspension Solution Flotation
Wind	Deflation	Corrasion or abrasion	Attrition	Traction Saltation Suspension
Glaciers	Scouring Plucking or sapping	Corrasion or abrasion Gouging	Attrition	Traction Suspension

* Groundwater is not interpreted to include underground streams. The same processes would apply to them as to surface streams.

The outline indicates that there are four aspects of erosion, the acquisition of loose material by an erosional agency, the wearing away of solid rock by the impact upon it of materials in transit, the mutual wear of rock particles in transit through contact with each other, and transportation.

Hydraulic action is the sweeping away of loose material by moving water, as when a stream of water from a hose is turned on a sidewalk to clean the dirt off it. This process has also been called *fluviraption* (from *fluvius,* river, and *rapere,* to seize) by Malott (1928). The corresponding process as performed by wind is known as *deflation.* When achieved by ice moving over a land surface it is commonly called *scouring.* Bedrock surfaces may be eroded by rock debris in transit. The removal of bedrock particles by the "tooling action" of transported material is known as *corrasion* or *abrasion*. Pothole drilling is a special type of corrasion or abrasion. The removal of material by solution is called *corrosion*.

Plucking, sapping, and gouging are erosional processes restricted to glaciers. *Plucking* refers to the acquisition of parts of the bedrock by a glacier when water enters cracks in the rock and subsequently freezes with resulting de-

FIG. 3.5.   Glacially scoured rock surface dotted with glacial boulders, near Tuolumne Meadows, Yosemite National Park.   (Photo by C. L. Heald.)

tachment of rock fragments as the ice moves forward.   The term *sapping* implies undermining and is used by some as synonymous with plucking, but by others it is restricted to detachment which takes place at the bottoms of crevasses.   The local basining of bedrock surfaces often effected by glacial erosion is sometimes called *gouging,* but this usage of the term is not widespread.

Attrition is the wear and tear that rock particles in transit undergo through mutual rubbing, grinding, knocking, scraping, and bumping with resulting comminution in size.

Transportation may be accomplished by the mobile agents in four ways. These are traction, suspension, solution, and flotation. *Traction* involves the partial support of the material being transported by the buoyancy of the water or air but consists chiefly of the rolling, pushing, and dragging along of rock particles which are too large to be lifted into the main body of the stream or current. Moving water can transport both small- and large-sized particles in this way but wind can transport only material of much smaller size because its density is much lower and the resulting buoyancy is less. At extreme velocities wind can move pebbles by traction. When movement is noticeably by intermittent leaps and bounds it is referred to as *saltation*.

Suspension involves the temporary support of rock particles by moving air or water. It is possible because the flow of air and water is mainly turbulent with upward currents which can lift and keep particles in suspension. It may not be strictly correct to speak of glaciers carrying material in suspension, but no more satisfactory term is available to describe how glaciers transport much of their load. A part of the load carried by moving water is in *solution* and becomes a part of the fluid; thus no extra expenditure of energy for its transport is involved.

Flotation is a minor transporting process. Some inorganic materials such as rock pumice or sheets of mica may be carried in this way. The rock debris on the surface of a glacier can hardly be said to be floating. It is merely resting upon a solid surface which carries it along as it moves.

Aggradation or Deposition

Aggradation is an inevitable concomitant of degradation and contributes to the general leveling of the earth's surface. Deposition, except where groundwater is involved, results from a loss in transporting power. Deposition from groundwater results from changes in conditions of pressure and temperature, or from the action of organisms which cause precipitation. Deposition by a glacier as it melts may be considered a special type of loss in transporting power.

More attention has been given to erosional than to depositional land forms, with the possible exception of those produced by glaciers. One reason for this may be that the sculptured forms are often more striking, but to some extent the depositional forms have been less studied because many of them have too little relief to be shown well on most topographic maps. Alluvial and glacial deposits have been fairly satisfactorily classified, but we still lack thorough and systematic classifications of the deposits made by wind, groundwater, and waves and currents.

Diastrophism

Diastrophism, and also vulcanism, are classed as hypogene or endogenous processes, because the forces responsible for them originate at some depth within the earth's crust. They elevate or build up portions of the earth's surface and thereby prevent the gradational processes from ultimately reducing the earth's land areas to sea level.

Diastrophic processes are usually classified as two types, *orogenic* (mountain-building with deformation) and *epeirogenic* (regional uplift without im-

FIG. 3.6.   Dust starting to rise from Sycamore Canyon slide, which resulted from aftershock of the Arvin-Tehachapi, California, earthquake of July 25, 1952.   (Photo by Robert C. Frampton.)

portant deformation). Orogenic movements are more localized than are epeirogenic and usually involve tangential forces with resulting compression or tension of rock strata. G. K. Gilbert (1890) was the first to point out the distinction between the two, and he cited the Colorado Plateaus as an example of a region which has undergone epeirogenic uplift in contrast to the Wasatch Range which has experienced block faulting associated with orogenic

FIG. 3.7. The Waterpocket fold, a monoclinal fold west of the Henry Mountains, in southern Utah. Triassic rocks are exposed at the left and Cretaceous rocks at the right. (Photo by J. S. Shelton and R. C. Frampton.)

movements. The distinction between the two seems simple enough in principle but is actually not so in practice, for, as Gilluly (1950) has pointed out, the fault zone which bounds the Wasatch Range continues southward as the boundary of the Wasatch Plateau, which is usually considered a typical epeirogenic feature. Furthermore the Colorado Plateau, another epeirogenic feature, is bounded by the Hurricane and Grand Wash faults.

It is commonly believed that periods of mountain-building (orogenies) are episodic in occurrence, widely spaced in geologic time, and world-wide in extent. It is also believed that between orogenies there are long periods during which the earth's crust is relatively stable or subject only to slow epeirogenic uplift or subsidence. According to this viewpoint, we are now living in the dying phase of an orogenic episode or have actually entered an anorogenic period, during which present mountains and plateaus will be destroyed by erosion and the earth's surface reduced to one of low relief.

Gilluly (1949, 1950) has challenged this concept of diastrophism. He neither believed in the world-wide nature of orogenies nor in their contemporaneity. In his opinion, present geologic time is little different from most of geologic time and mountain-building is going on as rapidly now as it ever did. He doubted that epeirogenic movements as usually defined are sharply distinguished from orogenic movements in geologic time.

Obviously these contrasting viewpoints with respect to the nature of diastrophism will profoundly influence certain geomorphic interpretations. The concept of the geomorphic cycle was to a large degree postulated upon a belief that there are long anorogenic periods during which the earth's crust is relatively stable. Those who doubt the validity of such anorogenic periods will be inclined to doubt the validity of geomorphic cycles.

Vulcanism

Vulcanism includes the movement of molten rock or magma onto or toward the earth's surface. It is beyond the scope of geomorphology to explain the

FIG. 3.8. A lava flow near Grants, New Mexico. (Photo by Fairchild Aerial Surveys, Inc.)

complex changes within the earth which produce vulcanism. *Volcanism* includes the various ways by which molten rock is extruded. This may be through centralized vents called volcanoes or through extended openings or fissures as mass eruptions. The topographic effects of extrusions are direct and immediate. The effects of igneous intrusions are more commonly indirect and delayed. They consist of: rock deformation resulting in domal folds; disturbance of overlying rock strata; or the intrusion into older rocks of igneous masses, which, when subsequently exposed by erosion, give rise to topographic forms different from those developed upon the enclosing strata.

Impact of Meteorites

Probably the most unusual land forms are those that originate from the impact of meteorites. Such forms are rare but a few well-substantiated ones such as Meteor Crater, Arizona, exist. Their uniqueness lies in the fact that they were produced by extraterrestrial agents, although the earth's gravity was responsible for the infall of the meteorites. No generally acceptable term has been proposed yet for this process of land form development, although geobolism and meteoritism have been informally suggested.

Topographic Effects of Organisms

Organisms, including man, should not be overlooked as geomorphic agents. Locally man-made quarries, road cuts and fills, and many other types of excavations profoundly modify the earth's surface. Bomb craters may become a distinctive and prevalent type of land form if modern warfare continues. Reefs built by corals and other organisms are widespread and striking features of tropical seas. Beaver dams with lakes and meadows back of them are found in many areas. Ants, termites, prairie dogs, gophers, birds, and other animals build mounds which locally may be conspicuous. Termite mounds as high as 25 feet have been described. Even vegetation may play a role in the development of land surfaces, for the final filling of a lake often is with vegetation, resulting in peat bogs and marshes.

CLIMATIC INFLUENCES UPON GEOMORPHIC PROCESSES

In Chapter 2, we briefly discussed the effects of varying climatic conditions upon the operation of geomorphic processes. These relationships are so important that they merit further discussion. This phase of geomorphology has been neglected in the United States, but there are signs that its significance is beginning to be realized. Most American textbooks either largely ignore the interrelationships of climate and land forms or treat them incidentally.

Europeans have not been so negligent.   There has been too much of a tendency to standardize the geomorphic cycle as it proceeds under temperate humid conditions without recognizing that there are many other sets of climatic factors under which the relative importance of the individual geomorphic processes may vary greatly.

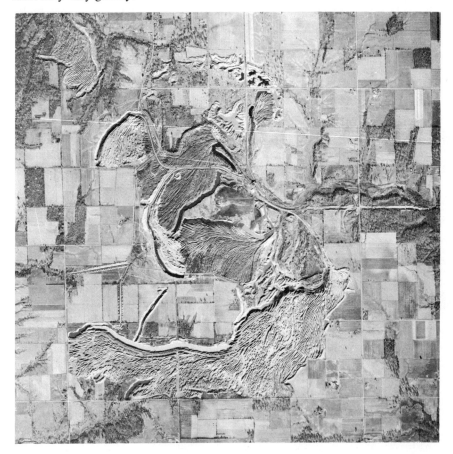

Fɪɢ. 3.9.   Vertical photo of strip mine topography, Clay County, Indiana.   (Production and Marketing Administration photo.)

Any one who has observed land forms produced under widely different climatic conditions can hardly fail to note significant differences in the landscapes and a certain degree of harmony between climate and landscapes, particularly with respect to the lesser topographic features.   A corollary of the principle that certain aspects of the landscape are to be correlated with climatic factors is that the effects of climatic changes may be recognizable in specific features of the landscape.

Krynine (1936) in reviewing the conclusions of Sapper regarding geomorphic processes in the humid tropics pointed out some significant differences between these areas and the humid temperate lands. Deep chemical
decay of rocks is usually considered one of the outstanding features of humid

FIG. 3.10. A termite mound built up a tree, Kabondo, Belgian Congo, Africa. (Photo
by L. C. King.)

tropical regions. This is found as expected on the flatlands, but in mountainous areas also as much as 20 feet of reddish soil mantle may be encountered. Yet within the same mountain area there may be found steep-
walled gullies and canyons which hardly fit in with our usual concept of the
type of topography to be associated with deep residual soils. The role which
the heavy tropical rain-forest vegetation plays has to be appreciated in order
to reconcile the presumed incompatibility of deep residual soils and deep
canyons. Heavy vegetation may grow in the humid tropics on slopes as great
as 70 degrees. Under its protective cover and with the existing temperature

conditions, mechanical weathering is insignificant and sheet wash and soil erosion may be negligible. Vegetation may extend up to the very banks of streams and prevent lateral erosion from becoming a significant process in the shaping of the valley form. Although the tropical rain-forest effectively protects the soil beneath from erosion, when it is destroyed the effects of gulleying may be striking. Large runoff is typical of these areas because soil and subsoil are nearly always saturated with moisture. Sapper (1935) considered vertical erosion to be the dominant process in the humid tropics and to be even more strikingly displayed than in arid and semiarid regions. Slumping, landslides, and avalanches of saturated weathered soil masses contribute significantly to the steepening of canyon walls. During periods of long-continued rain, clay masses produced by deep chemical weathering become saturated and highly mobile, and flow out from beneath the roots of the vegetation to produce mudflows. Friese (1935) has also attached great significance to these "underground earthflows." Wentworth (1928) has described deep, steep-walled canyons on the humid northeastern side of Oahu, in the Hawaiian Islands, which he attributed to the fact that lack of freezing temperatures and wide temperature ranges along with high annual temperature and high annual rainfall accentuate the importance of chemical weathering and reduce mechanical weathering to a minimum.

Other observers (Chamberlin and Chamberlin, 1910) have noted striking differences between humid tropical land forms and those of the middle latitudes. One striking difference noted was the lack of an accumulation of rock fragments at the base of valley sides. Valley sides characteristically rise sharply from the valley floor and their slopes are clean of fragmented rock material. This was attributed to: (1) lack of freezing and thawing, which would produce talus; (2) the presence of deep chemical decay; and (3) the restraining effects of vegetation.

The relative resistance of limestone to the gradational processes is an interesting example of how climatic differences may be significant. In humid regions limestone is usually considered a "weak" rock. Areas underlain by limestone are generally lower than surrounding areas. This is the result not so much of the physical weakness of limestone as of its susceptibility to solution. In arid regions, however, where moisture is deficient and solution insignificant, we frequently find that limestone is a "strong" rock and commonly is a cliff or ridge former.

Persons who compare the topography of arid regions with that of humid regions are usually impressed by the greater angularity which the topography displays in arid regions. Humid regions most typically exhibit smooth, flowing slopes instead of the sharp and abrupt changes of slopes which are so common in arid regions. Although not the only reason for this difference,

FIG. 3.11. The angularity of arid-land topography versus the rounded slopes of humid-land topography. *A*. View looking northwest from Lipan Point toward Walhalla Plateau, Grand Canyon, Arizona. (Fairchild Aerial Surveys, Inc. photo.) *B*. The Smoky Mountains near Brevard, North Carolina. (Photo by Elliot Lyman Fisher.)

certainly one of the major ones is the unimportance of downslope movement of material by creep in arid regions. Abundant moisture is essential to most movements of soil and subsoil downslope, and where it is deficient the smoothing effects of the mass movement of weathered materials are lacking, with the result that the varying resistances of the rock materials beneath the slopes are sharply reflected in the topography. Other contributing factors, related to climatic differences, are the slower rate of soil formation and the lack of a continuous vegetal cover.

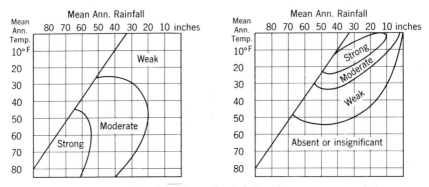

FIG. 3.12. Diagrams suggesting the effects of rainfall and temperature variations upon the relative importance of chemical (*left*) and mechanical (*right*) weathering. (After Louis Peltier.)

Even within a given climatic region there may be local variations in climatic factors which affect significantly geomorphic processes. Differences in such factors as altitude and exposure to moisture-bearing winds and insolation are important. It has been noted by various observers that some south-facing slopes of east-west valleys in the northern hemisphere are less steep than adjacent north-facing slopes. Variations in the microclimatology of the two slopes are generally believed to be responsible for this. North-facing slopes have a snow cover longer; experience fewer days of freeze and thaw; retain their soil moisture longer and probably have a better vegetal cover, all of which are likely to result in less active erosion on these slopes than those facing the sun. Exposure may also significantly affect the size of glaciers in mountain areas because of its influence upon the rate of evaporation and melting of their snowfields.

Intensity of chemical weathering depends to a large degree upon an abundance of water and high air temperatures. It will be at a minimum in arid regions where moisture is scarce and in cold regions where reduced rates of chemical reactions result from low temperatures and low moisture conditions (low at least as far as its availability for chemical reactions because water is in the frozen state much or all of the time). Mechanical disintegration of rocks also depends to a large degree upon the presence of water, but it is most rapid where there is repeated freezing and thawing of water in rocks. It will thus be apparent that there are two types of climatic regions under which mechanical weathering is at a minimum: those regions where temperatures are too high for freezing to take place and those where it is so

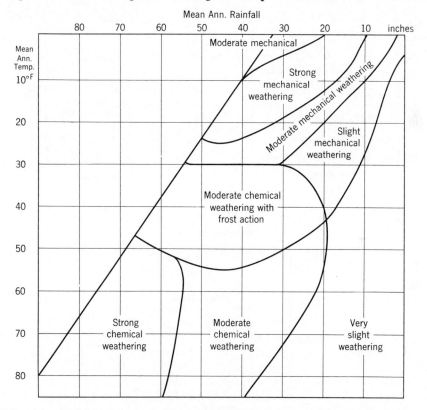

FIG. 3.13. Diagram suggesting the relative importance of various types of weathering under varying temperature and rainfall conditions. (After Louis Peltier.)

cold that water rarely thaws. The relationships between the two climatic factors of temperature and rainfall and the relative intensity of chemical and mechanical weathering are graphically suggested in Figs. 3.12 and 3.13. Other geomorphic processes will each be at a maximum under a certain set of climatic conditions and at a minimum under another set of conditions.

The rapidity and type of erosion by running water are affected by such factors as: permeability of the soil and bedrock (permeability may be determined by whether the ground is frozen or saturated with water as well as by such factors as porosity and arrangement of materials); attitude of the beds; degree of induration of the bedrock; amount and kind of vegetation; rate of evaporation and transpiration; intensity of precipitation; and the frequency of rain-producing storms. The last four factors are either directly or indirectly related to climatic factors, and to a considerable degree per-

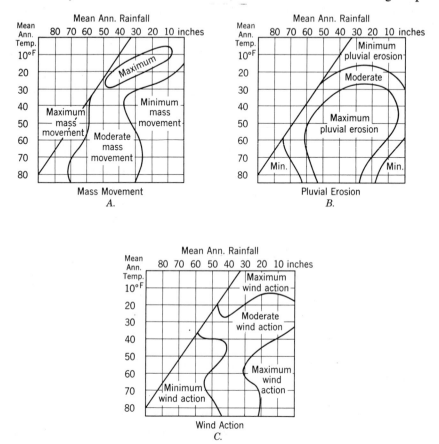

FIG. 3.14. Diagrams suggesting the relative importance of mass-wasting, stream erosion, and wind erosion under varying climatic conditions. (After Louis Peltier.)

meability of the soil is also. As shown in Chapter 11 differences in these last four factors especially are responsible for differences in the landscapes of arid and humid regions rather than difference in the geology.

THE CONCEPT OF MORPHOGENETIC REGIONS

If it is recognized that different geomorphic processes produce different land forms, it follows that the characteristics of the topography should to a

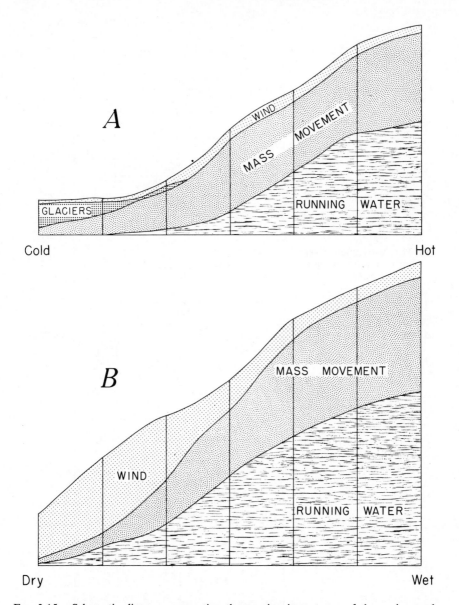

FIG. 3.15. Schematic diagrams suggesting the varying importance of the major gradational processes in relation to variations in *A*, temperature, and *B*, effective precipitation. (After D. I. Blumenstock and C. W. Thornthwaite in *Yearbook of Agriculture
for 1941*.)

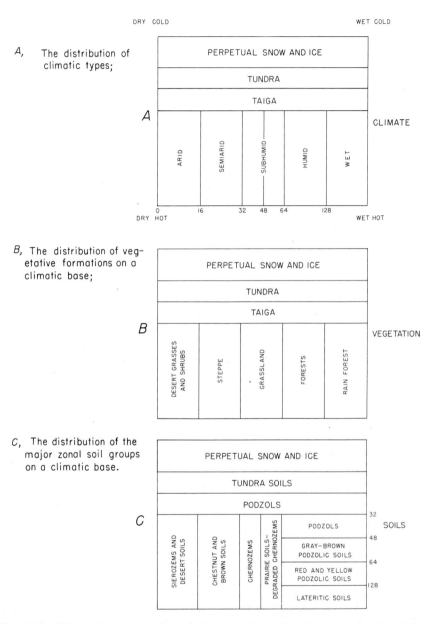

A, The distribution of climatic types;

B, The distribution of vegetative formations on a climatic base;

C, The distribution of the major zonal soil groups on a climatic base.

FIG. 3.16. Schematic representation of the interrelationships between climatic, vegetation, and soil types. (After D. I. Blumenstock and C. W. Thornthwaite in *Yearbook of Agriculture for 1941*.)

certain degree reflect the climatic conditions under which the topography developed. Thus a particular climatic regime is marked by a particular assemblage of geomorphic processes which in turn will have their own topographic expression. If the geologic materials upon which geomorphic processes operated were the same everywhere, relationships between topography and climate would, of course, be much more striking than they are. But lithology and geologic structure are not the same everywhere, nor have processes everywhere been operating upon the terrain for equal lengths of time. Thus one has difficulty in attributing as much significance to climate as did Sauer (1925) when he stated, "We may ··· assert that under a given climate a distinctive landscape will develop in time, the climate ultimately concealing the geognostic factor [the kind and attitude of materials in the earth's crust] in many cases." We can, perhaps, agree with him that "it is geographically much more important to establish the synthesis of natural landscape forms in terms of the individual climatic area than to follow through the mechanics of a single process, rarely expressing itself individually in a land form of any great extent."

Geographers, more than geologists, are likely to recognize the significance of the similarity in the world distribution patterns of physiographic regions, soil groups, vegetation types, and climatic regions. The common denominator which helps to explain these similarities is climate. In Europe, Büdel (1944, 1948) has suggested the existence of *formkreisen* or what may be called *morphogenetic regions,* and Peltier (1950) has put forth a tentative list of such regions. The concept of a morphogenetic region is that under a certain set of climatic conditions particular geomorphic processes will predominate and hence will give to the landscape of the region characteristics that will set it off from those of other areas developed under different climatic conditions. Peltier postulated nine morphogenetic regions and attempted rough quantitative definitions of them in terms of temperature and moisture conditions and suggested the dominant geomorphic processes in each.

Figure 3.17 graphically suggests the climatic attributes of the morphogenetic regions of the accompanying outline.

Except for the suggested boreal and maritime morphogenetic regions, all those listed in the outline have to some degree been recognized. Davis (1899) in his development of the idea of the "normal cycle" applied it chiefly to those areas which Peltier classed under the moderate morphogenetic region. Davis further recognized to some extent the importance of climatic differences in classing under "climatic accidents" the arid (1905) and glacial (1909) cycles. Cotton (1947) suggested that perhaps there are sufficient differences in the relative importance of deflation and lateral erosion by streams in arid, semi-arid, and savanna regions to justify recognizing each as a distinct geomorphic

MORPHOGENETIC REGIONS

(After Peltier)

Morpho-genetic Region	Estimated Range of Average Annual Temperature, degrees	Estimated Range of Average Annual Rainfall, inches	Morphologic Characteristics
Glacial	0–20	0–45	Glacial erosion Nivation Wind action
Periglacial	5–30	5–55	Strong mass movement Moderate to strong wind action Weak effect of running water
Boreal	15–38	10–60	Moderate frost action Moderate to slight wind action Moderate effect of running water
Maritime	35–70	50–75	Strong mass action Moderate to strong action of running water
Selva	60–85	55–90	Strong mass action Slight effect of slope wash No wind action
Moderate	38–85	35–60	Maximum effect of running water Moderate mass movement Frost action slight in colder part of region No significant wind action except on coasts
Savanna	10–85	25–50	Strong to weak action or running water Moderate wind action
Semiarid	35–85	10–25	Strong wind action Moderate to strong action of running water
Arid	55–85	0–15	Strong wind action Slight action of running water and mass movement

region. Many workers in Europe and America have called attention to the distinctive processes in periglacial regions as related to the climatic characteristics of such regions (see p. 411). Men like Sapper (1935) and Friese (1935) have emphasized the greater intensity of chemical denudation in the humid tropics (selva) than in the humid middle latitudes and the importance

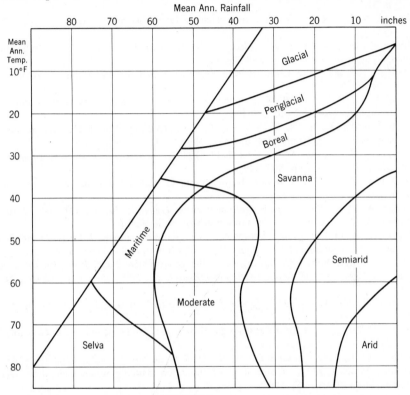

FIG. 3.17. Diagram indicating possible climatic boundaries of morphogenetic regions. (After Louis Peltier.)

of the heavy rain-forest vegetation in the maintenance of steep slopes. Thus the basic idea of morphogenetic regions is not entirely new. What Peltier and others suggest is the need for greater emphasis upon climatic control of the gradational processes than has generally been given. Certainly in the past more emphasis has been placed upon processes per se than upon processes as conditioned by climate. The concept of process as developed by Davis and his followers should be expanded to include processes as controlled by particular climatic regimes, which, as Peltier stated, "represent the sum of all geomorphic processes in those peculiar proportions which characterize a particular climatic region."

REFERENCES CITED IN TEXT

Blackwelder, Eliot (1925). Exfoliation as a phase of rock weathering, *J. Geol., 33,* pp. 793–806.

Blackwelder, Eliot (1933). The insolation hypothesis of rock weathering, *Am. J. Sci., 226,* pp. 97–113.

Bowen, N. L. (1922). The reaction principle in petrogenesis, *J. Geol., 30,* pp. 177–198.

Büdel, J. (1944). Die morphologischen Wirkungen des Eiszeitklimas im gletscherfrein Gebiet, *Geol. Rundschau, 34,* pp. 482–519.

Büdel, J. (1948). Die Klima morphologischen Zonen der Polarländer, *Erkunde, 2,* pp. 25–53.

Chamberlin, T. C., and R. T. Chamberlin (1910). Certain valley configurations in low latitudes, *J. Geol., 18,* pp. 117–124.

Chamberlin, T. C., and R. D. Salisbury (1904). *Geology,* Vol. 1, Geologic processes and their results, p. 2, Henry Holt and Co., New York.

Chapman, R. W., and M. A. Greenfield (1949). Spheroidal weathering of igneous rocks, *Am. J. Sci., 247,* pp. 407–429.

Cotton, C. A. (1947). *Climatic Accidents,* pp. 11–100, Whitcombe and Tombs, Ltd., Wellington.

Davis, W. M. (1899). The geographical cycle, *Geog. J., 14,* pp. 481–504. Also in *Geographical Essays,* pp. 249–278, Ginn and Co., New York.

Davis, W. M. (1905). The geographical cycle in an arid climate, *J. Geol., 13,* pp. 381–407. Also in *Geographical Essays,* pp. 296–322, Ginn and Co., New York.

Davis, W. M. (1909). Complications of the geographical cycle, *Proc. 8th Int. Geol. Cong.,* pp. 150–163. Also in *Geographical Essays,* pp. 279–295, Ginn and Co., New York.

Farmin, Rollin (1937). Hypogene exfoliation in rock masses, *J. Geol., 45,* pp. 625–635.

Friese, F. W. (1935). Erscheinungen des Erdfliessens im Tropenwalde, *Z. Geomorph., 9,* pp. 88–98.

Gilbert, G. K. (1890). Lake Bonneville, *U. S. Geol. Survey, Mon. 1,* pp. 340–345.

Gilluly, James (1949). Distribution of mountain building in geologic time, *Geol. Soc. Am., Bull. 60,* pp. 561–590.

Gilluly, James (1950). Reply to discussion by Hans Stille, *Geol. Rundschau, Band 38, Heft 2,* pp. 103–107.

Goldich, S. S. (1938). A study in rock-weathering, *J. Geol., 46,* pp. 17–58.

Jahns, R. H. (1943). Sheet structure in granites: Its origin and uses as a measure of glacial erosion, *J. Geol., 51,* pp. 71–98.

Krynine, P. D. (1936). Geomorphology and sedimentation in the humid tropics, *Am. J. Sci., 232,* pp. 297–306.

Lyon, T. L., and H. O. Buckman (1943). *The Nature and Properties of Soils,* 4th ed., The Macmillan Co., New York.

Malott, C. A. (1928). An analysis of erosion, *Proc. Indiana Acad. Sci., 37,* pp. 153–163.

Matthes, F. E. (1937). Exfoliation of massive granite in the Sierra Nevada of California, *Geol. Soc. Am., Proc. for 1936,* pp. 342–343.

Peltier, Louis (1950). The geographic cycle in periglacial regions as it is related to climatic geomorphology, *Assoc. Am. Geog., Ann., 40,* pp. 214–236.

Reiche, Parry (1950). *A Survey of Weathering Processes and Products,* rev. ed., University of New Mexico Press, 95 pp.

Sapper, K. (1935). *Geomorphologie der feuchten Tropen,* B. G. Teubner, Berlin.

Sauer, C. O. (1925). The morphology of landscapes, *Univ. Calif. Publs. Geog., 2,* pp. 19–53.

Sharpe, C. F. S. (1938). *Landslides and Related Phenomena,* Columbia University Press, 137 pp.

Taber, Stephen (1930). The mechanics of frost heaving, *J. Geol. 38,* pp. 303–317.

Wentworth, C. K. (1928). Principles of stream erosion in Hawaii, *J. Geol., 36,* pp. 385–410.

Additional References

Barrell, Joseph (1917). Rhythms and the measurement of geological time, *Geol. Soc. Am., Bull. 28,* pp. 745–914.

Chamberlin, T. C. (1909). Diastrophism as the fundamental basis of correlation, *J. Geol., 17,* pp. 3–59.

Griggs, D. T. (1936). The factor of fatigue in rock exfoliation, *J. Geol., 44,* pp. 783–796.

Scott, H. W. (1951). The geological work of the mound-building ants in western United States, *J. Geol., 59,* pp. 173–175.

Strahler, A. N. (1940). Landslides of the Vermilion and Echo Cliffs, northern Arizona, *J. Geomorph., 3,* pp. 285–301.

4 · Weathering, Soil Processes, and Mass-Wasting

GEOMORPHIC SIGNIFICANCE OF WEATHERING

Some of the products of weathering, such as soils, are not strictly geomorphic phenomena, yet may have great geomorphic, geologic, and economic significance. Rock weathering will be considered under four headings: (*a*) as an aid to mass-wasting and erosion; (*b*) as a factor in the general lowering of land surfaces; (*c*) as it contributes to the creation and modification of land forms; and (*d*) as a major process involved in the formation of regolith and soils.

Weathering as an aid to mass-wasting and erosion. The combined weathering processes result in either weakening, fragmentation, or decomposition of bedrock at and near the earth's surface. The portion of the earth's crust in which this happens is called the *zone of weathering.* It cannot be said with certainty to how great a depth weathering may extend. There are weathered materials as much as 100 feet thick in the Piedmont region of the southeastern United States, and greater thicknesses have been reported from humid tropical areas. The position of the water table and the length of time that weathering has been in operation have an important influence upon the depth to which some of the weathering processes may extend. Most active chemical weathering takes place in the zone of aeration above the water table. Oxidation effects are seldom conspicuous below this level, but solution and hydration may extend below it.

Erosional agents, particularly if carrying rock debris, can break off and remove solid bedrock, but this is more easily done if the bedrock has been previously weakened by weathering. Hydraulic action, deflation, and glacial scouring are most effective on unconsolidated materials, although some hydraulic action is possible on thinly laminated and weakly indurated materials. It is impossible to give any quantitative estimate of the extent to which rock weathering accelerates erosion, but it most certainly contributes significantly to the effectiveness of most erosional processes.

Lowering of surfaces by weathering. In certain areas, particularly those underlain by limestone, dolomite, or gypsum, solution may produce a lowering en masse of topographic surfaces. As noted before, limestone areas under humid climates usually are lower than adjacent areas of clastic rocks, but in arid regions limestone and dolomite usually are resistant rocks. The possibility of the solutional lowering of extensive areas needs to be kept in mind in the interpretation of topographic surfaces at different levels. Even in areas of insoluble rocks differential mass weathering may be significant, unless the lithology is remarkably uniform.

Creation and modification of land forms. Some of the products of differential weathering are hardly to be classed as land forms but rather are geologic features which add variety to topographic surfaces. Examples of such are stone lattice, honeycombed rocks, and stratification ribs produced by differential etching of thinly bedded rock. Other features such as hollows and niches in rock walls and their reciprocal features, projecting ledges, juts, and other lesser prominences, are part of the sculptural detail on larger land forms.

Weathering pits are common on limestones and often resemble potholes produced by abrasion. They are also found in granites and similar rocks. Smith (1941) has described circular or elliptical weathering pits in the southern Piedmont of South Carolina which are 10 to 40 feet in diameter. They are most common on flat surfaces, and most of them contain weathered granite in varying degrees of decomposition. Algae and mosses seem to have contributed to their formation through their ability to retain moisture and add carbon dioxide to it. Striking weathering pits in granite near Llano, Texas, which are surrounded by raised annular ridges have been described by Blank (1951). Similar pits have been described in granite from Georgia, Bear Mountain, New York, and Yosemite National Park.

The Great Stone Face, in New Hampshire, which Hawthorne made famous, is a jut or prominence produced by differential weathering. Features produced by weathering with profiles which if viewed with imagination may suggest humans or animals are found in numerous places. The Garden of the Gods near Colorado Springs is well known for such forms.

It is common to find at high altitudes accumulations of riven rocks known as *felsenmeere*. These boulder fields attest to the rapidity of the weathering processes at high altitudes. Many of the boulders are spheroidal in shape as a result of the exfoliation of massive rock blocks. Related to spheroidal boulders but on a larger scale are the exfoliation domes discussed in the preceding chapter. White (1945) concluded that Stone Mountain and many lesser so-called exfoliation domes in the southern Piedmont were not the result of exfoliation of thin sheets of rock but were produced mainly by granular disintegration accompanying chemical weathering. These monoliths seem to have developed chiefly on massive rocks such as granite and granite gneiss. Chapman (1940) pointed out that the Percy Peaks in New

Hampshire, similar granite monoliths, developed where the Conway granite lacked the complex jointing which would favor weathering of the rock into blocks.

Differential weathering helps to develop and modify such upstanding topographic firms as columns, pillars, pedestal rocks, earth pillars, and toadstool

FIG. 4.1. Weathering pits in granite near Katemcy, Texas. (Photo by H. R. Blank.

rocks which, because of their bizarre shapes, are sometimes collectively classed as *hoodoo rocks*. Bryan (1925) concluded that those which he studied in southwestern United States were chiefly the work of rainwash plus the effects of mechanical and chemical weathering. In areas of sedimentary rock, pedestal rocks usually have a hard cap rock which has protected weaker strata beneath, but similar forms are found in massive igneous rocks of uniform lithology. The pedestal rocks of the southern Piedmont have attracted considerable attention and various explanations of them have been offered. Crickmay (1935) attributed the more rapid destruction of their basal portions to granular disintegration produced through increase in volume by hydration.   He thought that hydration was at a maximum near their bases because there was less rapid evaporation here than on their more exposed tops.   Frost action and rainwash were believed to have been of minor importance.   White (1944) explained them as the result of breaching

Fig. 4.2.   A profile rock near Pennington Gap, Virginia, that is the product of differential weathering and rainwash on massive sandstone.

of surficial veneers of iron oxides formed by capillary action in the interstices of partially weathered granite.   These indurated veneers were breached by the abrasive action of running water or by solution by humic and carbonic acids produced by mosses.   After a breach was made in the indurated veneer the peculiar forms then developed.

Weathering contributes in an important way to the recession of cliffs and scarps.   Escarpments which develop where gently dipping sedimentary rocks of varying resistance are subjected to degradation are often called *weathering escarpments*.   The name is not too appropriate because it overemphasizes the role which weathering plays in their formation. Weathering does contribute to their development, but other processes, such as mass-wasting, sheet-wash, and stream erosion, are equally or more important. The backweathering of scarps has been considered particularly significant in arid regions and has been thought to be a major process involved in the formation of the pediments of arid regions (see Chapter 11). Weathering on valley sides con-

FIG. 4.3.   Erosion remnants of Dawson arkose in Monument Park, Colorado Springs, Colorado. (Photo by N. H. Darton, U. S. Geol. Survey.)

tributes significantly to valley widening in the early stages of a geomorphic cycle, and this process and mass-wasting become dominant in the final stage of the fluvial geomorphic cycle.

Talus slopes or *scree* are among the few land forms which are produced mainly by weathering aided by mass-wasting. Talus is generally considered to be largely a product of frost action aided by gravity, but hydration may contribute more to the spalling off of rock than is generally realized. Blackwelder (1942) has called attention to the occasional erosive action of rolling debris or talus moving down steep slopes resulting in the formation of V-shaped gullies. Thus such an apparently simple process as the formation of talus involves mass-wasting as well as weathering and may under extreme circumstances be accompanied by erosion. Again the fundamental principle is illustrated that many land forms are the product of several inter-related processes rather than simple in origin. Because it is difficult to cite land forms produced solely by weathering, it should not be thought that weathering is of little geomorphic significance. Rather the process is so intimately related to other processes that its effects cannot be readily separated.

Fig. 4.4. A talus cone near Challis, Idaho. (Photo by C. L. Heald.)

SOILS

Soils are, next to water, man's most vital natural resource. Some plants are now grown from chemical solutions, but this method is not likely ever to produce much of our food supply. Hence, from an economic viewpoint the production of soils is the most significant result of rock weathering. Yet, if soils were significant only as an economic resource, a discussion of them could hardly be justified in a book on geomorphology. So important is a knowledge of soils that no geomorphologist today is adequately trained who lacks an appreciation of the soil-forming processes and a basic understanding of soil characteristics. The soils of every region reflect its geomorphic history and in some areas do so better than do the land forms.

No entirely satisfactory definition of soil has been formulated even by the pedologists or students of soils. Perhaps the following definition patterned upon one given by Bushnell (1944) will suffice. *Soil* is a natural part of the earth's surface, being characterized by layers parallel to the surface resulting from modification of parent materials by physical, chemical, and biological processes operating under varying conditions during varying periods of time.

Formation of soils. It is now commonly agreed that there are five major factors which condition the development of soils. These five factors are:

a. Climate—particularly temperature and the amount and kind of precipitation.

b. Topography—especially as it affects both external and internal drainage.

c. Soil biota—including both vegetative cover and organisms within the soil.

d. Parent material—including the texture and structure of the material as well as its mineralogic and chemical composition.

e. Time—the length of time that pedologic processes have been operating.

Three of these at least, topography, parent material, and time, are geologic in nature. Climate is perhaps more geographical in character but to a considerable degree is influenced by topographic features.

That soils owe some of their characteristics to the kind of parent materials from which they were derived is generally agreed, but pedologists disagree about the extent to which this is true. Soil classifications once were largely geologic in nature. Soils were commonly divided into two major groups, residual and transported. Residual soils were classified according to the type of rock from which they were derived. Thus, limestone soils, sandstone soils, gravelly soils, granitic and other soils were recognized. Transported soils were classified as glacial, alluvial, colluvial, eolian, lacustrine, and marine, according to the way in which the materials were deposited. Such a classification may have value, but it fails to take into account that there are factors other than parent material which contribute to the differentiation of soil types.

We owe to Dokuchaiev, a Russian, and his followers the realization of the importance of climate and soil biota, particularly the type of vegetation, in the differentiation of soil types. Geologists at first were skeptical of the claim made by Russian pedologists, beginning about 1870, that, regardless of what the parent material was, with similar topographic, climatic, and vegetative conditions and after equal time intervals the soils produced were essentially the same. This is probably not strictly true but it is nearer to the truth than many geologists realized.

When a vertical section is made through soil it is found that it shows a series of more or less distinct layers or zones which, although differing in their chemical and physical characteristics, are related genetically to each other. These successive vertical layers constitute the *soil profile*. A simple profile has three distinct divisions, which are designated by pedologists as the *A*, *B*, and *C* horizons. Actually each of these three horizons may be subdivided into several distinct layers. The *A* horizon or group of layers lies immediately beneath the surface and is often described as the *eluvial* (washed out) layer because material has been removed from it and carried downward into

FIG. 4.5. A soil profile in Illinoian till. The man's hand rests at the contact between the *A* and *B* horizons, and his foot is at the contact between the *B* and *C* horizons. (Photo by Engineering Experiment Station, Purdue University.)

the *B* horizon. Some material was leached out of the *A* horizon and carried down in solution and precipitated in the *B* horizon, but the fine colloidal clays were mechanically removed by descending soil water. Although the *A* layer has lost material to the *B* horizon, it has also had varying amounts of organic materials added to it. The *B* layer is designated as the *illuvial* (washed in) layer because of the presence in it of minerals which were carried down into it. Iron and aluminum minerals including clays accumulate here. The *B* layer is commonly called the subsoil. In it hardpan and claypan layers develop. In arid regions calcium carbonate and other salts may accumulate just below the *B* horizon in a horizon designated as the *Cca* layer. The *A* and *B* layers together constitute the *solum* (soil) and are the parts of the soil profile in which the soil-forming processes have greatly modified the original materials. The *C* layer or parent material below does not exhibit this high degree of alteration. The *C* horizon may display evidence of weathering, particularly of oxidation, hydration, and lime accumulation, but weathering has not progressed to the point at which the original characteristics of the parent materials are no longer recognizable.

Geologic age affects the thickness and the degree of maturity of a soil profile. A mature soil profile exhibits well-developed horizons. If the soil-forming processes have not been in operation long, the soil profile will be immature and the various layers and horizons may be indistinct or lacking. Geologic time also affects the depth to which the soil-forming processes have penetrated and hence affords a possible method of determining the relative ages of soils. Comparative depths of leaching and thickness of *B*-zone development have been utilized extensively by glacial geologists to determine the ages of glacial deposits.

Topography profoundly affects the soil profile. On steep slopes, the profile may never become mature because erosion removes weathering products as rapidly as they form. Under these conditions we find *truncated profiles* in which one or both of the upper horizons have been removed, if they were ever present. Topography affects the amount of water penetrating the soil through its control of runoff and infiltration of water. It further influences the position of the water table and thereby the depth of penetration of most chemical weathering processes. To a considerable degree, the amount of relief governs the circulation of water through the soil and parent material and thus determines whether the soil profile develops under conditions of good or poor drainage.

As indicated by Lyon and Buckman (1943) three main ideas are involved in the present-day interpretation and classification of soils: first, that soils are everywhere the product of certain chemical and physical processes; secondly, that, as a result of variations in climate, soil biota, parent material, age, and topography, soils differ from place to place, particularly in the later stages of their development; and thirdly, that as soils mature the original differences related largely to parent materials become less obvious and climate becomes a dominant factor in differentiating soils. Climate acts directly through temperature and rainfall variations and indirectly through its influence upon the amount and kind of vegetation. Mature and old soils in areas that are climatically alike are strikingly similar, and it is possible to classify them in soil groups that developed under similar climatic conditions. Soils that can be so grouped are called *zonal soils.* Within most areas where a zonal grouping of soils is possible there are likely to be some soils which reflect more the effects of local conditions than broad climatic controls. Such soils are classed as *intrazonal.* Soils of bog areas and alkali flats are examples of these. Locally there may also be soils that lack definite profiles. They are called *azonal soils.* Stream alluvium, dry sands, and colluvium are examples of azonal soils.

Description of major soil groups. The following brief outline of the major soil groups of the world follows that of Lyon and Buckman (1943) and indicates the climatic and vegetative conditions under which they formed as well as certain significant characteristics of the soil profiles in each group.

Tundra soils develop under the tundra type of vegetation at high altitudes and high latitudes. Drainage conditions are poor and usually boggy. A

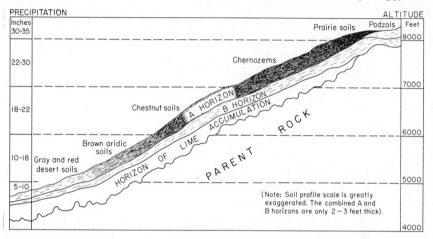

FIG. 4.6. A schematic diagram showing the climatic zonation of soils from the arid Bighorn Basin at the left to the humid Bighorn Mountains at the right. (After James Thorp.)

permanently frozen substratum underlies them. The profile is shallow, and much undecomposed organic material is found at the surface.

Podzol soils possess well-developed *A, B,* and *C* horizons. The surface material is organic matter, and under it is a whitish or grayish layer which gives them their name, which comes from two Russian words meaning under and ash. Below the gray layer there is a zone in which iron and aluminum minerals accumulate. The *A* and *B* layers are strongly acid. They develop under coniferous and mixed hardwood forests.

Laterites are soils which form under hot and humid climatic conditions and under forest vegetation. They have a thin organic layer over a reddish leached layer, which in turn is underlain by a still deeper red layer. Hydrolysis and oxidation have been intense. They are rather granular soils. They are confined mainly to tropical and subtropical regions, but some soils in the middle latitudes have been described as lateritic.

Chernozems originate under tall-grass prairie vegetation. The name is from the Russian word for black earth and suggests the color and high organic content of the *A* horizon. The *B* layer exhibits an accumulation of calcium carbonate rather than leaching. Columnar structure is common. They are among the most fertile soils.

Chestnut soils are brown or grayish brown soils that developed under short-grass vegetation in areas slightly drier than those that produced chernozems. Secondary lime is found near the surface. The soil profile is weakly developed.

Brown aridic soils are found around the margins of deserts and semiarid regions. They have a low organic content and are highly calcareous.

Gray desert soils (Sierozems) and *reddish desert soils* develop under desert or short-grass vegetation. Calcium carbonate accumulates near the surface. The sierozems are found in the continental deserts, and the reddish desert soils in what are called the subtropical deserts.

Non-calcic brown soils are soils that form in areas which originally had forest or brush vegetation. Weak podzolization makes the surface layer slightly acid.

Soil series. Within any group of soils there are many varieties related to local variations in topography, parent material, and climate. For purposes of detailed mapping soils are divided into soil series. It is with soil series that the geologist is most likely to work. To him the concept of a soil series is as important as that of the soil profile.

A *soil series* consists of a group of soils having horizons of similar origin, character, and arrangement in the soil profile which were derived from similar parent material. Except for differences in the texture of the surface soil and upper subsoil, the members of a soil series have essentially the same color, structure, and drainage conditions. The individual members of a soil series are called soil types and are differentiated from each other by slight differences in the lithology of the parent material and the topographic conditions under which they developed. A common soil series in the middle western states is the Miami series. The soils of this series were derived from glacial till belonging to the Tazewell substage of the Wisconsin glacial stage. Included in this series are such soil types as the Miami fine sandy loam, Miami loam, Miami silt loam, and Miami silty clay loam. It should be apparent that because the members of a soil series were derived from common parent material they delineate mappable geologic units.

Probably even more useful to a geomorphologist than the concept of a soil series is that of a soil catena. A *soil catena* consists of a group of soils within a particular soil region which developed from similar parent material but differ in the characteristics of their profiles because of the varying topographic and drainage conditions under which they formed. The topographic conditions under which a catena develops may vary through flat uplands, slight slopes, moderate slopes to steep slopes. In addition, on upland flats there

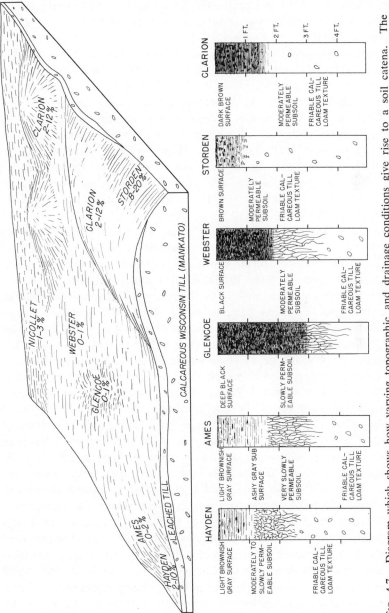

FIG. 4.7. Diagram which shows how varying topographic and drainage conditions give rise to a soil catena. The varying degrees of black indicate the relative abundance of organic matter in the soil profiles. Percentages indicate the steepness of slope. (After W. H. Scholtes, R. V. Ruhe, and F. F. Rieken.)

may be shallow to deep depressions. Each of these topographic settings affects the character of the soil profile chiefly because of the varying surface and internal drainage conditions. Thus a soil catena has both lithologic and topographic implications.

A common soil catena found on the till plains of Ohio, Indiana, and Illinois consists of the Hennepin, Miami, Crosby, Bethel, Brookston, Kokomo, and Carlisle soils. All these soils were derived from pebbly clay till belonging to the Tazewell substage of the Wisconsin glacial stage, but they exhibit notable differences in their profiles because of the different topographic and drainage conditions under which each evolved. The Hennepin soils are found on steep slopes (20–55%); the Miami soils on moderate slopes (4–15%); the Crosby soils on slight slopes (1–2%); the Bethel soils on flat rises (0–1%); the Brookston soils in slight depressions (0–1%); the Kokomo soils in deep depressions (0–1%); and the Carlisle soils (mucks) in the deepest depressions (0–1%).

Complex or polygenetic soils. Some soils have profiles that could not have developed under a single set of soil-developing controls, but consist of a younger profile developed under existing topographic and climatic conditions superposed upon an older profile formed under different conditions. Such soils have been called *polygenetic* or *complex soils* (Bryan and Albritton, 1943). Two types of change produce a polygenetic soil—climatic change with attendant vegetation changes and change in topographic and associated drainage conditions. It is generally recognized that important changes occurred during the Pleistocene as conditions changed from glacial to interglacial. Extensive areas have experienced cooler and rainier climates alternating with warmer and drier. World climates have undergone significant changes since the recession of the last ice sheet. Examples of polygenetic soils produced as the result of these changes have been observed. Some of the so-called degraded chernozems and gray podzols of Russia have been interpreted as having been originally true chernozems, developed when the steppe vegetation extended farther poleward than it does now, which have undergone alteration through podzolization as the forests shifted southward. Examples have also been described in Russia of chernozem and podzol profiles developed over red residuum thought to be lateritic in nature. They suggest that warmer and more humid conditions than those at present prevailed at the time that the lower lateritic profile formed. Bryan and Albritton (1943) cited examples in Minnesota where subsoils on both till and gravel show marked limonitic staining which at present is being replaced by caliche, indicating a postglacial change toward more arid conditions. They also described a complex soil profile in the Trans-Pecos region of Texas, in which there are three layers of caliche suggestive of three periods of aridity with accompanying caliche deposition.

Change in topography accompanying stream dissection may result in modification of both the external and internal drainage conditions under which a soil profile has developed. Leighton and MacClintock (1930) have described complex soil profiles upon Illinoian tills which show silttil characteristics above gumbotil below and which they thought resulted from improved drainage conditions accompanying dissection of the till plain. *Gumbotil profiles* develop under conditions which exist on flat and poorly drained till plains.

Fig. 4.8. A paleosol or buried soil profile on Illinoian till beneath Wisconsin till. The dark band above the maddox is organic material in the *A* horizon of the paleosol.

In the gumbotil profile the *B* horizon is composed of a sticky clay mass produced from advanced chemical weathering. Under conditions of good drainage the *B* horizon in Illinoian till is composed of a much looser silt-like material. The profile so developed has been called a *silttil profile*. Intermediate profiles between gumbotil and silttil are called *mesotil profiles*. Geomorphologists have barely begun the study and interpretation of polygenetic or complex soil profiles, but it should be apparent that they may throw considerable light upon the recent geomorphic history of an area.

Composite soil profiles. A soil profile may develop through two distinct parent materials and be composite in nature. Recognition of this may aid in the interpretation of the geologic and geomorphic history of a region. Composite profiles are occasionally found in glaciated areas. A soil profile may start to develop upon glacial till and become buried with a thin veneer of loess, windblown sand, or outwash sand and gravel. At first the old profile would be classed as a fossil or buried soil, but when the new profile developing downward through the covermass reaches the buried soil the profile becomes composite. It will usually be possible to distinguish the two parts of the profile, but if loess is over till it may be difficult because of the similarity in the soils formed from the two materials. It may be only by careful petrographic study that the composite nature of the profile is recognized.

Fossil soils. *Fossil soils* or *paleosols* are those which have been buried subsequent to their formation. Especially important in this class are the soils which formed during interglacial ages and were subsequently buried by glacial till, outwash, wind-blown sands, and loess. Many examples of such soils are known, and they, along with associated organic deposits, throw light upon the climatic and topographic conditions under which they formed.

Geomorphic Applications of Soil Studies

Soils are the result of biochemical and physical processes operating upon earth materials under various topographic and climatic conditions. They reflect as much as do land forms the climatic and geomorphic history of the region in which they evolved. This is particularly true of the *B* horizon of the soil profile. Wooldridge (1949) has stressed the point that present-day topography contains many facets of varying ages which have been exposed to the soil-forming processes for different lengths of time and in some areas under markedly different climatic conditions. These varying histories are reflected in the soil profiles. He cited examples of three distinct facets of the landscape of the chalk region of England which are sharply and readily distinguishable from the differences in the soils upon each of the topographic surfaces.

Paleosols and polygenetic soils may furnish information from which it is possible to reconstruct past climatic and topographic conditions. Paleosols may be found on buried landscapes, on exhumed portions of ancient landscapes, or upon features of the present topography which are relics of previous geomorphic cycles.

Wahlstrom (1948) has described an ancient weathered zone in the foothills of the Colorado Front Range near Boulder, Colorado, which is developed upon Pre-Cambrian granodiorite and lies below the Fountain arkose of Pennsylvanian age. The weathered zone is as much as 80 feet thick. An *A* horizon is lacking, but *B* and *C* horizons are well preserved. The *B* horizon consists of red lateritic-like material. The nature of the paleosol indicates that it was formed upon a topographic surface of low relief but with sufficient drainage to permit deep weathering. The climatic conditions suggested are those of high humidity and temperature such as characterize humid tropical or subtropical regions today. The soils that are forming today upon the granodiorite are the typical brownearths which characterize such semiarid regions as Colorado.

What have been called *duricrusts* have been described in Australia by Woolnough (1930) upon topographic surfaces which he thought had formed during a Miocene erosion cycle. These duricrusts are concentrations in the upper part of the soil profile of aluminous, siliceous, ferruginous, or calcareous materials. They were thought to have been formed under a semiarid climate with alternating wet and dry seasons by means of capillarity which brought into the upper soil during the dry season minerals which had been dissolved during the wet season. Under the present arid conditions which prevail today duricrusts are not forming in a significant amount but rather are undergoing destruction. Dixey (1941) has described masses of chalcedony from 1 to 6 feet thick in paleosols in south-central Africa which he thought were formed on erosion surfaces of middle to late Tertiary age. He considered them to be evidence of intense and widespread aridity at that time.

Bretz and Horberg (1949) have described polygenetic soil profiles developed on limestone gravels in the Llano Estacado and Pecos depression of New Mexico, which they thought recorded the effects of alternating wet and dry periods.

Caliche layers in the soil were interpreted as having been formed during periods of relative aridity. The effects of wetter periods are indicated by solution cavities in the caliche and solution pits in the limestone pebbles through which the soil profiles are developed. Soil profiles containing caliche are found on at least four different topographic surfaces, and the soil profiles on these surfaces exhibit differences in depth and maturity related to the ages of the surfaces upon which they exist. They concluded that detailed study of the soil profiles along with further geomorphic studies may give a fairly complete record of the climatic changes which have taken place in the area since early or middle Pliocene time down to the present.

Hunt and Sokoloff (1949) have described a paleosol which is widespread in the Rocky Mountain region. Its relations are such as to indicate that it is older in age than Lake Bonneville and the youngest moraines in the Wasatch Range but younger than the oldest moraines. Wherever it is found this paleosol is associated with a distinct topographic unconformity. The topography upon which it formed seems to have had as great relief as the present topography but slopes were much smoother than at present. Climatic conditions different from those of late Wisconsin and postglacial times are indicated by the fact that the soil-forming processes kept ahead of erosion.

More ancient paleosols include the weathered zones developed upon land surfaces and then buried under sediments which were subsequently lithified. These ancient weathered zones are usually associated with unconformities and have long been recognized for what they are, but modern pedology enables us to interpret more accurately their geomorphic and climatic implications.

MASS-WASTING AND ITS GEOMORPHIC SIGNIFICANCE

The processes involved in mass-wasting have been discussed briefly in the preceding chapter. We are concerned here with the topographic effects of mass-wasting. Generally speaking, the geomorphic effects and forms produced are not large-scaled although certain ones may be striking. The really

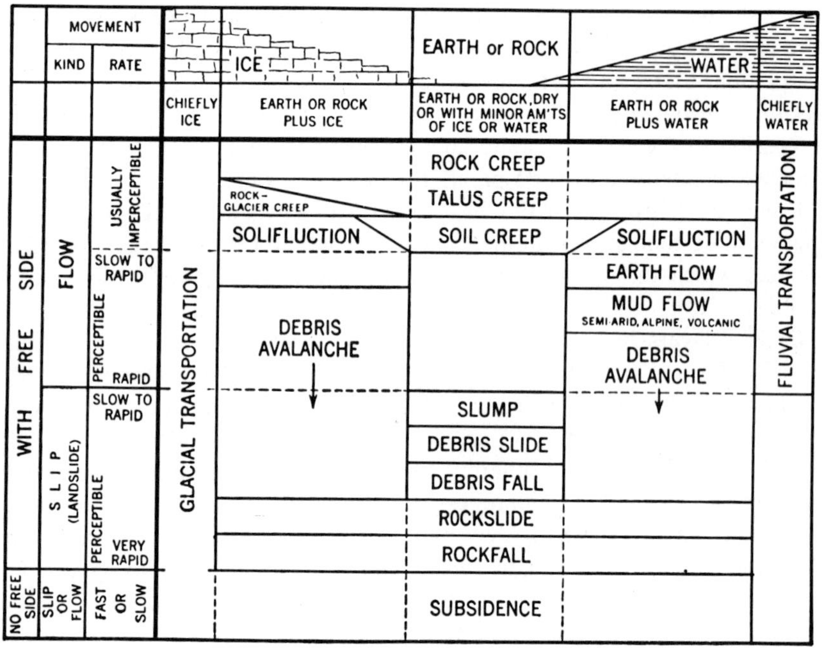

MOVEMENT		CHIEFLY ICE	EARTH OR ROCK PLUS ICE	EARTH OR ROCK, DRY OR WITH MINOR AM'TS OF ICE OR WATER	EARTH OR ROCK PLUS WATER	CHIEFLY WATER
KIND	RATE	(ICE)		(EARTH or ROCK)	(WATER)	
WITH FREE SIDE — FLOW	USUALLY IMPERCEPTIBLE			ROCK CREEP		
			ROCK-GLACIER CREEP	TALUS CREEP		
			SOLIFLUCTION	SOIL CREEP	SOLIFLUCTION	
	SLOW TO RAPID				EARTH FLOW	
	PERCEPTIBLE		DEBRIS AVALANCHE		MUD FLOW — SEMI ARID, ALPINE, VOLCANIC	
	RAPID				DEBRIS AVALANCHE	
WITH FREE SIDE — SLIP (LANDSLIDE)	SLOW TO RAPID			SLUMP		
				DEBRIS SLIDE		
	PERCEPTIBLE			DEBRIS FALL		
				ROCKSLIDE		
	VERY RAPID			ROCKFALL		
NO FREE SIDE — SLIP OR FLOW	FAST OR SLOW			SUBSIDENCE		

(Left side spanning label: GLACIAL TRANSPORTATION; right side spanning label: FLUVIAL TRANSPORTATION)

FIG. 4.9. A classification of types of mass-wasting. (After C. F. S. Sharpe, by permission of Columbia University Press.)

significant role of mass-wasting is in the unspectacular contribution that it makes to the slow reduction of land masses.

Creep and Solifluction

Creep. Sharpe (1938) recognized four types of creep, soil creep, talus creep, rock-glacier creep, and rock creep. The effects of soil creep are usually not particularly apparent except upon vegetation and man-made structures. It is made evident through such features as curved trees and tilted fence, telephone, and telegraph poles. Soil creep does aid in slope wash and sometimes produces miniature scarps and shallow, channelless depressions on slopes. Frost heaving (Taber, 1930) is probably the most important contributory process to soil creep, although heating and cooling and wetting and drying of the mantle along with the wedging action of root growth aid it.

Talus or scree is one of the conspicuous products of the weathering of cliffs, scarps, and hill- and mountain-sides. Talus is composed mainly of large-sized materials and can maintain steep slopes. Behre (1933) observed that talus slopes in the Rocky Mountains of Colorado characteristically had

FIG. 4.10. Rock stream in Silver Basin, San Juan Mountains, Colorado. (Photo by Whitman Cross, U. S. Geol. Survey.)

slopes varying between 26 and 35 degrees. Talus creep is especially rapid where there is frequent freeze and thaw. It grades more or less imperceptibly into rock-glacier creep. Striking examples of rock glaciers, rock streams, or rock rivers have been described by Howe (1909) in the San Juan Mountains of southwestern Colorado, by Capps (1910) in Alaska, and by Kesseli (1941) in the Sierra Nevada of California. They are stream-like arms of talus extending down valleys. Difference of opinion exists in regard to whether they are genetically related to former glaciers. Capps thought that those in Alaska were products of a dying stage of glaciation and that all gradations existed between true glaciers and rock glaciers. Kesseli concluded that the rock glaciers of the Sierra Nevada were essentially fossil glaciers. Howe at first thought that the rock glaciers of the San Juan Mountains were formed by rock falling onto glaciers, but he later decided that they were produced by rockfalls from glacially oversteepened cliffs. Sharpe (1938) concluded that a rock glacier might form from the material in a dying glacier but he did not consider this essential. He believed that the wrinkles or corrugations often observed on the surface of a rock glacier were evidence of true flowage of the rock mass of the type designated by him as talus creep.

FIG. 4.11. Rock creep near Steamboat Point, Bighorn Mountains, Wyoming. (Photo by C. L. Heald.)

The movement of individual rock blocks down slopes was classed by Sharpe as rock creep. Large, detached boulders are often found at considerable distances from cliff or scarp faces. This type of creep is likely to be most noticeable where the rocks are massive sandstones and conglomerates or granitoid rocks with widely spaced joints. Masses of such detached rock blocks are sometimes called "rock cities" because of the imagined similarity of the individual blocks to the buildings in a city square and of the spaces between them to streets.

Solifluction. Andersson (1906) from a study of the "mud glaciers" of Bear Island, in the north Atlantic, and the "stone rivers" of the Falkland Islands, in the south Atlantic, concluded that they were a product of "slow flowing

from higher to lower ground of masses of soil or earth saturated with water."
This method of gravitative transfer of mantle rock he called solifluction
(*solum,* soil + *fluere,* to flow). Features similar to those on Bear Island
and the Falkland Islands have been described in other high-latitude areas such
as South Georgia, Graham Land, Spitzbergen, and Scandinavia. Solifluction
is also significant at high altitudes. There are four conditions which promote
it: (*a*) a good supply of water from the melting of snow and ground ice; (*b*)
moderate to steep slopes relatively free of vegetation; (*c*) the presence beneath the surface of permanently frozen ground or *tjaele,* as it is called in
Scandinavia; and (*d*) rapid production by weathering of new rock waste.
Solifluction differs from mudflow (see below) in that: it is a slower and more
continuous movement; it is not confined to a channel as a mudflow usually is;
and it develops under severe subarctic or alpine climates rather than under
arid or semiarid climates as do mudflows. During the summer months the
surface thaws to a depth of several feet and there develops over the still-
frozen tjaele a water-saturated mass of soil and rock debris which flows downslope en masse. The topographic effects of solifluction are usually not striking because it operates over the whole surface rather than being concentrated
in channels. Locally it may result in terrace-like forms upon slopes or the
filling of smaller basins.

Soil structures. Closely related to solifluction are certain features that are
variously referred to as *soil structures,* soil patterns, or structure soils. It will
be evident from the ensuing discussion that soil is not used here in a pedologic
sense, for the various structures include rock debris of varying sizes, exhibit
various degrees of weathering, and usually show no evidence of a soil profile.
Included in this category are forms which commonly go by the names of
stone nets, stone rings, stone stripes, earth stripes, earth hummocks, and block-
fields or felsenmeere. Space will permit but the barest outline of their characteristics and the theories as to their origin. For further information concerning them the reader is referred particularly to discussions by Sharp
(1942*a*) and Sharpe (1938).

Not all soil structures are, strictly speaking, solifluction features, for downslope movement may be either lacking or of minor importance. They are
phenomena which develop in either high latitudes under arctic or subarctic
climatic conditions or at high altitudes. Conditions of these environments
which contribute most to their development are intensity and frequency of
freeze and thaw and the existence of permanently frozen ground. Various
names have been applied to permanently frozen ground. The name tjaele,
mentioned above, is more widely used in Europe than in the United States.
More commonly it is referred to as *permafrost,* despite the fact that the word
suggests the trade name for a refrigerator and is not etymologically sound.

Much interest in permanently frozen ground developed during World War II as a result of the building of the Alcan and Alaskan highways and construction of air bases in Alaska and northern Canada. This interest is likely to continue with the increasingly strategic importance of the subpolar portions of North America.

Soil structures are believed to be caused by repeated freezing and thawing with consequent acceleration of mechanical weathering, growth of ground

Fig. 4.12.   Polygonal ground near Churchill, Canada.   (Royal Canadian Air Force photo.)

ice, pronounced volume changes in the surficial materials, differential and progressive movement of rock debris leading to crude assortment, and mass movement of material downslope.

Stone nets are three-dimensional soil structures having centers chiefly of clay, silt, and gravel and roughly circular or polygonal borders of coarse stones. An isolated example is called a *stone ring*. Other names such as *stone polygons, frost polygons, polygonal ground* or *soil* have been applied to stone nets. Their net-like arrangement may extend downward as much as 2 feet, and they may range in diameter from a few feet to more than 30 feet. Stone nets are found upon flat or nearly flat ground; upon slopes of moderate inclination (slopes of 5° to 15° according to Sharp) they will be drawn out downslope by solifluction into tongue-like or elliptical shapes known as *stone garlands.* Stone garlands show assortment between finer materials at the center and coarser at the edges but the assortment is not so sharp as in stone nets. On still steeper slopes (slopes of 5° to 30°, according to Sharp) stone garlands give way to *stone stripes,* which are parallel stony and earthy bands. There is considerable overlap in steepness of the slopes upon which stone garlands and stone stripes develop, but stripes characteristically form on steeper slopes than do garlands.

FIG. 4.13. Stone rings on an old pond near Summit Lake, British Columbia. (Photo by H. M. Raup.)

Earth stripes or *soil stripes* are similar to stone stripes except that they have finer textures. They are low ridges of clay and silt a few inches high which rise above broader pebbly bands. They sometimes form within the finer-textured portions of stone stripes and garlands. On surfaces covered by a good growth of tundra vegetation low, rounded mounds composed of fine materials are often found. They are called *earth hummocks* or *palsen.* They usually have a central core of fine material capped with vegetation.

Numerous theories have been proposed to explain the many types of structure soils but none is generally accepted. Sharp (1942a) has presented an excellent summary of some of the main theories. Perhaps the favored theory is that of Högbom. His theory as summarized by Sharp * is as follows:

* Quoted by permission of Columbia University Press from R. P. Sharp in *J. Geomorph.,* 5, pp. 284–285.

He assumes an initial surficial accumulation of heterogeneous rock debris of all sizes from clay to boulders and of any origin, glacial, alluvial, or residual. The significant features of the theory are: (1) inhomogeneous debris undergoes inhomogeneous freeze and thaw; (2) areas of fine material are essentially pressure centers, for they contain more water and are able to draw moisture from surround-

Fig. 4.14.   Close-up of a stone stripe. (Photo by R. P. Sharp.)

ing materials as they freeze; (3) repeated freeze and thaw sorts the coarse and fine elements of the debris in the following manner. On freezing the stones and adjacent fines are shoved upward and outward from the pressure center or area of concentrated fines. On thawing the fines contract and are pulled back farther than the stones owing to greater mutual cohesion. This leaves the stones in a new position relative to the surrounding material and repetition of the process eventually produces the sorting observed in stone nets.

Frost heaving is an essential part of most of the theories which have been proposed but the exact way in which it operates has not been agreed upon.

It seems likely that the many variations in soil structures are related to local variations in other factors such as topography, drainage, vegetation, types of surface debris, and perhaps the depth to frozen ground.

On many high mountain summits and on subpolar islands are found notable concentrations of rock blocks. These rock fields are known as *block fields* or *felsenmeere*. They are believed to be caused by the vigorous frost riving which characterizes such locations. They are found especially on massive rocks with widely spaced joints and fractures. Some undoubtedly were formed in place but others were derived from glacially transported boulders. Some may be stablized in position but others show downslope movement by solifluction. When they become concentrated into stream-like masses moving downslope, they become "stone rivers." The soil structures discussed above have come to be considered diagnostic features of periglacial regions. The significance of such features in regions now remote from ice caps will be discussed in Chapter 16.

Rapid Flowage Types of Mass-Wasting

Earthflows. Earthflows are likely to be confused with slumping and mudflows. They are often accompanied by slumping, but they differ from slumping in that there is no backward rotation of the mass. They differ from mudflows in that: (1) they are slower, seldom being perceptible to the eye except to observations extending over several hours or days; (2) they are not confined to channels as are mudflows; (3) there is a lower water content than in mudflows; and (4) they are not specifically characteristic of dry regions, as are mudflows, but are more common in humid areas. Earthflows form on terraces and hillsides where earth materials capable of flowage when saturated with water lie beneath mantle rock or artificial fill. They are common in the Appalachian Plateau region (Sharpe and Dosch, 1942), where they are called "slides." Shallow scars mark their points of origin, and hummocky benches similar in profile to lateral moraines mark their termini.

Mudflows. Mudflows move rapidly enough to be perceptible to the eye, have a higher water content than earthflows, and are usually confined to channels. Blackwelder (1928), who emphasized their geomorphic importance in arid and semiarid regions, listed the following conditions as favoring their formation: (1) unconsolidated materials at the surface which will become slippery when wet; (2) steep slopes; (3) an abundant but intermittent supply of water; and (4) sparsity of vegetation. Mudflows are characteristic of the drier regions because in such areas the vegetation is sparse and the infrequent rains are often torrential. Mudflows commonly follow the channels cut by streamfloods across alluvial fans at the bases of mountains. They consist largely of mud but may transport boulders weighing many tons. Many of the large isolated boulders seen on alluvial fans at considerable distances from mountains were thought by Blackwelder (1928) to have been carried by mudflows, which, except for the unassorted nature of the deposits, have lost most of their original characteristics. The edges of mudflows may be marked by sharp linear ridges called *mudflow levees* (Sharp, 1942*b*). Sharpe (1938) recognized three types of mudflows, the semiarid, the alpine, and the volcanic type. A famous alpine mudflow is the Slumgullion mudflow (Howe, 1909) which dammed the Lake Fork of the Gunnison River, in western Colorado, and formed Lake Cristobal. The flow is about 6 miles long. A similar mudflow, in 1914, partially covered the mining town of Telluride on the west slope of the San Juan Mountains. Ancient and dissected mudflow deposits in mountain valleys may be mistaken for glacial deposits because of their lack of assortment and the presence in them of striated rocks. Although pebbles and boulders carried by mudflows may become striated, they lack the facets which distinguish glacial boulders.

Debris avalanches. Debris avalanches are found in humid regions and are much like snow avalanches except that rock debris rather than snow makes up the bulk of their mass. It would be difficult in some cases to distinguish them from debris slides (see below) except at the time they take place, as the main distinguishing difference is that debris avalanches have more water in them than debris slides. Many of the elongate and narrow scars on the forested slopes of the White and Green Mountains probably were made by debris avalanches.

Landslides

Landslide was used by Sharpe as a group name for several types of mass movement of rock debris. As a class, landslides are distinguished from the preceding groups in that there is more rapid movement and the moving mass has less water in it. Five types of landslides are recognized: slump, debris slide, debris fall, rockslide, and rockfall.

Slump usually takes place as an intermittent movement of earth or rock masses for a short distance and typically involves a backward rotation of the mass or masses involved, as a result of which the surface of the slumped mass often exhibits a reversed slope. Frequently slumping takes place as several small independent units; the result is that the surface of the slumped mass has a number of step-like "terracettes." The loess deposits of the Mississippi Valley often display these small terraces, which have been called catsteps. Undercutting of slopes by streams, waves, and man are the most common causes of slumping.

What were called *debris slides* by Sharpe are often described as earth slides or soil slips. They differ from slumping in that the movement does not exhibit backward rotation. Instead there is a sliding or rolling motion. The amount of water is usually small; otherwise it would be a debris avalanche. *Debris falls* differ from debris slides in that the material falls from a vertical or overhanging cliff or bluff. Debris falls are especially common along the undercut banks of streams.

FIG. 4.15. A landslide near Gorman, California. (Photo by J. S. Shelton and R. C. Frampton.)

Rockslides are masses of bedrock sliding or slipping down what are usually bedding, joint, or fault surfaces. Two well-known rockslides were the Gros Ventre slide, in Wyoming, in 1925 and the Turtle Mountain slide at Frank, Alberta, in 1903. It was estimated that there were some 50,000,000 cubic yards of material in the Gros Ventre slide. Worcester (1939) interpreted it as a combined rockslide and debris slide, which is probably correct. The Turtle Mountain slide partially overwhelmed the town of Frank, Alberta, resulting in the loss of 70 lives. Rockfalls take place when recently detached rock blocks, usually small, move precipitously down a steep cliff or rock face. They are most frequent in mountain areas and during the spring months when there is repeated freezing and thawing. They may involve large rock masses and result in severe loss of life and damage to property.

FIG. 4.16. *A.* Breccia flow in Mojave Desert, 30 miles southeast of Victorville, California. The lobe is about 2½ miles across at its forward edge. *B.* View of surface of breccia flow, showing the many corrugations upon its surface. (Photos by J. S. Shelton and R. C. Frampton.)

94

Subsidence Forms

Subsidence, in contrast to other types of mass movements, does not take place along a free surface but is a downward settling of material with little horizontal motion. The most common cause is the slow removal of material beneath the subsiding mass. Common examples of land forms produced in this way are the kettles which form as glacial drift settles over melting ice blocks (see Chapter 16) and the sinkholes of limestone regions (Chapter 13). Removal of material at depth by mining, the settling of faulted masses, overloading of peat and muck areas with highway fills, and removal of fluid lava beneath a solid crust are other causes of subsidence.

CONCLUSIONS

Weathering and mass-wasting contribute to the gradation of the earth's surface to a much greater degree than is usually realized. These processes, along with sheetwash, are important not only in the general lowering of the land but in the shaping of the details of the topography of interstream areas. Rich (1951) has well stated that land sculpture has been thought of too much in terms of stream erosion and not enough in terms of the slow etching of the earth's surface by differential weathering, mass-wasting, and sheetwash. These processes, he believed, were particularly significant in areas of diverse rocks. What he referred to as the etching concept was described as follows:

Wherever the earth's crust is composed of diverse materials varying in resistance to weathering and corrasion, and lying in various attitudes, differential weathering, followed by removal of weathered products, mainly by sheetwash and creep (except in arid lands where winds are important), so etches the surface that areas underlain by the more resistant rocks are brought into relief as the less resistant are lowered. Even small differences in structure and composition are thus revealed for interpretation by the geomorphologist.

A fundamental basis for the etching concept is the extreme contrast in rate of weathering exhibited by different rocks. Some waste away very quickly, others extremely slowly. Herein lies a possible field of investigation for those interested in the quantitative approach to geomorphic problems. Here also is a fundamental reason why much of the geomorphology of the past has been ineffective. Far too little consideration has been given to the nature of the underlying rocks.

Differential weathering is extremely selective. Wherever the products of weathering can be removed as rapidly as formed, even slight differences in resistance are expressed in the topography. But even where removal is slow, the soil cover deep, and vegetation dense, the weak rocks still form lowlands and the resistant rocks stand in relief.

Under the etching concept of erosion, interstream degradation is recognized as being of extreme importance. Streams serve mainly to cut down narrow notches which expose the rocks to the other agencies; they cut laterally to some extent; and they serve as sewers to carry away the material delivered to them by creep and sheetwash. They directly affect only a small proportion of the total land surface.

FIG. 4.17. Trench formed on a sill of basic igneous rock in limestone as a result of differential weathering and erosion. An example of etching as conceived by Rich.

The influence of creep in moist, vegetation-covered areas seems to have been much underestimated. In such areas, it is by far the most effective agency for the removal of the products of weathering, whereas sheetwash and deflation play the leading roles in arid regions.

REFERENCES CITED IN TEXT

Andersson, J. G. (1906). Solifluction, a component of subaerial denudation, *J. Geol.*, *14*, pp. 91–112.

Behre, C. H., Jr. (1933). Talus behavior above timber line in the Rocky Mountains, *J. Geol., 41*, pp. 622–635.

Blackwelder, Eliot (1928). Mudflow as a geologic agent in semiarid mountains, *Geol. Soc. Am., Bull. 39,* pp. 465–480.

Blackwelder, Eliot (1942). The process of mountain sculpture by rolling debris, *J. Geomorph., 5,* pp. 325–328.

Blank, H. R. (1951). "Rock doughnuts," a product of granite weathering, *Am. J. Sci., 249,* pp. 822–829.

Bretz, J. H., and Leland Horberg (1949). Caliche in southeastern New Mexico, *J. Geol., 57,* pp. 491–511.

Bryan, Kirk (1925). Pedestal rocks in the arid southwest, *U. S. Geol. Survey, Bull. 760,* pp. 1–11.

Bryan, Kirk, and C. C. Albritton (1943). Soil phenomena as evidence of climatic change, *Am. J. Sci., 241,* pp. 469–490.

Bushnell, T. M. (1944). The story of Indiana soils, *Purdue Univ. Agr. Expt. Sta., Spec. Cir. 1,* Lafayette.

Capps, S. R., Jr. (1910). Rock glaciers in Alaska, *J. Geol., 18,* pp. 359–375.

Chapman, R. W. (1940). Monoliths in the White Mountains of New Hampshire, *J. Geomorph., 3,* pp. 302–310.

Crickmay, G. F. (1935). Granite pedestal rocks in the southern Appalachian Piedmont, *J. Geol., 43,* pp. 745–758.

Dixey, F. (1941). The age of the silicified surface deposits in northern Rhodesia, Angola and the Belgian Congo, *Geol. Soc. S. Africa, Trans. 44,* pp. 39–49.

Howe, Ernest (1909). Landslides in the San Juan Mountains, including a consideration of their causes and classification, *U. S. Geol. Survey, Profess. Paper 67,* 58 pp.

Hunt, C. B., and V. P. Sokoloff (1949). Pre-Wisconsin soil in the Rocky Mountain region, *U. S. Geol. Survey, Profess. Paper 221-G,* pp. 109–123.

Kesseli, J. E. (1941). Rock streams in the Sierra Nevada, California, *Geog. Rev., 31,* pp. 203–227.

Leighton, M. M., and Paul MacClintock (1930). Weathered zones of the drift-sheets of Illinois, *J. Geol., 38,* pp. 28–53.

Lyon, T. L., and H. O. Buckman (1943). *The Nature and Properties of Soils,* The Macmillan Co., New York.

Rich, J. L. (1951). Geomorphology as a tool for the interpretation of geology and earth history, *Trans. N. Y. Acad. Sci.,* Ser. 2, *13,* pp. 188–192.

Sharp, R. P. (1942a). Soil structures in the St. Elias Range, Yukon Territory, *J. Geomorph., 5,* pp. 274–301.

Sharp, R. P. (1942b). Mudflow levees, *J. Geomorph., 5,* pp. 222–227.

Sharpe, C. F. S. (1938). *Landslides and Related Phenomena,* 137 pp., Columbia University Press, New York.

Sharpe, C. F. S., and E. F. Dosch (1942). Relation of soil-creep to earthflow in the Appalachian Plateau, *J. Geomorph., 5,* pp. 312–324.

Smith, L. L. (1941). Weather pits in granite of the southern Piedmont, *J. Geomorph., 4,* pp. 117–127.

Taber, Stephen (1930). The mechanics of frost heaving, *J. Geol., 38,* pp. 303–317.

Wahlstrom, E. E. (1948). Pre-Fountain and recent weathering on Flagstaff Mountain near Boulder, Colorado, *Geol. Soc. Am., Bull. 59,* pp. 1173–1190.

White, W. A. (1944). Geomorphic effects of indurated veneers on granites in the southern states, *J. Geol., 52,* pp. 333–341.

White, W. A. (1945). Origin of granite domes in the southeastern Piedmont, *J. Geol., 53,* pp. 276–282.

Wooldridge, S. W. (1949). Geomorphology and soil science, *J. Soil Sci., 1,* pp. 31–34.

Woolnough, W. G. (1930). The influence of climate and topography in the formation and distribution of products of weathering, *Geol. Mag., 67,* pp. 123–132.

Additional References

Baldwin, Mark, C. E. Kellogg, and James Thorp (1938). Soil classification, *Yearbook Agr.,* pp. 979–1001.

Bryan, Kirk (1946). Cryopedology—The study of frozen ground and intense frost-action with suggestions of nomenclature, *Am. J. Sci., 244,* pp. 622–642.

Bushnell, T. M. (1942). Some aspects of the catena concept, *Soil Sci. Soc. Am., Proc. 7,* pp. 466–476.

Byers, H. G., C. E. Kellogg, M. S. Anderson, and James Thorp (1938). Formation of soil, *Yearbook Agr.,* pp. 948–978.

Glinka, K. D. (1927). *The Great Soil Groups of the World and Their Development* (translated by C. F. Marbut), 235 pp., Edwards Bros., Ann Arbor.

Jenny, Hans (1941). *Factors of Soil Formation,* 281 pp., McGraw-Hill Book Co., New York.

Jenny, Hans (1946). Arrangement of soil series and types according to functions of soil-forming factors, *Soil Sci., 61,* pp. 375–391.

Joyce, J. R. F. (1950). Stone runs of the Falkland Islands, *Geol. Mag., 87,* pp. 105–115.

Nikiforoff, C. C. (1943). Introduction to paleopedology, *Am. J. Sci., 241,* pp. 194–200.

Scheffel, E. R. (1920). "Slides" in the Conemaugh formation near Morgantown, West Virginia, *J. Geol., 28,* pp. 340–355.

Strahler, A. N. (1940). Landslides of the Vermilion and Echo Cliffs, northern Arizona, *J. Geomorph., 3,* pp. 285–301.

Thorp, James (1949). Interrelations of Pleistocene geology and soil science, *Geol. Soc. Am., Bull. 60,* pp. 1517–1526.

Wentworth, C. K. (1943). Soil avalanches on Oahu, Hawaii, *Geol. Soc. Am., Bull. 54,* pp. 53–64.

5 · The Fluvial Geomorphic Cycle

The topical heading of this chapter is not entirely satisfactory, for it fails to suggest that weathering, mass-wasting, and unconcentrated runoff contribute to the production of many of the features to be discussed. Davis and others have included the processes discussed in this chapter under the "normal cycle," but that designation is not particularly appropriate for all geomorphic processes are normal. Since emphasis is upon the work of streams, it seems that "the fluvial cycle" suggests more nearly the nature of the topics to be discussed.

Throughout most of the world, runoff waters are the dominant geomorphic agency. Except for areas now covered by glaciers there are few places where rainfall does not have an opportunity to perform geomorphic work. In the most absolute deserts it rains, even though infrequently, and, to a large degree, as we shall see in a later chapter, desert land forms are the work of running water. In areas that were covered by ice sheets during the Pleistocene and were veneered with glacial deposits having a unique topographic expression, streams and sheetwash have already made an impression upon the landscape and are rapidly changing it.

FUNDAMENTAL CONCEPTS PERTAINING TO THE FLUVIAL CYCLE

Implications of a fluvial cycle. In Chapter 2, we introduced the concept of a geomorphic cycle. Let us now consider more specifically the implications of a fluvial cycle. The following ideas seem to be implied.

1. Uplift of the land provides initial units of geologic materials which are subjected to sculpturing and reduction. These earth units have their own lithologic, structural, and geomorphic aspects. They may be such forms as: recently uplifted sections of the continental shelf; domes and basins; alternating anticlines and synclines of folded mountain structures; faulted structures; complex mountain structures; volcanic mountains or plateaus; ice-scoured plains; glacial till plains and associated features; lacustrine plains; areas of sand and loess accumulation; areas previously reduced to low relief by fluvial erosion (peneplains), and other types.

2. In the fluvial cycle, sculpture and reduction of the initial land mass take place chiefly through the combined action of weathering, mass-wasting, and erosion by both concentrated and unconcentrated runoff waters.

3. Reduction may take place under varying diastrophic conditions. It may take place after or during uplift. Uplift may be continuous or it may be intermittent. Davis usually assumed rapid uplift followed by a long period of stillstand, but it must be recognized that this is by no means the only condition that may prevail. Some would argue that it is not even the normal one.

4. Sculpture and reduction may be only partially completed during a period of stillstand before uplift inaugurates a new cycle, and thus partial and multiple cycles result. Completed cycles are the exception rather than the rule.

5. Stages of progress in land sculpture and reduction may be recognized and designated by the metaphorical stages of youth, maturity, and old age. Substages may be suggested by such prefixes as early, middle, and late.

6. As a cycle progresses a systematic sequence of land forms ensues, and each stage is characterized by a particular group of forms in harmony with its place in the cycle.

7. The land forms produced during a cycle are dependent upon and explainable in terms of structure, process, and stage.

8. It is possible for an uplifted land mass to be reduced to that ultimate limit of land reduction called base level.

9. Nearly complete consumption of an uplifted land unit by the combined processes of weathering, mass-wasting, and erosion by runoff waters gives rise to a topographic surface of low relief called a peneplain.

STREAMS AND VALLEYS

Valleys are so common on the earth's surface that we seldom bother to define them. This lack of definition leads to the application of the term to features that really are not valleys. True valleys are present over most of the land surface of the earth. They are known by such names as gully, draw, ravine, gulch, hollow, run, arroyo, gorge, canyon, or by more poetic forms such as vale, glen, and dale, but the one thing that is common to all is that they were cut by running water. They are negative land forms of varying size and shape occupied by either perennial or intermittent streams. The term valley should be restricted to features of such origin. It is too much to suppose that we shall cease to speak of Death Valley, the Great Valley of California, the Vale of Chile, and the Jordan Valley, all features largely of diastrophic origin, but we should at least be cognizant of the fact that they owe their existence to diastrophism rather than stream erosion.

We frequently use carelessly the terms valley and stream as if they were practically synonymous. We speak of young streams, mature streams, and old streams when we really mean young, mature, and old valleys. It is possible to have an old stream in a young valley or a young stream in an old valley, for young, mature, and old as applied to a valley have no time implications but rather imply certain characteristics of the valley which are diagnostic of its stage of development.

FIG. 5.1. Sheetwash merging into gulleying, near Redfield, South Dakota. (Soil Conservation Service photo.)

Most large valleys or valley systems have had a complicated history which cannot always be determined in detail. The later geologic history of a valley usually is suggested by its form or by lesser topographic features within it. Hence a comprehension of the significance of valley forms is of utmost importance to the geomorphologist.

Valley development. A valley takes form through the operation of three concomitant processes, valley deepening, valley widening, and valley lengthening. Valley deepening is effected by several processes (discussed in Chapter 3). They are: (1) hydraulic action; (2) corrasion or abrasion on the floor of the valley; (3) pothole drilling along the valley floor and at the base of waterfalls (this is really a special type of corrasion or abrasion but is perhaps unique enough to merit special designation); (4) corrosion; and (5) weathering of the stream bed (in the case of an intermittent stream) plus subsequent removal of weathered material by hydraulic action.

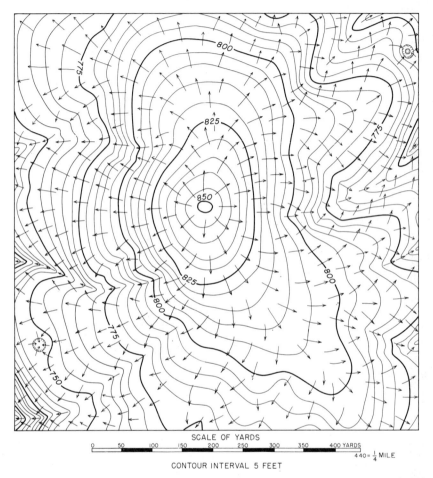

SCALE OF YARDS

CONTOUR INTERVAL 5 FEET

FIG. 5.2. Topographic map of Sheets Hill, north of Bloomington, Indiana, showing the transition from unconcentrated to concentrated runoff. (After C. A. Malott.)

Valley width is the linear distance between valley sides and is expressed in terms of the cross section of a valley. The upper and outer parts of many valley sides are indefinite and must be delimited arbitrarily. The width of the valley floor is more definitely determinable, and we commonly think of valley width in terms of it. Valley widening may be accomplished in a number of ways. (1) Lateral erosion or planation by the stream in a valley may remove material from the base of the valley side through hydraulic and corrasive action. This results in local oversteepening of the valley side with attendant undercut slopes, which favors slumping of materials into the stream. This process may operate during any stage of valley development

FIG. 5.3. A youthful valley which is being deepened through weathering on its floor plus subsequent removal of the weathering spalls by hydraulic action. (Photo by P. B. Stockdale.)

but is most noticeable during maturity and old age because then valley deepening has essentially ceased and the effects of lateral erosion are more obvious. (2) Rainwash or sheetwash on valley sides contributes in an important way to valley widening. (3) Gulleying on valley sides is another method by which valley widening is effected. Although more spectacular than sheetwash, it is doubtful if it is as important. (4) Weathering and mass-wasting may contribute to valley widening both directly and indirectly. Some valleys have such steep sides that little mantle rock remains on them. Under such conditions weathering may loosen material which moves directly downslope into the stream channels; more commonly weathering acts indirectly through production of mantle rock which then is moved downslope by creep, slump, other types of mass-wasting, and by sheetwash. (5) Incoming tributaries contribute to valley widening even though they be no more than over-

Fig. 5.4. Valley widening by slumping along the Rio Grande River near Taos, New Mexico. (Photo by J. S. Shelton and R. C. Frampton.)

grown gullies. Valleys commonly widen noticeably where tributaries join them because the valley wall here is being attacked from two directions.

Valley lengthening may take place in three ways. (1) Valleys may be extended by the process of headward erosion. This is particularly significant in the growth of lesser valleys. It would be erroneous to assume that a major stream like the Mississippi River started as a short stream heading near the sea and then extended itself by headward erosion to its present source. No great valley was ever produced in this way, but many minor tributary valleys were so formed. The most common type of headward erosion involves the extension of a ravine by incoming sheet waters with attendant weathering and slumping at the ravine head. A second type of headward erosion is effected by spring sapping. This takes place where a spring emerges at the head of a valley. The overlying rock is undermined by solution and weathering and subsequent slumping above the spring may produce an abrupt valley head. Valleys may extend headward into or across swampy areas accompanying filling and drainage of such areas or after deepening of the valley downstream from the swamp has lowered the water table sufficiently to permit valley incision across it. (2) Valleys may be lengthened through increase in size of their crooks or meanders. As long as a meandering stream is confined by valley walls increase in its crookedness will augment the valley length. (3) Valleys also may lengthen at their termini. Uplift of the land or lowering of sea or lake level will result in extension of the valley form across the newly exposed land. Many of the valleys across the Atlantic and Gulf coastal plains have been extended in this way, as have valleys which empty into the present Great Lakes at altitudes lower than their Pleistocene predecessors. Streams may be lengthened at their termini by seaward extension of deltas. The lower Mississippi well illustrates this type of lengthening, as its mouth at the beginning of the Pleistocene was approximately 125 miles inland from its present position. It is doubtful, however, whether the trench of the Mississippi across its delta should be called a valley.

Base level and its varieties. It is generally recognized that there is a downward limit to valley deepening. This limitation on vertical erosion is known as *base level*. When Powell (1875) proposed the concept of base level he introduced an idea which probably has been more abused than any other concept except that of the peneplain. After 75 years, we cannot be certain what geologists mean when they use the term. They cannot even agree upon how to spell it. For discussions of the many usages that have been made of the term the reader is referred to articles by Davis (1902) and Malott (1928). Powell introduced the idea of base level in the following words:

We may consider the level of the sea to be a grand base level, below which the dry lands cannot be eroded; but we may also have, for local and temporary purposes, other base levels of erosion, which are the levels of the beds of the principal streams which carry away the products of erosion. '(I take some liberty in using the term level in this connection, as the action of a running stream in wearing its channel ceases, for all practical purposes, before its bed has quite reached the level of the lower end of the stream. What I have called base level would, in fact, be an imaginary surface, inclining slightly in all its parts toward the lower end of the principal stream draining the area through which the level is supposed to extend, or having the inclination of its parts varied in direction as determined by tributary streams.) Where such a stream crosses a series of rocks in its course, some of which are hard, and others soft, the harder beds form a series of temporary dams, above which the corrasion of the channel through the softer beds is checked, and thus we may have a series of base levels of erosion, below which the rocks on either side of the river, though exceedingly friable, cannot be degraded.

It would seem that Powell intended to define base level as a limit of land reduction and not as a "graded plain," a "peneplain," a "plane," "the condition of a river," "a stage in river history," or a "geomorphic form," as it has been variously interpreted.

To oversimplify somewhat, we may say that the major contention has been over whether the ultimate surface of a region reduced to base level should be characterized as a plain or a plane, and whether Powell envisaged two or three types of base level. Malott (1928) interpreted Powell's statement as implying three types of base level, *ultimate, local,* and *temporary.* Davis (1902) seems to have thought likewise, for in referring to Powell's ideas he stated:

Base level, as thus defined, seems to include three ideas: first, the grand or general base level for sub-aerial erosion is the level of the sea; second, a base level is an imaginary, sloping surface which generalizes the faint inclination of the trunk and branch rivers of a region when the erosion of their channels has practically ceased; third, local and temporary base levels are those slow reaches in a river which are determined by ledges in its course farther downstream.

Further on in the same article, however, Davis indicated that there are only two types of base level, general or permanent and local or temporary. One of the major differences of opinion is whether the expression of Powell's "for local and temporary purposes" was meant to imply local *and* temporary or local *or* temporary base levels. There is also difference of opinion about the nature of the "grand base level." Johnson (1929) was one of the staunchest supporters of the idea that the ultimate level of land reduction is sea level and that ultimate base level is the *plane* surface of the sea extended under the lands. This to the author seems a theoretical abstraction which has little practical application except for areas immediately adjacent to the sea. Even if we grant the theoretical possibility of continental masses being worn down to sea level (a possibility open to question), the probability of the land remaining stationary long enough for this to be consummated is so unlikely, in view of our knowledge of diastrophism, that sea level has little significance as a control level of land reduction for areas remote from it. The control level at present for land reduction in Minnesota or Colorado and other areas thousands of miles by stream distance from the ocean is certainly not the sea. If the concept of base level is to have practical value in the interpretation of land forms, it needs to be based upon something less theoretical than the reduction to sea level of areas subcontinental in size. It is not implied that there is no such thing as an ultimate base level of erosion. There is such, and it would be at or perhaps slightly below sea level, for streams can and do cut below sea level at their mouths. After a region has attained such low relief that little mechanical load is being carried by streams, the ocean may encroach upon the land and reduce it not to sea level but to wave base (see Chapter 17). That local areas have been eroded to wave base seems likely, but it is difficult to cite examples of subcontinental tracts so reduced.

The more conventional viewpoint is that Powell intended to suggest that such features as hard-rock barriers or lakes athwart a stream course could act as local or temporary base levels for portions of the stream above such barriers. Johnson referred to them as "temporary local base levels" and thus combined the two into one idea. We cannot judge what Powell's intentions were, but a case can be made for the view that there are both local and temporary base levels. If these two are distinguished, *local base level* would be the downward limit of land reduction for any region as controlled by the level of streams across the region which are graded to the sea.

This somewhat unorthodox idea of local base level has much to recommend it to one interested in the practical interpretation of land forms, although it will probably be opposed by those who cling to the abstraction that land remote from the sea will eventually be worn down to sea level. Certainly the control level of land reduction now for areas remote from the sea is not sea level, but for each valley it is the present level of the valley to which it is tributary. If we assume that the river systems are graded to the sea (i.e., a state of equilibrium between erosion and deposition exists), then by projecting the profiles of these graded river systems beneath unreduced interstream uplands we obtain a measure of the limits of land reduction for each particular locality.

It is generally agreed that such features as hard-rock barriers or lakes along a stream may act as temporary base levels for the areas above them. The only question is whether they should also be considered as local base levels. It seems appropriate to describe a region as being at *temporary base level* whenever it is graded toward some level which in turn is not graded to the sea. Used in this sense, temporary base level is different from local base level.

The graded stream. Davis, building upon Gilbert's ideas, developed the concept that a stream rather early in the geomorphic cycle attains that slope or gradient which under existing conditions of discharge and channel characteristics is just sufficient for transportation of its load. Such a stream is said to be *graded* or *at grade*. Grade is not to be confused with gradient; all streams flow over slopes and thereby possess gradients which may be expressed in degrees, per cent, or perhaps in feet per mile. A graded stream is not a stream which is loaded to capacity as often stated, for streams are probably never loaded to capacity. Neither is it a stream which is neither eroding nor depositing, as sometimes said, for erosion in one part of a stream channel and deposition in another part are as characteristic of a graded stream as of a non-graded one. Although a stream at grade has attained a particular gradient, it is not gradient alone that determines the graded condition. Other factors are involved, such as velocity of the stream, which is not entirely determined by gradient, channel characteristics, and the caliber of the material that the stream has to transport.

The concept of the graded stream has been challenged by Kesseli (1941), who would prefer to have graded stream imply only an absence of rapids and waterfalls, but it seems that his objections came mainly from a misinterpretation of what is implied in the idea of a graded stream. Mackin (1948) has vigorously defended the concept and done much to clarify its meaning. His definition of a graded stream is somewhat involved but, inasmuch as it expresses the essential ideas so well, it is here quoted:

A graded stream is one in which, over a period of years, slope is delicately adjusted to provide, with available discharge and with prevailing channel characteristics, just the velocity required for transportation of the load supplied from the drainage basin. The graded stream is a system in equilibrium; its diagnostic characteristic is that any change in any of the controlling factors will cause a displacement of the equilibrium in a direction that will tend to absorb the effect of the change.

The graded condition does not imply necessarily either a high or low gradient. Some streams may be graded and still have high gradients. Mackin (1948) cited the Shoshone River, east of Cody, Wyoming, with an average gradient of over 30 feet to the mile as an example of a high-gradient, graded stream and the Illinois River, with a gradient of less than 2 inches per mile, as an example of a low-gradient, graded stream. The Shoshone River is required to transport boulders 8 to 12 inches in diameter, whereas the Illinois River is transporting largely silt and clay-sized sediments.

The longitudinal profile of a graded stream is referred to as a *profile of equilibrium*. It is generally pictured as a smoothly concave-upward hyperbolic curve which decreases in slope gradually and systematically downvalley. Although this is true in a general way, it is by no means so in detail. Actually streams may have both graded and ungraded sections along their courses, but even the longitudinal profile of a stream that is graded throughout its length is seldom a smooth curve decreasing progressively in slope downstream. Noticeable changes in gradient may take place below junctions of major tributaries related to the quantity and caliber of material brought into the main stream by the tributaries. A steepening of the gradient of the Missouri River below the junction of the Platte River takes place primarily because the Platte brings into the Missouri considerable material of gravel size, which necessitates a steeper gradient for its transport. Between major tributaries there is frequently a gradual decrease in slope because of reduction in size of material by attrition. The factors influencing the variations in gradient along a graded stream were stated by Mackin (1948) as follows:

The longitudinal profile of a graded stream may be thought of as consisting of a number of segments, each differing from those that adjoin it but all closely related parts of one system · · ·.

Each segment has a slope that will provide the velocity required for transportation of all of the load supplied to it from above, and this slope is maintained without change as long as controlling conditions remain the same. The graded profile is a slope of transportation; it is influenced directly neither by the corrasive power of the stream nor bedrock resistance to corrasion.

Some changes from segment to segment in factors controlling the slope of the graded profile are matters of geographic circumstance that are not systematic in any way; these include the downvalley increase in discharge, and downvalley decrease in load relative to discharge, that characterize trunk streams flowing from highland areas through humid lowland. Other changes, as downvalley decrease in caliber of load by reason of attrition, may be more or less systematic between tributary junctions. Still others, as change in channel characteristics, are partly dependent on changes in load and discharge; the channel characteristics are determined chiefly by caliber of load and rate of lateral shifting of the channel, and the rate of channel shifting is itself dependent upon velocity and erodibility of the banks.

These changes are usually such as to decrease slope requirements in a downvalley direction but, because none of them is systematic, the graded profile cannot be a simple mathematical curve in anything more than a loose or superficial sense. We can proceed toward an understanding of the graded profile, not by "curve matching," but by rigorous analysis of adequate sets of data for unit segments of natural and laboratory streams numerous and varied enough to reveal the effect of variation of each of the factors separately. An essential prerequisite for efficiency in the gathering and analysis of the data is recognition of the difference between the graded profile that is maintained without change, and the ungraded profile that is being modified by upbuilding and downcutting.

The attainment of a profile of equilibrium does not mean necessarily that the stream has attained the lowest slope or gradient over which it will ever be able to flow. Actually its gradient may be slowly modified as conditions change during the progress of the geomorphic cycle, but this usually takes place at such an imperceptibly slow rate that, as pointed out above, the profile of a graded stream remains the local base level of erosion for the land area adjacent to it. Decrease in the amount and caliber of load resulting from lowering of interstream areas by slope wash and mass-wasting as the cycle progresses permits a lesser gradient for the trunk stream, but these changes take place concomitantly and the profile of the graded trunk stream remains the control level toward which local land reduction proceeds.

As stated above, there may be both graded and ungraded stretches along a stream course. The breaks between these stretches consist not of minor inflections in the profile as observed along graded streams but rather of sharp changes in gradient marked by rapids or waterfalls. A stream profile marked by such abrupt changes in slope is called an *interrupted profile*. The main problem which an interrupted profile presents is in determining whether it reflects the local effect of resistant bedrock along the valley or whether it has cyclical implication and was produced by diastrophic or eustatic changes in sea level. The sharp inflections in an interrupted profile are called *knick-*

Fɪɢ. 5.5. Cataract Falls, near Cloverdale, Indiana. The falls is over a resistant bed of sandstone and marks the headward limit of post-Illinoian valley deepening. (Photo by C. A. Malott.)

punkte or, in English, *nickpoints*. It is maintained by some that they are always explainable in terms of varying rock hardness along the stream course and hence have no cyclic significance. A temporary base level, such as a hard-rock barrier or lake, may exist along a stream and cause a section of the stream above it to attain a graded condition while below the barrier the stream remains ungraded. Such a nickpoint is a temporary thing and usually has no cyclical implication. Along some streams, however, nickpoints are found which do not seem to be related to more resistant rock. They have been interpreted as the headward limits of successive periods of base leveling in multicyclic valleys. This origin for nickpoints, found along valleys cut in homogeneous rocks, would seem likely. For that matter those associated with resistant rocks may represent points attained in the up-valley migration of a nickpoint, for the resistant rock may cause a nickpoint to remain fixed until the hard-rock barrier has been breached.

Cross profiles of valleys. The cross profile of a valley may shed considerable light upon its geomorphic history, as well as indicate the influence of local geologic and climatic controls. It is sometimes carelessly stated that during youth valley deepening exceeds valley widening. In a sense valley deepening is the most conspicuous effect of stream activity during youth, but it seldom exceeds valley widening in amount. The V-shaped cross profile, which is usually considered most typical of youthful valleys, will nearly always have greater width, measured at the top, than depth. The Grand Canyon is as much as 15 miles wide at its top as compared with a depth of 1 mile. Only where special geologic conditions exist, such as nearly vertical beds or rocks which are especially resistant to weathering and mass-wasting, is the gorge profile in which depth exceeds width likely to develop. It is true, however, that in youth a stream is more engaged in vertical than in lateral erosion, for then much of the valley widening is effected by weathering, sheetwash, and mass-wasting on the valley sides.

Although stream meandering and lateral erosion may take place in youth, they become particularly significant processes after a stream has attained a profile of equilibrium, for then downcutting becomes almost imperceptibly slow and valley widening becomes dominant. This change in characteristic stream activity may be said to mark the transition from youth to maturity. The interlocking spurs of youth are trimmed, sharpened, and removed with resulting development of a freely meandering stream, rather than one confined by valley walls, as in youth. Lateral planation produces the valley flat with its incipient floodplain or veneer of alluvium. Continued valley widening produces the wide, open valley of old age which has a width several times that of the meander belt (roughly the distance between the arcs of successive meanders).

Most textbook diagrams follow Davis's diagrams and show the slopes of the valley sides as becoming more and more gentle as a cycle proceeds. That this is always true may be seriously doubted. Probably the most important result of Walther Penck's challenge to some of Davis's ideas (see Chapter 8) has been to call our attention to our lack of information on methods of slope retreat (Bryan, 1940).

The cross profile of a valley is often asymmetrical, and recognition of the causes of asymmetry is essential to a proper interpretation of valley history. There are several causes which may produce asymmetrical cross profiles.

1. Probably the most common one is the production of *undercut* and *slip-off slopes* attendant upon meandering and lateral erosion. The valley side on the outside of a meander often displays an oversteepened and undercut slope, whereas the opposite valley side descends more gradually from the upland to the valley floor to form what is called a slip-off slope.

2. An asymmetrical cross profile may reflect structural controls. Valleys cut in alternating strong and weak strata having an essentially horizontal attitude will frequently exhibit *structural benches* upon the strong strata. These benches are usually not continuous for long distances and may be present on one side but lacking on the other side of a valley. They must be distinguished from alluvial terraces and strath terraces (bedrock surfaces marking former base levels of erosion) which have different implications as far as the history of the valley is concerned. Structural benches have no implication of former base level control or of partial erosion cycles, as do strath terraces, nor do they represent a stage of aggradation followed by partial excavation of the valley fill, as do alluvial terraces. Usually it is not difficult to recognize structural benches, for it will often be evident that their existence is dependent upon specific lithologic conditions.

3. Faulting may in some cases be responsible for asymmetry of the cross profile. This can happen where a valley follows along or close to a fault which has brought into juxtaposition rocks of varying lithology, such as weak rock against strong, or massive rock against highly fractured rock, in which case erosion will proceed more easily along one valley side than the other.

4. Where valleys roughly parallel the strike of inclined strata it is commonly observed that the stream has a marked tendency to shift its position down the dip of the beds, even following the surface of a particularly resistant bed. This leads to an asymmetrical cross profile, and the process responsible for it is referred to as *homoclinal shifting*.

5. An east-west valley may display marked asymmetry in cross profile as a result of the direct or indirect effects of the climatic differences on the two sides of the valley. As yet few quantitative studies have been made of this factor in valley development. The asymmetry results largely from the effects of differences in exposures upon the rates of weathering, mass-wasting, and erosion. A south-facing slope (in the northern hemisphere) will receive more direct sunshine, have a higher evaporation loss, experience more frequent freezes and thaws and retain a snow cover for a shorter period than will a north-facing slope. As a result of the higher temperatures and lower soil moisture on the south-facing slope there usually will be less vegetation (this is particularly noticeable in semiarid regions where moisture is a more critical factor than in humid regions). Hence weathering, sheetwash and mass-wasting will go on more rapidly and the slope of the valley side will be less steep than that of the north-facing side. Where moisture-bearing winds are prevailingly from the same direction, as in the trade wind belt, there may result striking asymmetry of the windward and leeward slopes because of the variation in the amount of rainfall on the two sides of a valley.

Classifications of Valleys

Many classifications of valleys and associated streams have been proposed. Johnson (1932) summarized some of the ways of classifying valleys, and his classification is followed below with certain emendations and additions. It should, perhaps, be emphasized that we are attempting to classify valleys and not streams, although Johnson argued somewhat illogically that streams and valleys are so interrelated that the words can be used interchangeably.

According to stage in the geomorphic cycle. The much-used metaphorical classification of valleys developed by Davis is familiar to most students of geology. According to it, valleys are classed as young, mature, or old, depending upon characteristics developed at different stages in their evolution. This classification has much to recommend it provided we keep in mind that no time implications are intended. The three stages are not of equal length. Johnson (1932) assigned 5 per cent of the total cycle to youth, 25 per cent to maturity, and 70 per cent to old age. Although these are purely arbitrary values, they at least suggest the decreasing rate at which valley modification proceeds.

Genetic classification. Davis, again, was largely responsible for this method of classifying valleys. Powell, in 1875, had introduced the idea of consequent valleys and to this Davis later added such names as subsequent, obsequent, insequent, resequent, and others which today are rarely used. A *consequent valley* is one whose course was determined by the initial slope of the land. The only valleys that we can feel certain are consequent valleys are those which have developed upon newly created land surfaces such as alluvial plains, glacial plains, lava cones or plains, or recently uplifted coastal plains. Valleys whose courses align themselves with the regional slope or dip of the rock are often inferred to be consequent valleys, but this cannot always be demonstrated conclusively. *Subsequent valleys* are those whose courses have been shifted from the original consequent ones to belts of more readily erosible rocks. They represent structurally adjusted stream courses. Because of the coincidence of subsequent valleys with belts of weak rock it is usually concluded that any valley which follows such a course is a subsequent valley. This may not be true, for the valley may have been on the weak rock from the beginning. With lesser tributary valleys which definitely are on weak rock we are probably correct in assuming that shifts in valley positions have taken place to produce this relationship. It will be evident that most subsequent valleys follow the strike of the rock and hence may also be designated as *strike valleys*. They are also sometimes called *longitudinal valleys*.

Insequent valleys are those whose courses are controlled by factors which are not determinable. They show no apparent adjustment to structure or initial slopes and seemingly have developed where they are by chance. This undoubtedly was not so but the controlling factors escape detection. Much of the drainage of homogeneous sedimentary and igneous rock areas may have to be classed as insequent, for there is seldom a detectable reason why a particular valley is where it is rather than elsewhere. Thus to classify valleys as insequent tells little more than that there is a notable lack of structural or lithologic control of their location.

Obsequent valleys drain in a direction opposite to the original consequent valleys, and *resequent valleys* drain in the same direction as the original consequent drainage but are at lower topographic levels and have developed with respect to new base levels of erosion. Designation of valleys as obsequent or resequent may be easily subject to error, for it involves interpretations of former drainage directions not always demonstrable. Hence, some geomorphologists make little use of these two terms, but they are convenient and useful designations for those valleys where it is possible to reconstruct the drainage relationships which they imply.

Classification of valleys according to controlling structures. It is frequently possible to classify valleys on the basis of the types of geologic structure which have controlled their development. Thus we may recognize homoclinal, anticlinal, synclinal, fault, fault-line, and joint valleys. *Homoclinal valleys* (commonly in the older literature called monoclinal) are strike valleys which follow beds of weaker rock along the flanks of folds and on homoclinal structures where alternating weak and strong strata having moderate to high dips have been truncated. They are usually subsequent valleys. They are especially common in the folded Appalachians, the Jura Mountains, and along the flanks of the Rocky Mountains.

Anticlinal valleys follow the axes of breached anticlines, and *synclinal valleys* the axes of synclines. Valleys whose positions are determined by faults may be of two types. They are *fault valleys* if the streams follow depressions consequent upon faulting and are *fault-line valleys* if they are subsequent valleys following a fault line. Most fault-line valleys are second or nth cycle valleys. Some valley courses or portions of valley courses are controlled by major joint systems and may be classed as *joint valleys*. They are likely to be minor valleys or sections of valleys, although in certain areas, as in the Adirondacks, it has been claimed that many valley courses are joint-controlled.

Valleys transverse to structure. Valleys whose courses cut across geologic structures may be classed as *transverse valleys*. Four types of transverse valleys have been recognized. First, there are those developed across the folded or faulted structures of tectonic basins whose lowest outlets extend

FIG. 5.6.   *A*. Initial stage in the uncovering of a pre-Karoo quartzite of the Witwatersrand system by drainage lines superposed from younger rocks of the Karoo system. View is near Benoni, Transvaal, South Africa.   *B*. A later stage in which the beds of the Karoo system have been removed and a new topography has developed upon the beds of the Witwatersrand system, with a quartzite in this system forming a ridge. Both exhumation and differential erosion of the undermass are involved.   View is north of Johannesburg, Transvaal, South Africa.   (Photos by J. H. Wellington.)

across the structures.   Secondly, there are those developed by headward erosion along zones of weakness, such as tear faults, which are transverse to the major structures.   These represent a special type of subsequent valleys.

PRE-CAMBRIAN   MESOZOIC   TERTIARY VOLCANICS

FIG. 5.7.   *A*. View of the Black Canyon of the Gunnison River, western Colorado. (Photo by G. A. Grant, National Park Service.)   *B*. Diagram to show how the Gunnison River was superposed from Tertiary volcanics.   (After W. W. Atwood, Sr., and W. W. Atwood, Jr.)

The term transverse valley may be applied also to two other types of valleys which cross geologic structures: superposed and antecedent valleys. Powell proposed the terms superimposed and antecedent to describe certain types of valleys transverse to geologic structures.   Superimposed was later shortened by McGee to superposed, and this spelling is now more generally used. A *superposed valley* extends across a geologic structure which antedates the valley, but which was not exposed at the time cutting of the valley began but was buried beneath a cover mass such as a sedimentary rock cover, glacial drift, or lava flows. The stream originated upon the cover mass, perhaps as an ordinary consequent stream, but as it cut its valley downward

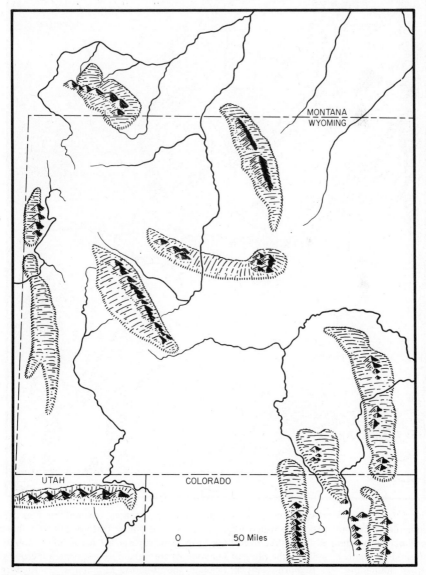

FIG. 5.8. The late Tertiary topography of western Wyoming and adjacent areas. (After W. W. Atwood, Sr., and W. W. Atwood, Jr.)

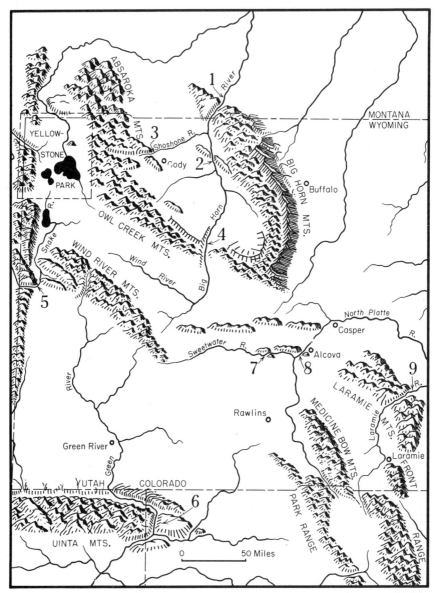

Fɪɢ. 5.9. Diagram showing the present topography in the area covered by Fig. 5.8. Numerous mountain structures which were largely buried under Tertiary alluvium have been partially exhumed, and stream courses have been superposed across these mountains. The numbers indicate the sites of canyons where stream courses are superposed across mountains. 1. Bighorn Canyon through the Bighorn Mountains; 2. Bighorn River near Greybull; 3. Shoshone Canyon; 4. Wind River Canyon; 5. Snake River Canyon; 6. Lodore Canyon of the Green River; 7. Devil's Gate on the Sweetwater; 8. North Platte River near Alcova; and 9. Laramie River through the Laramie Mountains. (After W. W. Atwood, Sr., and W. W. Atwood, Jr.)

it encountered the buried transverse structure and continued to cut its valley into it. An *antecedent valley,* however, antedates the structure across which it is cut, and the stream in it has been able to keep pace in downcutting with transverse doming, upfolding or upfaulting.

It was formerly common practice to interpret most gorges transverse to mountain structures as portions of antecedent valleys. Powell's example of an antecedent gorge was the Lodore Canyon of the Green River across the Uinta Mountains. It is now becoming evident that many gorges formerly thought to be antecedent are really superposed, which seems to be the correct interpretation of the Lodore Canyon of the Green River. The work of the Atwoods (1938) has shown that many, if not most, of the gorges in the Rocky Mountain region are the result of superposition. A most striking example is the gorge of the Sweetwater River in Wyoming known as the Devil's Gate. Here, the Sweetwater River cuts through the end of a mountain range, whereas by merely detouring a fraction of a mile it could have avoided the mountain structure completely. Such a situation seems logical only if we conceive of the valley as being let down from a cover mass onto a buried structure, under which condition the river had little chance to pick a course which would avoid the hidden mountain structure.

It is not meant to infer that there are no antecedent valleys. Such valleys do exist, but they are probably largely confined to regions of recent mountain movements, such as the circum-Pacific Tertiary mountain belt. The gorge of the Columbia River through the Cascade Mountains and the Santa Ana Valley across the north end of the Santa Ana Mountains in southern California are most likely antecedent. It requires detailed study of the geologic and geomorphic history of a region before it can be stated with assurance whether a valley is antecedent or superposed, particularly when remnants of any possible cover mass are lacking.

Valleys classified according to effects of change of base level. Many valleys show the effects of change in base level resulting from either diastrophic or eustatic rise or fall of sea level. Rise of sea level results in the development of *drowned valleys.* Chesapeake Bay is a broad estuary which was produced by the drowning of the lower part of the Susquehanna Valley. Drowning has resulted in the dismemberment of the lower part of the former valley system. The Delaware, Rappahannock, James, and Potomac valleys were formerly tributaries of the Susquehanna Valley. Drowned valleys are common today because of the postglacial rise of sea level which resulted from the return of vast quantities of melted ice to the oceans. Lowering of sea level (and other causes to be discussed later) may result in *rejuvenated valleys.* It is really the river which is rejuvenated, but it is not illogical to think of the valley as having been rejuvenated too, because the effect has been to impose youthful characteristics upon it.

DRAINAGE PATTERNS AND THEIR SIGNIFICANCE

Drainage pattern refers to the particular plan or design which the individual stream courses collectively form.   A distinction can and in some instances should be made between the patterns of the individual streams and their spatial relationships to one another.   Certain so-called drainage patterns might better be termed *drainage arrangements* since they refer more to the spatial relationships of individual streams than to the over-all pattern made by the individual drainage lines.   It is generally recognized that drainage patterns reflect the influence of such factors as initial slopes, inequalities in rock hardness, structural controls, recent diastrophism, and the recent geologic and geomorphic history of the drainage basin.   Because drainage patterns are influenced by so many factors they are extremely helpful in the interpretation of geomorphic features, and study of them represents one of the more practical approaches to an understanding of structural and lithologic control of land form evolution.

Types of drainage patterns.   The most commonly encountered drainage patterns are the dendritic, trellis, barbed, rectangular, complex and deranged. Of these, *dendritic patterns* are by far the most common.   They are charac-

Fig. 5.10.   Sawtooth Range, Idaho, viewed southward from McGowan Peak.   The topography is that developed upon a large batholith.   (Photo by A. A. Monner, U. S. Forest Service.)

terized by irregular branching of tributary streams in many directions and at almost any angle, although usually at considerably less than a right angle. They develop upon rocks of uniform resistance and imply a notable lack of structural control. Dendritic patterns are most likely to be found upon nearly

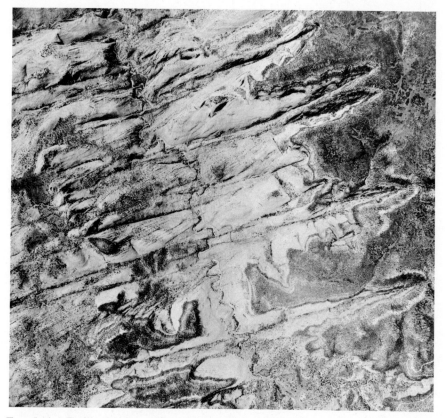

FIG. 5.11. Trellis drainage pattern controlled by jointing, near Zion Canyon, Utah. (Soil Conservation Service photo.)

horizontal sedimentary rocks or in areas of massive igneous rocks, but may be seen on folded or complexly metamorphosed rocks, particularly when imposed upon them through superposition. A special dendritic pattern is the *pinnate*. The tributaries to the main stream are subparallel and join it at acute angles. It is believed to represent the effect of the unusually steep slopes on which the tributaries developed.

Trellis patterns display a system of subparallel streams, usually aligned along the strike of the rock formations or between parallel or nearly parallel topographic features recently deposited by wind or ice. The major streams

Dendritic pattern
Virginia, Ill., quadrangle

Trellis pattern
Monterey, Va., quadrangle

Rectangular pattern
Elizabethtown, N.Y., quadrangle

Radial pattern
Katahdin, Me., quadrangle

Scale

0   1   2   3   4 Miles

FIG. 5.12. Types of drainage patterns.

frequently make nearly right-angled bends to cross or pass between aligned ridges, and the primary tributary streams are usually at right angles to the main stream and are themselves joined at right angles by secondary tributaries whose courses commonly parallel the master stream. Trellis patterns reflect marked structural control of most stream courses, except perhaps the trunk streams. The tributary valleys are usually subsequent strike valleys. Trellis patterns are particularly well displayed in the folded Appalachians where alternating weak and strong strata have been truncated by stream erosion. They may also be found on a limited scale around the flanks of mountains, such as the Colorado Front Range, where steeply tilted strong and weak rocks rest against a core of crystalline rocks. A variety of trellis pattern is the *fault trellis pattern,* which may be found where a series of parallel faults have brought together alternating bands of strong and weak rock. Usually a trellis pattern is largely subsequent in origin but there are areas where the streams are largely consequent ones. This may be encountered in certain glaciated areas where parallel hills known as drumlins give rise to this pattern and also to some degree in areas of parallel sand dunes, if surface drainage lines exist on such permeable materials.

Barbed drainage patterns usually have only local extent and will be found at or near the headwater portions of drainage systems. The tributaries join the main stream in "boathook bends" which point upstream. Most barbed patterns are the result of stream piracy which has effected a reversal of the drainage of part of a separate river system. Less commonly, the drainage reversal may have been effected by warping or tilting of the land or may represent drainage changes effected by glaciation.

In *rectangular drainage patterns,* both the main stream and its tributaries display right-angled bends. They reflect control exerted by joint or fault systems. Such patterns are especially well developed along the Norwegian coast and in portions of the Adirondack Mountains. Although there has probably been a tendency among some geologists to attribute undue importance to the control of joint and fault systems upon stream courses, it is certainly true that locally they may determine stream locations. A variant of rectangular drainage is the *angulate pattern.* It develops where faults or joints join each other at acute or obtuse angles rather than at right angles.

Some drainage patterns show such variations between component parts that it is impossible to describe the over-all patterns other than as *complex.* This is especially true in areas of complicated geologic structure and geomorphic history. The drainage of some of the more recently glaciated areas might be thought to belong in this category, but the fundamental difference between the drainage here and complex drainage is that there is usually complete lack of structural and bedrock control. The preglacial drainage has been effaced, and the new drainage has not had time to develop any significant degree of integration. This type of drainage pattern is commonly referred to as *deranged*. It is marked by irregular stream courses which flow into and out of lakes and have only a few short tributaries. Much of the inter-

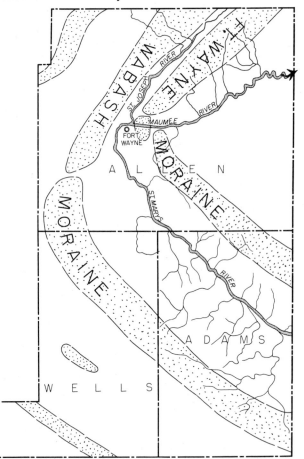

Fɪɢ. 5.13. A barbed drainage pattern determined by two end moraines.

stream area is swampy and frequently the streams are mere threads of water through the swampy areas.

Other drainage patterns which are encountered locally are the centripetal, radial, parallel, and annular. *Centripetal patterns* show drainage lines converging into a central depression. They are found on sinkholes, craters, and other basin-like depressions. *Radial patterns* have streams diverging from a central elevated tract. They develop on domes, volcanic cones, and various other types of isolated conical or subconical hills. *Parallel patterns* are

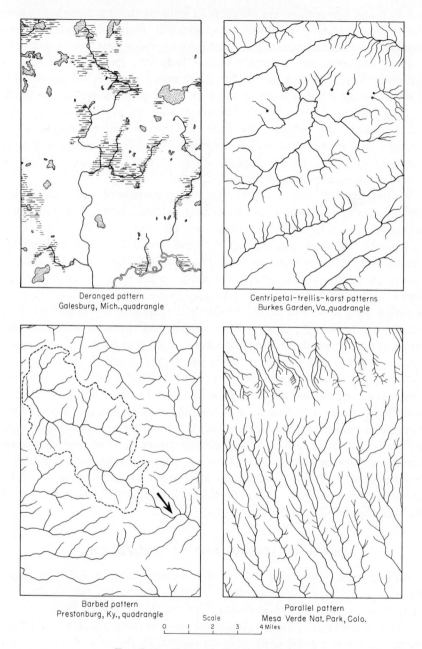

Deranged pattern
Galesburg, Mich.,quadrangle

Centripetal-trellis-karst patterns
Burkes Garden, Va.,quadrangle

Barbed pattern
Prestonburg, Ky., quadrangle

Parallel pattern
Mesa Verde Nat. Park, Colo.

Scale
0   1   2   3   4 Miles

FIG. 5.14. Types of drainage patterns.

usually found where there are pronounced slope or structural controls which lead to regular spacing of parallel or near-parallel streams. *Annular patterns* may be found around maturely dissected domes which have alternating belts of strong and weak rock encircling them. They have ring-like plans such as that of the Race Track or Red Valley which nearly encircles the Black Hills. It might be argued that three of the four above patterns are not so much drainage patterns as they are spatial relationships of adjacent streams. In radial patterns, for example, the individual stream patterns may well be dendritic or pinnate, and radial designates more their arrangement with respect to each other than it does the stream pattern. The same thing may be said of the centripetal and parallel patterns.

We frequently have occasion to refer to the patterns of individual stream courses in contrast to the over-all drainage pattern. Some of the common patterns displayed by individual streams are the straight, crooked, contorted, meandering, anastomosing, rectangular, braided, and deltaic or distributary. Streams are seldom straight for long distances, but where they are there is usually an implication of structural control or the effect of initially steep slopes upon homogeneous rocks. Varying degrees of crookedness may be exhibited, such as sinuous curves, broad, open meanders, contorted meanders, and the anastomosing pattern with its winding and tortuous intercommunicating channels, sloughs, and oxbow lakes. *Braided patterns* are marked by the streams being split into a number of intertwinned channels separated from each other by islands or channel bars. The braided condition is generally believed to indicate that a stream is unable to carry all of its load. It may result from excessive contribution of load to the main stream, from sudden decrease in stream gradient when a stream debouches from mountains onto lowlands with resulting loss of transporting power or from loss of volume through seepage, evaporation, or diversion. Where streams enter their deltas they frequently split into a number of independent channels across the delta known as *distributaries*. This pattern is well displayed by such streams as the Mississippi, Indus, and Rhine.

DRAINAGE TEXTURE AND ITS IMPLICATIONS

An important geomorphic concept is *drainage texture,* by which we mean the relative spacing of drainage lines. Horton (1945) has pointed out that what we commonly refer to as drainage texture really includes both drainage density and stream frequency. He defined *drainage density* as the total length of the streams in a given drainage basin divided by the area of the drainage basin, and *stream frequency* as the total number of streams in a drainage basin divided by the area of the drainage basin. Drainage texture does not refer to such factors as steepness of slope, amount of relief, or stage in the geomorphic cycle. We may have fine or coarse drainage texture in regions of low or high relief, in regions of gentle or steep slopes, or in young or old age topography.

FIG. 5.15. Vertical photo of a portion of Brown County, Indiana, showing medium-fine drainage texture on siltstones and shales. (Production and Marketing Administration photo.)

As yet we have no generally applicable quantitative definitions of the commonly used expressions fine, medium, and coarse drainage texture. Such attempts as have been made to define drainage texture in quantitative terms (Smith, 1950) have arrived at purely empirical formulae. In fact, it seems inevitable that any quantitative definitions will necessarily be of such a nature and applicable largely to specific areas. By their very nature fine, medium, and coarse are relative terms and not subject to arbitrary definition.

Much work remains to be done before we understand fully all the factors that influence drainage texture, but we do recognize the major controls. The following discussion does little more than point these out, for as yet the quantitative importance of each factor is not thoroughly understood. Climate affects drainage texture both directly and indirectly. The amount and type of precipitation influence directly the quantity and character of runoff. In areas where the precipitation comes largely as thundershowers, a larger percentage of the rainfall will run off immediately and, other factors being equal, there will be more surface drainage lines. Climate indirectly affects drainage texture by its control upon the amount and kind of vegetation through their influence upon the amount and rate of surface runoff. Climate affects the capacity of the soil to absorb rainfall by determining whether the soil is frozen and whether it is nearly saturated with moisture. It is probably true that with similar conditions of lithology and geologic structure semiarid regions have finer drainage texture than humid regions, although major streams may be more widely spaced in semiarid than in humid regions. The reason for this lies chiefly in the less extensive vegetal covering in dry regions and the larger percentage of runoff.

What Horton (1945) has called infiltration capacity, or what is more commonly referred to as the permeability of the mantle rock and bedrock, is probably the most important single factor influencing drainage texture. It is commonly observed that drainage lines are more numerous over impermeable materials than over permeable ones. In fact, areas covered by such permeable materials as sands and gravels may practically lack surface drainage lines. Infiltration capacity is influenced by a large number of factors. Among the more important ones are: soil texture or size of the individual particles; soil structure or the mode of arrangement of the individual particles; amount and kind of vegetal cover; biologic structures in the soil such as root perforations, animal burrows, humus and vegetal debris; moisture content of the soil; condition of the soil surface as determined by whether it is newly plowed, baked, or sun-cracked; and the temperature of the soil, particularly whether frozen or not.

Drainage texture is influenced by the amount of initial relief, for drainage lines will develop in larger number upon an irregular surface than upon one which lacks conspicuous relief. Rock structure, including such factors as massiveness and the abundance of joints and faults, may also exert an influence upon the drainage texture. *Available relief* (Glock, 1932; and Johnson, 1933), or the vertical distance from initial upland flats to the level of adjacent graded valleys, is significant, for high available relief will permit the cutting of deep valleys and thereby favor the development of many headward-eroding streams. It is not clear to what extent stage in an erosion cycle affects drainage texture but it appears that drainage texture is finest during maturity when drainage lines are most numerous. Whether a region is in its first or *n*th cycle of erosion may possibly be another factor affecting drainage texture. It would seem likely that regions in a first cycle might have coarser texture than

Fig. 5.16. Vertical photo of a part of the Wabash Valley and adjoining upland near Clinton, Indiana, showing difference in the drainage texture upon a sand and gravel terrace and on a till upland. (Production and Marketing Administration photo.)

those which have passed through several cycles but this is largely supposition. The relative importance of sheetwash and creep as compared with the transportation of rock waste by streams is another factor which may have significance. Where sheetwash and mass-wasting are most effective there will be correspondingly fewer stream courses.

Badland topography illustrates one set of conditions which leads to fine drainage texture. Impermeable clays and shales, sparse vegetation, and a prevalence of dash rainfall have been responsible for the extremely fine drainage texture. Coarse drainage texture is particularly well displayed on the sand and gravel outwash plains and valley trains of glaciated areas. Gravel plains have fewer surface drainage lines on them than adjacent till plains underlain by relatively impermeable clay till. Many of the small streams which descend from till plain uplands onto gravel terrace remnants of valley trains cease to flow as surface streams upon reaching the permeable terrace materials.

STREAM MEANDERING AND LATERAL EROSION

Lateral erosion was discussed incidentally in a previous section (p. 102), but it merits further attention because of its importance in the later stages of the fluvial cycle. The *valley flat* is the fundamental land form produced by lateral erosion. This is a bedrock erosional surface which may go unrecognized, because it is commonly veneered with alluvium. The first stage in the development of a valley flat involves primarily the elimination of the spurs between stream meanders. This process is frequently referred to as spur trimming. This is effected by lateral erosion as a stream flows against the upside and downstream sides of spurs in going around meanders. Continuation of this process will reduce the width of a meander spur until only a narrow *meander neck* separates the stream on the two sides of the meander. Continued impingement of a stream against the two sides of a meander neck will result in the stream cutting through it and forming a *meander cutoff*. A remnant of the meander spur will then be left as a *meander core,* and the abandoned route around it becomes a *meander scar*. The stage which immediately precedes the formation of a detached meander core is that in which a stream flows across a meander neck beneath a *natural bridge* or *arch*. The famous natural bridges of Utah are examples. Meander cutoffs are characteristic of late youth and are expectable features attendant upon the valley widening which marks the initiation of maturity. Various stages in meander cutoff development ranging from incipient ones to multiple completed cutoffs are shown along the course of the Kentucky River on the Frankfort, Lockport, and Burnside, Kentucky, topographic sheets. Topographic maps frequently show meander cores, meander scars, and meander scarps perched above present valley floors which attest to cutoffs made when the streams flowed at higher levels.

With the beginning of a valley flat we get the initiation of a *floodplain,* the deposit of alluvium which covers a valley flat. It may be impractical to distinguish valley flat from floodplain, yet it should be kept in mind that along many valleys the floodplain is but a veneer of alluvium over a bedrock floor which has been cut by lateral erosion. The initial floodplain is different in many respects from the final floodplain of old age. Its beginnings appear in youth before streams have ceased downcutting as crescent-shaped or slightly sinuous strips of coarse alluvium along the convex banks of stream meanders. These deposits have been called *floodplain scrolls*. They are subject to repeated reworking as the stream cuts laterally and downward in the process of attaining a profile of equilibrium. By the time a graded condition is attained the thickness of alluvium usually has become so great that little if any bedrock is exposed along the channel bottom. Although scour and fill, particularly during flood stage, may continue to rework the alluvium, there does develop over the bedrock floor of the valley a laterally and upward-expanding cover of alluvium which is laid down chiefly when the stream overflows its banks. Upper floodplain materials consist primarily of finer alluvium and may have a thickness which exceeds the depth of the river channel, but beneath this finer alluvium there usually will be found coarser materials representing channel deposits made at the various positions occupied by the stream as it migrated laterally over its valley flat.

Two significant achievements mark the attainment of valley maturity. The stream in it has attained a profile of equilibrium and is no longer actively deepening its valley, and the valley flat at least equals the meander belt in width, so that the stream meanders are free meanders and are not inclosed by valley walls. Downstream meander sweep now becomes easy and rapid, since it involves only the removal of alluvial materials rather than bedrock, as in youth. Lateral sweep of a meander belt causes the stream to impinge upon valley sides and continue the process of valley widening. Meander cutoffs may accompany this process but they will differ from the cutoffs along incised streams in that there will be no bedrock meander cores to mark the sites of former meander necks. Meander scars on the surface of the flood-plain will mark abandoned channels, and the deeper ones may be sites of oxbow lakes, bayous, or swamps.

Opinion differs as to whether there is a limit to the extent to which lateral erosion may continue to widen valleys and reduce interstream upland tracts. Davis attributed the reduction of interstream areas more to downwasting than to lateral planation. He thought that the gently undulating topography which presumably develops toward the end of a fluvial cycle, to which he gave the name *peneplain,* was only in part a result of lateral erosion. Floodplains were incidental as compared with the more extensive interstream areas which were *worn down* rather than *planed off*. Johnson (1931) and Crickmay (1933) in particular have attached much greater significance to lateral plana-tion in the process of peneplain development. (See Chapter 8.) If rivers that flow across floodplains many times wider than their meander belts are observed, it will be found that in relatively few places are the streams actually against and undercutting the valley sides. This suggests at least that there may be a limiting width of valley flat beyond which lateral erosion becomes insignificant.

The valleys of many, if not most, of the world's large rivers are so deeply filled with alluvium that it may seem inappropriate to consider their flood-plains as veneers over bedrock valley flats. The alluvial fills in such valleys

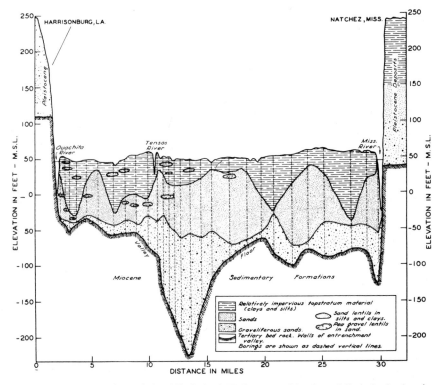

Fig. 5.17. Cross section of the Mississippi Valley near Natchez, Mississippi, showing the alluvial fill in the valley and the configuration of the pre-Pleistocene bedrock valley. The vertical scale is greatly exaggerated. (After H. N. Fisk.)

as those of the Mississippi, Missouri, and Ohio in places are several hundred feet thick. The unusually deep alluvium in these valleys is largely the result of rising sea levels which accompanied deglaciation, discharge of large quantities of glacial outwash down the valleys, and local downwarping of the earth's crust. Fisk (1944) has described the buried valley systems which underlie the Mississippi alluvial plain and has shown that the bedrock floor beneath the alluvial fill exhibits notable relief. The alluvial fill increases in thickness from about 160 feet near St. Louis to over 600 feet near the Gulf of Mexico. Although the present floodplains of most of our great rivers are much more than alluvial veneers over an erosional bedrock surface, the fact still remains that floodplains many miles wide could not have been built up through aggradation had not the rivers previously by lateral erosion opened up wide valleys.

INFLUENCES OF HOMOCLINAL STRUCTURES UPON TOPOGRAPHY

Extensive areas of the earth's surface are underlain by sedimentary rocks which have uniformly gentle to steep dips. Such structures are commonly referred to as homoclinal structures. They may be carelessly described as monoclinal structures, but this term should be restricted to local flexures which are frequently imposed upon broader regional structures. If the rocks in a homoclinal structure are beds varying in resistance to weathering and erosion, there will develop a number of topographic features whose major differences are related to the steepness of dip.

One of the common regional expressions of gently or moderately dipping rock is the *cuesta*. This has an abrupt escarpment or in-face on the up-dip side and a more gentle backslope or dip slope extending in the direction of the regional dip. The scarp which constitutes the front slope of a cuesta is called by some geologists a *structural escarpment*. Such a designation is undesirable because it may carry with it the implication of being the result of the more intense types of deformation which produce folds and faults. We may avoid such possible confusion by calling it a *cuesta scarp*.

One of the simplest situations under which cuestas may develop is represented by a recently emerged coastal plain underlain by seaward-dipping weak and strong strata. The Gulf coastal plain through Alabama and Mississippi is a good example of such a situation. Here a series of cuestas has developed with intervening lowlands or vales to form what is called a *belted coastal plain*. The lowland which exists between the Tertiary and Cretaceous sediments of the coastal plain and the old land against which they abut is called an *inner lowland* or *vale*. The cuestas which flank the Paris Basin comprise a well-known belted coastal plain and their scarps played a strategic role during World War I, as many of the famous battles of that war were fought for control of them.

Not all cuestas are associated with coastal plains but some are found on the flanks of geologically old folds or domes. The well-known Niagara escarpment which extends east-west across New York State and continues around the Great Lakes until it disappears beneath glacial drift in northern Illinois is an example of this type. The so-called Catskill Mountains of southeastern New York are the dissected front of a cuesta, as is the Allegheny front in Pennsylvania. The Dripping Springs escarpment of Kentucky, which continues northward into Indiana, where it is called the Chester escarpment, marks the edge of a cuesta on the west flank of the Cincinnati arch. The various wolds of southern England are similar features.

The scarps which form the front slopes of cuestas may be straight and imposing topographic features to begin with but as weathering, mass-wasting, and erosion attack them they recede down the dip of the strata and become

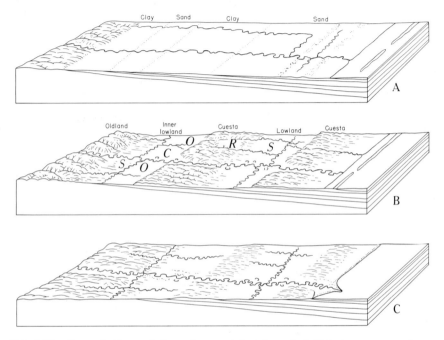

Fig. 5.18. Stages in the evolution of a coastal plain. *A.* Initial stage: a recently uplifted portion of the continental shelf. *B.* Mature stage of development with a series of cuestas and lowlands. *C.* Late mature or old-age stage with cuestas largely obliterated. *S,* subsequent stream; *C,* consequent stream; *O,* obsequent stream; and *R,* resequent stream. (After A. N. Strahler, *Physical Geography,* John Wiley & Sons.)

sinuous or irregular in outline. As indicated in Chapter 4, such old cuesta scarps are sometimes called weathering escarpments but this term is not particularly appropriate, for it overemphasizes the role played by weathering in causing scarp recession. During recession of a cuesta scarp portions are likely to be detached and isolated from the cuesta proper and become *outliers.* In some areas, the strata underlying a cuesta were deposited upon an oldland of igneous or metamorphic rock having an irregular surface. In the process of valley cutting across the cuesta some of the buried hills on the oldland surface may be exposed as *inliers.* They are sometimes referred to as *mendips* from the Mendip Hills of England, which were of such origin.

It is frequently inferred that streams that flow down the front slope of a cuesta are obsequent streams and those which flow down the more gentle back slope are consequent or resequent streams, as if obsequent streams are those that flow against the dip of the rock and consequent streams or resequent streams are those that flow with the dip. This is not necessarily true, and we should hesitate to designate them as such unless the geomorphic evidence sup-

Fig. 5.19. View of the Orange Cliffs, Utah, near Millard Canyon, showing numerous erosion outliers. (Photo by G. A. Grant, National Park Service.)

ports such a conclusion. Streams flowing down both slopes of a first cycle cuesta may well be consequent streams.

All gradations are to be found between hogbacks, homoclinal ridges, cuestas, and mesas. Their fundamental differences are related to the steepness in dip of the resistant beds which are responsible for them and to their geographic extent. Depending upon whether rock attitudes are nearly vertical, moderately dipping, gently dipping or nearly horizontal, we may expect to find hogbacks, homoclinal ridges, cuestas, or mesas. It may be difficult to distinguish sharply immediately adjacent members of the series, but there are significant differences between the extremes of the series, hogbacks and mesas. *Hogbacks* are sharp-crested ridges which develop where the rock dips are steep, roughly in excess of 45 degrees. They have limited area, although they may have considerable length. Rock dips in cuestas are seldom more than a few degrees. Although the front slope of a hogback may be slightly steeper than the back slope, there is not the marked difference in the two that there is in a cuesta. Furthermore, hogbacks remain rather well fixed in position. A slight shifting may take place as the landscape is lowered, but it is likely to be a matter of feet rather than miles, as may happen with cuesta scarps. Such features as dike ridges (see p. 511) should not be called hogbacks, even though they may resemble them in appearance, for they are

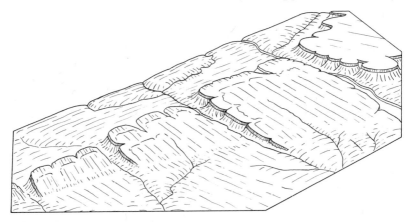

Fig. 5.20. Diagram showing the transition from a hogback through a homoclinal ridge and cuesta to a mesa as controlled by varying rock dip. (Modified after W. M. Davis.)

features developed upon igneous intrusions rather than upon steeply dipping stratified rocks.

Homoclinal ridges (Cotton, 1948) develop in areas of moderately dipping strata, and there is a notable difference in the steepness as well as in the length of the front and back slopes. A homoclinal ridge is not so sharply defined as is a hogback nor does it have the areal extent of a cuesta. Although a hogback is a homoclinal ridge, there does seem to be a need for a term which describes those intermediate forms between hogbacks and cuestas which are found in so many areas.

Mesas are isolated tablelands capped with a protective covering which is essentially horizontal in attitude. Features similar in character but with more limited summits are usually called *buttes*. Some mesas owe their existence to the protective cover of sedimentary rocks, particularly sandstones, but others may be capped with lava flows or gravels and boulders. Commonly a mesa is rimmed by a scarp formed by the resistant cap rock, but at times the term is extended to include tablelands which are not completely isolated. Some mesas have such extensive regional development that they may be better described as *structural plateaus.* The Kaibab Plateau in the Colorado plateaus area, the Great Sage Plateau of southeastern Utah and southwestern Colorado, the Edwards Plateau of Texas and the Drakenburg Plateau of southern Africa are well-known examples of structural plateaus of great regional extent.

AN IDEALIZED FLUVIAL CYCLE

An attempt is made here to indicate the changes in topographic conditions that accompany the progress of a somewhat idealized geomorphic cycle. It should be recognized that complications of geologic structure, diastrophic history, and climatic conditions may result in notable departures from the sequence of events suggested and may be just as "normal" as those indicated. Let us start with a region that has recently been uplifted from beneath the sea. Let us further assume that the geologic structure is simple and that there are sedimentary rocks of varying hardness having moderate seaward dips uncomplicated by any pronounced geologic structures such as folds and faults. We shall further assume that this newly uplifted land mass remains stationary with respect to sea level long enough for a geomorphic cycle nearly to run its course. What then would be the characteristics of the landscape during each of the stages of the cycle under these conditions?

Youth

1. There will be a few consequent trunk streams but few large tributaries. Numerous short tributaries and gullies will be extending themselves by headward erosion and developing dendritic valley systems.

2. Valleys will have V-shaped cross profiles and will be shallow or deep depending upon the height of the region above sea level.

3. There will be a general lack of floodplain development except along trunk streams, and valley sides will rise from near the streams' edges.

4. Interstream tracts may be extensive and poorly drained. Lakes and swamps may exist in the interstream areas if these areas are not well above the local base level.

5. Waterfalls or rapids may exist where stream courses cross beds of particularly resistant rock. They are most typical of early youth and will have disappeared before maturity is attained.

6. Stream divides will be broad and poorly defined.

7. Stream meandering may exist in youth, but the meanders are those on a flat and undissected initial surface or are closely confined meanders in valleys incised below the upland surface.

Maturity

1. Valleys have extended themselves so that the region now has a well-integrated drainage system.

2. Adjustment of streams to such lithologic variations as exist will be evident in the existence of some longitudinal tributaries along belts of particularly weak rock.

3. Stream divides will be sharp and ridge-like resulting in a minimum of interstream uplands.

4. A profile of equilibrium has been attained by the master streams, but many of the tributaries may still be ungraded.

Fig. 5.21. Youthful drainage on the Pliocene Chanac formation in the San Joaquin Valley southeast of Bakersfield, California. (Photo by J. S. Shelton and R. C. Frampton.)

5. Any lakes or waterfalls that existed in youth have been eliminated.

6. Floodplain tracts constitute a considerable portion of the valley floors.

7. Meanders may be conspicuous but, in contrast to those of youth, they are free to shift their positions over the floodplains.

8. The widths of the valley floors do not greatly exceed the widths of the meander belts.

9. The maximum possible relief exists.

10. The topography consists not so much of valley bottoms and upland tracts as it does of slopes of hillsides and valley sides.

Old age

1. Tributaries to trunk streams are usually fewer in number than in maturity but more numerous than in youth.

FIG. 5.22. Topography in late youth to early maturity. The Lammerlaws near Otago, New Zealand. (Photo by Whites Aviation, Ltd.)

FIG. 5.23. White River Valley near Shoals, Indiana, a valley in late maturity. (Indiana State Geol. Survey photo.)

2. Valleys are extremely broad and gently sloping both laterally and longi-tudinally.

3. There is marked development of floodplains over which streams flow in broadly meandering courses.

4. Valley widths are several times those of the meander belts.

5. Interstream areas have been reduced in height, and stream divides are not so sharp as in maturity.

6. Lakes, swamps, and marshes may be present but they are on the floodplains and not in the interstream areas as in youth.

7. Mass-wasting is dominant over fluvial processes.

8. The adjustments of stream courses to varying lithology which were apparent in maturity may now be obscure.

9. Extensive areas are at or near the base level of erosion.

It is recognized that the preceding outline is based upon certain assumptions that are not everywhere attained, particularly homogeneous rocks, lack of pronounced structural features, and stability of the land mass. It might be argued that it represents an exception rather than the rule. Even so, it does suggest changes in the characteristics of valleys and interstream areas which may be expected with modifications under different conditions. It should be kept in mind that the changes from one stage to another are transitional rather than abrupt. The outline attempts to give the regional picture and not that of individual valleys. A region may be classed as mature and have within it individual valleys which are youthful or in early old age, or it may be in old age and still have within it mature valleys. Usually we are more concerned with the topographic age of areas than of individual valleys, and it is the picture that the total assemblage of land forms presents which determines this. Complications that may affect the cycle will be discussed in the following chapter.

REFERENCES CITED IN TEXT

Atwood, W. W., Sr., and W. W. Atwood, Jr. (1938). Working hypothesis for the physiographic history of the Rocky Mountain region, *Geol. Soc. Am., Bull. 49,* pp. 957–980.

Bryan, Kirk (1940). The retreat of slopes, *Assoc. Am. Geog., Ann., 30,* pp. 254–268.

Cotton, C. A. (1948). *Landscape,* pp. 128–129, Whitcombe and Tombs, Ltd., Wellington.

Crickmay, C. H. (1933). The later stages of the cycle of erosion, *Geol. Mag., 70,* pp. 337–347.

Davis, W. M. (1902). Base-level, grade and peneplain, *J. Geol., 10,* pp. 77–111; also in *Geographical Essays,* pp. 381–412, Ginn and Co., New York.

Fisk, H. N. (1944). *Geological Investigation of the Alluvial Valley of the Lower Mississippi River,* pp. 11–16, Miss. River Comm., Corps of Engineers, War Dept.

Glock, W. S. (1932). Available relief as a factor of control in the profile of a land form, *J. Geol., 40,* pp. 74–83.

Horton, R. E. (1945). Erosional development of streams and their drainage basins; hydrophysical approach to quantitative morphology, *Geol. Soc. Am., Bull. 56,* pp. 275–370.

Johnson, D. W. (1929). Baselevel, *J. Geol., 37,* pp. 775–782.

Johnson, D. W. (1931). Planes of lateral corrasion, *Science, 73,* pp. 174–177.

Johnson, D. W. (1932). Streams and their significance, *J. Geol., 40,* pp. 481–497.

Johnson, D. W. (1933). Available relief and texture of topography: a discussion, *J. Geol., 41,* pp. 293–305.

Kesseli, J. E. (1941). The concept of the graded river, *J. Geol., 49,* pp. 561–588.

Mackin, J. H. (1948). Concept of the graded river, *Geol. Soc. Am., Bull. 59,* pp. 463–512.

Malott, C. A. (1928). Base-level and its varieties, *Indiana Univ. Studies, 82,* pp. 37–59.

Powell, J. W. (1875). *Exploration of the Colorado River of the West,* p. 203, Smithsonian Institution, Washington.

Smith, K. G. (1950). Standards of grading texture of erosional topography, *Am. J. Sci., 248,* pp. 655–668.

Additional References

Baulig, Henri (1950). Le notion de profil d'équilibre, Essais de géomorphologie, pp. 43–86, *Publs. de la faculté des lettres de l'université de Strasbourg, Paris;* also in *Compt. rend. congr. inter. géogr., Le Caire* (1925), *3,* pp. 51–63.

Challinor, J. (1930). The curve of stream erosion, *Geol. Mag., 67,* pp. 61–67.

Chamberlin, R. T. (1930). The level of baselevel, *J. Geol., 38,* pp. 166–173.

Davis, W. M. (1899). The geographical cycle, *Geog. J., 14,* pp. 481–504; also in *Geographical Essays,* pp. 249–278, Ginn and Co., New York.

Davis, W. M. (1923). The scheme of the erosion cycle, *J. Geol., 31,* pp. 10–25.

Glock, W. S. (1931). The development of drainage systems: a synoptic view, *Geog. Rev., 21,* pp. 475–482.

Holmes, C. D. (1952). Stream competence and the graded stream profile, *Am. J. Sci., 250,* pp. 899–906.

Johnson, D. W. (1933). Development of drainage systems and the dynamic cycle, *Geog. Rev., 23,* pp. 114–121.

Macar, P. F. (1934). Effects of cut-off meanders on the longitudinal profiles of streams, *J. Geol., 42,* pp. 523–536.

Malott, C. A. (1928). The valley form and its development, *Indiana Univ. Studies, 81,* pp. 3–34.

6 · Complications of the Fluvial Cycle

INTERRUPTIONS OF THE CYCLE

Rejuvenation. Multicyclic evolution of landscapes is more common than the monocyclic development discussed in the preceding chapter. Mature or old-age topography is likely to have superposed upon it youthful features as a result of rejuvenation. This results from any change which causes streams that had previously attained profiles of equilibrium, or were aggrading, to engage again actively in valley deepening. Rejuvenation may result from causes which are dynamic, eustatic, or static in nature.

Dynamic rejuvenation may be caused by epeirogenic uplift of a land mass, with accompanying tilting and warping. Such movements may be rather localized and associated with neighboring orogenic movements or they may be, as thought by some, world-wide in nature. Localized downtilting, warping, or faulting of a drainage basin will result in a steepening of stream gradients followed by downcutting by the streams which now have transporting power in excess of that required for transport of their loads. The effects of seaward tilting would presumably be felt along the entire stream course and immediately reflected in deepening of the valley by the stream in it. Its effects will be felt immediately, however, only by those streams whose courses roughly parallel the direction of tilting. A stream flowing at right angles to the direction of tilting will respond to rejuvenation only after the stream which it joins has deepened its valley so as to leave the tributary out of adjustment. Even then the effect is felt only at the mouth of the tributary rather than along its entire length.

Eustatic rejuvenation results from causes that produce world-wide lowerings of sea level rather than localized changes in base level. Two types of eustatism have been recognized. *Diastrophic eustatism* is change of sea level resulting from variation in capacity of the ocean basins, whereas *glacio-eustatism* refers to changes in sea level produced by withdrawal or return of water to the oceans, accompanying the accumulation or melting of successive ice sheets. Eduard Suess was the father of the idea of diastrophic eustatism. Baulig (1935) has been one of its modern proponents. He recognized the importance of glacio-eustatism but thought that there is evidence of world-wide lowerings of sea level during the Pliocene and Pleistocene of a greater magnitude than would have been possible through glacio-eustatic changes. He thought these lowerings of sea level were the result of epeirogenic movements to be associated with orogenic movements and a consequence of them rather than independent of orogenic movements, as thought by some geologists. As large segments of the earth's crust were up-arched through tangential forces, isostatic adjustments entailed subcrustal flow from beneath the ocean basins toward the rising blocks. Removal of earth material from beneath the ocean basins caused their floors to sink with resulting fall in world sea levels.

Eustatic lowering of sea level will cause rejuvenation of a stream at its mouth. Regrading of a stream toward the new base level will progress up-valley. The result is an interrupted profile, with the point of intersection of the two base levels being marked by a nickpoint which gradually proceeds up stream as the new base level is extended headward. It should not, however, be assumed that every break in a stream profile is to be interpreted as a nickpoint produced by rejuvenation. We have previously seen (p. 108) that along a graded stream there are numerous changes in gradient related to local factors which affect the slope of the profile of equilibrium. Where nickpoints exist, however, that cannot be accounted for by lithologic or structural controls, along with marked topographic differences in the portions of the valley below and above them (youthful characteristics downstream and old age or mature features upstream), we are probably safe in attributing them to rejuvenation.

The most recent eustatic changes of sea level were those produced by the alternations between the glacial and interglacial stages of the Pleistocene. The most commonly accepted estimate is that sea level today is about 300 feet higher than at the maximum extent of the Wisconsin ice sheets. During the low sea levels of the glacial ages streams presumably were rejuvenated in their lower courses. Baulig (1940) held that, during these periods of low sea level, waves of headward erosion were initiated which are still continuing inland and are marked by nickpoints in stream profiles. Johnson (1938) was skeptical of our ability to recognize such effects in present-day stream profiles. If nickpoints are produced in this manner along stream courses, they should be discernible along all streams rather than restricted to certain valleys, as those resulting from local diastrophism may be.

Streams may exhibit signs of renewed youth from changes which involve neither uplift of the land nor eustatic lowering of sea level. This has been called *static rejuvenation* (Malott, 1920). At least three changes may produce static rejuvenation. They are decrease in load, increase in runoff because of increased rainfall, and increase in stream volume through acquisition of new drainage by stream diversion or derangement (see p. 152).

Rejuvenation because of decrease in load has taken place during post-glacial times along many valleys that formerly received large quantities of glacial outwash. With change to non-glacial conditions, stream loads decreased and as a result valley deepening ensued, leaving the former valley floors partially preserved as gravel terraces.

A change to greater rainfall may cause an increase in stream volume without corresponding increase in load, as result of which streams find themselves able to transport their loads over more gentle slopes. Valley incision may then result. Opinion differs as to whether the effects of increased rainfall are always as suggested. This point is further discussed on p. 150.

Stream rejuvenation because of an increase in stream volume through diversion of drainage from one river system to another is well illustrated along the Ohio valley. The preglacial Ohio was a much shorter stream than the present river. It headed in either southeastern Indiana or southwestern Ohio. The present upper Ohio drainage, including that of the Kanawha, Monongehela, and Allegheny flowed northward in preglacial time. This drainage was blocked, ponded, and diverted southwestward along the ice front into the preglacial Ohio to form the present stream. As a result of its increased volume the lower Ohio River proceeded to incise its valley. Malott (1920) estimated that as much as 90 feet of valley trenching in the lower Ohio Valley is attributable to this cause.

Topographic evidence of rejuvenation. Rejuvenation of a region is suggested when we have within it a *topographic unconformity* or *discordance*. By this is meant a lack of conformity between the topographic forms of the upper parts of valleys and adjacent uplands and those of the lower parts of valleys. The upland and upper valley forms are those of maturity or old age, whereas the lower valley forms are those of youth. A common example of this is the *valley-in-valley* cross profile, also described as a *two-story valley* or *two-cycle valley*. The chief characteristic of such a cross profile is a break in slope along the valley side marked by a shoulder separating the youthful valley form below from the older valley form above. This bedrock shoulder is a remnant of the valley floor formed during a cycle of erosion preceding the present. Such bedrock terraces are not to be confused with structural benches which correlate with and are caused by resistant strata. Rather than being structurally controlled levels they mark former base levels of erosion and extend across rocks of varying lithology.

The long profile of a valley may show evidence of rejuvenation, consisting of nickpoints along the valley which cannot be attributed to hard-rock barriers. It may even be possible in this way to detect evidence of rejuvenation in valleys which have not reached a sufficiently advanced stage of development to have had valley flats along them which would form terraces after rejuvenation.

Incised or inclosed meanders are usually considered evidence of rejuvenation. Unfortunately there is confusion in the usage of terms intended to describe valley meanders which are entrenched in bedrock, as well as in our ideas as to what significance such meanders have. Five terms, incised, entrenched, intrenched, inclosed and ingrown, have been applied to them with varying meanings. Our usage of these terms will have to be somewhat

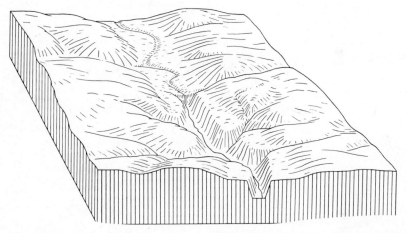

FIG. 6.1. Diagram of a two-cycle valley showing a nickpoint that marks the head of rejuvenation. (Drawing by William J. Wayne.)

arbitrary. It seems to be more common practice to use incised and inclosed as generic terms to include any meanders inclosed by rock walls. We shall therefore consider *incised meanders* and *inclosed meanders* as synonymous group names for all meanders set down in bedrock, regardless of the cross profile of the valley. Two types of incised or inclosed meanders are generally recognized: (1) *entrenched* or *intrenched meanders* (the difference is only in spelling), which show little or no contrast between the slopes of the two valley sides of a meander curve, and (2) *ingrown meanders* (Rich, 1914), which exhibit pronounced asymmetry of cross profile with undercut slopes on the outside of the meander curves and slipoff slopes on the inside. A good deal of the confusion in terminology arises from the fact that the various terms that have been used have so nearly synonymous meanings that they do not inherently carry the different connotations that have often been implied by those who have used them.

It is often inferred that incised meanders were inherited from a previous erosion cycle and are hence second-cycle features and evidence of rejuvena-

tion. Blache (1939) and Cole (1930) have challenged this conclusion.
Some incised meanders may be inherited but this is not necessarily so.
Mahard (1942) has reminded us of the possibility of the development of
incised meanders during an initial cycle of erosion. Certainly, ingrown
meanders may develop during a single cycle through accentuation of initial
sinuosities of a stream by lateral corrasion and accompanying downcutting.

FIG. 6.2. The entrenched meanders of the North Fork of the Shenandoah River, west
of Massannutten Mountain, Luray quadrangle, Virginia. (Photo by John L. Rich,
courtesy American Geographical Society.)

Although it would seem likely that first-cycle incised meanders will usually
be the ingrown type, it does not follow that all ingrown meanders are first-
cycle meanders. Ingrown meanders are probably the expectable type of incised meanders, even when inherited, and entrenched meanders are likely to
develop under special conditions. It is not clear, however, why in one area
downcutting is accompanied by significant lateral erosion and in another area
not. The problem is further complicated where both entrenched and ingrown meanders are found along different stretches of the same valley. This
suggests that the two types may be related to local differences in geologic
structure or lithology. Moore (1926) in discussing the inclosed (incised)
meanders of the Colorado plateau concluded that whether entrenched or ingrown meanders develop depends largely upon the character of the inclosing
rock and the amount of stream load. He found entrenched meanders in hard-

rock areas and ingrown meanders or open valleys in areas of weak rock. He attached considerable significance to whether the stream carried a large or small load and concluded that heavily loaded streams are more likely to cut laterally as they incise their valleys than those with lesser loads and would thereby develop ingrown meanders rather than entrenched.

Rich (1914) attributed much importance to the rate of uplift in determining whether entrenched or ingrown meanders develop from inherited meanders. He thought the entrenched type will probably develop if the rate of uplift is rapid, for then downcutting will be rapid in comparison to lateral erosion and meander sweep; if, however, the rate of uplift is slow, the ingrown type will develop because lateral erosion and meander sweep proceed apace with downcutting.

It is sometimes assumed that, if a region having meandering streams is vertically uplifted, the effects of the new base level will be felt only at the termini of the valleys and the new base level will migrate upstream as nickpoints following the meander curves. Geomorphologists who have difficulty in believing in nickpoints in homogeneous rock have equal difficulty in explaining so simply entrenched meanders. We may conclude, in the first place, that the reasons for the difference between ingrown and entrenched meanders are still obscure. Secondly, we should be cautious in taking incised meanders as prima-facie evidence of rejuvenation. Along with other supporting evidence, however, they may strengthen the arguments for rejuvenation.

Aggradation. We have considered dynamic, eustatic, and static causes which may effect rejuvenation of streams. For each type the effects are the same as those that would result from an actual lowering of base level. It will be evident that in each of the three types of change the effects may be of such a nature as to cause streams to aggrade rather than deepen their valleys. Sinking of the land may result in the drowning of the portions of valleys adjacent to the sea, which will be accompanied by alluviation or valley filling upstream. Actual drowning of a valley is not necessary to cause a stream to become an aggrading one. Downward tilting of the land inland will produce the same effect as soon as a stream's gradient becomes too low for transport of its load. Aggradation for this reason will become obvious first along streams which were graded, but it may also affect non-graded streams provided downtilting inland is sufficient.

The effect on streams of rise of sea level is the same as if the land sank. Fisk (1939) has interpreted the geomorphic history of the lower Mississippi Valley as one of alternation between alluviation during interglacial high sea levels and valley trenching during glacial low sea levels. He has described four terraces which he thought represent the tops of valley fills made during times of high sea levels. He believed that, because there was continuing uplift of the Mississippi Valley inland from Baton Rouge to isostatically com-

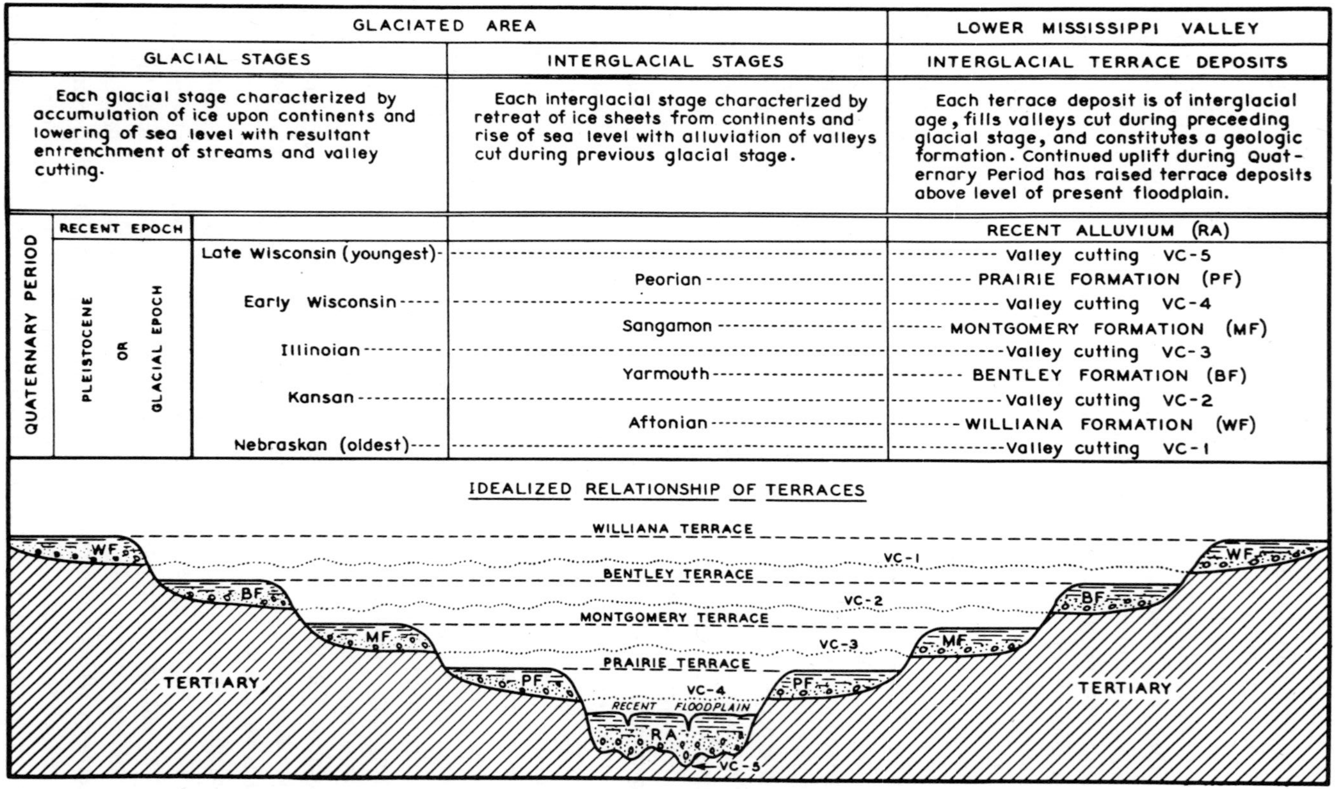

	GLACIATED AREA		LOWER MISSISSIPPI VALLEY
	GLACIAL STAGES	INTERGLACIAL STAGES	INTERGLACIAL TERRACE DEPOSITS
QUATERNARY PERIOD — PLEISTOCENE OR GLACIAL EPOCH / RECENT EPOCH	Each glacial stage characterized by accumulation of ice upon continents and lowering of sea level with resultant entrenchment of streams and valley cutting.	Each interglacial stage characterized by retreat of ice sheets from continents and rise of sea level with alluviation of valleys cut during previous glacial stage.	Each terrace deposit is of interglacial age, fills valleys cut during preceding glacial stage, and constitutes a geologic formation. Continued uplift during Quaternary Period has raised terrace deposits above level of present floodplain.
	Late Wisconsin (youngest)		RECENT ALLUVIUM (RA)
		Peorian	-- Valley cutting VC-5
	Early Wisconsin		-- PRAIRIE FORMATION (PF)
		Sangamon	-- Valley cutting VC-4
	Illinoian		-- MONTGOMERY FORMATION (MF)
		Yarmouth	-- Valley cutting VC-3
	Kansan		-- BENTLEY FORMATION (BF)
		Aftonian	-- Valley cutting VC-2
	Nebraskan (oldest)		-- WILLIANA FORMATION (WF)
			-- Valley cutting VC-1

IDEALIZED RELATIONSHIP OF TERRACES

WILLIANA TERRACE — BENTLEY TERRACE — MONTGOMERY TERRACE — PRAIRIE TERRACE — RECENT FLOODPLAIN

VC-1, VC-2, VC-3, VC-4, VC-5

WF, BF, MF, PF, RA

TERTIARY

FIG. 6.3. An interpretation of the Pleistocene history of the lower Mississippi Valley. (After H. N. Fisk.)

pensate for the great mass of delta deposits, each successive interglacial high sea level was slightly lower than the one preceding it. Although there was trenching of the interglacial valley fills during glacial stages when sea level was lowered, the fills were not completely removed. Remnants of these fills constitute the four terrace levels.

Considerable skepticism exists among glacial geologists in the upper Mississippi Valley as to the validity of these conclusions. The evidence there seems to indicate that glacial ages (times of low sea levels) were characterized not so much by valley trenching as by filling with glacial outwash, although it is recognized that during a glacial age there were alternating episodes of deposition and erosion. Valley trenching, rather than filling, marked interglacial time (time of high sea level), for then large quantities of glacial outwash were not being carried by streams and the gradients which had been necessary for the transport of this outwash were no longer required.

It is possible that these two sharply contrasting viewpoints are not entirely incompatible. Stream behavior in the lower Mississippi Valley may have been dominated by the effects of eustatically lowered and raised sea levels, whereas the upper Mississippi Valley was too remote from the sea to have felt the influence of sea-level changes of such short duration and hence alternated between aggradation during glacial times and erosion during interglacial times.

At least two changes which are static in nature may cause streams to change from erosion to aggradation. Pleistocene glacierization caused many of the valleys beyond the margins of the ice sheets in North America, Europe, and elsewhere to receive quantities of glacial outwash beyond the transporting capacities of the streams in them. Valley aggradation from this cause is widely observed throughout the Mississippi-Ohio-Missouri drainage basins. There are numerous stretches along valleys where bedrock may not be exposed for many miles. The lower Ohio Valley has been aggraded as much as 200 feet. Not only were the valleys which carried glacial outwash aggraded but their tributaries were also, for the rapid aggradation of the main valleys resulted in the tributary valleys which were not receiving glacial outwash being ponded by the valley fills in the trunk valleys and becoming areas of deposition.

Although a large part of the fill in such valleys as the Mississippi consists of glacial outwash, there has been another factor, static in nature, which has contributed to aggradation of the Mississippi Valley and its tributaries. The delta of the Mississippi River has been extended approximately 125 miles since the beginning of the Pleistocene. The mouth of the Mississippi at the beginning of the Pleistocene was near where the Red River at present joins the Mississippi. This point, which was originally at sea level, is now about 50 feet above sea level. If the preglacial Mississippi and its major tributaries were graded streams, it is evident that at least 50 feet of aggradation in their valleys has resulted from the seaward extension of the Mississippi delta (Malott, 1922).

The idea that fluctuations between drier and wetter periods, particularly in arid and semiarid regions, lead to alternating periods of valley aggradation and trenching was first fully expounded by Huntington (1914). He held that the balance between deposition and erosion is largely determined by the amount of vegetal cover. Decrease in rainfall would presumably cause a decrease in the amount of vegetation, and hence more debris would be shed into streams from surrounding slopes and the stream aggradation would result. On the other hand, an increase in rainfall would increase the amount of vegetal cover and less debris would reach stream courses and the streams would have ability to erode their channels.

A different viewpoint (Antevs, 1951) is that valley trenching takes place during drier times when the plant and soil mantles are reduced, permitting the waters of intense rains to run off rapidly and form torrents which excavate valley floors. Valley filling would characterize more humid times when increased vegetal cover slows down the rate of runoff and causes streams to drop their loads.

Thornthwaite (1942) and associates maintained, however, that the alternation in southwestern United States between periods of cut and fill do not necessarily reflect fluctuations between semiarid and arid climates. They concluded that the evidence from tree rings, lake levels, and the history of the Pueblo peoples suggested that there had been no significant climatic changes in southwestern United States in the last 2000 years. They thought that the times of cutting and filling could be correlated with times when storms were more or less intense, the times of intense storms being marked by erosion and those of less intense storms being marked by valley filling.

These contrasting viewpoints indicate the complexity of the problem and suggest that no generally applicable conclusion can be drawn as to the effects of increased or decreased rainfall upon the behavior of streams. Each area must be considered separately in terms of the specific factors which affect stream activity.

Termination of the fluvial cycle. Because of the length of time required for the completion of a fluvial cycle there is always the possibility that a geomorphic cycle instead of merely being interrupted may actually be terminated. Davis foresaw the possibilities of this and applied the somewhat inappropriate term "accident" to the termination of a cycle. He recognized two types of accidents which might end the so-called normal cycle: volcanism and climatic change with ensuing glaciation or aridity. To these we could add submergence beneath the sea of a land mass and possibly complex diastrophism.

Covermasses of lava or glacial drift may bury completely or partially obscure the land forms produced by streams but following termination of glaciation or volcanism the fluvial cycle will reestablish itself. Change to aridity does not completely destroy the regime of streams, but changes in their characteristics and relative importance are sufficient to justify considering an arid cycle as distinct from the humid-land fluvial cycle. Submergence of continental areas beneath the sea has taken place many times in geologic history, but, since the middle of the Tertiary, North America, and the other continents for that matter, have been mostly above sea level and inundations have been restricted to narrow marginal strips and nowhere have been extensive enough to completely submerge a major drainage basin. In areas where repeated folding and faulting are taking place, or in what we may designate as areas of complex diastrophism, it may be questionable whether anything approaching cyclical development of land forms is approached. It may be going too far to say that the fluvial cycle is terminated under these conditions, but at least its operation in the sense of sequential land-form development is suspended. A sizeable group of geologists believes that the fluvial cycle can make little headway until the diastrophic clock has run down, so to speak, and the conditions of stability essential for its progress are attained.

SHIFTING STREAM DIVIDES

As a fluvial cycle progresses there are significant changes in the character and position of stream divides. In youth, stream divides are often indistinct, but as valley systems extend themselves there is an increasing number of stream divides as well as sharper definition of them. By maturity, they are usually sharp although not necessarily fixed in position. Divide migration may take place slowly, or there may be rapid divide migration because of stream diversion or derangement.

Slow divide migration. It is commonly observed that, in the later stages of the fluvial cycle, stream divides for streams of similar size are somewhat uniformly spaced and stream gradients are roughly equal on opposite sides of divides. This is not always true in the earlier stages of the cycle, for then it is not unusual to find unequal stream lengths and gradients on the two sides of a divide. It was recognized by Gilbert (1877) in his classical study of the Henry Mountains of Utah that a stream flowing down the steeper slope of an asymmetrical ridge erodes its valley more rapidly than one flowing down the more gentle slope and as a result the divide line migrates away from the more actively eroding stream toward the less actively eroding one. This principle has been referred to as the *law of unequal slopes*. In a region of gently or moderately dipping weak and strong sedimentary rocks, cuestas or homoclinal ridges will form with steep front slopes facing up the dip and more gentle back slopes extending down the dip, under which conditions divides will migrate down the dip of the rock, and the shorter and steeper streams down the front slope will extend their drainage at the expense of the lower-gradient streams down the backslope.

Unequal rainfall on two sides of a divide may contribute to divide migration, especially where winds are prevailingly from one direction, as in the trade wind belts.   This effect is well illustrated on the island of Oahu, in the Hawaiian group, where much heavier rainfall on the northeast slope has caused a shifting of the divide southwestward with accompanying development of a steep-faced scarp on the northeast side known as the Pali.

Rapid shifting of divides.   Rapid divide shifting may result from either stream diversion or derangement.   *Stream diversion,* as here used, refers to the various ways by which streams may actively effect drainage changes in contrast with *stream derangement* in which drainage changes are accomplished by agents other than streams.   Stream diversion may come about either through stream piracy or aggradation.   *Stream piracy* involves active erosional attack by one stream upon the drainage of another with resulting diversion of part of its waters to the pirating stream.   The most common cause of stream piracy is the ability of one stream to maintain and extend its valley at a lower level than that of an adjacent stream.   Stream piracy may take place by abstraction, headward erosion, lateral planation, and subterranean diversion.   The term *abstraction* is often applied to the simplest type of piracy, which results from competition between adjacent consequent gullies and ravines.   This is an unspectacular but important process whereby innumerable minor drainage lines become integrated into a few dominant stream courses. It takes place chiefly at the upper ends of drainage lines and can be observed in progress among gullies on almost any steep slope.   It is of importance in the ultimate location of the divides between the lesser tributaries of a river system.

Stream piracy by headward erosion may arise from either of two causes. One stream may have an advantage over another because it is cutting its valley in more easily erosible rocks, or piracy may take place because one stream has a much steeper gradient than another.   Numerous examples of stream piracy because of variation in rock resistance along competing streams are to be found in the folded Appalachian region of eastern United States. Here a few streams have courses transverse to the geologic structure, but most streams flow in strike valleys parallel to the geologic structure.   Transverse streams are limited in the rate at which they can deepen their valleys by their ability to erode the more resistant rocks athwart their courses, whereas the longitudinal streams are usually flowing upon belts of weak rock and hence are able to erode their valleys more rapidly.   One of the best-known examples of piracy in this area is that which resulted in the formation of Snickers, Ashby, and Manassas gaps. This represents an illustration of progressive stream piracy by headward erosion. These gaps are now *wind gaps,* a term applied to former *water gaps* through which streams no longer flow, but formerly a stream flowed eastward through the Blue Ridge in each of them. A tributary of the Potomac, the Shenandoah River, cut headward along a belt of weak rock, back of the Blue Ridge, and successively captured the headwaters of the streams flowing through Snickers, Ashby, and Manassas gaps. The Shenandoah River was able to do this for two reasons: first, it was eroding its valley in weak rock, whereas the three transverse streams were retarded in downcutting by the harder rock of the Blue Ridge; secondly, the Shenandoah was at a lower level than the three other streams because the Potomac, of which it is a tributary, is a large stream and hence able to cut more rapidly downward through the Blue Ridge than lesser streams. The altitudes of Snickers, Ashby, and Manassas gaps are approximately 1100, 1000, and 900 feet, respectively. This shows that there was a lowering of Ashby Gap of 100 feet between the time of capture of the stream through Snickers Gap and the capture of the stream which flowed through Ashby Gap, and that there was still another 100 feet of downcutting of Manassas Gap before the stream through it was captured. This type of capture is encountered most commonly in areas of folded rocks or where there are steeply dipping rocks around the flank of a domal or anticlinal uplift. The point at which the capture is effected is known as the *elbow of capture,* since it is commonly marked by a right-angled turn in the present stream. The stream which has lost part of its drainage is said to have been *beheaded.* Most commonly the pirating stream is a subsequent stream, and the pirated stream is a transverse one.

Unequal gradients along two opposed and competing streams commonly are found where there are cuestas, plateaus, or tilted fault blocks with steep front and gentle backslopes. There need be no difference in rock resistance along the two competing streams for piracy to take place. The classic locality for this type of piracy is along the front of the Catskill Mountains where Kaaterskill Creek and other streams that flow down the steep east-facing front have captured some of the headward drainage of the westward-flowing streams. This type of stream piracy results fundamentally from the fact that the pirating streams have lower local base levels of erosion and consequently steeper gradients in their headward portions at least. Stream piracy of this type is usually marked by a barbed drainage pattern in the headwater portion of the pirating stream. This is so common that wherever the barbed pattern is seen the possibilities of piracy should be considered.

An example of stream capture because of unequal stream gradients but not inherently related to topographic contrasts on two opposing slopes is that described by Mackin (1936), involving the Greybull River in the Bighorn Basin of Wyoming. The Greybull River heads in the Absaroka Mountains

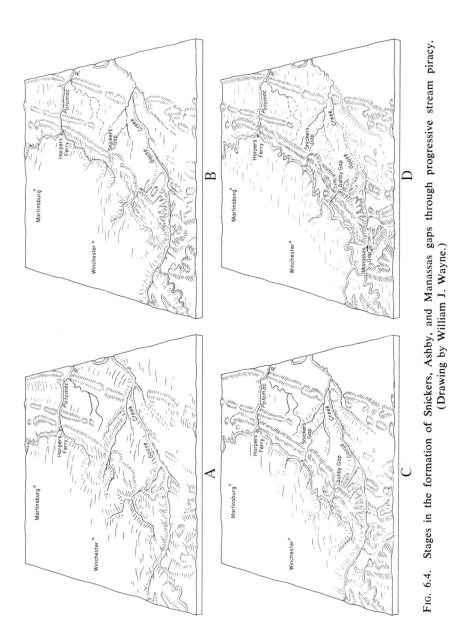

Fig. 6.4. Stages in the formation of Snickers, Ashby, and Manassas gaps through progressive stream piracy. (Drawing by William J. Wayne.)

and has a rather high gradient of 40 feet to the mile because of the large-sized materials which it is required to transport. Its captor, a shorter stream, which headed in the Bighorn Basin, had, because of the smaller caliber of its load, a gradient of only 3 feet per mile. Both the Greybull River and its captor were tributaries of the Bighorn River, and, because the Greybull River had a much steeper gradient than the other stream, its valley floor at the point of capture was higher than that of its lower-gradient rival. This permitted capture of the Greybull and its diversion to the competing stream.

In areas where limestone or other soluble rock lies above the base level of erosion, subterranean stream piracy may take place. The condition that permits it is that a surface stream be perched upon soluble rocks adjacent to stream valleys at lower levels. Under this condition portions of the perched drainage may be diverted through underground routes to adjacent entrenched streams. An interesting example of this type of piracy, described by Beede (1911), is found west of Bloomington, Indiana, where about 15 square miles of the drainage of Indian Creek were diverted through subterranean piracy to adjacent Clear and Richland creeks.

Stream piracy may also result from lateral erosion or planation. This type of piracy is to be expected after streams have become graded and lateral erosion has become dominant. The master stream may through lateral erosion cut through the ridge separating it from a tributary and divert that part of the tributary above the point of spur intercision to the main stream and leave a short and diminished stream to occupy the lower valley portion of the beheaded stream. An example of this type of piracy is shown on the Lancaster, Wisconsin, topographic sheet. Here little Maquoketa Creek, which formerly joined the Mississippi River through Couler Valley, has been diverted by lateral planation to a more directly eastward route, leaving Couler Valley with a much reduced stream in it.

Stream diversion not involving piracy may take place when, after rapid aggradation of its valley, a stream spills out of the valley into a lower route. This is illustrated by the lower Hwang Ho in China which has repeatedly shifted its course from the north to the south side of the Shantung Peninsula. A notable change in the course of the lower Ohio River has resulted from aggradation. Extending across southern Illinois is the Cache Valley, which is recognized as an abandoned former route of the Ohio River. The probable cause for its abandonment was that during Wisconsin glaciation the floor of the Ohio Valley was aggraded with glacial outwash until finally the river spilled over a divide to its present more southerly route.

Stream derangement. Stream derangement has little in common with stream piracy except that changes in stream courses are involved. Derangement may result from glaciation, wind deposition, or diastrophism. Glaciation has probably more often been responsible for stream derangements than any other single cause. The present course of the upper Missouri River is largely the result of ice sheets advancing across the preglacial course of the Missouri River to Hudson Bay and causing the Upper Missouri River to establish a route to the Mississippi which roughly parallels the glacial boundary. This type of stream derangement is further discussed in Chapter 16. The Calumet River in northwestern Indiana, which flows for a distance parallel to the south shore of Lake Michigan before emptying into the lake, does so because its former more direct route to the lake was blocked by sand dunes. King (1942) stated that the drainage of the Nyanza region in east-central Africa formerly flowed into the Congo River, but downsinking of the area produced a lake which now empties into the Nile. The extreme branching of Lake Kioga in the same part of Africa has been attributed to a backward tilting of the headwaters of the Kafu River, which formerly flowed westward.

MISFIT RIVERS

It has been observed that some streams are not proportionate in size to the valleys that they occupy. These have been called *misfit rivers*. If a stream, or more correctly the size of the stream meanders, is too small for the size of the valley, the stream is said to be *underfit* (Davis, 1913); if too large, it is referred to as *overfit*. It is difficult to cite examples of overfit rivers, or streams with floodplains too small for the size of the stream. Hence there may well be a question whether overfit streams exist. The reason for their scarcity or non-existence may be that a stream cannot long remain overfit, for an increase in volume will be accompanied by increased erosive power and rapid adjustment of valley size to river size. The underfit condition can persist indefinitely; hence many examples of such streams exist.

Just because a valley is large and the river in it is small does not necessarily mean that the river is underfit, for a small stream can by lateral erosion cut a fairly wide valley. Evidence of pronounced shrinkage in meander radius is probably the strongest evidence for the underfit condition. If present stream meanders have a radius of curvature much less than those of abandoned ones on the valley floor or those indicated by the curvatures of the valley walls, the conclusion that the river is underfit seems justified. It is also unlikely that any scarps along the valley sides of an underfit river will be undergoing active undercutting but rather will be in the process of being obscured by mass-wasting. Extensive alluvial fans where tributaries join the main stream are also suggestive that the trunk stream because of its shrunken size is no longer capable of carrying all the debris contributed to it by its tributaries.

A misfit condition commonly results from drainage changes effected by stream piracy or derangement. However, another common cause in areas that have been glaciated is the marked changes in stream volumes which accompanied the waxing and waning of the ice sheets. Many valleys which acted as drainage ways for glacial meltwaters carried for a time large volumes of waters and then were abandoned as outlets for glacial waters. The streams in these valleys today are commonly underfit.

RIVER TERRACES AND THEIR SIGNIFICANCE

River terraces are topographic surfaces which mark former valley floor levels. In the main they are vestiges of former floodplains, although some may have little or no alluvium on them and may thus be classed as *bedrock terraces* in contrast with *alluvial terraces,* which consist of gravel, sand, and finer alluvium. As Gilbert (1877) has stressed, river terraces are largely the products of stream erosion and not of stream deposition. Although it is not unusual for the surface of a terrace to be a depositional surface, the terrace itself came into being only through the development by erosion of another valley flat below the top of the former valley floor. Alluvial terraces are particularly prominent along valleys which carried glacial outwash.

Bedrock terraces may have a thin veneer of alluvium upon them, but it is usually inconspicuous. They are essentially remnant valley flats which were produced in most instances through lateral erosion by graded streams. They are not to be confused with structural benches (see p. 112) which carry no implication of former local base levels of erosion. The slope of a structural bench will be controlled by the local dip of the rock, whereas bedrock terraces, assuming lack of deformation, will reflect the longitudinal profile of the stream which formerly flowed over them.

River terraces are produced by surges of erosion along river valleys and hence reflect periods of rejuvenation which have affected streams. Cotton (1940) has suggested that river terraces may be classified as either cyclic or non-cyclic. *Cyclic terraces* represent former valley floors formed during periods when valley deepening had largely stopped and lateral erosion had become dominant. The vertical distance that a terrace is above the present floodplain or another terrace below it represents the amount of valley deepening which took place after rejuvenation. The distinguishing feature of cyclic terraces is that, for any set of terraces, remnants on opposite sides of the valley will be paired or will correspond in altitude along any particular valley stretch. This implies that there was not continuing, concomitant downcutting and lateral erosion, but that valley deepening essentially ceased for a time and allowed lateral erosion to form an extensive valley flat at the local base level of erosion. Cyclic terraces may imply partial erosion cycles related to intermittent uplift, or they may represent alternations between periods of aggradation and degradation related to static changes. *Non-cyclic terraces,* according to Cotton (1940), will be non-paired terraces. They imply that there was continued downcutting accompanied by lateral erosion. Under these conditions the meander belt of the stream shifted back and forth over

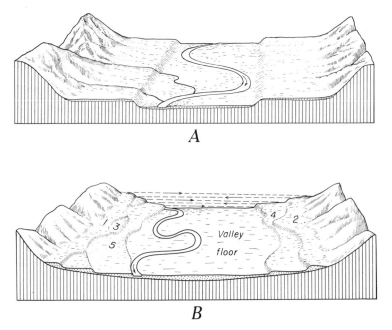

FIG. 6.5. Diagrams showing the difference between: *A,* paired terraces, and *B,* non-paired terraces. (After Longwell, Knopf, and Flint, *Physical Geology,* John Wiley & Sons.)

the valley and, by the time that it moved from one side of the valley to the other, the valley floor had been lowered somewhat and terraces left on opposing valley sides were at different altitudes. Non-paired terraces imply continued slow uplift of the land or other changes that would have a similar effect upon a stream.

In some valleys, where the vertical range between successive terraces is not great and there has been appreciable modification of the terrace surfaces, it may be difficult or impossible to say whether the opposing terraces are to be classed as paired or non-paired. For that matter, both types may exist in a valley, with some representing halts in downcutting and others representing preserved cusps of stream meanders (Lobeck, 1939) formed as a stream shifted back and forth across its valley.

Alluvial terraces may be better preserved in some parts of a valley than in others because of the protective effect of buried bedrock beneath an alluvial cover. These have been called *rock-defended terraces*. Alluviation may bury bedrock spurs which, when encountered during ensuing valley excavation, will deflect the stream from that part of the valley and protect the alluvial fill

Fɪɢ. 6.6. Bedrock terraces thinly veneered with gravels along Cola Creek, Castle Hill Basin, New Zealand. The terraces may be called rock-defended terraces, for a hard-rock barrier downstream has restricted lateral planation. The terraces probably resulted from alternating phases of aggradation and erosion related to Pleistocene climatic changes. (Photo by Maxwell Gage.)

both upstream and downstream from the bedrock buttress from removal by lateral erosion.

Terrace surfaces and slopes. The cross profile of a terrace is seldom flat. Numerous processes operate to destroy its flatness, if it ever possessed it. Chief among them are slope wash from the valley side against which the terrace abuts and deposition of alluvium upon it by streams which debouch from the adjacent upland onto it. Mackin (1937) has shown to what extent slope wash has obscured the terraces along the Shoshone River in Wyoming. Two terraces along this valley, called the Powell and Cody terraces, have a minimum vertical distance between them of 90 feet, yet in places slope wash has completely obscured the scarp which separates the two terraces. Along some valleys wind-blown silts and sands may mask a terrace surface or add to its irregularity. Even where an initial terrace surface is well-preserved there may be considerable relief. Such features as stream bars, swales representing former channels of braided streams, meander scars, and natural levees may give relief to its surface. Terrace treads are level only in the broad sense of the word, and many exhibit marked differences in altitude between their distal portions and the parts that abut against the upland or terrace above them. This should be kept in mind when attempts at terrace correlations are made.

Terraces underlain by sand and gravel are among the most persistent. Sand and gravel are resistant geologic materials because of their great permeability. They may be as well- or better-preserved than bedrock terraces because the rain that falls on them and the streams that debouch onto them may be largely absorbed and thus reduce stream dissection to a minimum. Their permeable character does not, however, prevent them from having their original surfaces obscured by slope wash (colluvium), eolian deposits, and alluvium.

Normally a terrace level rises up-valley, although by no means necessarily at the same rate as the present floodplain. Warping or tilting of an area after terrace development may in rare instances cause terraces to slope up-valley, but they will more likely lessen or increase their down-valley slopes. Warped terraces are useful in determining the amount and nature of recent deformation. It may not be possible to determine the absolute amount of deformation that has taken place, for the initial slope of the terrace will not be known, but a rough estimate of the original terrace gradient can be attained by assuming that it was about that of the present floodplain or by using the gradient of undeformed terraces in near-by valleys which presumably are correlative.

Where two or more terrace levels are present along a valley it may be observed that the terrace surfaces do not parallel each other or the present valley floor but instead may become closer together or farther apart vertically as they are followed down-valley. If the terrace surfaces come nearer together, there is *terrace convergence;* if they become farther apart, there is *terrace divergence.* Terrace convergence may suggest a decreasing rate of uplift or longer periods of time between successive rejuvenations, either of which would permit streams to develop lower-gradient valley floors than had been possible during the preceding partial cycle. By the same line of reasoning, terrace profiles which diverge down-valley may be attributed to either an increased rate of uplift inland as compared with that nearer the sea or to a shortening of the intervals between successive rejuvenations, as a result of which there would be successively less opportunity for the streams to develop low-gradient valley floors before uplift started a new cycle. This would result in progressively steeper gradients in going from the uppermost terrace level down to the present valley floor and would constitute terrace divergence. More commonly lack of parallelism of terrace profiles with present stream profiles simply reflects a change in the regimen of the stream related to some cause such as climatic change, but in the tectonically active regions it is more likely to be the result of tilting produced by faulting or localized folding.

Correlation of terraces. The problem of terrace correlation is, as Johnson (1944) has stressed, often a difficult one. Even where terraces can be established unequivocally as river terraces, there still remain two major problems. If more than one terrace level exists along a valley, there is the problem of determining to which level any particular terrace tract belongs. There is also the more difficult problem of correlating terraces from one valley to another.

Some of the more common methods of correlation, such as identification by fossils, degree of dissection, comparative decomposition of boulders, lithology of the terrace materials, and depth of weathering of terrace materials, may not be usable because the time intervals between the formation of successive terraces are usually too short for these tests to be applicable. In that event, near identity of altitude of terraces in a particular valley, permissible changes in altitude in going up- or down-valley being taken into account, must remain the only safe method of correlation. Correlations made in this way should, if possible, be strengthened by corroborative evidence obtained by one or more of the previously mentioned methods. If the vertical interval between successive terraces is large, correlation by relative elevations will usually be satisfactory, but, if it is small, there may be difficulties in making exact correlations. The smaller the vertical interval is between terraces, the less is the permissible variation in the altitudes of terrace tracts considered to be equivalent.

Correlation of terraces from one valley to another presents an even more difficult problem, for absolute altitudes are not likely to be the same even in adjacent valleys because of variations in stream profiles. The only positive method of correlating terraces in different valleys is to trace them until they join each other at the junction of the valleys or join with a similar terrace in a trunk valley to which the two valleys are tributary. If valleys are so widely separated that neither of these methods is possible, extreme caution should be exerted in correlating valley terraces on the basis of similarity in altitudes.

A similar sequence of terraces with similar vertical intervals between them may be strongly presumptive evidence of their correlation but does not necessarily prove it.

REFERENCES CITED IN TEXT

Antevs, Ernst (1952). Arroyo cutting and filling, *J. Geol., 60*, pp. 375–385.

Baulig, Henri (1935). *The Changing Sea Level,* 46 pp., George Philips and Son, Ltd., London.

Baulig, Henri (1940). Reconstruction of stream profiles, *J. Geomorph., 3*, pp. 3–15.

Beede, J. W. (1911). The cycle of subterranean drainage as illustrated in the Bloomington, Indiana, quadrangle, *Indiana Acad. Sci., Proc. 20*, pp. 81–111.

Blache, Jules (1939). Le problème des méandres encaissés et les rivières Lorraines, *J. Geomorph., 2*, pp. 201–212.

Cole, W. S. (1930). The interpretation of intrenched meanders, *J. Geol., 38*, pp. 423–436.

Cotton, C. A. (1940). Classification and correlation of river terraces, *J. Geomorph., 3*, pp. 27–37.

Davis, W. M. (1913). Meandering valleys and underfit rivers, *Assoc. Am. Geog., Anns., 3, pp. 3–28.*

Fisk, H. N. (1939). Depositional terrace slopes in Louisiana, *J. Geomorph., 2*, pp. 385–410.

Gilbert, G. K. (1877). *Report on the Geology of the Henry Mountains,* pp. 120–127, U. S. Geographical and Geological Survey of the Rocky Mountain Region (Powell), Washington.

Huntington, Ellsworth (1914). The climatic factor as illustrated in arid America, *Carnegie Inst. Wash. Pub., 192*, 341 pp.

Johnson, D. W. (1938). Stream profiles as evidence of eustatic changes of sea level, *J. Geomorph., 1*, pp. 178–181.

Johnson, D. W. (1944). Problems of terrace correlation, *Geol. Soc. Am., Bull. 55*, pp. 793–818.

King, L. C. (1942). *South African Scenery,* pp. 179–183, Oliver and Boyd, Ltd., London.

Lobeck, A. K. (1939). *Geomorphology,* pp. 238–241, McGraw-Hill Book Co., New York.

Mackin, J. H. (1936). The capture of the Greybull River, *Am. J. Sci., 231*, pp. 373–385.

Mackin, J. H. (1937). Erosional history of the Big Horn Basin, Wyoming, *Geol. Soc. Am., Bull. 48*, pp. 813–894.

Mahard, R. M. (1942). The origin and significance of intrenched meanders, *J. Geomorph., 5*, pp. 32–44.

Malott, C. A. (1920). Static rejuvenation, *Science, 52*, pp. 182–183.

Malott, C. A. (1922). The physiography of Indiana, *Indiana Dept. Conserv., Pub. 21*, pp. 139–141.

Moore, R. C. (1926). Origin of inclosed meanders on streams of the Colorado Plateau, *J. Geol., 34*, pp. 29–57.

Rich, J. L. (1914). Certain types of streams and their meaning, *J. Geol., 22*, pp. 469–497.

Thornthwaite, C. W., C. F. S. Sharpe, and E. F. Dosch (1942). Climate and accelerated erosion in the arid and semi-arid southwest with special reference to the Polacca Wash drainage basin, Arizona, *U. S. Dept. Agr., Tech. Bull. 808*, 134 pp.

Additional References

Challinor, J. (1932).   River-terraces as normal features of valley development, *Geography, 17,* pp. 141–147.

Cotton, C. A. (1945).   The significance of terraces due to climatic oscillation, *Geol. Mag., 82,* pp. 10–16.

Crosby, I. B. (1937).   Methods of stream piracy, *J. Geol., 45,* pp. 465–486.

Lewis, W. V. (1944).   Stream trough experiments and terrace formation, *Geol. Mag., 81,* pp. 241–253.

Malott, C. A. (1921).   Planation stream piracy, *Indiana Acad. Sci., Proc. 30,* pp. 249–260.

Peltier, L. C. (1949).   Pleistocene terraces of the Susquehanna River, Pennsylvania, *Penn. Geol. Survey, 4th Ser., Bull. G 23,* 158 pp.

Rubey, W. W. (1952).   Geology and mineral resources of the Hardin and Brussels quadrangles (in Illinois), *U. S. Geol. Survey, Profess. Paper 218,* pp. 129–136.

Wooldridge, S. W., and J. F. Kirkaldy (1936).   River profiles and denudational chronology in southern England, *Geol. Mag., 73,* pp. 1–16.

7 · Stream Deposition

It is certainly true, as Price (1947) has stated, that geomorphology stresses the sculptural more than the constructional effects of geomorphic processes. For this there are several reasons. In the first place, erosional forms generally possess greater relief and are more striking and until rather recently were better shown on topographic maps. Another factor, perhaps an important reason for the stressing of erosional geomorphology, is that fluvial cycles and partial cycles have been used, along with other features, in the interpretation of the recent diastrophic histories of regions. Stratigraphers, as yet, have not shown too much interest in the depositional record of erosion cycles. It is further true that erosional forms comprise a larger part of the earth's surface than do those produced by deposition but it is doubtful that they exist in a ratio commensurate with the importance attached to each. Mapping of areas on 1- and 5-foot contour intervals has brought out an amazing amount of detail on topographic surfaces as supposedly featureless as floodplains and deltas. The study of aerial photographs by stereoscopic methods further permits better recognition of slight relief.

CAUSES OF STREAM DEPOSITION

The two most fundamental causes of stream deposition are loss of transporting power and inability of a stream to transport all the material brought into it by its tributaries. The following outline attempts to show the more common and immediate causes of stream deposition.

A. Loss of transporting power resulting from:
  1. Decrease in velocity produced by:
     a. Decrease in gradient because of:
        1. Passage from upland to lowland areas.
        2. Diastrophic warping or tilting.
        3. Delta extension of a trunk stream.
        4. Increase in valley crookedness.
     b. Increase in spread of stream waters.
        1. Where they debouch from mountain valleys on lowlands.
        2. Where they overflow banks during flood stage.

    c. Obstructions caused by:
        1. More rapid aggradation of main stream than by tributaries.
        2. Boulder fans built by steep tributaries into main stream.
        3. Landslide dams.
        4. Lava flow dams.
        5. Sand dune dams.
        6. Timber rafts.
        7. Artificial dams.
        8. Minor obstructions in stream channel.
    d. Decrease in volume as a result of:
        1. Decreased runoff because of climatic change.
        2. Evaporation losses.
        3. Seepage losses.
        4. Diversion of waters by stream piracy.
        5. Diversion of waters by man.
  2. Cessation of flow in:
    a. Local pools of standing water along stream courses.
    b. Standing backwaters of flood stage.
    c. Standing bodies of waters such as lakes and oceans.
B. Excessive load produced by:
  1. Great quantities of glacio-fluviatile outwash.
  2. Increased erosion in drainage basin because of reduction of vegetal cover caused by:
    a. Man's activities.
    b. Climatic change.
  3. Excessive load contribution from:
    a. Tributary valleys being cut in easily erodible materials.
    b. Man's activities.

LAND FORMS AND GEOLOGIC FEATURES RESULTING FROM STREAM DEPOSITION

The floodplain and associated features. Just as the valley is the fundamental feature resulting from stream erosion, the floodplain is the essential product of stream deposition. A floodplain, like a valley, exhibits much variety in all gradations from the initial floodplain scrolls of coarse alluvium to the broad floodplain of old age with its varied features. Many individual features add diversity to the surface of a mature or old floodplain. Along the stream channel are likely to be found numerous deposits of sand and gravel commonly designated by such names as channel bars, meander bars, and delta bars, depending upon their position and origin. *Channel bars* are located in the stream course and are perhaps most characteristic of braided streams, although they are by no means restricted to them. Objection has been raised to the term braided stream (Melton, 1936) on the grounds that it is a rather superficial term which is applicable to streams only when they are at low-water stage.   On the other hand, some, like Fisk (1947), recognize a fundamental difference between a braided and a meandering stream in that a braided stream is one that has an excessive load and is not capable of carrying on lateral erosion as is a meandering stream.   In the more common

FIG. 7.1.   The braided Rakaia River, New Zealand.   The valley is slightly over 1 mile wide.   (Photo by V. C. Browne.)

terminology of floodplain features, *meander bar* is used to designate the bars that form on the inside of meanders and extend into meander curves.   *Delta bars* are formed by tributary streams building deltas into the main stream channel.   It is doubtful whether they merit the designation bar.

Other floodplain features are natural levees, floodplain scour routes, floodplain lakes and swamps, of which the oxbow type is perhaps most distinctive, and meander scars.   *Natural levees* are low ridges that parallel a river course; they are highest near the river and slope gradually away from it.   They may be a mile or more in width, and they owe their greater height near the stream channel to the cumulative effect of sudden loss in transporting power when a river overspreads its banks.   They cause the present meander belt of a river to stand up above the floodplain as a low alluvial ridge.   Natural levees marking former river positions may be present on a floodplain, although usually in a fragmentary state. There are shallow channels across floodplains which are used only during floods which may be called *floodplain scour routes.* They may represent either abandoned courses or initial stages in the development of new courses. Scattered over the floodplain are numerous *meander scars* representing meanders abandoned by cutoffs. They may be the sites of oxbow lakes or swamps, or they may be recognizable only by dense vegetation because of their unsuitability for cultivation.

The terminology of the preceding paragraphs has been used for many years to describe the topographic details of floodplains. Extensive studies of floodplain deposits and land forms have been carried on by the Mississippi River Commission of the Corps of Engineers of the War Department, particularly in the lower Mississippi Valley, under the leadership of Fisk and Russell, and as a result of this work there has emerged a terminology for alluvial deposits which has much merit. Because of the significance of the suggested terminology and because of the importance of the Mississippi Valley, both geologically and geographically, a summary of part of one of the more recent reports on the Mississippi River (Fisk, 1947) follows.

The original meandering course of the Mississippi River from Cairo, Illinois, to the Gulf of Mexico was over 1100 miles long. This has been shortened by man-made cutoffs until it is now 970 miles. The alluvial valley begins at Cape Girardeau, Missouri, and is approximately 600 miles long. The valley varies in width from a maximum of 125 miles at Helena, Arkansas, to a minimum of 25 miles at Natchez, Mississippi, and it descends from an altitude of approximately 300 feet at Cairo, Illinois, to sea level at the Gulf of Mexico, with an average gradient of 0.8 foot per mile. Its gradient is not everywhere the same and the longitudinal profile of the valley shows a series of flatter stretches separated by shorter and steeper risers, especially at points where alluvium from tributary valleys comes into the trunk valley.

The Mississippi River apparently has attained a profile of equilibrium so that its volume is properly adjusted to the load which it has to carry and hence is a graded stream. Its load consists primarily of silt and clay. It was the opinion of Fisk that pronounced meandering became characteristic rather recently and succeeded the braided stream condition which prevailed during late-glacial if not much of postglacial time. Abandonment of a braided condition for one of meandering and confinement to one channel has resulted in deeper scouring action, periodic flooding, and rapid channel shifting. The idea commonly held that the river consists of a continuous series of meanders is not entirely correct for there are numerous stretches where the river's course is relatively straight.

The alluvial deposits of the lower Mississippi Valley were classified by Fisk (1944, 1947) as follows:

A. Graveliferous deposits.

B. Non-graveliferous deposits.
    1. Meander belt deposits.
        a. Point bar deposits.
        b. Abandoned channel fillings.
        c. Natural levee deposits.
    2. Backswamp deposits.
    3. Braided stream deposits.
    4. Deltaic plain deposits.

Fisk estimated that about 45 per cent of the Recent alluvium is graveliferous and 55 per cent non-graveliferous. The graveliferous deposits form the basal

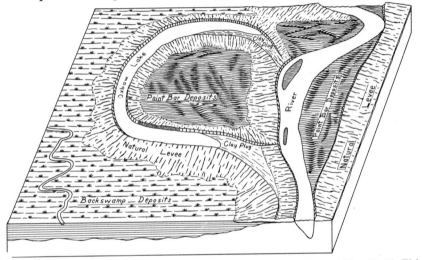

Fɪɢ. 7.2. Types of alluvial deposits in the lower Mississippi Valley. (After H. N. Fisk. Drawing by W. C. Heisterkamp.)

portion of the alluvial fill. Their character and distribution are known largely from the logs of thousands of wells. These show that the gravels extend away from the mouths of tributary valleys as a series of alluvial fans. The overlying non-graveliferous deposits consist of a lower part of pervious sands and a top stratum composed of varying proportions of sand, silt, and clay.

Meander belt deposits. The term *point bar* was applied to the bar that develops on the inside of a meander bend and grows by the slow addition of individual accretions accompanying migration of the meander. It is roughly equivalent to what has been called a meander bar, meander scroll (Davis, 1913), or scroll meander (Melton, 1936). As a point bar extends itself into a meander curve a series of alternating arcuate ridges and sloughs marks the individual accretions. The sloughs may become sealed off and become sites of lakes or swamps which gradually fill with finer sediments deposited during flood stages.   Point bars stand out on contour maps with a small contour interval because of their alternating low ridges and swales, but they show even better on aerial photographs because of the contrasts in color which the ridges and sloughs usually exhibit.

F𝐼𝐺. 7.3.   Vertical photo of portion of White River Valley, Greene County, Indiana, showing details of a floodplain.   Several meander cutoffs are shown, and point bar deposits are shown by the many light lines.   (Production and Marketing Administration photo.)

Channel fillings are deposits made in abandoned cutoff channels.   Two types of cutoffs were recognized, chute cutoffs and neck cutoffs.   A *chute cutoff* takes place through one of the sloughs of a point bar resulting in isolation of a part of the bar as an island between two river channels.   As this is a low-angle cutoff, part of the river may continue to follow its old route until it becomes filled with a *sand plug* at its upper end.   A *neck cutoff* takes place in the final stage of meander-loop development when a meander neck is cut through. It is a high-angle cutoff and results in shortening of the river course and local steepening of stream gradient. The channel abandoned during a neck cutoff will in time be converted into an *oxbow lake,* which as the new stream course migrates away from its former site becomes farther and farther removed from the river. The abandoned route will in time be filled with silt and clay materials and form what was called a *clay plug* in the alluvial valley. Most chute cutoffs and neck cutoffs show distinctly on topographic maps and aerial photographs, although older ones may be obscure. Some chute cutoffs may not be well-defined on topographic maps, but even after the abandoned channels are no longer sites of lakes or swamps they may stand out on aerial photographs because of their differences in vegetation or lack of cultivation. Abandoned channels are characterized by such trees as the cypress, tupelo gum, and water elm in contrast to the oaks, red gum, and pecans of the natural levees.

Natural levee deposits along present and former stream courses stand out not only because of their height and slope away from the river but also because their silty character and better drainage favor cultivation. They are in many places the only cultivated land. On aerial photographs they are usually lighter in color and display a fine network of drainage lines leading radially down their slopes away from the river. *Crevasses* or breaks through natural levees frequently cause the contours to display numerous inflections marking the materials carried down the slopes of the natural levees.

Backswamp deposits. Backswamp deposits are those that were laid down in the flood basins back of natural levees. They consist of extensive layers of silt and clay. They are rare along the Mississippi Valley north of Helena, Arkansas, but are extensive south of Vicksburg, Mississippi, as far as Donaldsonville, Louisiana, where they are replaced by deltaic plain sediments. Areas of backswamp deposits usually are marked by slight relief, usually less than 5 feet, and by drainage networks which reflect the position of older drainage lines, although successive floodings may eventually obscure these.

Braided stream deposits. Four portions of the Mississippi alluvial plain between Cairo and northern Louisiana have been interpreted as having been built by braided rather than meandering streams. The braided pattern was thought to have been the result of the large amount of glacial outwash which entered the Mississippi Valley; according to Fisk, it persisted until about 5000 years ago when sea level became fixed and the present meandering condition of the river was established. Pronounced lenticularity of sand and gravel distinguishes braided stream deposits from those of a meandering stream. There is a notable lack of silts and clays in the finer part of the alluvium, suggesting that the braided stream was able to carry the finer materials to the sea. On some aerial photographs, the braided pattern is still discernible but it becomes less evident southward as it becomes buried beneath backswamp deposits.

Deltaic plain deposits. South of Donaldsonville, Louisiana, the Mississippi River flows across delta deposits. These thicken seaward until in the Gulf of Mexico, according to seismic data, they attain a thickness of over 30,000 feet. This great mass of sediments is not all Recent alluvium, but rather it represents deposition which has taken place since the beginning of

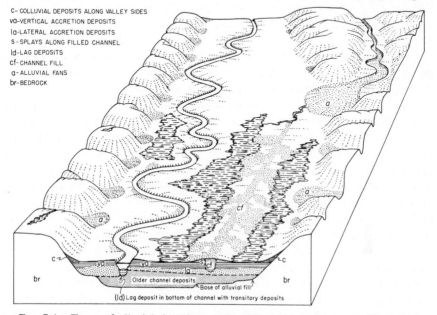

c- COLLUVIAL DEPOSITS ALONG VALLEY SIDES
va-VERTICAL ACCRETION DEPOSITS
la-LATERAL ACCRETION DEPOSITS
s-SPLAYS ALONG FILLED CHANNEL
ld-LAG DEPOSITS
cf- CHANNEL FILL
a- ALLUVIAL FANS
br-BEDROCK

Older channel deposits
Base of alluvial fill
(ld) Lag deposit in bottom of channel with transitory deposits

FIG. 7.4. Types of alluvial deposits. (After Happ, Rittenhouse, and Dobson.)

Mesozoic time. Deltaic plain deposits consist of alternating clay and silty clays which locally are sandy. Shells and organic oozes are common among them. The delta plain is marked by its near-sea-level altitude and by the branching stream patterns of present and ancient channels of the Mississippi River across it.

Whether or not we agree with all of the interpretations which have been made by Fisk and his associates regarding the geomorphic history of the Mississippi Valley (see p. 415), we are indebted to them for a systematic classification of alluvial deposits which has ready geomorphic application.

Happ, Rittenhouse, and Dobson (1940) proposed a somewhat different classification of alluvial deposits which took into account that they consist most commonly of the coarser sands and gravels of stream channels overlain by finer silts and clays that were deposited when streams overflowed their banks. They recognized six types of materials which may underlie the alluvial floor of a valley. These were: (*a*) *channel fill deposits,* which are largely bed-load materials; (*b*) *vertical accretion deposits,* consisting largely of suspended-load materials; (*c*) *floodplain splays,* a term applied to materials spread over the floodplain through restricted low sections, through breaks in natural levees, or along distributary channels; (*d*) *deposits of lateral accretion,* or materials deposited at the sides of channels where bed-load materials are being moved toward the inner sides of meanders (usually they are later covered with vertical accretion deposits); (*e*) *lag deposits,* or coarse materials which have been sorted out and left behind on the bed of the stream; and (*f*) *colluvial deposits,* consisting of debris carried by slope wash into the valley and mixed with varying amounts of talus. These different types of deposits are found in four distinct associations: normal floodplain or valley flat, alluvial fan or alluvial cone, valley plug and delta associations.

River deltas. Along with floodplains, river deltas have played an important role in providing sites for many of man's early civilizations. Not all rivers build deltas, nor are all deltas shaped like the Greek letter from which Herodotus in the fifth century B.C. coined the name for the deltaic plain at the mouth of the Nile. Deltas may be lacking at the mouths of streams which enter the sea because their mouths are so exposed to wave and current action that the sediments are removed as rapidly as they are deposited. Strong tidal action through bottle-necked bays, as at the mouth of the Hudson, may sweep material seaward as rapidly as it accumulates. Some streams lack deltas because they carry so little load. This is particularly true of streams which flow out of lakes before entering the sea. In some localities, the ocean bottom may be subsiding as rapidly as or more rapidly than a stream is extending its alluvium seaward, which will at least prevent the delta from becoming a surface feature. Although each delta has its own individual form as far as details are concerned, at least four forms are commonly encountered. These may be called the true delta form (in the sense of simulating the Greek delta), the arcuate or fan-shaped delta, which is the most common form, the digitate form, and the estuarine form. The delta of the Nile is an example of the first; that of the Rhine is an example of the arcuate; that of the Mississippi, of the digitate; and that of the Susquehanna, of the estuarine form.

What has come to be called delta structure is produced by the three sets of beds often observable in a delta. *Bottom-set beds* consist of the finer materials carried farthest seaward and laid down on the floor of the embayment into which the delta is being built. *Fore-set beds* are somewhat coarser and they represent the advancing front of the delta and the greater part of its bulk. They usually have a distinctly steeper dip than the bottom-set beds over which they are slowly advancing. *Top-set beds* lie above the fore-set beds and are in reality a continuation of the alluvial plain of which the delta is the terminal portion. Not all deltas display distinctly these three sets of beds. They are most conspicuous in deltas built by small streams flowing into lakes or protected embayments. Large rivers may carry such fine sediments and spread them over so large an area that there may be no sharp contrast in the dips of the bottom-set and fore-set beds. Wave and current action may also obscure the contact of the two.

Deltaic plains frequently are densely populated, and the problem of confining the river to permanent channels becomes an important one. Commonly, a river crosses its delta through a number of channels known variously as *distributaries, mouths,* or *passes.* Extensions of these mouths with their associated alluvial ridges may enclose former arms of the sea and form *delta lakes* such as Lake Ponchartrain on the Mississippi delta or the Zuyder Zee on the Rhine delta. Aggradation of one route across a delta to the point where it is higher than other possible routes may cause the river to break through its levees and follow a lower route.

Alluvial fans and bajadas. Somewhat analogous to deltas are the deposits on land known as *alluvial fans.* Where a heavily loaded stream emerges

Fig. 7.5. Braided drainage upon alluvial fans at base of the San Gabriel Mountains, north of Cucamonga, California. (Photo by J. S. Shelton and R. C. Frampton.)

from hills or mountains onto a lowland there is a marked change in gradient with resulting deposition of alluvium, apexing at the point of emergence and spreading out in fan-like form onto the lowland. If the slope of the surface is steep, as it is likely to be where minor streams descend from upland areas, the feature may be called an *alluvial cone*. The material comprising a fan varies in texture from coarse boulders and pebbles at its head to finer material down its slope. A series of adjacent fans may in time coalesce to form an extensive *piedmont alluvial plain* or *bajada,* which may extend for several miles away from a mountain front. The building of fans takes place largely during flood times, when great volumes of water with accompanying alluvium de-

FIG. 7.6.   A fanhead trench in an alluvial fan at base of San Gabriel Mountains, southern California.   (Photo by J. S. Shelton and R. C. Frampton.)

bouch onto them. During much of the time the stream channels across the fans are dry, or, if permanent streams occupy the channels, much of their waters is likely to sink into the coarse alluvium near the apices of the fans. The channels across the fans commonly are called *washes*. The deeper ones near the heads of the fans are better designated as *fan-head trenches*.

The growth of a series of alluvial fans to form a bajada takes place by repeated lateral shifting of the stream courses across the fans. As one channel is aggraded until it is higher than the adjacent parts of the fan the stream will spill out of it into a lower route. This shifting is aided by the frequent development in the shallow washes of *gravel plugs* (Eckis, 1928) formed when the sudden change in gradient at the foot of the mountain causes rapid deposition of the coarser debris. The braided stream pattern is characteristic of streams across alluvial fans, and as a result of repeated channel shifting streams at some time or other flow down almost every possible radius of the fan.

Bajadas present geologic conditions that are excellent for obtaining ground-water in large quantities from wells sunk into their permeable materials. Water infiltrates readily in the coarse materials at the heads of the fans and moves down the bajada under hydrostatic head. This source of groundwater is strikingly illustrated in the Los Angeles Basin where a dense population, along with the intensive growth of citrus fruits, vegetables, and other crops, has led to an enormous demand for water. Here a series of over thirty distinct water-producing basins has resulted from irregularities in the basement complex, which forms the floor upon which the alluvial deposits have been laid down, and from the faulting which has involved the water-bearing gravels. About 90 per cent of the water originating within this area comes from wells in these various groundwater basins.

Growth of bajadas in various directions into an enclosed basin will result in its partial filling. Such deposits are often referred to as *basin fillings*. They are especially characteristic of the many individual basins which col-lectively form the Great Basin region of western United States. The alluvium in the great structural trough called the Valley of California belongs in this category. The Vale of Kashmir is a well-known interior basin whose floor consists of alluvium washed down from the surrounding uplands.

Plains of older alluvium. Certain areas have many of the features of alluvial plains but are covered with alluvial deposits showing greater degrees of induration than is characteristic of Recent alluvium. The High Plains region east of the Rocky Mountains is an example of such an area. The materials which cap the interstream areas were laid down by streams which carried the waste from the Rocky Mountains during Tertiary erosion cycles. In many areas, stream dissection has so little affected the surfaces of the Tertiary deposits that they retain the essential features of alluvial plains; hence they may aptly be described as *plains of older alluvium*. The Pampa region of Argentina is another example of such an area.

REFERENCES CITED IN TEXT

Eckis, Rollin (1928). Alluvial fans of the Cucamonga district, southern California, *J. Geol., 36,* pp. 224–247.

Fisk, H. N. (1944). *Geological Investigation of the Alluvial Valley of the Lower Mississippi River,* pp. 17–20, Miss. River Comm., Corps of Engineers, War Dept.

Fisk, H. N. (1947). Fine-grained alluvial deposits and their effects on Mississippi River activity, Waterways Expt. Station, Vicksburg, 82 pp.

Happ, S. C., Gordon Rittenhouse, and G. C. Dobson (1940). Some principles of accelerated stream and valley sedimentation, *U. S. Dept. Agr., Tech. Bull. 695,* pp. 22–31.

Melton, F. A. (1936). An empirical classification of flood-plain streams, *Geog. Rev., 26,* pp. 593–609.

Price, W. A. (1947).   Geomorphology of depositional surfaces, *Am. Assoc. Petroleum Geol., Bull. 31,* pp. 1784–1800.

Additional References

Challinor, J. (1946).   Two contrasted types of alluvial deposits, *Geol. Mag., 83,* pp. 162–164.

Fenneman, N. M. (1906).   Floodplains produced without floods, *Geol. Soc. Am., Bull. 38,* pp. 89–91.

Happ, S. C. (1950).   Geological classification of alluvial soils, *Geol. Soc. Am., Bull. 61,* p. 1568.

Rubey, W. W. (1952).   Geology and mineral resources of the Hardin and Brussels quadrangles (in Illinois), *U. S. Geol. Survey, Profess. Paper 218,* pp. 122–129.

Russell, R. J. (1936).   Physiography of the Lower Mississippi delta, *Louisiana Conserv. Dept., Bull. 8,* pp. 3–199.

8 · The Peneplain Concept

HISTORICAL SKETCH

The term *peneplain* was introduced by W. M. Davis in 1889 to describe the low and gently undulating plain which the processes of subaerial erosion * presumably develop at the penultimate stage of a geomorphic cycle. The idea of the peneplain was a direct and logical outgrowth of his earlier proposal of the erosion cycle. The peneplain concept has had a rather controversial history which continues down to the present, and probably more has been written pro and con regarding it than about any other geomorphic idea.

Davis has given in one of his papers (1896) a somewhat detailed discussion of the earlier ideas which were held regarding the origin of plains of denudation. He pointed out that they were recognized as early as 1846 by the English geologist Ramsay and explained by him as the result of marine denudation. There soon developed what may be designated as the English school of marine denudationists. Richthofen in Germany was a leading supporter of this school of thought. Their ideas dominated geomorphic thinking from roughly 1846 down to about 1875.

As pointed out in Chapter 1, the bases for the development of the American school of subaerial erosion were laid by such men as Powell, Dutton, and Gilbert in their work in the arid and semiarid western United States. From their work evolved the concepts of base level, the geomorphic cycle, and later the almost inevitable corollary, the peneplain. The term peneplain as used by Davis was applied only to plains resulting from subaerial denudation. Perfect planation was not regarded as a probability; hence the word peneplain (almost a plain) was coined. Among the first erosional surfaces described as such were the Rocky Mountain peneplain of Marvine and a buried peneplain in the Colorado Plateau described by Powell. Subsequent work by Davis, Willis, Hayes, Campbell, and others soon led to the widespread acceptance by American geologists of the validity of peneplanation. In fact peneplains were described in such numbers that the author of the idea had to plead for caution in the interpretation of topographic surfaces as such.

---

* The expression "subaerial erosion," which refers to the various erosional processes operating upon land beneath the atmosphere in contrast to marine erosion beneath oceanic waters, stems from the days when there was a decided difference of opinion as to which was largely responsible for wearing down of the continents.

English geologists were slow to accept the validity of the subaerial cycle and its near-end product, the peneplain, because of their earlier, well-ingrained ideas regarding plains of marine denudation. In Germany and France, men like A. Penck and De Lapparent were favorable to the idea of subaerial planation.

Topographic surfaces were described so freely as peneplains that toward the close of the nineteenth century a reaction set in against the validity of the concept. Tarr (1898), Shaler (1899), Daly (1905), and Smith (1899) argued for other possible explanations for features which had been called peneplains. Further doubt as to the validity of peneplanation was raised in 1913, when Barrell revived the idea of marine denudation as an explanation for topographic surfaces in the Appalachians which had been called peneplains. Support to the idea of plains of marine denudation was given by Douglas Johnson in a paper appearing in 1916, in which he also suggested that the spelling be peneplane, a proposal which has received little support except from his students.

The rise of the German school under the leadership of W. Penck during the 1920's and 1930's was to a considerable degree a challenge to the peneplain concept, although Penck and his followers did not completely negate the existence of peneplains. In more recent years some geologists from the Pacific coast region have looked with considerable skepticism upon the validity of erosion cycles and peneplains. Although the peneplain still remains an important concept with most geomorphologists, it is now recognized that many topographic surfaces have been erroneously called peneplains.

ARGUMENTS FOR THE PENEPLAIN CONCEPT

If the geomorphic cycle is accepted as valid, the peneplain is a necessary corollary to it, in theory at least. It can hardly be argued that the idea of sequential land form development is valid in the earlier stages of the cycle but not in the later. Assuming stability of the earth's crust for a sufficient length of time, it seems inevitable that a land mass will eventually be reduced to base level. The main argument is whether stillstands of the earth's crust are ever of sufficient duration for the penultimate stage of the geomorphic cycle to be attained. It should, perhaps, be further emphasized that a peneplain is the near-end product of an erosion cycle and not the end product and that chemical weathering and mass-wasting are largely responsible for land reduction in the later stages of its development. A great deal of confusion regarding the nature of such a topographic surface will be eliminated if this is kept clearly in mind. Davis recognized that complete reduction of a land mass to base level was so unlikely that it was largely a theoretical abstraction.

Admittedly peneplanation is a long and slow process, and one's attitude toward the possibility of sufficiently long stillstands of the earth's crust for a geomorphic cycle to nearly run its course is likely to be influenced by the diastrophic conditions in the area in which he lives. Davis never pictured a peneplain surface as flat, as some seem to do, but rather visualized it as an area of low undulating relief with convex, gently graded interstream tracts sloping down to broad valley floors.

FIG. 8.1. The Sherman peneplain of Wyoming. (Photo by T. S. Lovering, U. S. Geol. Survey.)

The fact that buried topographic surfaces of sufficiently low relief to be considered peneplains exist was considered by Davis as one of the strong arguments for peneplanation. The Pre-Cretaceous erosion surface beneath the Cretaceous and Tertiary sediments of the Atlantic coastal plain is a well-known example. As previously stated, Wahlstrom (1948) has described an old erosion surface in the foothills of the Colorado Front Range, developed upon Pre-Cambrian rocks and overlain by the Fountain formation of Pennsylvanian age, which exhibits low relief and has a mantle of residual soil as much as 80 feet deep. The well-known Algonquin wedge in the Grand Canyon of the Colorado, where the Grand Canyon series rests upon the beveled surface of the Vishnu schist and Cambrian strata lie upon the truncated surface of the Grand Canyon series, is bounded by erosion surfaces which, according to Sharp (1939), are excellent examples of buried peneplains. The topographic surface in Wyoming called the Sherman peneplain may in part at least be a resurrected peneplain (Atwood and Atwood, 1948).

It is more difficult to cite good examples of extensive present-day peneplains at the present base level of erosion. Examples have been cited in Siberia south of the Ural Mountains, in central-western Missouri and adjacent

F<small>IG</small>. 8.2. The Gangplank between Cheyenne and Laramie, Wyoming. This name has been applied to this area because the materials removed in the process of forming a peneplain are in contact with the peneplain. The foreground is upon Tertiary alluvial deposits which extend up to the Sherman peneplain on granite in the background. (Photo by H. S. Palmer.)

southeastern Kansas, and portions of the lower Mississippi Valley. Some have even considered such an area as the Canadian Shield of Canada to be a peneplain surface that has persisted from Pre-Cambrian time. It is hardly compatible with our knowledge of diastrophism to expect such stability in any part of the earth's crust, and we may well be skeptical of any peneplains, except buried ones, which are dated back of the Tertiary. It is possible that the Canadian Shield was a peneplained surface when the Pleistocene ice sheet extended over it, but it has undergone so much modification by glacial processes that it is extending the concept too far to call it a peneplain. What has just been said applies equally well to portions of Scandinavia and Finland which have been described as peneplain surfaces. Admittedly there are few good examples of peneplains at the present base level of erosion, but their scarcity may be attributed to Pliocene-Pleistocene diastrophism. Locally, limited areas have been reduced to or nearly to base level, but they can hardly be called more than local or incipient peneplains.

Probably the strongest argument for the validity of peneplanation is the fact that in many areas the topographic features have characteristics and relationships that are most logically interpreted by assuming the existence of peneplains which have been uplifted and dissected. Remnants of what are believed to be uplifted and partially destroyed peneplains have been described in the Appalachian Mountains, the Rocky Mountains, the Mississippi Valley region, the mountains in Germany, the Central Plateau of France, South Africa, and elsewhere. Because much of the argument over the reality of peneplains pertains to the interpretation of topographic surfaces above the present base level of erosion, we may well consider the criteria which can be applied to the identification of uplifted peneplains before enumerating further objections to the peneplain concept.

CRITERIA FOR IDENTIFYING UPLIFTED PENEPLAINS

No single test can be made which will prove the existence of a former peneplain. Some criteria have more force than others, but it is through the application of several tests that we may come to the conclusion that the interpretation of an upland topographic surface as a former peneplain best meets all the facts. The more of these tests it can meet, the greater the likelihood will be that such a designation is correct.

Accordant interstream levels and summit areas. If a former peneplain had regional extent, it is to be expected that its remnants would exist as summit areas of low relief which would exhibit a subequality of levels (disregarding for the present the possibility of warping or differential faulting of the surface during uplift). Application of this test should be made with the realization that no peneplain was ever flat, because it must have had slopes both laterally toward the major axial streams and seaward with these streams. It should further be recognized that accordance of interstream summit areas can be produced in various ways, but at least this test must be met before the possibility of peneplanation is even considered. Just how much relief can be tolerated upon a topographic surface designated as a peneplain is a debatable question. In addition to the regional relief that is to be expected on an undulating surface, there may exist local erosion remnants which stand conspicuously above the general level. These isolated hills are called *monadnocks* from Mt. Monadnock, New Hampshire, and they commonly owe their existence to the more resistant rock of which they are composed, but some may have survived because of their remoteness from streams. Monadnocks of the latter type have been called *mosores* by Penck. Occasionally the unreduced remnants of the original surface rather than existing as single isolated hills form sprawling masses or groups. Such forms have been called *unakas* from the Unaka Mountains of North Carolina, which are believed to be of such an origin.   Accordance of divides should by no means be predicated upon the suggestion of such to the eye, for such impressions are untrustworthy, but should be based upon quantitative evidence from topographic maps or other sources.   If uplift of a peneplain took place in very recent geologic time, accordant interstream areas should be more extensive than if the uplift was at an earlier date.   Incised streams will suggest recent uplift.

Fɪɢ. 8.3.   Big Cobbler and other monadnocks standing above the Harrisburg peneplain in the Piedmont region of Virginia.   (Photo by John L. Rich, courtesy American Geographical Society.)

To lend strong support to the peneplain origin of accordant upland tracts, they should be traceable over wide areas and not be local in extent.   This does not imply that they must everywhere be at the same altitude.   It is to be expected that they will rise in accordance with their remoteness from the sea, any differential uplift that may have affected the region being taken into account.

Topographic unconformities.   In an area where an uplifted peneplain exists, it is to be expected that topographic unconformities of two sorts may be present.   The valleys cut in the uplifted peneplain should display interrupted profiles or nickpoints where they pass from old valley levels to the more recently incised valleys.   Unless the former peneplain has been practically destroyed, we should expect old-age conditions at the heads of the valleys rather than youthful characteristics.   Likewise, there should exist upland interstream areas with old age characteristics within which are cut valleys with youthful features, resulting in a sharp contrast between the cross profiles of the upland topography and that of the valleys below.

Truncation of rocks of varying resistance. The most crucial test that can be applied to an erosion surface thought to be a former peneplain is that its low-relief surface truncate strata of varying resistance to weathering and erosion. Unless this test can be met one should hesitate to call it a peneplain. Although the effects of varying geologic structure and rock resistance are never completely obscured, in old age they are less distinct and a peneplain surface should extend across geologic structures. It is, of course, conceivable that in a region of gently dipping rocks the base level of erosion locally might have coincided essentially with the dip of the rock but this is unlikely over an extensive area. Truncation of strata is almost inevitable on a peneplain because of the fact that its surface slopes both laterally and seaward.

Presence of a thick zone of deeply weathered rock debris. If extensive upland tracts remain, it is to be expected that they may retain some of the deep residual soil which forms on old-age surfaces. This mantle may consist of laterite or other clays or it may be composed predominately of relatively insoluble materials such as chert and quartz. Such old-age residuums are likely to be patchy, for erosion will have removed much of the material. Buried peneplains may lack them entirely because of removal or reworking by a transgressing sea or ice sheet.

As mentioned in Chapter 4, Woolnough (1930) has described some products of advanced chemical weathering on erosion surfaces in Australia to which he applied the name duricrusts and which he believed were developed on a nearly perfect peneplain of Miocene age under climatic conditions marked by alternating wet and dry seasons. These duricrusts consist of three types of materials: (1) aluminous and ferruginous lateritic materials upon feldspathic rocks; (2) siliceous crusts upon argillaceous and arenaceous rocks; and (3) calcareous and magnesian travertine or caliche upon sediments rich in lime. During the dry season capillarity brought to the surface the solutions formed during the wet seasons and concentrated them as a duricrust layer. He thought such crusts formed only at the peneplain stage, for so long as there was much relief there was sufficient downward and lateral circulation of groundwater to remove the materials. In his opinion these duricrusts are the correlatives of the deep residual soils which form on peneplain surfaces in humid climates lacking notable seasonal distribution of rainfall. Alternating wet and dry seasons are essential to their development since continued rainfall would remove the soluble salts rather than permit their concentration at the surfaces through capillarity. Dixey (1941) has described large masses and nodules of chalcedony from 1 to 6 feet thick in South Africa upon topographic surfaces interpreted as peneplains. He concluded that they formed in late Tertiary time under the conditions of aridity that then prevailed.

Remnants of former alluvium. As conceived by most geomorphologists, peneplains are not produced primarily by lateral stream erosion but by down-wasting of interstream tracts. Hence, alluvium is not extensively present on peneplains. It is limited to the relatively narrow strips comprising the flood-plains of former streams. The finding of patches of alluvium on upland tracts will, however, strengthen the case for the existence a peneplain provided such patches are not too common and extensive, in which event another possible origin should be considered.

ARGUMENTS AGAINST THE PENEPLAIN CONCEPT

The most frequent argument against peneplanation is that periods of still-stand of land masses are not of sufficient duration to permit reduction of wide areas to base level. No one has been able to give good quantitative data on how long a time would be required for a peneplain to develop, but it has been estimated that at the present rate of reduction North America could be base-leveled in some 15,000,000 years. Many geologists consider a stillstand of such duration unlikely. Some (Gilluly, 1949) even question the commonly accepted idea that earth diastrophism may be periodic and world-wide and may punctuate long periods of relative stability (see p. 52). This line of reasoning denies the attainment of the ultimate or even penultimate stages in the geomorphic cycle. In support of this viewpoint, it is argued that no plains of wide extent exist at the present base level of erosion but rather that most of the present-day topography is in one of the earlier stages of the geomorphic cycle. Actually it may not be necessary for there to be a complete stillstand. Peneplanation can proceed during slow uplift provided the rate of uplift is slower than the rate of degradation.

At least two explanations have been offered for the commonly observed subaccordance of hilltops and mountaintops which explain it without assuming any base level or structural control. Shaler (1899) and Smith (1899) maintained that it would be an expectable result of the beveling of uplands by uniformly spaced streams. Tarr (1898) stressed the importance of the tree line in mountain areas in bringing about a subequality of altitude. Above the tree line erosion and weathering presumably go on more rapidly than on the areas below it with its protective cover. The result would be that areas above timber line would be worn down more rapidly than those below, and in time the topography would approach a uniformity of level. Although the assumption of more rapid denudation above the timber line than below it is probably valid, it should be kept in mind that the altitude of the timber line may vary greatly within relatively short distances, and it may be questionable whether this control factor would produce a topographic surface simulating a peneplain. It has also been argued that uniform erosion on rocks of equal resistance would equally well account for subequality of upland surfaces. This may be true locally, but it can hardly have general application.

It is often maintained that topographic surfaces that have been attributed to peneplanation often display a greater lack of accordance than is permissible on a peneplain (however much that may be) or can be ascribed to deformation of a peneplain surface. Usually part of this relief developed after uplift of the peneplain in response to differential erosion or mass-wasting. Lack of a proper appreciation of the differential lowering of peneplain surfaces has been responsible for multiplication of alleged peneplains. When one notes eight or ten peneplains described in an area, as has been done, he will not only be skeptical of there being so many but will probably wonder if even one is present. Ashley (1935) and Fenneman (1936) have stressed the likelihood of multiplication of surfaces through differential lowering of one surface after uplift. In discussing this possibility Fenneman stated:

It is to be observed that this process of surficial wasting does not destroy the horizontality of a crest but only lowers it. If the amount of such lowering were everywhere the same, the record of the cycle would not be defaced and the count of cycles and peneplains would not be confused. Confusion begins when one ridge has been lowered 30 feet and another 300 feet. Both crests received their flatness at the same time—i.e., when both summits were parts of the same peneplain. Neither one has at any time lost its flatness. Both are lowering now as fast as ever, and neither is at base level. Yet a casual view, and perhaps the present vogue, would assign them to different cycles, with the tacit implication that neither summit has been lowered since uplift and that the summit plane of each cuts the mass now just where it did when the peneplain was made.

Much ingenuity has been expended in depicting a series of base levels so that each mountain crest may fall in one of the assumed planes. When an equal amount of exact study shall have been given to correlating each height with the character of the rock and the breadth of the outcrop, the time will have come to decide how many base levels must be assumed.

The exact rate at which this general wasting proceeds is not a matter of primary concern. The suggested rate of 100 feet in a million years is somewhat slower than the general wasting of the Mississippi basin. The important consideration is that, if the rate of lowering is any considerable fraction of this amount, no elevated surface older than Pleistocene is properly interpreted without taking this factor into account. If all surfaces were lowered at the same rate, the remains of former peneplains would still stand in their true relative positions. But no one will assert that all surfaces waste equally. The result is that all correlations based on altitude are liable to error in proportion (among other things) to the antiquity of the surface concerned. Before correlation, all readings must be corrected by an amount dependent (to say the least) on time, slope, and resistance. This correction is not a minute Einsteinian matter. For features dating from Miocene time it may well run high into hundreds of feet. To compute such corrections for peneplains older than late Tertiary is an arduous task. To correlate without them should be classed by law as among dangerous occupations.

If all persons who have attempted to identify peneplains had kept these facts in mind not only would there have been many fewer peneplains described but there probably would be less difference of opinion about whether peneplains exist.

The differential lowering of topographic surfaces on belts of weak and strong rock is a matter that has not received as much attention as it merits. It becomes particularly significant where extensive limestone tracts are involved. A limestone area may be lowered in toto by solution to a much lower level than the original peneplain level but still retain its peneplain-like upland surface above the present base level of erosion and, as a result, be considered a distinct peneplain younger in age than the surface above it on non-soluble rocks. Ward (1930) has made a rather good case for this origin of the so-called Somerville peneplain in eastern Pennsylvania, where a limestone plain is some 200 feet below what is considered to be the Harrisburg peneplain on adjacent non-soluble rocks. Numerous other examples of such solutional lowering can be found in the Appalachian area. Trundle Valley east of Knoxville, which is 200 feet below adjacent ridge tops, probably represents the Harrisburg level in that area. Powell Valley in western Virginia, shown on the Middlesboro, Kentucky-Virginia, topographic sheet probably has a similar explanation. It is possible that the so-called Coosa peneplain of Alabama is another example. The extensive Mitchell Plain of southern Indiana is a former peneplain surface which is now lower topographically than the surface of the same age on adjacent clastic rocks.

Not always is the existence of topographic surfaces of the same age at different altitudes a result of differential reduction of the original surface after uplift. Warping or faulting accompanying uplift may result in displacement of what was originally a nearly horizontal surface and may make difficult regional correlations. It is even possible for peneplain surfaces developed during the same cycle and within the same drainage basin to develop at different altitudes in immediately adjacent regions where there is a marked difference in lithology in the two areas. This is well illustrated in the difference in the altitudes of the Lexington and Highland Rim peneplain surfaces in south-central Kentucky. These two surfaces are now believed to be of the same age, but the Highland Rim surface stands well above the Lexington peneplain because it is in an area of resistant sandstones, whereas the Lexington surface developed across more easily erodible shales and limestones.

An interesting relation exists in southern Virginia and North Carolina where the Blue Ridge scarp separates a lower peneplain east of the Blue Ridge in the Piedmont from a higher one west of the Blue Ridge. Davis (1903) concluded that the two peneplains were equivalent in age (presumably products of the Harrisburg cycle in two different drainage basins) and thought that they "broke joint" at the Blue Ridge scarp because the local base level for the shorter streams draining to the Atlantic was lower than that of the longer streams west of the Blue Ridge which flowed to the Gulf of Mexico. Thus the Blue Ridge was considered a retreating scarp between peneplains of equivalent ages but at different levels rather than a result of more resistant rock. Wright (1928) concluded that the Piedmont peneplain and Upland peneplain were of different ages and thought that the Upland peneplain owed its greater altitude in part to more resistant rock. He conceded that nearness to the sea helped to explain the greater perfection of the peneplain in the Piedmont but did not believe that it accounted for its lower altitude. White (1950) considered the two peneplains to the sides of the Blue Ridge scarp to be equivalent in age but thought that their different altitudes were the result of late Tertiary downfaulting on the east along what he called the Blue Ridge border fault. However, his evidence for this displacement was not very convincing.

The critical point in the argument is whether the two peneplains are of the same age. Until this is definitely decided no final conclusion can be drawn. Regardless of this, it seems that Davis's contention that peneplains in two different drainage basins may exist adjacent to each other at different levels because of varying distances to the sea in the two drainage basins has merit.

TOPOGRAPHIC SURFACES CONFUSED WITH PENEPLAINS

Accordant topographic surfaces may be produced in numerous ways other than by peneplation. Before considering them, it might be well to consider further what the term peneplain implies. It has been used by some in the sense of the penultimate product of any of the erosional agencies and thus we see references to fluvial, marine, eolian, and even glacial peneplains. Such an extension of the term was hardly intended by Davis, and it certainly does not make for clarity in our interpretation of landscapes if we so use it. The term peneplain should be restricted to those gently undulating landscapes which develop under a base level control toward the end of a humid fluvial cycle in part through lateral planation by streams but more through mass-wasting and sheetwash on interstream areas than by stream erosion. Used in this sense, it has a definite meaning and implication in relation to the geomorphic history of the region that attains such a condition. Topographic surfaces similar in some respects to peneplains may be developed by the other gradational agencies, but they should be given different names. At the time that the peneplain concept was evolving, distinctive names for end or near-end products of the other erosional agencies were lacking, but today satisfactory names are available and should be used.

Stripped or structural plains. We have seen that one of the essential tests of a peneplain is that it be developed across varying rock structures, although

FIG. 8.4. A dissected structural plain, north of Price, Utah. (Photo by Chicago Aerial Survey Co.)

it is conceivable that, locally at least, the base level of erosion might coincide with a resistant rock stratum. It has been argued that many so-called peneplains are merely *stripped* or *structural plains* produced by stripping or near-stripping of a resistant formation. Undoubtedly this is true in many instances. A stripped plain may present many of the basic features of a peneplain such as accordant summit levels, upland stream deposits, erosion remnants rising above it, incised streams below it, and deeply weathered soils upon its surface. Only the lack of truncation of rocks of varying resistance may prevent its being considered a peneplain.

Usually structural plains have local extent and can be easily recognized. Examples are common in the Colorado Plateau region. They most commonly are bench-like areas along valley sides or isolated buttes and mesas, but some have considerable areal extent. In most descriptions of stripped plains the conclusion is usually drawn that the stripping took place during the current erosion cycle and above the present base level of erosion and that hence they are in no sense related in position to a base level control. This is a conclusion which has been questioned by Fenneman (1936), who doubted that extensive stripped plains can long maintain themselves above base level, but rather thought that they were more likely to have formed near a former base level control and to that extent to simulate peneplains. Regarding this interpretation of stripped plains he stated:

There is such a thing as a plain of stripping, quite independent of base level, but the limitations of stripping at high levels are very severe and are often ignored. To do a clean job of stripping a horizontal bed is just as difficult as to make a perfect peneplain. The process is the same, and its last stages are just as slow in one case as in the other if equal areas be assumed ····. The result is that, while rock terraces and mesas of limited extent are common, extensive plains, stripped at considerable altitudes, are in all cases subject to dispute. The stripping is not denied; only the altitude at which it was done. The question at issue pertains to the ability of any hard stratum to maintain a local base level high above the sea, during the long time required for peneplanation.

A familiar example is the Edward Plateau in Texas, 1000 to 3000 feet above the sea and 400 to 1200 feet above its surroundings. The underlying limestone formations, collectively known as the Edwards, are relatively resistant in the climate of central Texas. No doubt this is the explanation of the plateau's present height above the lowlands on weaker rock. But it does not follow that the stripping was done at that altitude. The margins of this formation are being raked and shredded in a way to show the precarious position of the entire mass at its present altitude.

An even more striking illustration is seen in the "Great Sage Plain," 6000 to 7000 feet high, in southeastern Utah and southwestern Colorado. Its agreement with the surface of the Dakota sandstone is noteworthy. It is mentioned as one of the best American examples of a stripped plain, generally with the implication that the strong Dakota sandstone was, in this case, an adequate substitute for base level. However, it is only necessary to look at the sharp canyons, already branching in the interior, to be convinced that the Great Sage Plain cannot last long under present conditions.

It should be a safe principle that a peneplain cannot originate under conditions which make it essentially unstable. So slow a process as peneplanation must not be asked to run a race against so swift a process as dendritic dissection of a high plateau. Applying this principle to the Great Sage Plain, it is safe to conclude that, at the time of its development, the altitude was much less than at present, certainly not high enough above the local base level to make dissection possible.

Within the limits of altitude and of time, a strong stratum may become a substitute for base level, but these limits are, without doubt, much narrower than are implied by many casual descriptions of stripped plains.

It is perhaps significant that many of the best examples of stripped plains are found in semiarid regions where destruction of areas above base level may be effected by the retreat of frontal scarps as much as by over-all stream dissection, a point more fully discussed in Chapter 11.

Plains of marine denudation. As previously indicated, the concept of marine planation received an early impetus in England and to a certain

Fig. 8.5. A plain of marine erosion of Pliocene age with hills which stood as islands above it. (British Information Services photo.)

degree has persisted there as a more likely explanation of accordant summit areas than subaerial planation, although an English textbook (Wooldridge and Morgan, 1937) is favorably inclined toward the process of peneplanation. Barrell (1920) in this country described as many as eleven terraces in the northern Appalachians which he attributed to marine erosion. Meyerhoff and Hubbell (1928) described fourteen surfaces in New England which they called terraces rather than peneplains but thought they were formed subaerially. More recently, Olmsted and Little (1946) described what they believe to be a terrace of marine origin which extends across New England below an altitude of 600 feet and assigned a Miocene age to it. Its shore line serves as the inner boundary of the Seaboard lowland section of New England. There can be little question about the efficacy of the sea as a planation agent, but there does exist some doubt whether sea level remains fixed long enough for a marine plain of subcontinental extent to be developed. It undoubtedly would take longer for marine planation to reduce a given area to a plain-like condition than would be required by subaerial erosion because of the narrow zone which is subject to wave attack. Marine-cut plains are thus more likely to be limited to relatively narrow terrace strips around the margins of the continents than to extend far inland over continental interiors. Marine plains should terminate landward with sea cliffs and associated shore line features and should have scattered over them deposits of marine character. Unless, however, they are of very recent age these features may have been effaced by subaerial erosion and weathering and be difficult to recognize. That this is true is indicated by the two different interpretations that have been placed upon the terraces in New England.

Pediplains. The term *pediment* is used to designate the rock-floored plains adjacent to or interpenetrating mountain masses in desert regions which have developed by planation of the mountains at altitudes dependent upon those of the interior basins into which the mountains drain. The term *pediplain* is now most generally used to describe a series of coalescing pediments. These forms will be discussed in a later chapter and are mentioned here merely to indicate that some topographic surfaces which have been interpreted as uplifted and dissected peneplains may have actually formed as pediments. Howard (1941) has suggested that the Flattop and Rocky Mountain peneplains in the Rocky Mountain region may have developed in this way rather than as true peneplains. Rich (1938) also thought that this possible origin for these supposed peneplains should be given serious consideration.

Bradley (1936) in giving his reasons for interpreting the Gilbert Peak erosion surface on the north flank of the Uinta Mountains as a pediment rather than a peneplain remnant stated:

Because undisturbed remnants of this surface have gradients ranging from about 400 feet to the mile near the crest of the range to 55 feet to the mile 35 miles out in the basin, because island mounts of limestone rise rather abruptly above it, and because it apparently never had a soil mantle, but is covered in most places by conglomerate, this surface is interpreted as a pediment formed in a semiarid or arid climate.

Mackin (1947) came to a similar conclusion regarding what has been called the Subsummit peneplain in the Bighorn Mountains. This interpretation eliminates postulating 6000 feet or more of Pliocene and Pleistocene uplift in the Bighorns, for the pediments could have been formed according to his estimate within 2000 feet of their present altitudes.

A most sweeping application of the pediplain origin of erosional surfaces commonly thought to be peneplains was made by King (1950), when he interpreted widely separated surfaces in Africa, Asia, North America, Europe, South America, and Australia as ancient pediplains dating back as far as the Cretaceous. It is doubtful whether many geomorphologists accept his ideas either as to origin or age of all these topographic surfaces; certainly correlation between such widely separated areas is hazardous and may lead to skepticism in regard to the validity of more likely interpretations.

Panplanes. Crickmay (1933) proposed the term *panplane* for "a plain formed of flood-plains joined by their own growth." Such a plain is the

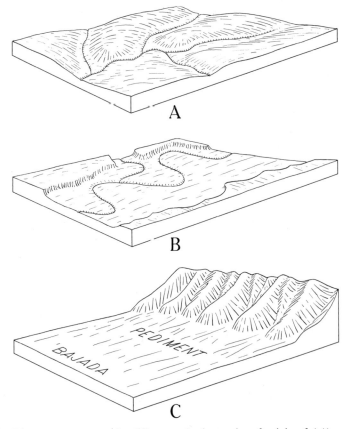

FIG. 8.6. Diagrams to suggest the differences in the modes of origin of (*A*) a peneplain, (*B*) a panplain, and (*C*) a pediplain. (Drawing by William J. Wayne.)

product of lateral erosion by streams with attendant floodplain construction and is not a peneplain according to the Davis concept, for a peneplain is primarily produced by the downwasting of interstream areas. According to

Crickmay "geographical old age and peneplanation rest on nothing but pure deduction by a few great masters of geography and geology and blind acquiescence by the rest of us." He believed that only the earlier stages of the erosion cycle are verified in nature and that the ability of mass-wasting to produce gentle slopes in old age is overemphasized. He would supplant the peneplain as produced by downwasting with the panplane and *panplanation.* The eventual joining of valley flats produces the panplane above which may rise monadnocks as envisaged on a peneplain. Even though we accept his thesis that lateral erosion has not received the attention that it merits, it would hardly seem necessary to discard the peneplain idea; at most it would involve a reevaluation of the processes responsible for its development. Many geomorphologists have difficulty in visualizing streams cutting away all of a land mass by lateral erosion. Floodplains constitute but a small portion of the total area of topographic surfaces now in the later stages of the erosion cycle. It may be admitted that many wide-open stretches along present-day valleys were produced mainly by lateral erosion, but they lack the geographic extent essential to being considered either peneplains or panplanes in the sense in which Crickmay used the latter term.

Etchplains. Wayland (1934), in discussing topographic surfaces in Uganda, Africa, introduced the concept of the *etchplain.* He described five topographic levels in Uganda, the uppermost of which he believed might be a true peneplain of Cretaceous age. The surfaces below it, however, he attributed to a process that he called etching, and he considered them etchplains rather than peneplains. The conditions under which they were thought to have developed are as follows: the original peneplain had attained a surface lacking any marked relief with the resultant development over it of a deeply weathered rock cover of *saprolite* extending down many tens of feet; uplift of the land supervened with the consequent removal of this thick zone of rotted rock; stability of the land followed with the development of another deep saprolitic cover which, following another uplift, was removed from part of the area; in this way the four topographic levels below the summit peneplain were believed to have been formed.

It is conceivable that the process designated by Wayland as etching might contribute to local differential lowering of a peneplain surface, just as solution and differential mass-wasting may, but it is difficult to visualize this process operating widely enough to produce etchplains of regional extent.

Resurrected or exhumed erosion surfaces. References to buried erosion surfaces are abundant in geologic literature. They constitute the unconformities or disconformities of stratigraphic geology. They vary in the amount of relief which they exhibit, but many of them possess slight enough relief to be ancient peneplains, although they may well be plains of marine erosion or pediplains. Most frequently they are seen in limited exposures with strata above them but in some places they are in the process of being exhumed or resurrected. Where they have not undergone distortion through folding or faulting they may retain their original characteristics after being exhumed, especially if the cover over them is relatively weak. Examples of such surfaces in the process of resurrection may be seen along the eastern edge of the Colorado Front Range, at the eastern edge of the Piedmont Plateau in eastern United States, in Brittany, the Central Plateau of France, and numerous other regions. The nearly flat surface of San Pedro Mountain in the Jemez Mountains of north-central New Mexico also has been interpreted as an exhumed erosion surface of early Tertiary age (Church and Hack, 1939).

Exhumed topographic surfaces retain their original peneplain-like features best when developed across rocks distinctly harder than those which covered them. Often these buried erosion surfaces are tilted with respect to erosion surfaces developed during later cycles of erosion. Davis (1911) applied the term *morvan* to:

> · · · a region of composite structure, consisting of an older undermass, usually made up of deformed crystalline rocks, that had been long ago worn down to small relief and that was then depressed, submerged, and buried beneath a heavy overmass of stratified deposits, the composite mass then being uplifted and tilted, the tilted mass being truncated across its double structure by renewed erosion, and in this worn-down condition rather evenly uplifted into a new cycle of destructive evolution.

The name comes from the Morvan Plateau of northeastern France which exhibits such relationships. Structurally the relationships described above are those of two intersecting erosional surfaces. The Fall Zone peneplain of the eastern Piedmont Plateau has been described as a morvan (Renner, 1927). Here a buried peneplain of probable Jurassic age is being resurrected, and its surface when projected inland intersects the late Tertiary Harrisburg peneplain at a distinct angle. Similar relationships exist in the Colorado Front Range between a partially resurrected pre-Pennsylvanian erosion surface and the Tertiary Rocky Mountain peneplain (or pediplain).

Where buried erosion surfaces have a considerable inclination with respect to the present base level of erosion only narrow strips or patches of them are likely to persist long after exhumation, but, if their surfaces should nearly parallel present base levels of erosion, more extensive tracts may be exposed which may be mistaken for peneplain surfaces of geologically recent age. Portions of Fennoscandia and Canada have been thought by some to be Pre-Cambrian peneplains which were exhumed and subjected to glaciation. Locally, exhumed topographic forms may constitute a part of our present landscape, but it is doubtful whether they exist on such a scale that they need to be confused with erosional surfaces of Tertiary or later age.

PARTIAL PENEPLAINS

How many peneplains may be represented in a region may properly be asked. If we were to apply the term peneplain in the sense of meaning complete reduction of an area to base level, the answer would of necessity be one, in which case there could be no such thing as a partial peneplain. It is a common thing, however, to find evidence of partially completed cycles in which the second cycle stopped short of the stage reached by the first and each later cycle fell short of the one immediately preceding it. Calling the remnants of the later erosion surfaces peneplains may lead to undue multiplication of them. Fenneman (1936) recognized the abundance of evidence for partial cycles when he stated:

The common experience is to find in a single area the partially destroyed, or newly begun, forms of several cycles, none of them carried to completion, unless it be the first. The work of the next cycle stopped somewhat short of the stage reached in the first, the third fell short of the second, and so on. Speaking only of those whose records remain, it follows that the cycles were of decreasing completeness and (presumably) in most cases of decreasing duration. The prevalence of this observation is sometimes noted as curious, as though implying that the earth's crust is becoming progressively less stable. Reflection shows, however, that the record could never have been otherwise, since only such cycles are recorded as were followed by others less complete.

It is obvious that we have need of a term which may be applied to a topographic surface which falls short of attaining the geographic extent and topographic condition to merit description as a peneplain. Various terms have been used such as *partial peneplain, incipient peneplain, local peneplain, berm,* and *strath.* None has proved completely satisfactory, for various meanings have been attached to them. It is to be expected that there will be all gradations from narrow valley strips reduced to base level to large expanses which approach obliteration of preexisting topography. Frequently the term berm or strath is applied to terrace-like remnants of former valley flats which have resulted from dissection following uplift of a region which had attained a stage of middle or late maturity. Bascom (1931) introduced the term berm in a geomorphic sense and suggested that it be applied to "those terraces which originate from interruptions of an erosion cycle with rejuvenation of a stream in the mature stage of development." Unfortunately she suggested that the term might also be applied to terrace remnants of a marine-formed abrasion platform.

Bucher (1932) suggested a restriction of the Scottish term strath to the "incipient peneplain" which is initiated "where the flat valley bottom is the result of degradation first by lateral stream cutting and later by whatever processes of degradation may be involved." Remnants of such a surface would then constitute *strath terraces.* It is well that this distinction be made, for the term strath is commonly used in the sense of strath terrace as suggested by Bucher, as is indicated by the frequent reference to the Parker strath in the Appalachian Plateau region. In general, it is true that the terms berm and strath terrace are restricted to planation surfaces confined to valleys, whereas such terms as partial or local peneplain are more likely to be applied to base-leveled areas not so limited in extent. Local peneplains or straths are usually found on areas of weak rock. The Somerville and Coosa peneplains, if not actually of the same age as the Harrisburg surface and lower because of solutional lowering of their limestone terrains, would be examples of such peneplains. The Scottsburg lowland of southern Indiana and the Black Belt of Alabama (Cleland, 1928) are probably good examples of local peneplains of Pleistocene age.

It often happens that a topographic surface is interpreted as evidence of a partial cycle when it is of the same age as a higher surface but at a lower altitude owing to either an originally lower altitude on belts of weak rock or to more rapid wasting of a weak rock area after uplift of an originally accordant surface. It requires careful discrimination to distinguish these surfaces from those produced by partial cycles, and to a considerable degree the failure to do so has led to the description of an excessive number of alleged peneplains.

THE AGES OF PENEPLAINS

In view of the increasing recognition of the recent age of most of our landscape features (Ashley, 1931), it seems likely that few peneplains, excluding buried and exhumed ones, date back of the Tertiary. There seems to be fairly good evidence for two peneplains in eastern North America, the Schooley and the Harrisburg. It is common practice now to consider the Schooley peneplain as early or mid-Tertiary age and the Harrisburg as of late Tertiary age. In the Rocky Mountain area the Flattop and Rocky Mountain peneplains, if they are such, probably have approximately corresponding age relationships. In a region where two peneplains exist, two age relationships may need to be recognized, the actual age and the comparative age of each surface (King, 1947). *Actual age* refers to the time of formation of the peneplain surface as indicated by fossils or sediments contemporaneous with its development. The actual age of a peneplain is usually difficult to determine; it can be done only for limited areas because different parts of the same peneplain may have varying actual ages. The peneplain stage will be attained in the lower part of a drainage basin long before it is attained in the upper part. A cycle may continue in interior areas long after it has been terminated near the sea by uplift or lowering sea level. King gave an interesting interpretation illustrating how this may lead to confusing relationships. Pleistocene faulting produced a series of lake basins in the interior of East Africa which were filled with Pleistocene sediments. A Pliocene erosion cycle is still advancing inland and will in time extend itself across these Pleistocene sediments. The actual age of the peneplain surface in that area will be post-Pleistocene yet it will correlate topographically with a surface near the coast which developed in Pliocene time.

Usually when we speak of the age of a peneplain we are considering its *comparative age*. By this we refer to its age in comparison with earlier or later peneplains in the same drainage basin; this age is arrived at by determining, as far as possible, during what period of geologic time the base level toward which it was being reduced was the current base level for that particular drainage basin. This is what we generally mean or should mean when we date a peneplain as of Miocene or Pliocene age. Comparative ages of erosion surfaces thus give the sequence of landscape evolution.

It is difficult enough to arrive at the actual or comparative age of a peneplain surface in a single drainage basin, but it is even more difficult and hazardous to try to correlate peneplains in different drainage areas. Their comparative ages may be ascertained more accurately, but even this may be difficult. Caution should be exercised in attempting precise correlation of peneplains in widely separated areas, for there is no assurance that diastrophism is sufficiently synchronized.

Conclusions

After realizing how much difference of opinion exists among professional geologists in regard to the reality of peneplains and how many types of topographic surfaces may be mistaken for them, the student is likely to be thoroughly confused. If so, he may take some consolation from the fact that he is not much more confused in his thinking than are geomorphologists. No final pronouncement can be made on the subject, but there are a few fundamental principles which, if kept in mind, will perhaps help to clarify our thinking.

The peneplain concept is fundamentally a sound one which has been greatly abused either through lack of proper appreciation of what it implies or through too hasty willingness to recognize peneplains. Topographic surfaces have many origins; peneplanation is only one possible origin. The term peneplain ought to be restricted to those surfaces that are the product of fluvial erosion plus mass-wasting of the interstream areas. It represents the penultimate stage of the fluvial cycle and not the ultimate; hence no peneplain was ever flat or plane-like. In its original condition it possessed slopes; it rose inland from the sea, and it had the slopes of the individual interstream areas toward adjacent streams. Deformation by warping or faulting may render difficult or impossible restoration of the original surface.

In the tectonically active belts, it is doubtful if the peneplain stage can be attained as long as mobility prevails, but there is a great deal of evidence to suggest that in some parts of the world there have been periods of sufficient crustal stability, not necessarily periods of complete stillstand, to permit extensive areas to be reduced to or nearly to a base level control.

Before interpreting any topographic surface as a peneplain the geologist should first take all possible precaution to make sure that the surface is not related to lithologic or structural controls or a product of some other geomorphic process. If this is done, many fewer peneplains will be suggested.

It cannot be said just how many peneplains are permissible in a region, but allegation of more than two should at least cause a person to be somewhat skeptical and to consider the probability that other explanations may apply to some of the alleged peneplains.

Extreme skepticism should prevail toward any peneplains that are dated back of the Tertiary, excluding buried and resurrected peneplains.

Any interpretation of topographic surfaces as peneplains should be based upon detailed quantitative studies and not mere subjective impressions of accordance of summit areas.

If these precautions are kept in mind, the peneplain concept becomes a useful tool in deciphering the geomorphic as well as the diastrophic history of a region.

WALTHER PENCK'S GEOMORPHIC IDEAS

The doctrines of Powell, Gilbert, Dutton, Davis, and others have become so firmly implanted in American books on geomorphology and physical geology that the beginning student is likely to get the idea that what has come to be called the "American or Davisian school of geomorphology" represents the thinking of all geomorphologists. It is true that the Davisian ideas are rather widely accepted, but there are dissenters even among American geomorphologists. Some British geologists are still inclined to look upon marine erosion as more significant than subaerial erosion or consider that land forms largely reflect the influences of varying rock resistance. Among some Germans, like the followers of Passarge, the idea that geomorphology should be primarily explanatory did not meet with ready acceptance. They were more inclined to follow the German penchant for detailed description of landscapes in empirical terms and leave the explanations to "armchair geologists." The German, Walther Penck, has come the nearest to offering a different philosophy of geomorphology, which has been considered by some as a real challenge to the teaching of Davis.

The author is not certain that he can do justice to Penck's ideas since his knowledge of them is largely second hand. Penck's *Die Morphologische Analyse* is in such tortuous and difficult German that even persons who are adept at that language have difficulty with it. Admittedly the objectives of Davis and Penck were different. Davis considered the major goal of geomorphic studies to be effective description of the earth's surface features and thought that explanatory description was to be preferred to purely empirical description. Penck approached geomorphology with the viewpoint of a geologist who seeks to find in it a tool for the more effective interpretation of diastrophic history. With his aim we certainly can have little quarrel.

Penck did not deny completely the validity of the geomorphic cycle as envisaged by Davis as a method of investigation, but he seems to have considered it a sort of special case rarely encountered, since he considered cyclical development of land forms dependent upon a stability of the earth's crust seldom attained. Although Penck recognized the value of considering the time factor in land-form development, he considered it more important for a proper interpretation of the land forms produced by the erosional processes to take into account the relative rate of uplift that the land mass was undergoing while it was being destroyed. This *principle of mobility* of the earth's crust is basic to any appreciation of Penck's ideas. We may say with some truth that he replaced the youth, maturity, and old age of the geomorphic cycle with three types of relative mobility: *aufsteigende Entwicklung*—waxing or accelerated development or uplift; *gleichförmige Entwicklung*—uniform development or uplift; and *absteigende Entwicklung*—waning or declining development or uplift.

According to Penck, the characteristics of valley side slopes are largely determined by which type of uplift has taken place. With aufsteigende entwicklung, valley deepening does not keep pace with uplift nor does valley widening with valley deepening; hence, the result is the development of convex side slopes. Gleichförmige entwicklung results in the maintenance of uniform relief and valley side slopes which are straight, whereas absteigende entwicklung produces concave slopes. There are probably few geologists who would be willing to accept the validity of this deduction. Johnson (1940), in referring to it, said, "Penck's conception that slope profiles are convex, plane, or concave, according to the circumstances of the uplifting action, is in my judgment one of the most fantastic errors ever introduced into geomorphology." Valley slopes are undoubtedly affected by various factors such as lithology, structure, climate, size of debris produced by weathering and stage of development, and it is too much to assume that they reflect mainly the difference in the type of uplift which affected the region in which they developed.

Yet, despite the probably erroneous conclusions drawn by Penck regarding the diastrophic significance of varying cross-valley profiles, it is likely that his most significant contribution to geomorphology has been that he has caused geomorphologists to realize how inadequate have been the usual explanations of how slopes develop and recede. According to Penck, there are two basic elements in any valley side slope formed under *absteigende entwicklung*—the

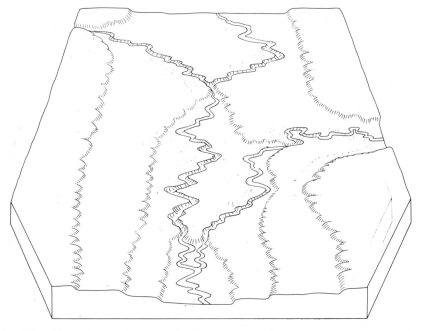

FIG. 8.7. Three stages of terrace development, according to the treppen concept of upstream migration of nickpoints and parallel retreat of slopes. (After H. A. Meyerhoff and M. Hubbell.)

upper, relatively steep *bösche* or *steilwand* or what Meyerhoff (1940) has called the *gravity slope;* and the lower, more gentle *haldenhang* or *wash slope* of Meyerhoff. These two slopes usually meet at a distinct angle. The angles of the two slopes depend upon the size of the material that is moved over them and the amount of water available to carry this material. Large-sized material requires a steeper slope for its transport than smaller particles, and the slope decreases with increase in volume of water moving over it. In general the haldenhang will be steeper in arid regions than in humid regions because there is less water available to move rock debris over it.

Particularly with respect to the changes that slopes undergo as dissection of a region proceeds are Penck and Davis in serious conflict. Penck contended that, once the bösche has developed its characteristic angle, it retreats parallel to its former positions maintaining the same angle that it had at the beginning and that its characteristic steepness is not lost until all the land mass above the level of the haldenhang is consumed. This is the principle of what has come to be called *parallel retreat of slopes*. In contrast, Davis contended that the steep slopes of youth become progressively flatter as the cycle proceeds until they are replaced by "the smooth and gentle slopes of old age." Thus the consumption of a land mass is conceived by Penck as largely the result of *backwasting* and by Davis as the result of *downwasting*. It is probably too much to say that one is right and the other is wrong. Certainly, the factors influencing the retreat of slopes are far more complex than is generally realized and there may be an element of truth to Penck's contention that, once slopes have become graded, they maintain a constant angle as they retreat. Even Davis in his later writings modified his views somewhat with respect to the development of slopes in arid regions, and in Chapter 11 we shall see that he nearly accepted the idea of parallel retreat of slopes under arid conditions. This may be true to greater degree than we realize in humid regions. Not enough information is yet at hand to say which idea is the more nearly correct. Bryan (1940) concluded that: (1) slopes develop certain characteristics dependent upon lithology and climate (and we might perhaps add processes involved in their development); (2) once slopes attain their characteristic angle, they maintain this inclination throughout their retreat; and (3) they disappear only after all the rock mass above the encroaching foot slope has been removed. Whether this be true or not it certainly represents a shift toward the ideas of Penck as far as slope development is concerned.

THE PRIMÄRRUMPF CONCEPT OF PENCK

The *primärrumpf* concept of Penck has been called by one of his adherents "one of Penck's magnificent generalizations." Some would not be so complimentary regarding it. It was a necessary concept according to Penck's thinking, because he considered that the diastrophic conditions necessary for the progress of an erosion cycle with its resultant peneplain represent a special and uncommon case. Penck recognized the existence of what were apparently remnants of former denudational surfaces, but he thought it unlikely that there were many uplifted and dissected peneplains. Although Davis recognized full well the possibilities of various types of diastrophic uplift, he commonly assumed rapid uplift as the more typical example of how a new geomorphic cycle was inaugurated, and furthermore he thought that this represented a simple condition which, if properly conceived, could be used as a point of departure for consideration of the complications of the geomorphic cycle.　It is perhaps only fair to Davis to say that his critics have insisted too vigorously that he believed that rapid uplift followed by long stillstand was the "normal" procedure.　Penck, however, believed that more commonly uplift of a land mass started very slowly, so slowly with respect to the rate of denudation that there would be no actual net rise of the surface nor increase in its relief.　Such a condition would give rise to a low, rather featureless plain which he called a *primärrumpf*.　No good English equivalent is available for this term.　Sauer has used the expression primary peneplain to express its implications, but the use of the word peneplain, even if prefixed by a qualifying adjective, is misleading.　It seemed to Penck that with slowness of initial uplift, regardless of what the geologic structure might be, degradation would keep pace with uplift resulting in the formation of a primärrumpf which would be the universal initial geomorphic unit for all the topographic sequences that followed.　After its formation, the subsequent history of a primärrumpf depends upon whether it undergoes aufsteigende entwicklung, gleichförmige entwicklung, or absteigende entwicklung.　With aufsteigende entwicklung (waxing or accelerated uplift) it may be elevated high above sea level, but it remains a primärrumpf even though it undergoes degradation.　Sauer, who has been most receptive to Penck's ideas, thought that the Mesa Grande and Julian Mesa in the Peninsular Range of southern California are examples of summit surfaces of the type which Penck designated as primärrümpfe.　Regarding them Sauer (1929) stated:

　　. . . the two mesas described form a summit peneplain of the types characterized by Penck as a primary peneplain (Primärrumpf).　We have here not, in so far as is known, a surface once worn down to a low level and then uplifted, but an assemblage of forms, which though at summit position, is in the process of reduction in relief · · ·.　In so far as we know the area has been subject to sub-aerial denudation indefinitely, perhaps since Mesozoic time.　It long ago became detached by uplift from any base-level of erosion extraneous to the local block, if such connection once existed.

　　It is difficult to see how a primärrumpf differs fundamentally from what Davis (1922) described as an *old-from-birth peneplain*.　He recognized the possibility that, if uplift were extremely slow, "the first-cut valleys may be deepened so little in excess of the down-wearing of the inter-valley uplands that the general expression of the surface will at once be 'old.' "　Thus the passage of a landscape through the stages of youth and maturity would be elided.　Penck called the terminal plain which resulted from the degradation of a land mass which has had high relief an *endrumpf*.　Thus the primärrumpf is the initial stage in a period of diastrophism marked by accelerated uplift, and the endrumpf is the end stage of a period of degradation marked by declining uplift (absteigende entwicklung).　It seems difficult indeed to see how these two could be distinguished from one another except in theory. Apparently the only difference between his concept of an endrumpf and Davis's concept of a peneplain is that Penck does not consider that the endrumpf has gone through any sequential stages in the process of its development.

It is difficult to see how Penck's concept of the primärrumpf is any less susceptible to the criticism of being a special case than is the claim that he made that Davis's postulation of rapid uplift followed by stillstand is such a case. Although Davis may have overworked the idea of rapid uplift followed by stillstand to permit the geomorphic cycle to run its course, he did at least recognize that there are many possible variations from this simplified case, even though he may have not stressed them enough. Whether we accept or reject Penck's concept of the primärrumpf and his ideas regarding the relationships between valley slopes and the type of diastrophic uplift, his challenge has had some salutary effects in that it has destroyed some of the complacency which had developed among geomorphologists and forced them to reexamine some of their basic assumptions. Modification may be found necessary to make the scheme of the geomorphic cycle fit certain circumstances, but it hardly seems logical to throw away a concept which has proved so useful because in certain areas it does not seem to apply.

THE TREPPEN CONCEPT OF PENCK

Probably the least favorably received of Penck's ideas is that which he proposed to explain the piedmont benchlands or piedmonttreppen which he encountered in his studies in the Black Forest (Schwarzwald) of Germany. The term *piedmonttreppen* has been used to describe the succession of step-like benches which are found around the flanks of the Black Forest and other mountains. According to more conventional interpretation, these benches or steps would be described as erosional surfaces representing a series of partial erosion cycles terminated by intermittent uplifts, but, according to Penck's interpretation, they are the result of continuous uplift of an *expanding dome* experiencing aufsteigende entwicklung (accelerated uplift). The summit area would thus be a primärrumpf. Penck thought that discontinuous erosional surfaces could develop in an area of homogeneous rocks as a result of accelerated uplift. A primärrumpf undergoing such uplift would experience a succession of land forms starting with "old age" forms, then "mature" forms and later "youthful" forms, in terms of conventional description, which is just the reverse of the sequence in the geomorphic cycle as pictured by Davis. Accelerated erosion would result in the development of nickpoints along the streams. Above each nick, degradation would proceed as if it were the local base level of erosion, and by the process of parallel retreat of slopes there would develop with respect to each nick a gradational surface extending back into the expanding dome. Thus the development of a series of nicks would in time be reflected in the topography by a series of piedmonttreppen. The reasoning by which Penck explains the development of nicks along stream courses in areas of homogeneous rocks is rather difficult to follow. Baulig (1939), although he rejected the explanation, summarized it as follows: *

The rivers being tacitly supposed to radiate from the center of the dome, their volume and erosive power increased outward. Moreover, through the dome-like uplift their gradients are increased in the peripheral more than in the central parts.

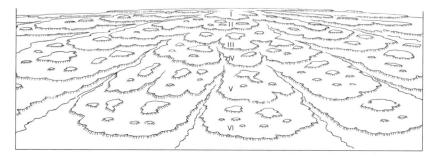

FIG. 8.8.   Diagram to explain the formation of piedmont benchlands, according to Penck's idea of an expanding dome.   (After W. M. Davis.)

For both reasons, they will rapidly and effectively react to uplift in their lower sections, while upper sections, where the current is feeble and the increase of slope small, suffer little change. As the dome constantly expands, the zone of greatest increase of slope and most active down-cutting of the river bed is constantly shifted outwards. Hence the river profile tends to divide into an upper, mature, concave section; a lower, vigorously rejuvenated tract likewise with a concave profile; and a middle, convex portion. As both volume and gradient are varying continuously both in space and in time, it seems as if the middle section ought to be relatively long and to extend both upstream through headward erosion and downstream in response to the shifting of the place of most rapid down-cutting. Penck, however, admits, in accordance with observable facts, that this section is short and recedes, as cyclic breaks do, through retrogressive erosion. He assumes, however, that for mysterious reasons, other breaks (or steps) will follow the first at shorter and shorter intervals, all receding, as the first does, upstream.

Few geomorphologists will doubt the reality of nickpoints along many stream courses, although they may differ as to whether they reflect the influence of harder rock or the effects of uplift, but there are probably few who

* Quoted by permission of Columbia University Press from *J. Geomorph.,* *2,* p. 300.

would admit the likelihood of their development under the conditions visualized by Penck. It may be pertinent here to call attention to the fact that the piedmonttreppen described by Penck bear similarity to features to be described as pediments in the chapter on the arid geomorphic cycle. The pediments, however, are marked by steep risers back of them, whereas the treppen have risers so indistinct as not to have been recognized by many workers.

REFERENCES CITED IN TEXT

Ashley, G. H. (1931). Our youthful scenery, *Geol. Soc. Am., Bull. 42*, pp. 537–546.

Ashley, G. H. (1935). Studies in Appalachian mountain structure, *Geol. Soc. Am., Bull. 46*, pp. 1395–1436.

Atwood, W. W., Sr., and W. W. Atwood, Jr. (1948). Tertiary-Pleistocene transition at the east margin of the Rocky Mountains, *Geol. Soc. Am., Bull. 59*, pp. 605–608.

Barrell, Joseph (1913). Piedmont terraces of the northern Appalachians and their mode of origin, *Geol. Soc. Am., Bull. 24*, pp. 688–690.

Barrell, Joseph (1920). The piedmont terraces of the northern Appalachians, *Am. J. Sci., 199*, pp. 227–258, 327–361, and 407–428.

Bascom, Florence (1931). Geomorphic nomenclature, *Science, 74*, pp. 172–173.

Baulig, Henri (1939). Sur les "Gradin de Piedmont," *J. Geomorph., 2*, pp. 281–304.

Bradley, W. H. (1936). Geomorphology of the north flank of the Uinta Mountains, *U. S. Geol. Survey, Profess. Paper 185-I*, pp. 163–199.

Bryan, Kirk (1940). The retreat of slopes, *Assoc. Am. Geog., Anns., 30*, pp. 254–268.

Bucher, W. H. (1932). "Strath" as a geomorphic term, *Science, 75*, pp. 130–131.

Church, F. S., and J. T. Hack (1939). An exhumed erosion surface in the Jemez Mountains, New Mexico, *J. Geol., 47*, pp. 613–629.

Cleland, H. F. (1920). A Pleistocene peneplain in the Coastal Plain, *J. Geol., 28*, pp. 702–706.

Crickmay, C. H. (1933). The later stages of the cycle of erosion, *Geol. Mag., 70*, pp. 337–347.

Daly, R. A. (1905). The accordance of summit levels among Alpine mountains, *J. Geol., 13*, pp. 105–125.

Davis, W. M. (1889). Topographic development of the Triassic formation of the Connecticut Valley, *Am. J. Sci.*, 3rd ser., *37*, p. 430.

Davis, W. M. (1896). Plains of marine and sub-aerial denudation, *Geol. Soc. Am., Bull. 7*, pp. 377–398; also in *Geographical Essays*, pp. 323–349, Ginn and Co., New York.

Davis, W. M. (1903). Stream contest along the Blue Ridge, *Geog. Soc. Philadelphia, Bull. 3*, pp. 213–244.

Davis, W. M. (1911). Relation of geography to geology, *Geol. Soc. Am., Bull. 23*, pp. 93–124.

Davis, W. M. (1922). Peneplains and the geographical cycle, *Geol. Soc. Am., Bull. 23*, pp. 587–598.

Dixey, F. (1941). The age of silicified surface deposits in Northern Rhodesia, Angola, and the Belgian Congo, *Geol. Soc. South Africa, Tran. 44*, pp. 39–49.

Fenneman, N. M. (1936). Cyclic and non-cyclic aspects of erosion, *Geol. Soc. Am., Bull. 47*, pp. 173–186.

Gilluly, James (1949). Distribution of mountain building in geologic time, *Geol. Soc. Am., Bull. 60,* pp. 561–590.

Howard, A. D. (1941). Rocky Mountain peneplains or pediments, *J. Geomorph., 4,* pp. 138–141.

Johnson, D. W. (1916). Plains, planes, and peneplanes, *Geog. Rev., 1,* pp. 443–447.

Johnson, D. W. (1940). Memorandum on the geomorphic ideas of Davis and Walther Penck, *Assoc. Am. Geog., Anns., 30,* p. 231.

King, L. C. (1947). Landscape study in southern Africa, *Geol. Soc. South Africa, Proc. 50,* pp. 23–52.

King, L. C. (1950). The study of the world's plainlands: a new approach in geomorphology, *Quart. J. Geol. Soc. London, 106,* pp. 101–127.

Mackin, J. H. (1947). Altitude and local relief of the Bighorn area during the Cenozoic, *Wyo. Geol. Assoc., Field conference in the Bighorn Basin, Guidebook,* pp. 103–120.

Meyerhoff, H. A., and M. Hubbell (1928). The erosional landforms of eastern and central Vermont, *Rept. Vt. State Geologist, 16,* pp. 315–381.

Meyerhoff, H. A. (1940). Migration of erosional surfaces, *Assoc. Am. Geog., Anns., 30,* pp. 247–254.

Olmsted, E. W., and L. S. Little (1946). Marine planation in southern New England, *Geol. Soc. Am., Bull. 57,* p. 1271.

Penck, Walther (1927). *Die morphologische Analyse,* Stuttgart.

Renner, G. T., Jr. (1927). The physiographic interpretation of the fall line, *Geog. Rev., 17,* pp. 278–286.

Rich, J. L. (1938). Recognition and significance of multiple erosion surfaces, *Geol. Soc. Am., Bull. 49,* pp. 1695–1722.

Sauer, Carl (1929). Land forms in the Peninsular Range of California as developed about Warner's Hot Springs and Mesa Grande, *Univ. Calif. Pub. Geog., 3,* pp. 212–215.

Shaler, N. S. (1899). Spacing of rivers with reference to hypothesis of baseleveling, *Geol. Soc. Am., Bull. 10,* pp. 263–276.

Sharp, R. P. (1939). Ep-Archean and Ep-Algonkian erosion surfaces, Grand Canyon, Arizona, *Geol. Soc. Am., Bull. 50,* p. 1933.

Smith, W. S. T. (1899). Some aspects of erosion in relation to the theory of peneplains, *Univ. Calif. Dept. Geol., Bull. 2,* pp. 155–178.

Tarr, R. S. (1898). The peneplain, *Am. Geol., 21,* pp. 351–370.

Wahlstrom, E. E. (1948). Pre-Fountain and recent weathering on Flagstaff Mountain near Boulder, Colorado, *Geol. Soc. Am., Bull. 59,* pp. 1173–1190.

Ward, Freeman (1930). The role of solution in peneplanation, *J. Geol., 38,* pp. 262–270.

Wayland, E. J. (1934). Peneplains and some erosional platforms, *Geol. Survey Uganda, Ann. Rept. Bull.,* pp. 77–79.

White, W. A. (1950). Blue Ridge Front—a fault scarp, *Geol. Soc. Am., Bull. 61,* pp. 1309–1346.

Wooldridge, S. W., and R. S. Morgan (1937). *The Physical Basis of Geography,* Chapter 13, Longmans, Green and Co., London.

Woolnough, W. G. (1930). The influence of climate and topography in the formation and distribution of products of weathering, *Geol. Mag., 67,* pp. 123–132.

Wright, F. J. (1928). The erosional history of the Blue Ridge, *Denison Univ. J. Sci. Labs., 23,* pp. 321–344.

Additional References

Bowman, Isaiah (1926). The analysis of land forms, *Geog. Rev., 16,* pp. 122–132.

Cotton, C. A. (1948). *Landscape,* 2nd ed., pp. 230–233, 270–296 and 357–361, Whitcombe and Tombs, Wellington.

Davis, W. M. (1932). Piedmont benchlands and primärrumpfe, *Geol. Soc. Am., Bull. 43,* pp. 399–440.

Johnson, D. W. (1930). Planes of lateral corrasion, *Science, 73,* pp. 174–177.

King, L. C. (1953). Canons of landscape evolution, *Geol. Soc. Am., Bull. 64,* pp. 721–752.

Sauer, Carl (1930). Basin and range forms in the Chiricahua area, *Univ. Calif. Publs. Geog., 3,* pp. 339–414.

9 · Topography on Domal and Folded Structures

Domes and basins underlie extensive portions of the earth's surface, if we place in this classification those broad warps in which the rocks possess slight dips and appear to the eye to be nearly horizontal. There are all gradations from gentle warps, such as the Cincinnati arch and Nashville dome, to those more definitely domal structures found in the Black Hills and Bighorns where the rocks have obviously steep dips. The term dome will be used in this chapter to include some structures that are more accurately described as elliptical anticlines, but for geomorphic purposes it is not too inaccurate to consider as domes such structures as the Kettleman Hills in California. Even gentle warps such as the Cincinnati arch possess the essential features of a dome, but the low dips around them cause the cuesta rather than the homoclinal ridge or hogback to be the characteristic flanking feature. Most large domal structures are of rather ancient geologic age. Some, like the Cincinnati and Nashville domes, were apparently initiated early in the Paleozoic. They have long been positive geologic structures and have undergone several periods of truncation by erosion. The only young domes are rather small ones, and even they display a considerable degree of dissection. Undissected domes are rare.

TYPES OF DOMAL STRUCTURES

Domes with cores of crystalline rocks. There are at least three types of domal structures in this class. The first is what for lack of a better name we shall call *domes with ancient crystalline cores*. These are often described as batholithic domes, but that name may be misleading, for a natural inference is that the intrusion of the batholith was responsible for the doming, whereas in fact the domal structure may long postdate the intrusion of the igneous core, which may not necessarily be batholithic in character. The Black Hills domal structure is an example of this type. Its crystalline core is of Pre-Cambrian age, whereas the doming did not take place until the Laramide Revolution at the close of the Mesozoic.

A second type of dome in this class is the *laccolithic dome*. It is produced by the intrusion of an igneous mass between the bedding planes of rock strata so as to form a lenticular mass convex upward. All gradations exist between

FIG. 9.1. Vertical photo of El Soltario dome, Presidio and Brewster counties, Texas. This domal structure is the result of intrusion of either a laccolith or small stock. (Photo by Edgar Tobin Aerial Surveys.)

laccoliths which produce doming and sills or intrusive sheets which produce little or no deformation of the intruded strata. Most laccolithic domes are likely to be small, and some of the larger so-called laccoliths are being classified as stocks. This seems to be true of the Henry Mountains of Utah, which have long been cited as classical examples of laccolithic mountains (Gilbert, 1877). Hunt (1946) has shown that the large peaks of the Henry Mountains, such as Mt. Hillers and Mt. Ellsworth, owe their existence to stocks rather than laccoliths.   There are laccoliths present in the Henry Mountains, but they are relatively small bodies and they exist as small tongue-like masses extending out radially from stocks.   The Judith Mountains of Montana, which also have been frequently cited as examples of laccolithic mountains, have been shown (Goddard, 1950) to consist of a group of stocks rather than

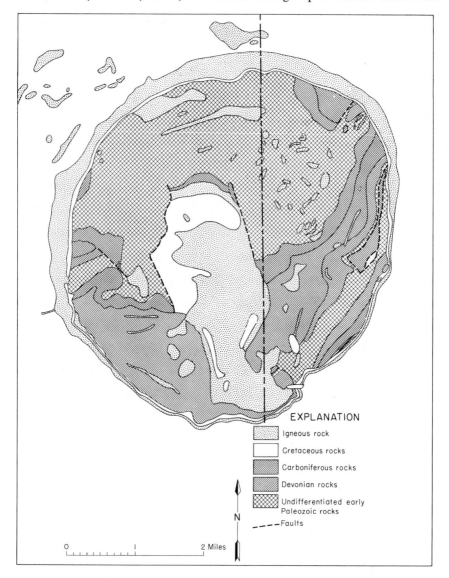

EXPLANATION

Igneous rock

Cretaceous rocks

Carboniferous rocks

Devonian rocks

Undifferentiated early
Paleozoic rocks

– – – – Faults

N

0 ⸻⸻⸻⸻ 2 Miles

FIG. 9.2.   Geologic map of El Soltario dome.   (Modified after E. H. Sellards et al.,
University of Texas Bull. 3232.)

laccoliths. Seemingly authentic examples of laccolithic domes are those of the Highwood Mountains in central Montana (Hurlbut and Griggs, 1939), where nine laccoliths exhibiting varying degrees of topographic expression and erosion have been described. Among them are Square Butte, which rises 2000 feet above its surroundings, and Round Butte or Palisade Butte, which

Fig. 9.3.   Green or Little Sundance Mountain dome, Crook County, Wyoming, a laccolithic dome.   (Production and Marketing Administration photo.)

is 800 feet high.   The Elk Mountains of western Colorado are believed to be laccolithic mountains.   Several laccolithic peaks are found around the Black Hills, among which are Crow Peak, Elkhorn Peak, and Crook Mountain.

A rare but unusually interesting type of dome is the *cryptovolcanic structure* (Bucher, 1936).   It is thought to be produced by the sudden release of volcanic gases at depth.   In a sense it represents an abortive attempt at the formation of a volcano.   This type of structure was first recognized in the Steinheim Basin in southern Germany.   Several have been described in the

United States, among which are Hicks dome in Hardin County, Illinois; Jeptha Knob dome in Shelby County, Kentucky; Kentland dome in Newton County, Indiana; Wells Creek dome in Houston and Stewart Counties, Tennessee; and Upheaval dome in San Juan County, Utah. All of them have several features in common, among which are: a nearly circular outline; a central uplifted portion which is usually marked by pronounced faulting; shattering and brecciation; and an absence of exposure of volcanic rocks. A typical cryptovolcanic dome is 2 to 3 miles in diameter. At the center of the Kentland dome in Indiana (Shrock and Malott, 1933) there is an upward displacement of strata of at least 1500 feet. Where the rocks are resistant, the central part of the dome is topographically as well as structurally high, as at Jeptha Knob, but, if the rocks are weak, as in the Well Creek dome, there is a central topographic basin. In some places, as at Hicks dome and Jeptha Knob, there are ring-like depressions around the dome developed upon weaker beds. The one around Hicks dome owes its existence to weak Devonian shale beneath it.

Salt domes.   Small domal structures produced by intrusion of masses of salt into rock strata, and hence called *salt domes,* have been known in this country since 1862 (DeGolyer, 1925). They have had economic significance since the development of the Spindletop oil pool on one of these in 1901. They are numerous in the Gulf coastal plain of Texas and Louisiana and are also found in the Harz Mountains, the North German plain, on the south flank of the Carpathian Mountains, in the Transylvanian portion of Rumania, in Egypt, Persia, Russia, Spain, Morocco, and Algeria. A typical salt dome is an anticlinal structure with a salt core which may or may not have a massive cap of anhydrite, gypsum, limestone, or dolomite. They vary in shape from sharp ridge-like flexures to shallow and flat circular domes. The latter form is particularly common in the Gulf Coast area. The diameter of a salt dome is usually from 1 to 4 miles.

Topographically, salt domes may correlate either positively or negatively with geologic structure, but more commonly they stand up from a few feet to a hundred feet or more above their surroundings. As a class they are the least dissected of domal structures, but the presence of even the smaller ones may be indicated either in the topography or associated vegetation. Before their origin was understood many were considered to be low monadnocks. The more dissected salt domes may be recognizable only from geophysical surveys. Their presence may be indicated by one or more of the following features (Powers, 1926): saline prairies or salt licks at their centers; lakes with or without surface outlets; slightly mineralized springs or wells; annular drainage about a central hill; radial drainage; tilted beds; exposure or presence at shallow depth of a cap rock or beds older than normal for the region; and soils and attendant vegetation that differ from those of their surroundings. In the Gulf Coast region, the salt domes farther inland tend to have circular basins at their centers enclosed by one or more rings of hills, whereas those nearer the coast are more likely to have a mound form with radial drainage on the larger ones.

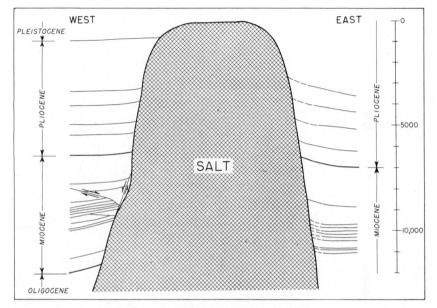

FIG. 9.4. Cross section of Avery Island salt dome. (After J. B. Carsey, *Am. Assoc. Petroleum Geol., Bull. 9.*)

Broad domal warps. Included in the class of broad upwarps are such structures as the Cincinnati arch, the Nashville dome, the Zuni Uplift of northwestern New Mexico, and the San Rafael Swell of Utah. In these the dip of the rock may be so gentle as to appear horizontal to the eye, in which case the domal structure is only apparent from the existence of rocks of older age at the center and progressively younger rocks away from the center and the development of a series of cuestas around the flanks of the dome. In none of the domes mentioned above is igneous rock exposed at the center of the dome, but in the Ozark, Wisconsin, and Ontario domes erosion has exposed Pre-Cambrian igneous rocks. Whether domes of this type are solely the result of broad epeirogenic warpings of the earth's crust or are partly the result of deep-seated igneous activity is difficult to say. The common view is that they are broad warps resulting from diastrophic causes. There is some evidence, in a few instances at least, to suggest that deep-seated igneous activity may have had something to do with their elevation. Suggestive of deep-seated igneous activity in connection with elevation of the Cincinnati arch and Nashville dome are the several cryptovolcanic structures around their flanks.

TOPOGRAPHIC EXPRESSION OF DOMES

The topographic expression of a domal structure depends primarily upon the length of time it has been undergoing dissection and the angle of dip

FIG. 9.5. Vertical photo of domed and faulted Cretaceous rocks in southwest Phillips County, Montana. (Photo by Jack Ammann, Photogrammetric Engineers.)

of the involved beds. Practically no large undissected domes exist, and the few examples of young undissected domes are mainly small laccolithic domes, salt domes, or domes resulting from localized late Pleistocene diastrophism, such as Western Coyote Hills dome in the Los Angeles Basin. Most domes exhibit at least a mature degree of dissection. The initial drainage on a small dome may well display a radial pattern, although, if there are through-going streams, this pattern may not be too evident. When dissection is well-advanced, annular or trellis patterns may become evident around the flanks.

If alternating weak and strong rock exist on the flanks of a dome, encircling ridges are likely to develop as dissection proceeds; they will take the form of hogbacks where dips are steep and of homoclinal ridges or cuestas where dips are moderate or slight. Between the hogbacks and homoclinal ridges formed by resistant rocks, there will be subsequent strike valleys on the belts

FIG. 9.6.   Hogbacks on the flanks of Hillers Peak, Henry Mountains, Utah.   (Photo by J. S. Shelton and R. C. Frampton.)

of weak rock. A belt of particularly weak rock may be reflected as a conspicuous lowland around a dome. Several alternating strong and weak beds will give rise to a series of concentric ridges and valleys with a combination of annular and trellis drainage patterns. In the course of headward extension of subsequent valleys, numerous instances of capture of the lower courses of the original radial streams may be expected. Their former routes will be suggested by wind gaps in the encircling hogbacks or homoclinal ridges. Short obsequent streams will descend the front slopes, and resequent streams the back slopes of the ridges to join subsequent streams in strike valleys.

Erosion of a dome may expose either strong or weak rock at its core. If weak rock is exposed, the central part of the dome may become a topographic basin; then we find an *inversion of topography* or a structural high expressed as a topographic low.   Although the idea of inversion of topography was applied originally by Davis to areas where the site of a valley later becomes a ridge, as may happen when a lava flow down a valley subsequently gives rise to a ridge, the expression has come to have a broader connotation and is now applied to any locality where geologic structure and topography do

FIG. 9.7.   A subsequent valley along the contact of granite and sedimentary rocks, Bighorn Mountains, Wyoming.   (Photo by C. L. Heald.)

not correlate.   Inversion of topography is common on many of the small salt domes.   If the rocks exposed at the center of a dome are more resistant than those that are being removed, the topography may become hilly or mountainous.   An initial stage of this condition is well-shown in the Ozark dome, where the St. Francois Mountains have developed upon a partially exposed granitic core.   The central portion of the Black Hills is another example of a mountainous domal core.

EXAMPLES OF DOMAL TOPOGRAPHY

West Point or Butler salt dome.   The West Point or Butler salt dome in southeastern Freestone County, Texas, is a good example of the type of topography that develops upon a salt dome.   Both DeGolyer (1919) and

Powers (1920) have described it as one of the most perfect and symmetrical domes both geologically and topographically in North America. It is about 8000 feet in diameter and almost perfectly circular in outline and is surrounded by a ring-shaped or annular valley formed by two streams which

FIG. 9.8.   The eroded Grenville dome, south of Sinclair, Wyoming.   (Photo by Jack Ammann, Photogrammetric Engineers.)

follow the strike of the rock around the dome.   The conspicuousness of this valley is increased by the fact that its vegetation is coarse marsh grass which allows open vistas.   This encircling valley is bounded on its outer margin by hills which present steep infacing scarps and have long, gentle, outward dip slopes.   The highest point on the dome is approximately 100 feet above the lowest point in the enclosing valley.   The top of the dome is flat, and drainage is by means of short channel-like ravines which join the annular valley at nearly right angles.   There are numerous springs and seepages around the valley. Some are briny and sulfurous, but most of them are moderately fresh. Salt incrustations mark their sites. They appear to be artesian springs from the Wilcox (Eocene) formation which caps the center of the dome. The time of the formation of the dome is uncertain. The oldest beds involved are probably of Cretaceous age, and the youngest beds are Eocene in age.

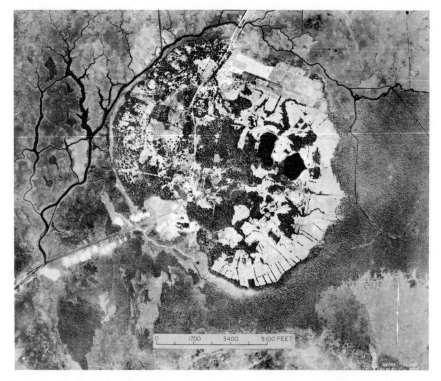

FIG. 9.9.   Vertical photo of Avery Island salt dome, about 10 miles southwest of New Iberia, Louisiana.   (Photo by Edgar Tobin Aerial Surveys.)

DeGolyer thought that its formation may have taken place in Quaternary or Recent time.

Kettleman Hills domes.   A good example of a young domal structure which is maturely dissected is found in the Kettleman Hills area at the west side of the San Joaquin Valley in California (Woodring, Stewart, and Richards, 1940). Structurally the Kettleman Hills consist of three elongated anticlines known as North, Middle, and South Domes, extending in a northwest-southeast direction and arranged en echelon, each fold being offset to the west of the one immediately north of it. Numerous faults, more than half of which are strike faults, are present. The total area involved in the folding is approximately 30 miles long and 5 miles wide. The folding began sometime during the Pleistocene, for the Tulare formation of upper Pliocene and Pleistocene (?) age was involved in it. Yet in the relatively short period of time which has elapsed since the folds were produced there has developed

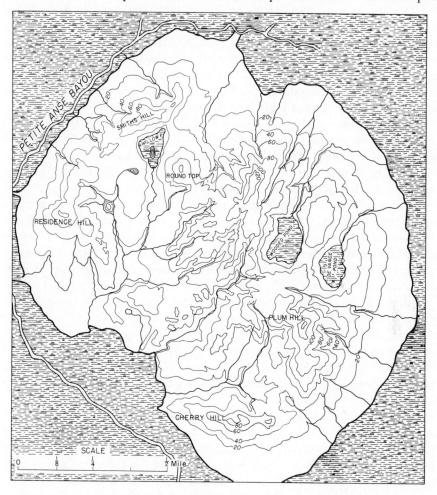

FIG. 9.10. The topographic expression of Avery Island salt dome. (After A. C. Veatch.)

a degree of dissection which has resulted in a nice adjustment of topography to geologic structure. There are two concentric rows of homoclinal ridges. Most of the streams are radial consequent streams which flow down the dip slopes roughly at right angles to the strike of the beds, but lesser strike valleys have developed on the weaker beds. One stream, however, flows through the hills at Avenal Gap between Middle and South Domes and is apparently an authentic example of an antecedent stream.   An old erosion surface at El Prado, near the north end of the Kettleman Hills, suggests that the area was probably base-leveled prior to its upfolding.   As a result of the lightness of the rainfall, most of the streams are intermittent and are forming alluvial fans cut by fanhead trenches where they empty onto adjoining plains.   Many of the valleys show an asymmetry of cross profile, even at this early stage of development, in which the north and northwest slopes are less steep than the opposite sides because of the difference in exposure.   Several instances of stream piracy are recognizable.   The most notable was the capture of the upper drainage of Arroyo Degallado by Arroyo Robador.

Nashville dome.   The Nashville dome, in central Tennessee, is a good example of the type of structure that we have designated as a broad domal warp.   It is especially interesting because it displays an inversion of topography.   The central part of the dome is now a topographic basin known as the Nashville Basin.   This development was made possible through removal by erosion of the resistant Mississippian sandstones and cherty limestones which once extended across the dome.   With their removal, following the Tertiary erosion cycle which produced the Lexington (Highland Rim) peneplain, there were exposed at the center of the dome weaker Ordovician limestones and shales on which the Cumberland River and its tributaries have developed a basin some 120 miles long and 60 miles wide.   This lowland is encircled by an escarpment known as the Highland Rim.   The central basin is by no means level.   Its floor varies in elevation by as much as 200 feet, and there are numerous monadnock-like hills which rise above its general level as much as 100 feet.   The floor of the Nashville Basin is probably best interpreted as a local strath terrace of probable Pleistocene age below the Lexington peneplain.   The Cumberland River at present is some 100 to 150 feet below this strath surface.

Black Hills.   The Black Hills are probably the best-known example of a maturely dissected domal structure with an ancient crystalline core around whose flanks are steeply dipping sedimentary rocks.   The crystalline rocks which form the core of the Black Hill uplift are of Pre-Cambrian age and far antedate the uplift that produced the present domal structure.   The rocks involved in the Black Hills uplift range in age from Pre-Cambrian to Cretaceous.   The uplift of the dome is believed to have begun in early Tertiary or possibly late Cretaceous time and to have continued intermittently down to late Quaternary time.   There were intrusions of igneous rocks during the Tertiary which produced several laccolithic domal structures, chiefly along the northern flank of the Black Hills, but these are lesser structures superposed upon the much larger Black Hills dome.

The Black Hills uplift (Darton and Paige, 1925) is elliptical in plan, being approximately 125 miles long and 65 miles wide.   The folding is

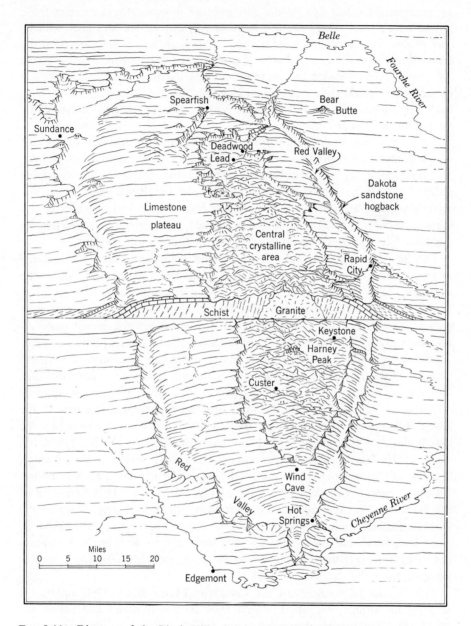

FIG. 9.11. Diagram of the Black Hills showing the main topographic regions on this broad domal structure. (After A. N. Strahler, *Physical Geography,* John Wiley & Sons.)

asymmetrical, for the dips on the east side are in general greater than those on the west flank.   There are five more or less distinct sections each with its own distinctive geologic and geomorphic characteristics.   These are from the center outward: (*a*) a central core of granitic and metamorphic rocks; (*b*) a limestone plateau outside the central core; (*c*) the Red Valley which nearly encircles the Black Hills but is best developed on the east side; (*d*) a hog-back ridge which lies outside the Red Valley; and (*e*) a belt of laccolithic mountains at the north edge.

The central core of crystalline rocks, which extends north-south slightly east of the center of the uplift, consists of a complex of Pre-Cambrian granites and metasediments which have been exposed by erosion of the overlying sediments.   Harney Peak, the highest peak in the Black Hills, is in this section and has an altitude of 7242 feet.   Most of the higher peaks in this section range from 5000 to 6600 feet in altitude, but between them are numer-ous park-like stretches above the canyons which cross this section.

The limestone plateau forms a rim around the granitic core but is much more extensive on the west than on the east side because of lower dips on the west side.   This plateau is developed mainly upon limestones of Mississippian, Pennsylvanian, and Permian age.   Its general altitude on the west is actually greater than that of the central core but its highest point does not quite at-tain the altitude of Harney Peak.   On the west, the limestone plateau is 15 to 20 miles wide and is relatively level and has the characteristics of a cuesta with an eastward-facing escarpment, but on the east side, where the dips are steeper, it is more a homoclinal ridge with a steep infacing escarpment.

The Red Valley is one of the most striking topographic features of the Black Hills uplift.   It encircles the Black Hills, but it is more striking on the east because of the steeper dips there.   It is developed upon the bright red shales of the Triassic Spearfish formation.   Its average width is about 2 miles, and the bright red soils and scarcity of trees make it a conspicuous feature. Its inner margin is marked by the dip slopes of the uppermost limestone of the limestone plateau, and its outer margin by the abrupt inner scarp of the Dakota hogback.   Major streams cross the Red Valley rather than follow it. Thus it appears that these streams have had little to do with its formation but that it is largely the product of erosion by intermittent subsequent streams working along the strike of the rock.

The Dakota hogback forms the outer rim of the Black Hills uplift.   It presents an abrupt inner face, which rises several hundred feet above the Red Valley, and a more gentle backslope, which descends to the surrounding plains.   On the west, where the dips are gentle, the hogback takes on the characteristics of a cuesta.   Numerous streams cross the Dakota hogback through conspicuous watergaps.

Numerous laccolithic mountains along the northern edge constitute a separate geologic, if not geomorphic, section. They are local domal structures which exhibit varying degrees of dissection. Green Mountain, or Little Sundance Mountain, has suffered so little dissection that the igneous core is not yet exposed; others, like the Devil's Tower or Mato Tepee, have only the igneous core left. It is not certain whether the Devil's Tower is a volcanic plug or a restricted laccolith. Its topographic form suggests a plug, but the vertical jointing has been thought by some to suggest a small laccolith. Dutton and Schwartz (1936) concluded that there are really three joint systems present and that their relationships were those that would be expected in a shaft-like mass such as a plug.

During the Tertiary, the Black Hills were undergoing erosion and were shedding their waste onto the adjacent plains. Intermittent upwarping took place, but there were periods of relative stillstand which are marked by erosional surfaces and terraces (Fillman, 1929). Three partial cycles at least are indicated in the present topography of the Black Hills and adjacent plains. These have been designated the Mountain Meadow, Rapid, and Sturgis cycles. The oldest or Mountain Meadow cycle is represented by numerous park-like areas in the central Black Hills. This erosional surface is one of late maturity rather than of old age. A mid-Oligocene age was ascribed to it because gravels found upon it contain fossils which have been correlated with those in the Oligocene White River beds in the Bad Lands to the east.

Uplift in the Oligocene varying from 100 feet in the plains to as much as 2500 to 3000 feet in the central Black Hills inaugurated the Rapid cycle. This cycle is represented by cobble and gravel-capped terraces about 100 feet above present stream floors. In the plains area to the east the Rapid surface has the characteristics of a local peneplain. Some 50 feet below the Rapid surface and a similar vertical distance above the present floodplain is another gravel and cobble-capped terrace which was cut during the Sturgis partial cycle. The Black Hills exhibit no evidence of Pleistocene glaciation.

The Weald. The Weald region of southeastern England and its continuation across the Straits of Dover into the Boulennais region of France is a well-known dome (Wooldridge and Morgan, 1937) which, as indicated by its accordant hilltops, is apparently in its second erosion cycle. Extending around the Weald is the more or less continuous Chalk escarpment known on the south as South Downs and on the north as North Downs. At the north and west a less conspicuous escarpment on the Greensand lies within the Chalk escarpment and between the two is a lowland known as The Vale. Within the hogback upon the Greensands is another lowland on the Weald clay, and the central area is a region of hills developed upon the Hastings sands of Jurassic age. The expected radial drainage pattern is more or less evident, and streams cut through the escarpments in a series of water gaps. Many of the tributary streams are subsequent streams which have developed along the strike of the beds.

THE GEOMORPHIC CYCLE ON FOLDED STRATA

In contrast to domal structures in which local folding of strata is either lacking or has had minor influence on the development of topography are those areas where the rocks have been intensely folded and exhibit true mountain structure. These are former geosynclinal regions in which sediments long accumulated and then were subjected to complex folding and, in many areas, complex faulting, following which they were uplifted and exposed to erosion. The following discussion pertains primarily to those areas in which close folding is dominant. Faulting may have accompanied folding, but it does not have major expression in the topography. The topographic effects of faulting will be discussed in the next chapter.

The type of geologic structure with which we are now concerned and its resultant topography are so well exemplified in the Ridge and Valley province of the Appalachian region of eastern United States that they are frequently described as Appalachian structure and topography. The outstanding feature of the structure is the existence of a series of roughly parallel folds varying from simple, open folds to complex, tightly compressed or overturned and faulted folds. Plunging anticlines and synclines normally alternate and are frequently arranged en echelon.

Initial stage. It is doubtful if the initial stage of topographic development upon folded structures can be depicted except in hypothetical terms, for all known examples of such topography have experienced previous cycles of erosion and are thus examples of multicyclic topography. Even in the Jura Mountains, where numerous examples of coincidence of structure and topography exist, the evidence seems to indicate that they are second-cycle mountains. The conventional idea that the original topography directly depicted the structure is predicated upon an assumption of rapid uplift of the folded structures with little or no accompanying erosion. More likely vast erosion accompanied uplift, and breaching of anticlines took place early in the history of mountain growth. The great original height sometimes attributed to folded mountains, which is obtained by projecting the strata involved in the folds upward, probably is not realistic since it ignores erosion during uplift. The Appalachian Mountains may never have been much higher than they are today. What we see today in most folded mountain structures are the roots of ancient folded structures which have been exposed by the removal of thousands of feet of strata during many erosion cycles.

Adjustment of topography to folded structures. The outstanding geo-morphic phenomenon encountered in folded mountains is the intimate way in which geologic structure, along with varying rock resistance, have controlled the details of the topography. This is most excellently illustrated in the folded Appalachians, and, because this region is so well-known, we shall use it as a type example.

Truncation of folds during previous erosion cycles has exposed rocks of varying resistance. The rocks involved in the folds are chiefly conglomerates, sandstones, limestones, and shales. Conglomerates and sandstones are the resistant rocks and ridge-makers. Certain geologic formations are notable in this respect, namely, the Clinch and Tuscarora sandstones of Silurian age, the Pocono sandstone of Mississippian age, and the Pottsville sandstone of Pennsylvanian age. Locally, other formations may be significant ridge-form-ers. Limestones and shales are the weak rocks and usually give rise to valleys. Because the strata are folded and truncated, the same formation may repeat itself in a direction transverse to the axes of the folds and thus be responsible for a number of ridges.

Differential degradation of strong and weak beds causes the topography (particularly the ridges) to show the positions and outlines of the folds. Plunging of anticlines and synclines gives rise to a zigzag pattern in the ridges. Where two or more resistant beds are involved in a fold the result is a series of zigzag ridges.

The six possible topographic expressions of geologic structure commonly encountered are anticlinal valleys, anticlinal ridges, synclinal valleys, syn-clinal ridges, homoclinal valleys, and homoclinal ridges. A valley developed along the axis of an anticline is called an *anticlinal valley.* If the topography and structure coincide an *anticlinal ridge* will develop along the axis of an anticline. A valley that follows the axis of a syncline is called a *synclinal valley,* and a ridge upon a syncline is a *synclinal ridge.* Anticlinal valleys and synclinal ridges represent inversion of topography. *Homoclinal ridges* develop upon the dipping beds on the flanks of anticlines and synclines. They are the most common and striking land forms encountered in areas of folded rocks. A belt of weak rock between resistant beds may give rise to a *homoclinal valley.* Frequently homoclinal valleys are not so conspicuous as either anti-clinal or synclinal valleys because they are likely to be minor rather than major drainage lines.

Major streams may have courses transverse to the regional structures, but their tributaries are more likely to flow in strike valleys with a resulting trellis drainage pattern. Where streams cut through ridges they do so in water gaps, such as the famous Delaware Water Gap, where the Delaware River cuts through the Kittatinny Mountain near Stroudsburg, Pennsylvania. Stream piracies have produced hundreds of wind gaps in the ridges through which transverse streams formerly flowed.   Geologic conditions are ideal for stream piracy because transverse streams flowing through water gaps in hard-rock ridges are retarded in downcutting, whereas subsequent streams on the belts of weaker rock are not so handicapped.

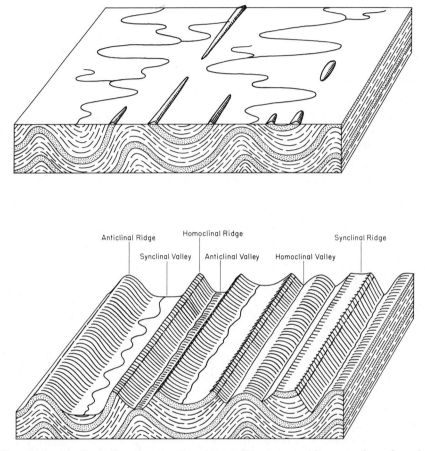

Fig. 9.12.   Idealized diagrams showing the possible topographic expression of eroded anticlines and synclines after the folds have been truncated.   (Drawing by W. C. Heisterkamp.)

   The topographic expression of a particular fold depends largely upon where the erosion surface which cuts across it passes with respect to the resistant and non-resistant beds involved in the fold.   If in a previous geomorphic cycle a peneplain has developed across an anticline in such a position that there are weak beds at the surface but a resistant bed near the surface, uplift of the peneplain will be followed by stripping of the weak beds and exposure of the resistant stratum.   An anticlinal mountain or ridge then will be formed on the hard rock.   Had, however, the strong bed been truncated previously, during the ensuing cycle there would be weak rock exposed along the axis of the anticline, and an anticlinal valley would have resulted.   By analogous reasoning it is evident that, if the erosion surface lay above a resistant bed in a syncline, with inauguration of a new cycle or erosion this resistant bed

Fig. 9.13.   The Delaware Water Gap through Kittatinny Mountain, near Stroudsburg, Pennsylvania.   (Photo by Chicago Aerial Survey Co.)

would give rise to a synclinal ridge.   If, however, erosion during the first cycle had removed the resistant bed in the trough of the syncline, conditions would then be favorable to the development of a synclinal valley in a later cycle.

CRITERIA FOR RELATING TOPOGRAPHY AND STRUCTURE

An attempt has been made above to sketch some of the more outstanding correlations which may exist between structure and topography in a region of folded rocks, but what has been said applies primarily to the broader aspects of the topography.   These relationships are often grasped only after a consideration of many details which may or may not be obvious.   An attempt is made below to formulate a number of criteria which may be applied singly and collectively in areas of folded rocks for the purpose of determining how geologic structures affect topographic form and, more particularly, how topographic forms may give clues to geologic structures. The criteria that are listed are presumably applicable to interpretation of topographic maps or aerial photographs. Some are simple restatements of facts already dis-

Fig. 9.14.   The Virgin anticline, southern Utah, a breached anticline marked by homo-clinal ridges and valleys.   (Photo by J. S. Shelton and R. C. Frampton.)

cussed, and others represent application of general principles previously considered. It is not presumed that all of them will be everywhere readily or invariably diagnostic.

1. Homoclinal ridges that develop after the breaching of a pitching anticline will converge in the direction in which the anticline plunges.

2. Homoclinal ridges formed by the erosion of a plunging syncline will converge in the direction opposite to that in which the syncline plunges.

3. Homoclinal ridges with infacing escarpments mark anticlines.   The plan made by these ridges is sometimes referred to as cigar-shaped.

4. Homoclinal ridges or plateau-like areas with outfacing escarpments mark synclines.   The outline made by such ridges is frequently described as canoe-shaped.

5. Alternating homoclinal ridges and valleys develop where there is a series of alternating strong and weak beds on the flanks of anticlines and synclines.

6. The noses of anticlines may be gentle or steep, depending upon their rates of plunge, but typically they are smooth and gently rounded.

7. The noses of pitching synclines are usually more abrupt, narrow, and scarp-like than those of anticlines.

8. Anticlinal mountains typically have rounded summits and smooth dip slopes.

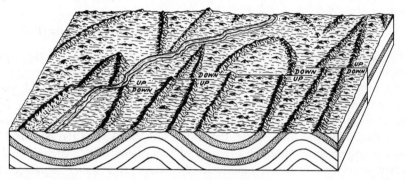

FIG. 9.15. Schematic diagram to show the effects of transverse normal faults upon the topographic expression of the noses of plunging anticlines and synclines. (Drawing by W. C. Heisterkamp.)

9. Synclinal mountains may vary from elongate narrow ridges to broad plateau-like expanses, but typically they are broader than anticlinal mountains and have steep scarp-like slopes on their flanks.

10. If there is a transverse fault across the nose of a fold and subequality of ridge summits on the two sides of the fault:

a. Lowering of the nose of a syncline will result in its being widened and extended at the fault and the homoclinal ridges marking the outline of the syncline will be "outset" at the fault.

b. Raising of the nose of a syncline will result in its being shortened and narrowed at the fault and the homoclinal ridges marking its outline will be "inset."

c. Lowering of the nose of an anticline will result in its being shortened and narrowed and the homoclinal ridges marking its outline will be "inset."

d. Raising of the nose of an anticline will result in its being lengthened and widened and the homoclinal ridges marking its outline will be "outset."

The reasons for the topographic effects of transverse faults indicated above are that, when folded strata are raised or lowered by faulting, the beds brought to the level of erosion are in a part of the fold where they are either farther apart or closer together.   Downfaulting of the nose of an anticline would mean that the upper part of the fold, where the strata are closer together, unless the fold is isoclinal, are brought down to the level of erosion.   The result will be a shortening of the nose of the anticline and insetting of the ridge marking its outline.   Conversely, upfaulting of an anticlinal nose will bring the broader part of the fold up to the level of erosion and will thus extend the nose of the anticline and cause "outsetting" at the fault of the ridge marking the anticline.   It should be obvious that the effects on a syncline will be the reverse.

GEOMORPHIC HISTORY OF THE FOLDED APPALACHIANS

Probably more has been written about the geomorphology of the Appalachian Highlands than about that of any other comparable area in the world, except possibly the Alps, and yet many of the major geomorphic problems remain unsolved.   What Fenneman (1938) called the Appalachian Highland division of the United States includes four physiographic regions known as the Piedmont, the Blue Ridge, the Ridge and Valley, and the Appalachian Plateaus provinces.   Although the type of topography with which we are at present concerned is found mainly in the Ridge and Valley province, the geomorphic history of the whole area is so interrelated that it is impossible to discuss one province without considering the others.   Hence a brief description of each of these provinces seems desirable.   The two easternmost provinces, the Piedmont and the Blue Ridge, are often referred to as the "Older Appalachians," and the two western provinces as the "Newer Appalachians." The validity of these designations is becoming less certain as geologic work progresses, and it may well be that the terms older and newer as now used designate areas of well-metamorphosed rocks versus areas of unmetamorphosed rocks rather than significant age differences.

The rocks of the Appalachian Plateaus range in age from Mississippian to Permian, and they are gently downwarped into an elongate syncline or synclinorium which parallels the more closely folded structures of the Ridge and Valley province to the east.   The topographic boundary between the two provinces is a prominent scarp known at the north as the Allegheny Front and at the south as the Cumberland Front.   In the Ridge and Valley province there are rocks representing all the Paleozoic systems from the Cambrian to the Pennsylvanian and having a total thickness of 30,000 to 40,000 feet. These rocks are intensely folded and there are also numerous low-angle thrust faults, particularly in the southern half.   The rocks of the Blue Ridge province are Lower Cambrian or older metasedimentary and metaigneous rocks.   The rocks of the Piedmont province are mainly metamorphic and plutonic rocks, but there are numerous areas of infolded metasedimentary rocks of early Paleozoic age, and in the northern part of the Piedmont there are several areas of downfolded and downfaulted Triassic sediments.

It is believed by many geologists that the Blue Ridge and Piedmont provinces are part of the ancient borderland of Appalachia which lay to the east of the Appalachian geosyncline and contributed sediments to it throughout the Paleozoic era. King (1950) has questioned the existence of such a persistent borderland. He believed that the evidence, in the southern Appalachians at least, argued against this concept and supported more the idea of intermittent deposition of sediments, derived more likely from fold ridges which arose within the Appalachian geosyncline as a result of successive periods of mountain building during the Paleozoic. Stratigraphic studies do not confirm the idea that the sediments always came from the east, for some formations thicken to the west. Neither did he believe that the deformation of the rocks in the Ridge and Valley province have resulted solely from the Appalachian revolution at the close of the Paleozoic. The orogeny at the close of the Paleozoic seemed to be only one of several which have affected the area. Mountain building seems to have affected the area as early as the Ordovician and to have continued intermittently throughout the Paleozoic, and the so-called Appalachian revolution was probably only the culminating phase of these various orogenies.

The Ridge and Valley province exhibits many outstanding geomorphic phenomena of which some of the more notable are: (*a*) marked parallelism of ridges and valleys in a general northeast-southwest direction; (*b*) a striking influence of alternating weak and strong beds upon topographic forms; (*c*) several major streams, such as the Susquehanna, Delaware, and Potomac with valleys transverse to the regional structure; (*d*) lesser stream valleys controlled by the geologic structure with striking development of trellis drainage patterns; (*e*) many ridges that display a noticeable accordance or subaccordance of summit levels, which suggests that their summits may be remnants of former erosional surfaces of cyclical origin; and (*f*) hundreds of wind gaps which attest to numerous drainage changes and in many instances exhibit a striking alignment suggestive of former transverse stream courses which have been dismembered.

Although there are many geomorphic problems in the folded Appalachians, and most of them contribute in some degree to the unraveling of their geomorphic history, we shall consider only a few of the more general problems. Any attempt at a synthesis of the geomorphic history of the area will of necessity try to answer the following questions. How many erosion cycles are represented in the area and what are their ages? What is the explanation of the aligned wind and water gaps? And, most difficult of all, how was the transverse drainage of the major streams reversed from an original northwestward direction into the Appalachian geosyncline to the present southeastward direction? This last problem is at the core of all attempts to interpret the region's geomorphic history. It is generally assumed that during the Paleozoic era drainage was from the continent of Appalachia to the east toward the Appalachian geosyncline to the northwest of it. The stratigraphy of the Mississippian and Pennsylvanian formations in the Appalachian geosyncline indicates that without much doubt these sediments came from the east, because the formations are thickest at the east and thin toward the west and also show a lateral change of facies from coarse materials on the east to finer sediments toward the west. As King has suggested, this may not have been so throughout the Paleozoic.

THEORIES TO ACCOUNT FOR REVERSAL OF DRAINAGE

Theory of Davis. In his classical paper "The rivers and valleys of Pennsylvania," published in 1909, Davis presented the first really systematic attempt to explain the reversal of drainage which is presumed to have taken place in the folded Appalachians. Although this paper dealt only with the northern part of the area, its principal thesis was believed to be applicable to other parts. He thought that the drainage during the Permian period was a consequent northwestward drainage which had developed across the growing folds produced by the Appalachian revolution. The major stream of the area was a stream which he called the Anthracite River. He thought that reversal of the upper part of this river to the southeast took place as the result of warping during the Triassic and postulated a series of captures of its headwaters by shorter, steeper-gradient streams flowing southeastward into the downwarped and downfaulted Triassic lowlands. Great significance was attached to the role that the Newark depression in New Jersey and Pennsylvania played during Triassic time in initiating a reversal of drainage to the southeast as Appalachia sank beneath the sea. According to Davis, the middle Susquehanna and its tributaries, and the upper portions of the Schuylkill and Lehigh Rivers are descendants of Permian rivers which flowed northwestward and were diverted by a series of involved stream captures to the shorter Triassic rivers flowing to the southeast. The lower portions of these rivers across the Triassic lowland and Piedmont province were believed to be of much younger age and were considered to be consequent stream courses initiated upon a former extension of a Cretaceous cover over a peneplain surface which he thought to be of Jurassic-Cretaceous age. Thus the portions of the streams across the Piedmont were superposed upon the complex structure of this region. The large westward-flowing streams in the Appalachian Plateaus were thought to be descendants of Permian streams which have suffered little change except for loss of their headwater portions through piracy to streams flowing eastward into the Triassic lowland. The region was thought to have been peneplained during a Jurassic-Cretaceous erosion cycle, and the higher mountain summits were considered to be remnants of this Cretaceous peneplain. He thought that this peneplain extended seaward beneath the Cretaceous and younger sediments of the coastal plain. Only one peneplain was recognized and it was called the Schooley peneplain. The Tertiary cycle was believed to have been incomplete and to have resulted only in the development of open valleys or Tertiary base level lowlands, as Davis called them (straths as we would now call them). Uplift at the close of the Tertiary inaugurated the Quaternary cycle which has produced the present entrenchment of streams. He admitted that his postulated series of stream captures left many events shrouded in doubt and that much further work was needed to fill out the series of events suggested by him.

Meyerhoff-Olmsted theory. The basic thesis of the suggested origin of Appalachian drainage as put forth by Meyerhoff and Olmsted (1936) is that the present southeastward-flowing streams are the direct lineal descendants of consequent streams initiated during the Permian period. The divide between the drainage to the northwest and the southeast presumably was established in the Ridge and Valley province in Permian time on great low-angle thrust sheets or overturned folds which apexed there. Eastward consequent drainage was established down the slopes of these thrust sheets and folds before the epoch of Newark faulting and folding by way of various structural sags and fault zones. Their contention is that, despite Triassic and post-Triassic adjustments which have taken place, the streams still essentially follow their early Mesozoic courses across the belt of folded rocks. Great stress was laid upon the supposed coincidence between present stream courses and structural sags and faulted zones. It was noted that each of the major streams crosses a belt of Triassic rocks on its way to the sea, which was thought to indicate that stream courses were established early through the Triassic belts and that the Triassic sediments were deposited by streams from the northwest. Mackin (1938) and Strahler (1945) have pointed out some of the difficulties encountered by this theory. Among the difficulties are: the doubt that exists in regard to whether the overthrusts, if they do exist, extended beyond the Blue Ridge; the doubtful Permian age of the overthrusts (an Ordovician age is indicated for some); uncertainty that the Triassic sediments were deposited by southeastward-flowing streams; the fact that the postulated divide was not far enough west to account for all the transverse drainage; and the question whether in an overturned fold or thrust sheet the slope would have been to the east and the head of the fold or thrust sheet the highest part.

Superposition theory of Johnson. The two theories outlined above have one thing in common. They assume that present Appalachian drainage has a direct lineal connection with stream courses dating back to the Permian, although each recognizes that to a greater or less degree modification of the Permian drainage has taken place. Johnson (1931) avoided most of the difficulties that these theories encountered by assuming that a Cretaceous marine transgression buried and destroyed the old drainage lines. The present streams were considered to be descendants of a system of southeastward-flowing consequent streams which developed upon a Cretaceous covermass following withdrawal of the Cretaceous sea. The present anomalous relationships between transverse streams and geologic structures were accounted for by assuming that the streams have been let down onto the structures which they cross. Some of the difficulties encountered by other theories are thus avoided, but lack of evidence of the covermass from which the drainage could have been superposed is a decided weakness in this theory. The stream captures and reversals postulated by Davis and others seemed improbable to Johnson. It further seemed to him that the peneplain surface which extends below the Cretaceous and Tertiary sediments of the coastal plain would if projected inland pass far above the summits of the highest mountains and that the Schooley peneplain of Davis was in reality a younger peneplain than that buried beneath the coastal plain. The relationship between the two surfaces is that which previously has been described as a morvan (see p. 194). The peneplain beneath the coastal plain was called the Fall Zone peneplain and assigned a probable Jurassic age, and the Schooley peneplain in the Appalachians was thought to be of Tertiary age. The probable steps in the evolution of Appalachian topography as outlined by Johnson were:

1. Formation in early Paleozoic time of the Appalachian geosyncline.

2. Intermittent accumulation in the Appalachian geosyncline during the Paleozoic era of sediments derived from Appalachia to the east.

3. The Appalachian Revolution. This probably started as far back as the Mississippian but culminated in the Permian period. It resulted in the folding, faulting, and uplift of the sediments in the Appalachian geosyncline.

4. Extensive faulting accompanied by igneous intrusions and extrusions during the Triassic period.

5. Erosion during the Jurassic period, culminating in the production of the Fall Zone peneplain.

6. Downtilting to the east of the Fall Zone peneplain and westward encroachment of the Cretaceous sea with resulting burial of the Fall Zone peneplain beneath a veneer of Cretaceous sediments.

7. Arching of the Fall Zone peneplain and its Cretaceous cover with resulting regional superposition of southeastward-flowing consequent streams.

8. A mid-Tertiary (probably Miocene and early Pliocene) erosion cycle which resulted in the formation of the Schooley peneplain.

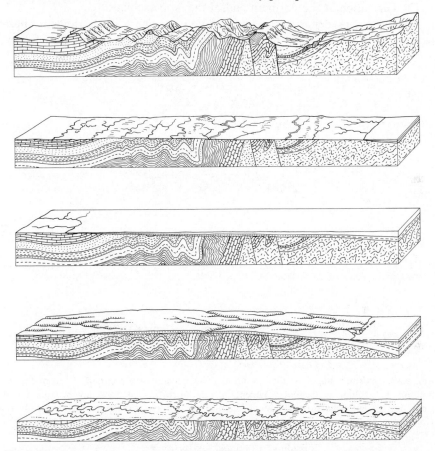

Fig. 9.16. Stages in the evolution of Appalachian topography, according to the theory of regional superposition of drainage. From top to bottom: 1. Rejuvenated Appalachians in post-Newark time. 2. Formation of Fall Zone peneplain. 3. Encroachment of Cretaceous sea over area. 4. Arching of Fall Zone peneplain and its Cretaceous cover and superposition of southeastward drainage. 5. Development of Schooley peneplain. (After Douglas Johnson, by permission of Columbia University Press.)

9. Uplift of the Schooley peneplain and inauguration of another erosion cycle with development in late Tertiary time of a second peneplain called the Harrisburg.

10. Uplift of the Harrisburg peneplain and ensuing dissection and formation of local peneplains or straths on belts of weak rock. (The Somerville peneplain of New Jersey is an example.)

11. Renewed uplift with inauguration of the present cycle of erosion and resulting stream entrenchment.

12. Glaciation at the northern end of the Appalachians.

No Cretaceous deposits have been found anywhere in the folded Appalachians, and according to this theory they should have extended as far in-

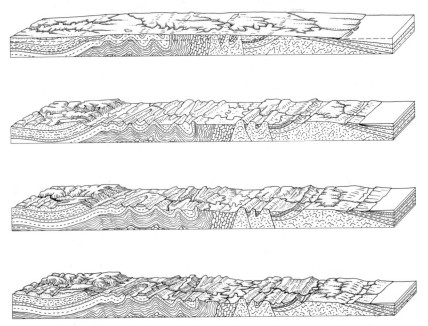

Fig. 9.17. Further stages in the evolution of Appalachian topography. From top to bottom: 1. Arching of Schooley peneplain. 2. Dissection of Schooley peneplain and formation of Harrisburg peneplain on belts of weak rock. 3. Uplift and dissection of Harrisburg peneplain and formation of local straths on weak-rock belts. 4. Renewal of uplift and dissection to give present topography. (After Douglas Johnson, by permission of Columbia University Press.)

land as the Appalachian Plateaus. The explanation offered by Johnson for their absence was that they were deposited upon a surface (the Fall Zone peneplain) which, if projected inland from the coastal plain, would rise far above the summits of the highest mountains and hence they were destroyed during the Schooley, Harrisburg, and later cycles. This is a logical explanation for the absence of Cretaceous deposits west of the coastal plain, but any interpretation of Appalachian geomorphic history which depends upon so questionable an assumption must of necessity remain a theory.

Von Engeln (1942) has suggested a partial modification of the events proposed by Johnson, so that following the development of the Fall Zone peneplain the sequence would be as follows: upwarping of the Fall Zone peneplain; production of the Schooley peneplain; submergence and covering of the Schooley peneplain with sediments; uplift of the covered Schooley peneplain; and superposition of streams upon this surface; and then the development of the Harrisburg peneplain and later erosional surfaces. Johnson considered this possible interpretation, but concluded that the complete lack of remnants of a sedimentary cover indicated that the cover must have been laid down on an erosional surface much older than the Schooley. It would seem that deposits laid down as recently as the Pliocene (assuming the Schooley surface to be of that age) might more likely be expected to have escaped erosion than those deposited during the Cretaceous, and hence their absence would cast even greater doubt upon the probability of the drainage having been superposed. The excellent adjustment of drainage to structure, except for major transverse streams, would seem to require a longer time than would be available after removal of the sedimentary cover over the Schooley peneplain. Furthermore, the many stream diversions which have taken place to form the present drainage lines might be expected to be more discernible than they are. Most of the drainage changes which have taken place to produce the present streams are rather vague, and it thus seems likely that the adjustments of stream courses to geologic structure have extended over a much longer period than post-Schooley time.

Progressive stream piracy theory of Thompson. Thompson (1939) believed that it was not necessary to resort to superposition to account for the reversal of drainage which most students of Appalachian geomorphic history presume has taken place. Rather, he thought that it could be explained by progressive stream piracy. He assumed that the drainage divide following the Appalachian revolution was in the Blue Ridge. Because of their shorter courses and steeper gradients, the streams which flowed to the Atlantic had an advantage over the streams which pursued much longer routes to the Gulf of Mexico. Progressive stream piracy was thought to have shifted the divide northwestward by amounts ranging from 10 to 20 miles at the south to as much as 80 to 100 miles in the northern Appalachians. The greater shifting of the divide at the north was thought to have been possible because the resistant rocks of the Blue Ridge are much narrower here than at the south and also because of the supposedly greater eastward tilting of the Atlantic border at the north. Thompson estimated that at its maximum this shifting represented a divide migration of only seven-tenths of a mile per million years.

He believed that present streams where they cross transverse structures do so to a large degree where the ridge-forming rocks are less resistant or narrower in width of outcrop and that there is not the striking lack of adjustment of transverse streams to structure which should be expected if their positions were chance locations resultant from superposition. The greatest difficulty encountered by this theory is to explain how a stream can effect piracy through a ridge of hard rock. Thompson recognized this and attempted to circumvent it by explaining the piracies as being effected not by direct attacks but more in the nature of flank attacks, whereby "the tributaries of the southeast-flowing streams undercut the little tributaries of their opponents." This was thought to have been possible because tributaries of a stream flowing by a roundabout route to the Gulf of Mexico would be at higher alti-

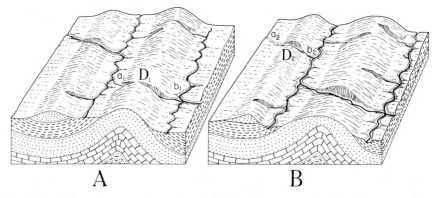

FIG. 9.18. Diagrams suggesting how stream piracy may have shifted westward the divide between drainage to the Atlantic and to the Gulf of Mexico and have thereby caused reversal of Appalachian drainage. (After H. D. Thompson.)

tudes than those of a stream flowing to the Atlantic. He further suggested that, where limestones were present, solution may have caused subterranean piracy.

Other theories. Bethune (1948) concluded that the geomorphic evidence in the Appalachians was such as to permit only two possible explanations of the transverse drainage, stream piracy or superposition. He discarded stream piracy as a possible explanation because the water and wind gaps do not correspond in position with points or zones of weakness caused either by faulting or highly jointed rock. He thought that Johnson's theory of superposition from a Cretaceous covermass rested upon negative evidence and suggested instead that the transverse drainage was superposed from extensive floodplains in downwarped areas on the Schooley peneplain. He believed that present streams are systematically consequent with respect to the slopes of the Schooley peneplain and thought that they originated at the time of upwarping of the Schooley peneplain. He further concluded that what has been called the Harrisburg peneplain was really parts of the Schooley peneplain which have been differentially lowered.

Still another possibility might be given brief consideration, although it has as yet only been barely suggested. If King (1950) is correct that Appalachia as a persistent borderland to the east of the Appalachian geosyncline is not in accordance with fact, then the possibility exists that the Susquehanna River may have flowed southeastward since the Triassic. His ideas regarding the nature of Appalachia were stated as follows: *

It seems doubtful now that such a persistent borderland ever existed, at least in the manner envisaged by its proponents. The main evidence adduced for it was the occurrence of great sheets of clastic sediments which spread northwestward across the area of the present Valley and Ridge province, from later Ordovician time onward. However, these clastic sediments were laid down intermittently and not continuously, and times of maximum deposition in one segment of the geosyncline do not correspond with maxima in others. More probably the clastic sediments were derived from fold ridges that arose in the interior zones of the Appalachian geosyncline during successive orogenic periods of Paleozoic time. These ridges may have been composed, not of basement rocks, but of materials from the inner zones of the geosyncline that were deformed and metamorphosed mainly in Paleozoic time.

Appalachia according to this concept may have consisted of large islands now here, now there, and if these islands of folded rock lay within the area of the folded Appalachians, then there is not the necessity of reversing the drainage from a land mass east of the folded Appalachians. Such a theory has not yet been formally proposed, but the idea is worth keeping in mind in attempting to arrive at a logical explanation of the evolution of Appalachian drainage.

PENEPLAINS OF THE APPALACHIAN REGION

There is much diversity of opinion as to the number and ages of peneplains in the Appalachian region. Ashley (1935) held that there is only one peneplain present and that the other topographic surfaces which have been called peneplains may be explained as the result of: differential lowering of that peneplain upon rocks of varying resistance, local base leveling of areas of weak rock, or stripping of areas of resistant rock. Bascom (1921) suggested that there are as many as five peneplains present and in addition five terraces below the youngest peneplain which have cyclical significance. Ages ranging from Jurassic to Pleistocene have been suggested for the supposed peneplains.

* Quoted by permission of the American Association of Petroleum Geologists from *Am. Assoc. Petroleum Geol., Bull. 34*, P. B. King, Tectonic framework of southeastern United States.

Probably the most widely held viewpoint is that there is good evidence for the existence of two peneplains to which the names Schooley and Harrisburg are most commonly applied. When Davis proposed the name Schooley for the older of these two peneplains, he assigned a Jurassic-Cretaceous age to it. Most students of the area are now inclined to doubt the existence of so ancient an erosion surface except for the peneplain beneath the coastal plain and the portion of it along the Fall Line which has been exhumed. The oldest peneplain remnants in the folded Appalachian region are generally considered to be of middle or late Tertiary age. The Kittatinny or Upland peneplain is frequently mentioned in discussions of Appalachian geomorphology. Some regard this as an older erosional surface above the Schooley peneplain, possibly equivalent to the Fall Zone peneplain beneath the coastal plain, but it is more commonly believed to be the equivalent of the Schooley. Wright (1942) concluded that the various terms such as Cretaceous, Kittatinny, and Upland which have been applied to supposed peneplains were probably equivalent to the surface that is now more commonly called the Schooley. It cannot be dogmatically stated that this is so, but preponderance of evidence seems to point toward this conclusion.

The Harrisburg cycle which followed the uplift of the Schooley peneplain was not a completed cycle or else evidence for the Schooley cycle would have been destroyed. East of the Blue Ridge, in the Piedmont, the Harrisburg peneplain is the most extensive peneplain surface, but within the folded Appalachians its development was largely restricted to belts of weak rocks. West of the Blue Ridge the Harrisburg peneplain is most extensively developed in the Great Valley, where it has considerable areal expanse, but along other valleys it is hardly more than a strath.

There is some evidence for a post-Harrisburg partial cycle because of the presence of terraces along stream courses and locally expansive lowland areas which have been interpreted by some as local peneplains on belts of weak rock. Examples of presumed local peneplains are the Somerville peneplain of New Jersey, the Coosa peneplain of eastern Tennessee, Georgia, and Alabama, and the peneplain in the Black Belt of Alabama. Two interpretations have been made of them. They have been considered post-Harrisburg local peneplains or straths upon belts of weak rock and lowlands produced by solutional lowering of the Harrisburg surface and hence not products of a later partial cycle. If they do represent local base-leveled plains or straths, their age is probably late Pliocene or early Pleistocene.

REFERENCES CITED IN TEXT

Ashley, G. H. (1935). Studies in Appalachian mountain structure, *Geol. Soc. Am., Bull. 46*, pp. 1395–1436.

REFERENCES 241

Bascom, Florence (1921). Cycles of erosion in the Piedmont province of Pennsylvania, *J. Geol., 29,* pp. 540–559.

Bethune, Pierre de (1948). Geomorphic studies in the Appalachians of Pennsylvania, *Am. J. Sci., 246,* pp. 1–22.

Bucher, W. H. (1936). Cryptovolcanic structures in the United States, *Rept. 16th Int. Geol. Cong.,* pp. 1055–1084.

Darton, N. H., and Sidney Paige (1925). Central Black Hills, *U. S. Geol. Survey, Folio 219,* 34 pp.

Davis, W. M. (1909). The rivers and valleys of Pennsylvania, *Geographical Essays,* pp. 413–484, Ginn and Co., New York. Also in *Nat. Geog. Mag.* (1899), *1,* pp. 183–253.

DeGolyer, E. L. (1919). The West Point, Texas, salt dome, Freestone County, *J. Geol., 27,* pp. 647–663.

DeGolyer, E. L. (1925). Origin of North American salt domes, *Am. Assoc. Petroleum Geol., Bull. 9,* pp. 831–872.

Dutton, C. E., and G. M. Schwartz (1936). Notes on the jointing of the Devil's Tower, Wyoming, *J. Geol., 44,* pp. 717–728.

Fenneman, N. M. (1938). *Physiography of Eastern United States,* pp. 121–123, McGraw-Hill Book Co., New York.

Fillman, Louise (1929). Cenozoic history of the northern Black Hills, *Univ. Iowa Stud. in Nat. Hist., 13,* No. 1, 50 pp.

Gilbert, G. K. (1877). Report on the geology of the Henry Mountains, *U. S. Geog. Geol. Survey Rocky Mt. Region,* pp. 18–98.

Goddard, E. N. (1950). Structure of the Judith Mountains, Montana, *Geol. Soc. Am., Bull. 61,* p. 1465.

Hunt, C. G. (1946). Guidebook to the geology of Utah, No. 1, *Utah Geol. Survey,* pp. 11–17.

Hurlburt, C. S., Jr., and D. T. Griggs (1939). Igneous rocks of the Highwood Mountains, Montana, Part I, The laccoliths, *Geol. Soc. Am., Bull. 50,* pp. 1043–1112.

Johnson, D. W. (1931). *Stream Sculpture on the Atlantic Slope,* Columbia University Press, New York, 142 pp.

King, P. B. (1950). Tectonic framework of southeastern United States, *Am. Assoc. Petroleum Geol., Bull. 34,* pp. 635–671.

Mackin, J. H. (1938). The origin of Appalachian drainage—a reply, *Am. J. Sci., 236,* pp. 27–53.

Meyerhoff, H. A., and E. W. Olmsted (1936). The origins of Appalachian drainages, *Am. J. Sci., 232,* pp. 21–41.

Powers, Sidney (1920). The Butler salt dome, Freestone County, Texas, *Am. J. Sci., 199,* pp. 127–142.

Powers, Sidney (1926). Interior salt domes of Texas, *Am. Assoc. Petroleum Geol., Bull. 10,* pp. 1–60.

Shrock, R. R., and C. A. Malott (1933). The Kentland area of disturbed Ordovician rocks in northwestern Indiana, *J. Geol., 41,* pp. 337–370.

Strahler, A. N. (1945). Hypotheses of stream development in the folded Appalachians of Pennsylvania, *Geol. Soc. Am., Bull. 56,* pp. 45–88.

Thompson, H. D. (1939). Drainage evolution in the southern Appalachians, *Geol. Soc. Am., Bull. 50,* pp. 1323–1356.

Von Engeln, O. D. (1942). *Geomorphology,* pp. 357–364, The Macmillan Co., New York.

Woodring, W. P., Ralph Stewart, and R. W. Richards (1940). Geology of the Kettleman Hills oil field, California, *U. S. Geol. Survey, Profess. Paper 195,* pp. 148–155.

Wooldridge, S. W., and R. S. Morgan (1937). *The Physical Basis of Geography,* pp. 240–247, Longmans, Green and Co., London.

Wright, F. J. (1942). Erosional history of the southern Appalachians, *J. Geomorph., 5,* pp. 151–161.

Additional References

Ashley, G. H. (1930). Age of the Appalachian peneplains, *Geol. Soc. Am., Bull. 41,* pp. 695–700.

Johnson, D. W. (1931). A theory of Appalachian geomorphic evolution, *J. Geol., 39,* pp. 497–508.

Renner, G. T. (1927). The physiographic interpretation of the Fall Line, *Geog. Rev., 17,* pp. 276–286.

Rich, J. L. (1939). A bird's eye cross section of the central Appalachian mountains and plateau: Washington to Cincinnati, *Geog. Rev., 29,* pp. 561–586.

Ver Steeg, K. A. (1942). A study in Appalachian physiography, *J. Geol., 50,* pp. 504–511.

Wright, F. J. (1925). The physiography of the upper James River basin in Virginia, *Va. Geol. Survey, Bull. 11,* pp. 11–55.

Wright, F. J. (1928). The erosional history of the Blue Ridge, *Denison Univ. J. Sci. Labs., 23,* pp. 321–344.

Wright, F. J. (1931). The older Appalachians of the South, *Denison Univ. J. Sci. Labs., 26,* pp. 143–269.

Wright, F. J. (1934). The newer Appalachians of the South (Part I), *Denison Univ. J. Sci. Labs., 29,* pp. 1–105.

Wright, F. J. (1936). The newer Appalachians of the South (Part II): south of the New River, *Denison Univ. J. Sci. Labs., 31,* pp. 93–142.

10 · Topography upon Faulted Structures

INTRODUCTION

Faults present many interesting and oftentimes complex geologic problems which constitute a considerable part of the field of structural geology. A geomorphologist is not concerned primarily with solving the stratigraphic and structural problems which faulting presents but rather with the topographic effects of faulting. Geomorphic features may aid, however, in the recognition of faults and, if carefully evaluated, may contribute to a correct interpretation of the diastrophic history of a region.

Faulting commonly has topographic expression because it elevates, lowers, tilts, or horizontally displaces blocks of the earth's crust along with their associated topographic features. Faulting may also create crushed or brecciated zones which are more easily eroded than surrounding rock and are thereby expressed topographically.

There are still two contrasting viewpoints as to whether faulting takes place rapidly enough to produce conspicuous land forms. One group of geologists is inclined to believe that degradation goes on more rapidly than faulting and that with the exception of such minor and ephemeral forms as *fault scarplets* erosion erases features produced by faulting about as rapidly as they form. According to this viewpoint, most topographic features that are associated with faults have resulted from erosion upon or along ancient or recent faults and thus were not directly fault-produced. Opposed to this view is the belief that faulting may go on rapidly enough so that weathering and erosion do not mask its effects and that there may exist major geomorphic features that are primarily fault-produced rather than the effect of differential erosion upon faulted structures. Actually, the two viewpoints are not quite so sharply opposed as the preceding statements would imply, but there are these two schools of thought regarding the interpretation of topography in faulted terrains. In general, it is the geologists in the stable regions who are skeptical of the existence of major land forms of tectonic origin, whereas those from the mobile belts have considered them common features. The major problem encountered in regions where faults exist is not so much recognition of the presence of faults as it is determining whether the land forms are tectonic forms produced directly by faulting or whether they owe their existence to erosion upon faulted structures.

Faulting may be either high-angle or low-angle in nature and in both types may be normal or reverse, although low-angle normal faulting is rather rare. In certain areas, there is evidence of a type of faulting now commonly called

FIG. 10.1. A fault cicatrice or scar in the Ruahine Range, New Zealand. The recently formed fault scarplet faces in toward the fault scarp and is about 20 feet high. (Photo by R. J. Waghorn.)

transcurrent in which there has been mainly strike-slip movement, resulting more in horizontal than in vertical displacement. In many respects the effects of high-angle faulting are much the same whether the faults are normal or reverse faults. Topographic expression is usually most striking for high-angle faults, and much of the literature dealing with topography upon faulted structures deals with high-angle faults. We should not overlook the fact that in some areas low-angle and transcurrent faulting influence or control topographic features.

TYPES OF SCARPS

A fault scarp is the most fundamental and diagnostic land form produced by faulting. The name *fault scarp* as applied to an abrupt cliff or scarp produced directly by faulting was apparently first used by Russell (1884) in describing features of this origin in the northwestern part of the Great Basin in southeastern Oregon. An escarpment is a common land form and may be produced in various ways, several of which are in no way related to faulting. Cuesta scarps are common in areas of gently tilted rocks; wave-cut escarpments often mark the positions of present and past shore lines; sides of glacial troughs and stream-cut valleys are often scarp-like in character, but they are paired and therefore easily distinguishable from fault scarps; and

FIG. 10.2.   Fault scarp northwest of Mount Tenabo, southwest of Elko, Nevada.   (Photo by J. S. Shelton and R. C. Frampton.)

fronts of lava flows and landslides may be locally scarp-like. Some mountain fronts are essentially escarpments which are receding as a result of long-continued degradation. This is particularly true in arid regions where bold mountain fronts may rise above broad rock-floored or alluvial plains at their bases.

A certain type of topography has come to be called *fault-block* or *block-faulted topography*. It is also frequently referred to as *basin and range topography* from the Great Basin region of western North America where it is excellently displayed. As originally used the term fault-block topography carried with it the implication that the land forms had been produced directly by relatively recent faulting. It has been shown, however, that many of the scarps have resulted from erosion long after the faulting. Thus the expression fault-block topography has come to refer to a particular type of topography developed upon faulted structures without specific implication as to whether the land forms were produced by faulting or by erosion along faults.

The type of topography found in the Basin and Range province is only one of several types of topography which may develop upon faulted structures. Probably too much emphasis has been placed upon this particular region, for, as Cotton (1950*a*) stated: "As a result of early discovery and recognition in a relatively waterless region, the fault scarp with such a specialized history became the type, especially after block diagrams of Basin Range scarps had been drawn by Davis, to be copied later by textbook writers. Fault scarps of the Basin Range type, as it may be called, are recognizable in some other parts of the world; but it is necessary to examine the development of tectonic forms in humid regions as well, where, primarily because of the vastly greater transporting capacity of rivers, erosional modification of scarps both during and after tectonic growth proceeds on different lines, ultimately producing a major landform with only a family resemblance to Basin Range fault scarps."

SCARPS ASSOCIATED WITH FAULTS

Davis (1913) was one of the first to attempt a clarification of the terminology applied to scarps resulting directly from faulting in contrast to those produced by differential erosion on two sides of a fault. This is the difference between a fault scarp and a *fault-line scarp*. A fault scarp as defined by Davis is a scarp which was *originally* produced by faulting, whereas a fault-line scarp was *originally* produced by differential erosion along a fault line. An understanding of the difference between the two is basic to an appreciation of much that follows. As Johnson (1939) has stressed, much of the confusion in usage of these terms has come from a lack of appreciation of the importance of the word "originally" in the above definitions. A scarp does not have to be at a fault line to be a fault scarp or even a fault-line scarp. The way in which the original scarp was initiated determines which type of scarp it is. A fault scarp originates by faulting and at its original position has a fault at or near its base, but there are mature and old fault scarps as well as young ones, and as weathering, mass-wasting, and erosion attack a fault scarp it undergoes change in both topographic appearance and position. As time passes the scarp loses much of its original abruptness and straightness and it becomes a denuded and battered fault scarp. Davis and Johnson both maintained that it still remains a fault scarp. To them it seemed as logical to recognize young, mature, and old fault scarps as to speak of young, mature, and old valleys, caverns, or mountains. A young scarp is likely to be situated near the fault that produced it, but a mature or old fault scarp may have receded from its originating fault.

A fault-line scarp originates by erosion along a fault line because there are rocks of varying resistance on the two sides of the fault line. Thus most fault-line scarps are second-, third-, or *n*th-cycle phenomena. However, a

FIG. 10.3. *A*, front, and *B*, back views of the Terlingua fault-line scarp and the Santa Helena Canyon of the Rio Grande River, Texas. The scarp is 1500 feet high. (Photos by L. V. Olson, International Boundary Commission.)

fault-line scarp may develop in the same cycle in which the original fault scarp was formed, but if it does the implication is that the fault scarp was first destroyed by erosion and later in the same cycle another scarp formed whose position was controlled by the fault line that marked the position of the original fault scarp. We may have young, mature, and old fault-line scarps, depending upon the degree of modification which they have undergone, and their positions with respect to associated faults may vary as much as do those of fault scarps. It should be obvious that the existence of a fault-line scarp implies a more complicated geomorphic history for the area in which it exists than does a fault scarp. Probably there are more fault-line scarps than fault scarps in nature. Most of the scarps in the stable portions of the earth's crust, which are associated with faults, are fault-line scarps. Some scarps associated with faults in the so-called mobile belts are also fault-line scarps.

It may seem that undue stress has been placed upon the importance of determining whether a particular scarp is a fault scarp or a fault-line scarp. This has been done because of the greatly differing geomorphic histories implied by the two types. Johnson (1929) has used the Colorado Plateaus area to illustrate this. Dutton (1882) interpreted numerous scarps in that region as fault scarps and assigned to them ages later than the peneplanation of the area, which he called the "Great Denudation." From this he concluded that the Colorado River is an antecedent stream which has maintained its course athwart block faulting. More recent work by Blackwelder (1934) and Longwell (1946) has indicated that most of the scarps are fault-line scarps and that the faulting was of ancient rather than recent age and thus antedated the period of peneplanation. It seemed more likely to them that the Colorado River was superposed across the faults from the peneplain which once extended across the region.

Davis recognized that a fault-line scarp may face in the same direction as the original fault scarp or in an opposite direction, depending upon the position of potential scarp-forming rocks with respect to the new base level of erosion and on which side of the fault they are. If resistant rock is on the upthrow side of a fault and above the base level of erosion with weak rock against it on the downthrow side, the fault-line scarp produced by erosion will be on the upthrow side of the fault and will essentially reproduce the original fault scarp. Davis defined such a scarp as a *resequent fault-line scarp*. If, on the other hand, weak rock is on the upthrow side and strong rock on the downthrow side at or above the new base level of erosion, an erosional scarp will form on the downthrow side of the fault and will face opposite the original fault scarp. Such a scarp is called an *obsequent fault-line scarp*. Recognition of an obsequent fault-line scarp is not too difficult, provided it is possible to determine the downthrow and upthrow sides of a fault, for a scarp on the downthrow side could not be a result of faulting. Distinguishing a resequent fault-line scarp from a fault scarp may be difficult, and success will depend upon the amount of geologic and geomorphic information available. However, this is a common problem which must be correctly answered if a proper interpretation of the geologic and geomorphic history of a region is to be made. Criteria for doing this will be given in a later section (see p. 252).

A single scarp may be in part a fault surface and in part an erosional surface. Such a scarp is called a *composite scarp* (Cotton, 1917). We can imagine two conditions under which a composite scarp may form. First, let us assume that a fault scarp has been formed and faulting has ceased. In the creation of the fault scarp weak beds were left at the base of the scarp on the downthrow side. Erosion of these weak rocks extends the scarp downward, resulting in a scarp that owes its upper part to faulting and lower portion to erosion. A composite scarp may be formed in another way. If, after a fault scarp has been destroyed by erosion, the region is uplifted and a resequent fault-line scarp formed, renewal of movement along the old fault plane will result in a composite scarp whose upper part is erosional in origin and whose lower part is of fault origin. It may seem that the circumstances just described are largely theoretical possibilities, but actually they are common in nature. Fault scarps and composite scarps may even exist in the same region. Sharp (1939), in describing the geology of the Ruby-East Humboldt Range of northeastern Nevada, came to the conclusion that the east face of the East Humboldt Mountains at the north end, at least, is a composite scarp. The total height of the scarp is approximately 4000 feet. He estimated that the lower 1000 feet were erosional in origin and the upper 3000 feet of the scarp were produced by faulting. On the other hand, he concluded that the east face of the Ruby Mountains has the appearance of a true fault scarp.

Cotton has in more recent years (1950*a,* 1950*b*) suggested that some of the controversy over fault scarps and fault-line scarps might be avoided by grouping scarps along faults into two classes, tectonic and non-tectonic. He classed as *tectonic scarps* fault scarps and fault-line scarps formed in the immediate postfaulting cycle of erosion and restricted fault-line scarps to second-cycle or multicycle scarps resulting from differential erosion. Thus, restricted fault-line scarps are structural features rather than tectonic. With reference to tectonic scarps he stated (1950*b*): "Purists in New Zealand · · · have made the very serious mistake of distinguishing far too meticulously between fault-line scarps and fault scarps in cases where the former have obviously been produced in the present major erosional (geomorphic) cycle by degradation of fault scarps. Australian geologists seem to have avoided this pitfall. The forms are tectonic, and must be recognized as such in any interpretation of geological history equally with less eroded fault scarps. It has been a mistake to insist on the present-day fault-line condition of many tectonic scarps, and it has been a greater mistake · · · to set up the category of 'composite' scarps, which are fault scarps in their upper and fault-line

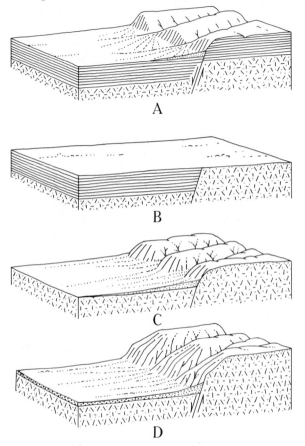

Fig. 10.4. Stages in the formation of a composite scarp. *A*. Formation of a fault scarp. *B*. Destruction of fault scarp by erosion. *C*. Renewal of erosion with development of a resequent fault-line scarp. *D*. Renewal of faulting with production of a composite scarp whose upper part is of erosional and lower part of fault origin. (After Douglas Johnson.)

erosional scarps in their lower parts. Clearly if any part of a scarp is a 'true' fault scarp · · · the feature as a whole is tectonic."

Not all scarps associated with faulting are most accurately described as fault, fault-line, or composite scarps. To these three types we may add *resurrected fault scarps* and *resurrected fault-line scarps*. It is possible, although probably rare, for a fault scarp to be buried under alluvium washed

down from the upthrow side of the fault and later exhumed. If we keep in
mind our definition of a fault scarp, one which was originally produced by
faulting, it follows that, even though it is buried and later exhumed, it remains a fault scarp, for removal of the alluvium did not produce a new scarp

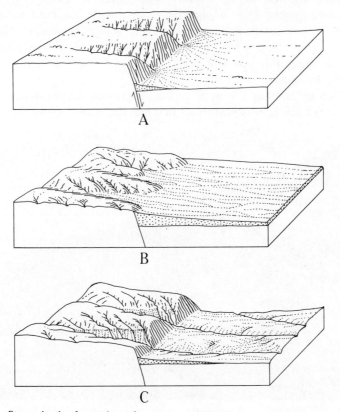

FIG. 10.5. Stages in the formation of a resurrected fault scarp. *A*. Creation of a fault
scarp. *B*. Burial of fault scarp with alluvium. *C*. Renewal of erosion accompanying
uplift with resurrection of the fault scarp. (After Douglas Johnson.)

but merely exposed a buried fault scarp. Baulig (1928) has cited as an
example of an exhumed fault scarp west of Clermont-Ferrand in the Central
Plateau district of France. Here a fault scarp was buried beneath alluvium,
after which lava from one of the Auvergne volcanoes spread across the
alluvium and buried it. Later, erosion removed most of the alluvium, but
that under the lava sheet was preserved to attest to the succession of events
that had transpired. Obviously, recognition of a resurrected fault-line scarp
is an even more difficult task and requires a full appreciation of the geologic
and geomorphic history of a region.

CRITERIA FOR DISTINGUISHING FAULT SCARPS AND FAULT-LINE SCARPS

That it is difficult to determine the origin of scarps associated with faults is evidenced by the varying interpretations given to them. For example, the imposing eastern front of the Teton Mountains has been variously described as a fault scarp, fault-line scarp, and composite scarp. Mere existence of a scarp along a fault is no proof that the scarp was produced by faulting, although this was long assumed to be so and apparently still is so considered by some.

Evidence of faulting. It is, of course, important that existence of a fault be established before it is postulated that a scarp is any one of the types that

FIG. 10.6. Slickensided fault face in quarry, Harrisons, Natal, South Africa. (Photo by L. C. King.)

may exist along faults. Geologic evidence for a fault consists of such phenomena as: actual exposure of a fault surface; presence of fault breccia or crushed zones; slickensided surfaces along a fault plane; zones of excessive jointing or fracturing; stratigraphic displacement of beds vertically, horizontally, or obliquely; and truncation of strata and geologic structure by a scarp. In areas of homogeneous and non-stratified rocks, the last two may be difficult to recognize. Even if all of them are present, they prove only that there is a fault and do not in themselves indicate whether a scarp was produced by faulting or differential erosion on its two sides.

It is desirable that we evaluate the various criteria which have been used to designate whether a scarp aligned with a fault is of diastrophic or erosional origin. With some emendations, the following discussion is based upon a

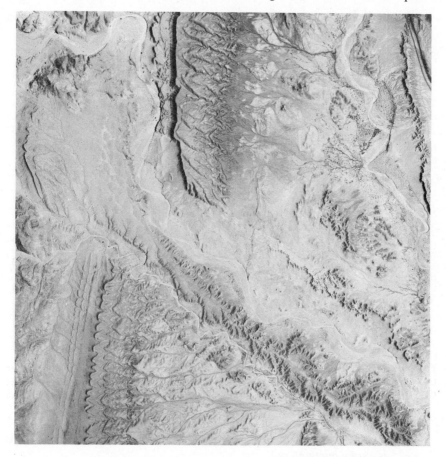

FIG. 10.7. Vertical photo showing displacement of a hogback by faulting, Emery County, Utah. (Soil Conservation Service photo.)

consideration of this problem by Blackwelder (1928*a*), in which he grouped various geologic and geomorphic phenomena found along fault-controlled scarps under three headings, those which are common to both fault and fault-line scarps, those which are positive evidence of fault scarps, and those which are restricted to fault-line scarps.

Features common to both fault scarps and fault-line scarps. 1. An abrupt and imposing front. This, of course, applies only to young scarps.

It is important that we recognize that an imposing front is not evidence that a scarp is a fault scarp, as so often has been assumed.

   2. Presence of a marginal fault as indicated by:

      a. Fault breccias, crushed zones, and zones of excessive fracturing and jointing.

Fig. 10.8. Displacement of beds and topographic features by step faulting, along lower Canyon Creek, near Glenwood Springs, Colorado. €S, Cambrian Sawatch quartzite; €D, Cambrian Dotsero dolomite; DC, Devonian Chaffee formation; ML, Mississippian Leadville limestone. (Photo by W. C. MacQuown, Jr.)

      b. Exposure of a fault surface with such features as slickensides. This is rarely encountered.
      c. Vertical, horizontal, or oblique displacement of beds.
      d. Truncation of rock strata and geologic structures along a scarp.

   3. Triangular facets on spur ends. Faceted spur ends terminating at or near a fault are commonly assumed to be proof of recent faulting. This is not necessarily so. Usually the slopes of spur-end faces are less than 30 degrees, whereas fault planes of normal faults are much steeper. Therefore, the spur ends have undergone considerable modification by weathering and erosion and do not mark the fault surface. Furthermore, it is entirely possible for faceted spurs to exist along a fault-line scarp as well as along a fault scarp, although they may be more common along fault scarps.

4. Linear base to a scarp. Most faults do not extend in straight lines but are sinuous in plan. However, in comparison with such scarps as cuesta scarps, those along high-angle faults are strikingly straight. Straightness, however, may characterize either fault scarps or fault-line scarps.

5. Sharp V-shaped canyons with bedrock floors extending down to a fault line. These are perhaps more common along fault scarps, but they may also develop along fault-line scarps or other types of scarps.

FIG. 10.9. View looking east at the scarp which marks the west edge of the Wasatch Mountains, Utah. Note faceted spur ends. (Photo by Chicago Aerial Survey Co.)

6. Increase in stream gradients as a fault line is approached. This along with feature 5 leads to so-called wine-glass valleys, which are frequently taken as evidence of recent faulting. Blackwelder thought that the same effect could be obtained by more rapid cutting of a stream channel in weak rock in contact with harder rock along a fault-line scarp.

7. Hanging valleys on the face of a scarp. Hanging valleys are common along fault scarps, but they may also be found along a fault-line scarp where there is notable difference in rock strength on the two sides of a fault.

8. Springs along the base of a scarp. Springs are sometimes found along faults but they are by no means limited to active or recently active faults.

9. Outflows of lavas along a fault trace. Even though the lavas can be shown to be of recent origin they are no proof that the associated scarp is of fault origin, for they may represent a renewal of vulcanism along a long inactive fault.

There are other topographic features which may suggest existence of a fault but do not have specific implication as to whether the fault, if present, is of recent or ancient age. Among these are such phenomena as:

1. Frequent landslides.
2. Alignment of notches, cols, and jogs in ridges showing no lithologic control.
3. Long, straight, parallel stream courses across rocks of various types and structures.
4. Nearly right-angled offsetting of stream courses. Unless the displacements are multiple and against the regional slope, they may be explained in other ways.

It seems possible for all of the above-mentioned features to be present along either fault-line or fault scarps; hence none should be considered diagnostic of a fault scarp. A combination of several of them may strongly suggest fault origin, but there should be other more conclusive evidence before a fault scarp is postulated.

Features diagnostic of fault scarps. Included in this class are various topographic and geologic features, some of which are positive evidence that the scarp with which they are associated was produced directly by faulting; others are merely presumptive evidence.

1. Poor correlation between rock resistance and topographic forms. If rocks on the scarp side of a fault are weaker than those on the downthrow side, or, if the materials on both sides are weak, e.g., unconsolidated sediments, then the evidence is positive that the scarp was produced directly by faulting, for along a fault-line scarp there will always be a close correlation between topography and rock strength. It is not unusual to find no notable difference in rock strength on the two sides of a fault, in which case this test can only be applied in a negative way.

2. Presence of "rift" features along a scarp. Various features such as small scarps, sag ponds, or basins, and small wedge-shaped hills in discordant relationships to each other have been called "rift" features from their prevalence along the famous San Andreas "Rift" of California. Where present, they are diagnostic of recent faulting, but more often they are lacking.

3. Ponded drainage up-valley from a scarp. If a scarp faces upstream and intersects a transverse valley to form a lake at its base, the evidence for recent faulting is strong.

4. Abnormally small alluvial fans at the base of a scarp. These may result from recent sinking of a fault block at the base of a scarp with accompanying lowering of the alluvium derived from the upthrown block. This criterion is more applicable in arid or semiarid regions, where permanent streams are few, than in humid areas where continued erosion along the base of the scarp may prevent the development of extensive alluvial fans.

FIG. 10.10. Vertical photo showing new alluvial fans along the front of a fault scarp in the Death Valley region. (Fairchild Aerial Surveys photo.)

5. Frequent severe earthquakes. Earthquakes of major magnitude are positive evidence that diastrophism is still active and thereby suggest that a fault scarp may be present.

6. Displacement of older topographic surfaces. If portions of mature or old-age erosion surfaces have been displaced with respect to one another, there is convincing evidence of recent faulting. Displaced topographic surfaces on the downthrown side of a fault are often buried under alluvium; therefore, care must be taken not to confuse a topographic level on the downthrown block which formed after faulting with older and higher topographic

forms on the upthrown side of the fault. Beheaded stream valleys on the backslope of an upthrown block are illustrative of this type of evidence. Where such valleys are cut off abruptly at the edge of a scarp there is a strong presumption of a fault scarp, although in rare instances they may be associated with a resurrected fault scarp.

7. Dislocation of recent or Pleistocene deposits. Dislocation of such youthful geologic materials points strongly to recent faulting, but this test is not always possible.

8. Warped terraces. Stream terraces may display abnormally steep slopes or even reversal of slope which can be attributed to faulting and thereby confirm the fault origin of a near-by scarp.

9. Presence of louderbacks. The term *louderback* was proposed by Davis (1930) for displaced segments of a lava flow on two sides of a fault. If it can be established that the lava flow is of late geologic age (preferably Pleistocene or Recent), there is justification for assuming that associated scarps were produced by faulting. This is really a special type of displaced older topographic surface.

10. Actual fault plane identified along a scarp. It rarely happens that the face of a fault plane is exposed or preserved along a fault scarp, for most fault scarps have undergone modification by weathering and erosion. However, if a fault plane surface is found, it strongly suggests recent faulting.

11. Scarps across alluvial deposits. *Alluvial scarps* are low scarps usually in alluvium, which are found at the bases of fault-block ridges. They range in height from a few feet to over 100 feet. Gilbert (1928) thought that they were produced largely by slumping accompanying movement along a fault plane and hence preferred to call them piedmont scarps rather than fault scarps. The Cucamonga scarp at the base of the San Gabriel Range in southern California is an excellent sample of an alluvial scarp (Eckis, 1928). It is 60 to 75 feet high in alluvium and 200 to 250 feet high where it separates alluvium from bedrock. The existence of such scarps is positive evidence of recent faulting, for they could not persist long in unconsolidated materials.

Strongly presumptive evidence of fault-line scarps. 1. Scarps situated on the downthrow side of a fault. Where this can be observed it is conclusive evidence, but it will be true only for obsequent fault-line scarps.

2. Close correlation between rock resistance, structure, and topography. This is one of the best tests that can be applied; it is applicable to both types of fault-line scarps.

3. A fault trace without marked topographic expression extending across lateral spurs rather than marking their ends. If the trace of a fault cuts across spurs extending out from a scarp rather than terminates them, the evidence

is strong for a fault-line scarp, for we should expect a fault line to cross spur ends if a scarp were a fault scarp.

4. Proof that a fault is pre-Pleistocene in age. If the assumption is made that most of the details of our present-day topography developed in post-Tertiary time, then it follows that any scarp associated with a pre-Pleistocene fault is likely to be a fault-line scarp. Some doubt the validity of this assumption and hence contend that fault scarps can persist for longer periods of time. Cotton (1950*a*) maintained that this criterion should be rejected because it practically assumes that there can be no fault scarps without active faulting along them.

5. Superposed drainage across a fault. Evidence is conclusive for a fault-line scarp if it can be demonstrated with certainty that drainage is superposed across a fault. However, the distinction between superposed and antecedent streams is one that is commonly difficult to make, and, unless there are actual remnants of the covermass from which the streams were superposed, the validity of the interpretation may be doubtful.

Because of the possibility of confusing fault and fault-line scarps, it is difficult to give specific examples of each type with assurance that they are correct. As examples of fault scarps, we may cite the east front of the Warner Range in northeastern California, the west face of the Schwarzwald on the east side of the Rhine Valley, and the east face of the Vosges Mountains to the west of the Rhine. The east front of the Sierra Nevada, California, and the west face of the Wasatch Mountains, Utah, are commonly cited as fault scarps, but they may be composite rather than fault scarps, as is true also of the east front of the Teton Mountains. Scarps which may be classed as fault-line with a reasonable degree of certainty are: the numerous scarps in the Connecticut Valley region on Triassic fault blocks as exemplified by Mount Holyoke, Mount Tom, and Mount Toby; the Ramapo scarp of New Jersey; and the Grand Wash Cliffs of Colorado.

BASIN AND RANGE TOPOGRAPHY AND ITS ORIGIN

The Basin and Range province of the western United States, bounded on the west by the Sierra Nevada and on the east by the Wasatch Mountains and Colorado Plateaus has become the type area for topography developed upon tilted fault blocks. Johnson (1929) has referred to the type of fault blocks found in this region as *tilted* or *monoclinal blocks*. These forms are also described as *block mountains,* although this term is used by some to include those upfaulted blocks which show little or no monoclinal tilting, the so-called horsts (see p. 264). Johnson designated the mountain blocks as *tilt block mountains* and the basins between them as *tilt block basins,* where they are direct products of faulting. If they are second- or later-cycle forms produced by erosion, the terms obsequent and resequent were used to indicate whether the present erosional topography coincides with or is opposite to the forms produced originally by faulting.   Thus, if an erosional topography essentially reproduces the faulted topography, the mountain blocks would be

Fig. 10.11.   View looking south at the tilt blocks which form the southern end of Steens Mountain in southeastern Oregon.   The downfaulted Broad Valley lies in the middle distance.   The mountain mass consists of lava flows and sedimentary rocks.
(Photo by R. E. Fuller.)

designated as *resequent tilt-block mountains* and the lowlands adjacent to them as *resequent tilt block valleys*.   The bounding scarps would be resequent fault-line scarps.   If on the other hand, the erosional topography is opposite to the original fault-produced topography, the mountain blocks would be *obsequent tilt block mountains,* the basins would be *obsequent tilt block valleys,* and the bounding scarps would be obsequent fault-line scarps.

Even though these terms may carry a clear implication of the geologic and geomorphic history of an area, there is always the difficulty of determining unequivocally whether the scarps are fault scarps or fault-line scarps.   It is largely because of this that there has developed wide divergence in the interpretation of basin and range topography. The following condensed review of the changing interpretations of basin and range topography has been given by Nolan (1943). King, in 1870, was the first to discuss the area, and he thought that the mountains were remnants of eroded folds. Gilbert, in papers in 1874 and 1875, first presented the idea that the mountains were essentially great blocks bordered on one or two sides by faults. Powell and Dutton recognized evidence for an ancient period of folding which was followed by peneplanation, but they thought that the period of faulting which was responsible for the present-day topography was of much later date. In general, a fault origin of the topography was rather generally accepted by geologists until Spurr dissented. Although recognizing that there are fault scarps in the Colorado Plateau area, he believed that the topography of the Basin and Range province had been produced largely by stream erosion since Jurassic time upon fault blocks and that relatively few recent faults have topographic expression. In other words, most of the scarps are fault-line scarps. He accounted for the great erosion by assuming a more humid climate during most of the time that the mountains were being formed. Baker, in 1913, was apparently the first to challenge the idea that the faults are normal faults and to suggest that they are reverse faults produced by lateral compression.

Louderback (1923) in discussing the Sierra Nevada, the greatest of the tilted blocks, concluded that:

1. The Sierra Nevada block has been uplifted 5000 feet or more at its eastern edge while its western edge remained near sea level.

2. This uplift was accompanied by much faulting, particularly on the east side of the block.

3. Faulting and elevation of the block took place during late Tertiary or/and Quaternary time and was thus not "relatively old."

4. Faulting has been directly responsible for the major topographic features, although it was recognized that they have been modified by erosion.

5. The eastern slope of the range was determined primarily by faulting.

6. Actual faults at the eastern base of the Sierra Nevada exist and are evidenced by such features as fault scarps, fault breccia, slickensided and striated surfaces, and displaced lava beds.

7. The faults are where they should be on the hypothesis that the topography was essentially fault-produced.

8. The valleys and basins to the east of the Sierra Nevada are basically depressed fault blocks and have been in the main areas of aggradation.

These conclusions of Louderback represent essentially the concepts of those who believe that Basin and Range topography is primarily a direct product of faulting.

Faults of different ages were recognized by Davis and used by him to account for differences in the geomorphic characteristics of various ranges. Gilbert in a posthumous paper (1928) presented additional information, largely geomorphic in character, to support his contention that the topography was largely fault-produced.

The idea that the faults are reverse faults continued to be expressed. Smith (1927) held this view, as did Lawson (1936), who believed that the eastern front of the Sierra Nevada is really the face of a steep thrust block. They did not, however, extend this concept to other blocks in the Basin and Range area.

Blackwelder (1928b) restated the belief that many of the intermontane basins were produced by erosion on belts of weak rock and that many of the scarps forming the mountain fronts are fault-line scarps. Thus the origin of basin and range topography remains a controversial matter. Most geologists are inclined to favor the view that much of the topography owes its characteristics to direct effects of faulting, but it is being recognized, by some at least, that not all the topography may fall into this category and each individual range needs to be considered a special problem. Preponderance of opinion is that most of the faults are normal rather than reverse faults. Considered judgment would seem to suggest that block faulting began perhaps as early as Oligocene time and has continued intermittently to the present. There may well be skepticism as to whether fault scarps formed in Oligocene time could persist to the present.

TOPOGRAPHY ALONG TRANSCURRENT FAULTS

As stated above, faulting in certain areas is of a type now commonly called transcurrent, in which strike-slip movement predominates. The San Andreas fault or fault zone (Willis, 1938a, 1938b) is the most famous example of this type. It has a known linear extent of 550 miles, but neither of its ends has been located, for at the north it passes under the waters of the Pacific Ocean and at the south it is buried under the sediments in Salton Basin. Evidence of recent movement consists of undissected fault scarplets, crush and gouge zones, pressure ridges, sag ponds, and variations in soils and vegetation on its two sides.

Movement along the fault has been such that it is described as *right-lateral* (across the fault the opposite side has moved to the right). This may be strikingly shown by offsetting of stream courses where they cross the fault. This is seen not only along the San Andreas fault but along other faults in southern California, such as the Whittier fault, along which there has been transcurrent movement.

Tectonic furrows, such as those along the San Andreas fault, have not been described elsewhere. Except in the San Andreas Valley, south of San Francisco, the fault furrow is not followed far by streams. More commonly streams cross the fault, but where they do so they follow it for a short distance and then leave it.

Associated with the San Andreas fault in southern California are numerous other faults characterized by a similar type of movement. Among these

FIG. 10.12. Lateral displacement of road near Arvin, California, caused by the Arvin-Tehachapi earthquake of July 25, 1952. (Photo by R. C. Frampton.)

are the San Jacinto, San Gabriel, Sierra Madre, Mission, and Whittier faults. Crowell (1952) has shown that movement along the San Gabriel fault, which parallels the San Andreas for 90 miles, has also been strike-slip in nature and has amounted to 15 to 25 miles of right-lateral displacement, as shown by offsetting of Miocene conglomerates and breccias on the two sides of the fault.

To a large degree the major topographic features of the Los Angeles region are delineated by a series of faults marked by strike-slip movement. The San Gabriel Mountains are bounded on the northeast by the San Andreas and San Jacinto faults and on the southwest by the Sierra Madre and Cucamonga faults. Several other faults coincide closely with the fronts of lesser mountain or hill tracts (see Fig. 10.14). These fault-bounded upland tracts have been shedding debris into a sinking basin at the south since at least early Miocene time, and sediments at least 25,000 feet thick have accumulated at the center of the basin.

Locally giant landslides are geomorphically significant along the flanks of the fault-bounded mountains as well as within the mountain masses them-

Fig. 10.13.   Right-lateral offsetting of drainage by San Andreas fault, 5 miles west of Taft, California.   (Photo by J. S. Shelton and R. C. Frampton.)

selves. Numerous streams within the mountains show alignment with faults. Some valleys are actually removed from the faults as if they had developed at higher levels and had been cut straight down, causing their present courses to be displaced from the positions of the locating faults.

HORSTS, GRABENS, AND RELATED FORMS

Blocks of the earth's crust may be relatively raised or lowered between more or less parallel faults without pronounced tilting. The relatively raised

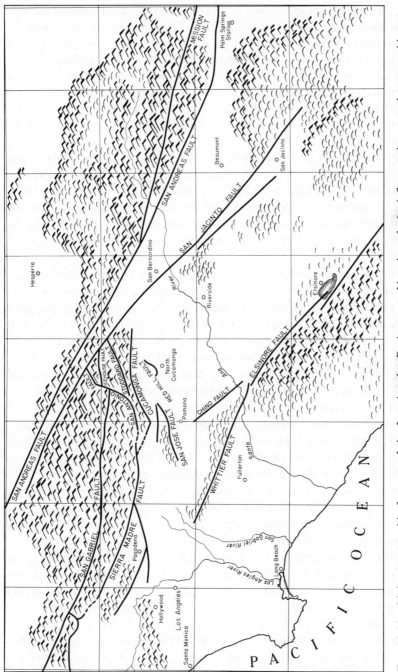

FIG. 10.14. Major topographic features of the Los Angeles Basin area. Note how faults influence the general topographic pattern. (After A. O. Woodford.)

blocks are commonly called *horsts* and the lowered blocks *grabens,* when they are the direct effects of faulting. Johnson (1929) preferred the name *rift blocks* for them, but there are objections to its use because rift is often applied to earthquake rifts such as the rift associated with the San Andreas fault in California, along which movement has been largely horizontal. Johnson would call an uplifted block a *rift block mountain* or horst and the lowered block a *rift block basin* or graben. The terms horst and graben are so firmly implanted in geologic literature that additional terminology will only be confusing.

Examples of what are generally considered grabens are: the Jordan-Dead Sea depression, Death Valley, the Rhine graben, great ocean troughs like the Bartlett trough off the coast of Cuba, and the "Rift Valleys" of eastern Africa. Examples of horsts are the Vosges Mountains to the west of the Rhine graben, the Schwarzwald or Black Forest plateau to the east of it, the Palestine Plateau to the west of the Dead Sea, and the Plateau of Transjordania to the east of it.

Terms are needed, however, to describe topographic forms which simulate horsts and grabens but owe their existence to erosion. Here Johnson's proposed terminology is helpful. The term rift valley as applied to the great troughs or grabens of East Africa and elsewhere is open to the criticism that the name valley should not be given to topographic forms produced by faulting but should be restricted to features produced by stream erosion. However, it has been applied so long to the deep trenches of eastern Africa and western Asia that it will probably continue to be used, but it should be kept in mind that it has a general rather than specific connotation.

If, after destruction by erosion of original horsts and grabens, a region is uplifted, a topographic lowland resembling a graben may develop on a block of weak rock bounded by parallel faults. Such a feature Johnson would call a *rift block valley* to distinguish it from an original graben produced by faulting. If the lowland occupies a position coinciding with a former graben, it would be described as a *resequent rift block valley.* The bounding scarps would be resequent fault-line scarps. If, however, an erosional lowland replaces a horst, it would be an *obsequent rift block valley* bounded by obsequent fault-line scarps. It is also possible that features resembling horsts may result from erosion. After a horst has been destroyed by erosion and the region is uplifted, erosion may essentially reproduce the original topography, provided that weak rocks lie below base level on the downthrow sides of the faults. The upstanding tract would then be called a *resequent rift block mountain* to distinguish it from the original horst, and the bounding scarps would be resequent fault-line scarps. If, however, a downdropped block of resistant rock which is bounded by weaker rock on either side is left through differential erosion above adjacent rift block valleys, it would be an *obsequent rift block mountain* and its bounding scarps would be obsequent faultline scarps. Obsequent rift block mountains and rift block valleys represent inversion of topography on faulted structures similar to that previously

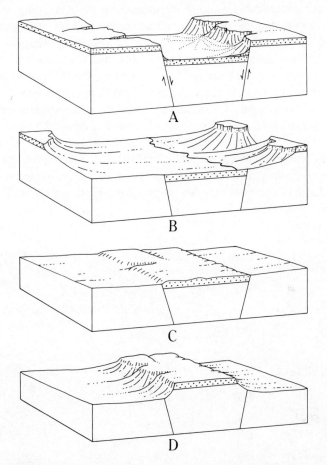

Fig. 10.15. Stages leading to topographic inversion on a graben. *A*. Formation of a graben. *B*. Destruction by erosion of the fault-produced topography. *C*. Renewal of erosion accompanying uplift. *D*. Development of an obsequent rift block mountain where the graben originally existed. (After Douglas Johnson.)

noted on folded structures. Good examples of such mountains are found in the central Texas uplift.

THRUST FAULTING

In the Alps, Appalachians, Rocky Mountains, the Highlands of northwestern Scotland, and elsewhere, compressive forces have produced thrust faulting with resulting complexity of geologic structure and related topographic forms.   Probably the most striking example of thrust faulting is found in the Alps, where great rock sheets have moved at low angles for many miles and created such complicated geologic structures that their mapping and interpre· tation have been among the most difficult problems that geologists have en- countered.   The thrust sheets are known as *decken* or *nappes*.   They are particularly associated with recumbent or inverted folds, and the complex of rock slices extending one over another has come to be designated as *imbricate structure* from its similarity to a number of overlapping shingles.

The Lewis overthrust of Montana has a length of approximately 135 miles (Billings, 1938) and a horizontal displacement of about 15 miles.   Its fault plane dips to the southwest at an angle of about 3 degrees.   The Bannock fault of southern Idaho (Mansfield, 1927) is 270 miles long and may have a horizontal displacement of as much as 35 miles.   High-angle thrust faulting is somewhat less common than low-angle; it is well displayed in the Sawtooth Mountains of Montana.

Fɪɢ. 10.16.   Small range-front graben at mouth of Wildrose Canyon at western edge of the Panamint Mountains, California.   (Photo by J. S. Shelton and R. C. Frampton.)

TOPOGRAPHIC EXPRESSION OF DIFFERENT TYPES OF FAULTS

High-angle normal faults. Topography associated with high-angle normal faults has been more frequently described than any other type, partly because of the spectacular nature of some fault and fault-line scarps and also because

FIG. 10.17. Frenchman Mountain near Las Vegas, Nevada. The topography is developed upon a tilted block consisting of alternating weak and strong sedimentary rocks. (Photo by Fairchild Aerial Surveys, Inc.)

of the attention that has been given to the Great Basin region of North America by many prominent geologists.

The topography resulting from normal faulting or from erosion along normal faults is not everywhere the same, even where it is on the grand scale found in western United States. Important differences may exist, depending upon whether the faulted rocks are massive, homogeneous rocks or stratified rocks of varying strength. In a region of massive, homogeneous rocks, the most striking effects will be a mountain mass elevated or tilted along boundary faults which are expressed as bold fault or fault-line scarps. Within the mountain mass, however, straight valley courses or aligned notches in ridges may reflect the position of other faults. Small range-front grabens are rather characteristically associated with many of the fault blocks in California. The Sierra Nevada of California and the Teton Mountains of Wyoming are examples of the topography that forms where massive crystalline rocks have been faulted along high-angle normal faults.

If the faulted blocks consist of stratified rocks which vary in susceptibility to erosion, their topographic expression may be quite different. If the block

FIG. 10.18.   Front view of a sheet of Mississippian Hannan limestone which has been thrust over Cretaceous shales and sandstones at head of No Business Creek, Saypo quadrangle, Montana. Ch, Mississippian Hannan limestone; Kk, Cretaceous Kootenai formation; Je, Jurrassic Ellis formation.   (Photo by C. F. Deiss.)

is a faulted anticline, such as the Santa Ana Mountains of southern California, the result is a series of hogbacks or homoclinal ridges, which show that the influence of the older structure was not eliminated by faulting. A simple example of normal faulting of stratified rocks may be seen in Frenchman Mountain, east of Las Vegas, Nevada. Here is a tilted block with moderate dips to the east. The block is marked by a marginal scarp and succession of hogbacks on tilted Paleozoic strata. Another range which combines the features of hogbacks and cuestas with marginal fault scarps is the southern Inyo Range between Owens and Saline valleys, California. A notable example of a tilted fault block consisting of lava flows and sediments is Steens Mountain (see Fig. 10.11) in southern Oregon (Fuller, 1931). Not only is a prominent marginal scarp present but step faults exist along its front.

High-angle thrust faults. High-angle thrust faulting may produce repetition of beds which when subjected to erosion will give a type of topography and drainage patterns that may easily be confused with those found in areas where folded or steeply dipping strata consisting of alternating strong and

FIG. 10.19. Topography upon a high-angle thrust sheet along north side of North Fork of Deep Creek, Saypo quadrangle, Montana. A similar sheet shows in the upper left background. Ch, Mississippian Hannan limestone; Kk, Cretaceous Kootenai formation; Je, Jurassic Ellis formation. (Photo by C. F. Deiss.)

weak rock have been truncated by erosion. Casual examination of the Saypo, Montana, quadrangle might very well suggest that the topography and structure here are similar to that found along the foothills of the Colorado Rockies, where resistant beds between weaker beds have given rise to parallel hogbacks. Yet, when the geology of the Saypo region is examined in the field, it is found (Deiss, 1943*a*, 1943*b*) that the parallel ridges are largely upon one geologic formation, the Hannan limestone of Mississippian age, which forms a series of imbricate slices which has been thrust over the shales and sandstones of the Cretaceous Kootenai formation. Valleys are mainly upon the shales in the Kootenai. This repetition of sharp, parallel ridges and valleys accounts for the mountains being known as the Sawtooth Mountains.

Low-angle thrust faults. Low-angle thrust faults are usually less conspicuously expressed in the topography than are high-angle faults.   If there is a fault or fault-line scarp marking the edge of a thrust sheet or nappe, it is

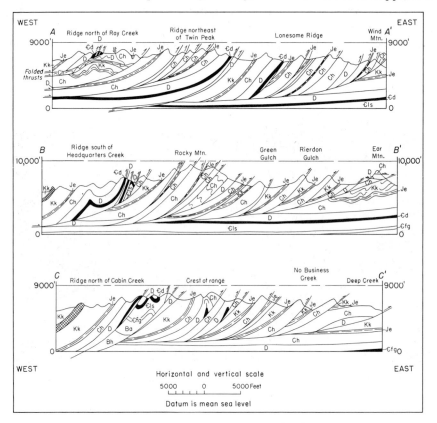

Fɪɢ. 10.20. Structure sections across the central Sawtooth Range, Montana, showing the imbricate structure which characterizes this range.   Kk, Cretaceous Kootenai formation; Je, Jurassic Ellis formation; Ch, Mississippian Hannan limestone; D, unnamed Devonian formation; Єd, Cambrian Devils's Glen dolomite; Єls, undifferentiated Cambrian limestone; Єfg, Cambrian Flathead sandstone and Gordon shale; Ba, Pre-Cambrian Belt Ahorn quartzite; Bh, Pre-Cambrian Belt Hoadley formation; cross-hatched pattern, diorite sill.   (After C. F. Deiss.)

not only likely to be less imposing than one associated with a high-angle fault but also it will usually be much more sinuous and irregular in plan.   In appearance the scarp will resemble more noticeably a cuesta scarp and in fact can probably not be distinguished from one except through field observations of the stratigraphic sequence.   After a nappe has been subjected to long erosion, portions of it may become isolated from the main part of the nappe.

These remnants are called *klippen* (singular, klippe). Chief Mountain, Montana, is a well-known example of a klippe, in which an isolated mass of Pre-Cambrian rock rests upon Cretaceous beds. It is an outlier of the Lewis overthrust mentioned above. Mythen Peak in the Alps is another example of a klippe.

It should be noted that the term klippe has a structural implication, that of older rock resting upon younger, and is not a topographic term. Topographically a klippe is a *nappe outlier* and from topography alone could not be distinguished from outliers of cuestas or plateaus. Study of the stratigraphic relationships is necessary for recognition of a klippe. There are important geologic differences between a nappe outlier and a cuesta outlier in that a nappe outlier has younger rocks surrounding it, whereas a cuesta outlier is surrounded by older rocks.

In the early stages of dissection of a nappe, erosion may penetrate the overthrust mass and locally expose patches of the younger rocks below it. Such exposures are called *fensters* or *windows*. Here again a particular structural or stratigraphic relationship is implied rather than a particular topographic form. Topographically fensters are likely to be basins or they may be restricted to narrow valley floors.

REFERENCES CITED IN TEXT

Baulig, Henri (1928). *Le Plateau Central de la France,* Armand Colin, Paris.
Billings, M. P. (1938). Physiographic relations of the Lewis overthrust in northern Montana, *Am. J. Sci., 235,* pp. 260–272.
Blackwelder, Eliot (1928*a*). The recognition of fault scarps, *J. Geol., 36,* pp. 289–311.
Blackwelder, Eliot (1928*b*). Origin of the desert basins of southwest United States, *Geol. Soc. Am., Bull. 39,* pp. 262–263.
Blackwelder, Eliot (1934). Origin of the Colorado River, *Geol. Soc. Am., Bull. 45,* pp. 551–566.
Cotton, C. A. (1917). Block mountains in New Zealand, *Am. J. Sci., 194,* p. 262.
Cotton, C. A. (1950*a*). Tectonic scarps and fault valleys, *Geol. Soc. Am., Bull. 61,* pp. 717–757.
Cotton, C. A. (1950*b*). Tectonic scarps and fault valleys, *Compt. rend. 16ᶜ congr. intern. géographie,* pp. 191–200.
Crowell, J. C. (1952). Lateral displacement on the San Gabriel fault, southern California, *Geol. Soc. Am., Bull. 63,* pp. 1241–1242.
Davis, W. M. (1913). Nomenclature of surface forms on faulted structures, *Geol., Soc. Am., Bull. 24,* pp. 187–216.
Davis, W. M. (1930). The Peacock Range, Arizona, *Geol. Soc. Am., Bull. 41,* pp. 293–313.
Deiss, C. F. (1943*a*). Stratigraphy and structure of southwest Saypo quadrangle, Montana, *Geol. Soc. Am., Bull. 54,* pp. 205–262.
Deiss, C. F. (1943*b*). Structure of central part of Sawtooth Range, Montana, *Geol. Soc. Am., Bull. 54,* pp. 1123–1168.

Dutton, C. E. (1882). Tertiary history of the Grand Canyon district, *U. S. Geol. Survey, Mon. 2,* pp. 61–121.

Eckis, Rollin (1928). Alluvial fans of the Cucamonga district, southern California, *J. Geol., 36,* pp. 224–247.

Fuller, R. E. (1931). The geomorphology and volcanic sequence of Steens Mountain in southeastern Oregon, *Univ. Wash. Publs. Geol., 3,* pp. 1–130.

Gilbert, G. K. (1928). Studies of basin range structure, *U. S. Geol. Survey, Profess. Paper 153,* pp. 1–92.

Johnson, Douglas (1929). Geomorphic aspects of rift valleys, *Compt. rend., 15ᵉ congr. intern. géol.,* Pt. 2, pp. 354–373.

Johnson, Douglas (1939). Fault scarps and fault-line scarps, *J. Geomorph., 2,* pp. 174–177.

Lawson, A. C. (1936). The Sierra Nevada in the light of isostasy, *Geol. Soc. Am., Bull. 47,* pp. 1691–1712.

Longwell, C. R. (1946). How old is the Colorado River? *Am. J. Sci., 244,* pp. 817–835.

Louderback, G. D. (1923). Basin Range structure in the Great Basin, *Univ. Calif. Pub. Geol. Sci., 14,* pp. 329–376.

Mansfield, G. R. (1927). Geography, geology and mineral resources of part of southeastern Idaho, *U. S. Geol. Survey, Profess. Paper 152,* pp. 150–159.

Nolan, T. B. (1943). The Basin and Range province in Utah, Nevada, and California, *U. S. Geol. Survey, Profess. Paper 197-D,* pp. 141–196.

Russell, I. C. (1884). A geological reconnaissance in southern Oregon, *U. S. Geol. Survey, 4th Ann. Rept.,* pp. 431–464.

Sharp, R. P. (1939). Basin-range structure of the Ruby-East Humboldt Range, northeastern Nevada, *Geol. Soc. Am., Bull. 50,* pp. 881–920.

Smith, W. D. (1927). Contribution to the geology of southeastern Oregon, *J. Geol., 35,* pp. 421–440.

Willis, Bailey (1938a). San Andreas Rift, California, *J. Geol., 46,* pp. 793–827.

Willis, Bailey (1938b). San Andreas Rift in southwestern California, *J. Geol., 46,* pp. 1017–1056.

Additional References

Butts, Charles (1927). Fensters in the Cumberland overthrust block, *Va. Geol. Survey, Bull. 28,* pp. 1–11.

Dixey, F. (1946). Erosion and tectonics in the East African rift system, *Quart. J. Geol. Soc. London, 102,* pp. 339–379.

Fuller, R. E., and A. A. Waters (1929). The nature and origin of the horst and graben structure of southern Oregon, *J. Geol., 37,* pp. 204–238.

Gardner, L. S. (1941). The Hurricane fault in southwestern Utah and northwestern Arizona, *Am. J. Sci., 239,* pp. 241–260.

Johnson, Douglas (1929). Block faulting in the Klamath Lakes region, *J. Geol., 26,* pp. 229–236.

Kenedy, W. Q. (1946). The great Glen fault, *Quart. J. Geol. Soc. London, 102,* pp. 41–72.

King, P. B. (1950). Tectonic framework of southeastern United States, *Am. Assoc. Petroleum Geol., Bull. 34,* pp. 635–671.

Longwell, C. R. (1930). Faulted fans west of the Sheep Range, southern Nevada, *Am. J. Sci., 220,* pp. 1–13.

MacQuown, W. C., Jr. (1945). Structure of the White River Plateau near Glenwood Springs, Colorado, *Geol. Soc. Am., Bull. 56,* pp. 877–892.

Quinn, A. W. (1933). Normal faults of the Lake Champlain region, *J. Geol., 41,* pp. 113–143.

Rich, J. L. (1933). Physiography and structure at Cumberland Gap, *Geol. Soc. Am., Bull. 44,* pp. 1219–1236.

Russell, R. J. (1928). Basin Range structure and stratigraphy of the Warner Range, northeastern California, *Univ. Calif. Pub. Geol., 17,* pp. 387–496.

Shand, S. J. (1936). Rift valley impressions, *Geol. Mag., 73,* pp. 307–312.

Wayland, E. J. (1929). Rift valleys and Lake Victoria, *Compt. rend. 15ᵉ congr. intern. géol.,* Pt. 2, pp. 323–353.

Willis, Bailey (1928). Dead Sea problem: Rift valley or ramp valley? *Geol. Soc. Am., Bull. 39,* pp. 490–542.

II · The Arid Cycle

CONTRASTS BETWEEN ARID AND HUMID REGIONS

It has been only within relatively recent times that arid lands have been studied in detail with a resulting recognition that there are significant differences in the evolution of land forms in arid and humid regions. When it is realized that about 30 per cent of the land may be classed as arid, it becomes evident that this phase of geomorphology has been neglected.

Climatic contrasts between arid and humid regions. No completely satisfactory definition of a desert has been proposed, but it is generally agreed that the most significant characteristics of deserts are scarcity of rainfall and vegetation. There is probably no place where it never rains, but there are some parts of the world where rainfall is exceedingly slight. At Iquique, Chile, in the Atacama desert, the mean annual rainfall over a period of 25 years was 0.05 inch. Calama, Chile, in the same desert, went for 13 consecutive years without rain. These are extreme examples, but they do show that there are places where practically no rain falls. A striking feature of most desert rainfall is its aperiodic nature. It is so irregular in occurrence that average values have little significance. One rain may equal or exceed the so-called mean annual rainfall, which is simply a mathematical mean arrived at from a number of very irregular and variable amounts of precipitation.

Desert rainfall is commonly described as torrential rainfall, and this has sometimes led to the belief that many of the world's heaviest rains fall in deserts. Examination of climatic data on extreme rainfalls will show, however, that these occur in humid areas and not in deserts. Rainfalls of 20 or more inches in 24 hours have been measured in the southeastern United States in contrast with 24-hour maxima of 4 inches or so in the arid and semi-arid west. Russell (1936) has questioned the validity of the two commonly held ideas that desert rainfall is typically torrential in character and comes most typically as afternoon thunderstorms. He pointed out that at Phoenix, Arizona, two-thirds of the summer rainfall comes at night and not as convectional afternoon thundershowers. What is significant, according to him, is that because of sparsity of vegetation the percentage of runoff is generally much greater than for rains of corresponding intensity in humid regions. It is also commonly believed that desert rainfall is much more spotty than that in humid regions. Spottiness of rainfall is characteristic of many humid

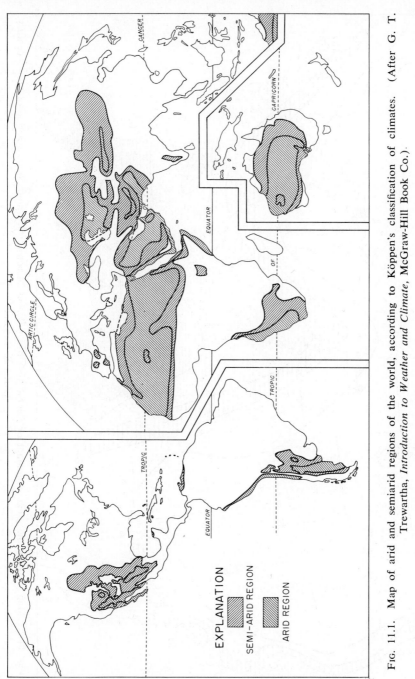

FIG. 11.1. Map of arid and semiarid regions of the world, according to Köppen's classification of climates. (After G. T. Trewartha, *Introduction to Weather and Climate*, McGraw-Hill Book Co.).

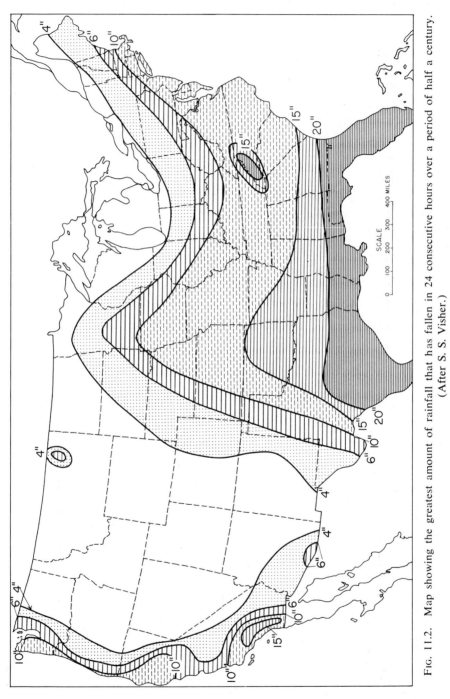

FIG. 11.2. Map showing the greatest amount of rainfall that has fallen in 24 consecutive hours over a period of half a century. (After S. S. Visher.)

regions, particularly of continental climates with convectional summer rainfall. Spottiness is characteristic of most precipitation and in general increases in proportion to the amount.

Large daily ranges of temperature characterize deserts. These result largely from low humidity and clear skies with consequent rapid heating by day and cooling at night. To some extent the barren nature of the ground in deserts contributes to their higher daily range of temperature. There is an extreme case from northern Tripoli of a maximum temperature of 99° F. and a minimum temperature of 31° F. during one 24-hour period. Annual range of temperature (the difference between the average monthly temperatures of the warmest and coldest months) is also greater in arid than in humid climates in comparable latitudes. The recognized intensity of solar radiation in deserts along with the large daily range of temperature were largely responsible, as we have previously seen, for the idea that exfoliation is produced mainly by alternate heating and cooling of rock surfaces.

Related to climatic differences, particularly rainfall deficiency and irregularity, are noticeable differences in the amount and kind of vegetation. Continuous cover of forest or grass is never found in deserts. Sparsity of vegetation results in extensive bare rock and soil surfaces that affect significantly the rapidity of operation of certain geomorphic processes.

Geologic differences between arid and humid regions. In most deserts, drainage is internal and does not reach the sea. There are exceptional streams which flow through deserts to the sea, the most notable being the Nile and Colorado rivers. These streams are exotic streams which originate outside the deserts in areas of moderate or heavy rainfall and have sufficient volumes when they enter the deserts which they cross to maintain themselves despite heavy losses by evaporation and seepage. The Nile is fed by heavy monsoonal rainfall in the Abyssinian Highlands and the Colorado by more moderate precipitation in the Rocky Mountains. The Nile through a distance of some 1200 miles of desert does not have a single incoming tributary.

Where drainage is internal, the concept of sea level as a base level control does not apply. Base level actually may be rising as a result of aggradation of interior basins. Even for the desert areas crossed by the Nile and Colorado rivers, sea level as the ultimate level of land reduction applies only to relatively narrow strips adjacent to each stream. It might be thought that absence of sea level control of erosion in mountain-rimmed basins with interior drainage represents only a temporary condition which will end with destruction of the mountain barriers and establishment of through streams to the sea. Actually, internal drainage is likely to persist so long that local and temporary base levels, for all practical purposes, control reduction of the upland areas.

In general, it may be said that mechanical weathering processes are relatively more important in arid than in humid regions, but it does not follow necessarily that mechanical weathering processes are dominant in arid regions. It is commonly stated that in arid climates mechanical weathering predominates over chemical weathering, but this is open to question. Considerable evidence indicates that spalling off of rock surfaces, commonly called exfoliation, results more from chemical weathering than from the effects of alternate heating and cooling of rock surfaces (Chapman and Greenfield, 1949). Davis (1938) concluded that many of the spheroidal boulders found in desert granitic areas are the result, in part at least, of chemical decomposition by what he called *subsoil weathering*. Evidence for this is found in the presence beneath the mantle rock of spheroidal boulders. Davis thought that their formation started with percolation of water along joints and that only after they were exposed through rainwash and creep did flaking or spalling off of their surfaces by diurnal temperature changes take place.

Rather than to say that chemical weathering is not important in deserts, it is probably more nearly correct to say that the advanced chemical processes which lead to the complete breakdown of the complex silicate minerals are lacking here. This does not preclude oxidation and hydration from being significant. Deeply weathered soils are rare, however, and *grus* (grit or slack) derived from granitoid rocks is a type of material from which arkoses are formed. Desert land forms typically display greater angularity than those of humid lands. This is partly the result of the reduced rate of desert weathering and partly because of the lesser importance of mass-wasting in arid regions.

Desert streams are with few exceptions notably intermittent, short, and discontinuous. Permanent streams largely rise outside deserts in areas of heavier rainfall. During most of the time desert stream courses are dry, although there may be a movement of water as underflow through the sands and gravels at the bottoms of their beds. What Davis (1938) has called *streamfloods* are characteristic of desert streams. Although stream channels may be dry most of the time, there are occasional floods of water down them after local heavy rains. Davis preferred to call them streamfloods rather than streams because of their "spasmodic and impetuous flow." A marked difference exists between floods in humid and arid regions. A flood on a river like the Ohio or Mississippi takes many days to attain its crest and to subside to normal flow. Desert streamfloods arrive suddenly and likewise abate rapidly. A person arriving at the bank of a desert streamflood may wait for the stream to "run out," for he is aware that in a short time the torrent will subside. This is particularly true of lesser stream courses, the *washes* as they are called in western United States or the *wadis* of the Saharan region. A visitor from a humid region is likely to be impressed by the number of bridges without streams beneath them and the number of dips in roads where they cross intermittent stream courses.

Perhaps even more important geomorphically than streamfloods are desert *sheetfloods* (McGee, 1897; Davis, 1938). This term was first used by McGee to designate broad sheets of storm-borne waters which move in a system of small, enmeshed channels rather than in definite stream courses. The conditions that seem most favorable to their development are: (*a*) rapid rainfall

FIG. 11.3. Vertical photo showing drainage lines on a piedmont slope, Pima Papago area, southern Arizona. (Soil Conservation Service photo.)

on a barren surface of detrital materials, whereby runoff waters become heavily loaded almost immediately, and (*b*) absence of low-water streams between periods of sheetfloods which would carve channels in the detrital slope and thus prevent the outspreading of sheetfloods. Sheetflood erosion is not lacking in humid regions, where it is more likely to be referred to as sheetwash or rill wash. The significant difference between the sheetwash in humid regions and sheetfloods in arid regions is that there is so much loose detrital material available in arid regions that the waters become almost immediately loaded with debris, whereas in humid regions vegetation helps retain the weathered mantle rock. Striking features of sheetfloods are their short distance of flow and brief duration. Fed as they are by short-lived downpours, they do not last long, and because they flow over both dry and pervious surfaces they do not persist for long distances. Blackwelder (1928) has pointed out that heavy rains in semiarid regions may produce either floods or mudflows (see p. 90), depending upon the type of material that the rain falls on. Locally mudflows are an important method of transport of debris in arid and semiarid regions.

Common misconceptions about deserts. A common fallacy about deserts is that they have little or no vegetation. Vegetation is present, although it is reduced in amount as compared with humid lands and different in character. Many people visualize deserts as consisting of mile after mile of drifting sand. Sand dunes may be striking features and may cover extensive areas, as in the Libyan portion of the Sahara, but dunes and related sand accumulations occupy a rather small proportion of a total desert area. A common misconception is that deserts are always hot. Tropical deserts do lack a cold season, but temperatures at night may be many degrees lower than those during daytime. Middle-latitude deserts, however, experience seasonal changes of temperature comparable to humid lands in the same latitudes. Probably the most common geomorphic misconception concerning deserts is the seemingly logical idea that the wind has been mainly responsible for the creation of desert land forms. As will be shown later, running water is a more important geomorphic agent.

ORIGINS OF DESERTS

Polar deserts. On the basis of origin, deserts may be grouped into three classes, polar, middle latitude, and tropical. Polar deserts do not concern us at the present at least. They include the areas of great ice sheets. The drought of polar deserts is what may be called *physiological drought*. Sufficient moisture is present in the ice sheets of Antarctica and Greenland for plant growth but it is not available because it is frozen.

Middle-latitude deserts. Middle-latitude deserts may be described as *topographic deserts,* because they owe their existence either to their location in the deep interiors of large continental masses or to the presence of high mountains across the paths of prevailing winds. An example of a desert area resulting from the first cause is the Desert of Turkestan, east of the Caspian sea. More commonly, middle-latitude deserts are a consequence of high mountain masses to the windward of the deserts. Examples of such deserts are the Great Basin Region of western United States and a lesser desert area to the east of the Andes in southern Argentina. The desert of Mongolia (Gobi) owes its existence to a combination of these two causes.

Low-latitude deserts. Low-latitude deserts, sometimes called tropical deserts, are products of still different conditions. They can be understood only by taking into account the general planetary circulation of the atmosphere. These deserts center around latitudes 20 to 25 degrees north and south and may extend approximately between latitudes 15 to 30 degrees north and south. These are the latitudes of the subtropical high-pressure belts or horse latitudes and of the trade winds. The subtropical high-pressure belts are more or less permanent areas of high pressure marked in general by relatively calm, subsiding air. As air is heated in descent, conditions are not favorable to precipitation. In the trade wind belts equatorward from the two high-pressure belts the general air movement is toward the equatorial low-pressure belt, from the northeast in the northern hemisphere and from the southeast in the southern hemisphere. Such movement is toward regions of slightly higher temperatures and is not conducive to precipitation, except where high mountains lie athwart the winds. Thus, the latitudes of the trade winds and the subtropical high-pressure belts are in general marked by deserts. Examples are: the Sahara in northern Africa; the Kalihari in southwest Africa; the Arabian Desert; the Thar of northwestern India, Afghanistan, and Baluchistan; the Great Australian Desert or the Victoria Desert; the Atacama Desert of Chile and Peru; and the Sonora Desert of northwest Mexico and southern Arizona and California.

MAJOR LAND FORMS OF ARID REGIONS

It has been emphasized by Blackwelder (1931) that plains are the most common features of deserts. He estimated that in the arid western United States, where fault-block mountains are conspicuous features, plains comprise more than three-fourths of the total area. He recognized five types of plains that are common in deserts: river floodplains, structural plains, playas, bajadas, and pediments. River floodplains are relatively rare in deserts because permanent rivers are uncommon. In the western United States only the Humboldt and Colorado rivers have significant floodplain areas. Structural plains, or dip slopes as Blackwelder called them, are striking features in such areas as the Colorado Plateaus, but, where the rocks are igneous or metamorphic, or where sedimentary beds have been complexly faulted or folded, structural plains are not likely to be extensive.

The term *bolson* is applied in the desert areas of western United States and Mexico to a basin more or less rimmed by mountains. Centripetal drainage is characteristic of a bolson and at or near its center there is commonly a level plain, or *playa,* which marks the site of a present or former lake. Most playas are dry except after the occasional desert shower, and their surfaces are often covered with glistening salts which have been precipitated

from the ephemeral lakes which exist on them.   Such salt-covered surfaces
are called *alkali flats*.   The waters of some playa lakes, such as Walker Lake
and Carson Lake in Nevada, are brackish or salty.   Those having a high concentration of salts are called *salinas*.

Mountains in arid regions are commonly bordered by smooth piedmont
slopes which extend downward to neighboring basin floors.   It was formerly
thought that a piedmont slope was entirely of aggradational origin, but it is

Fig. 11.4.   The Race Track Playa in the northern Panamint Mountain area, California.
(Photo by J. S. Shelton and R. C. Frampton.)

now realized that it actually consists of two parts, a lower part of aggradational origin, called a *bajada,* and an upper part which is really an eroded
bedrock surface, although it is commonly veneered with alluvium.   Although
various names have been suggested for the bedrock portion of a piedmont
slope, it is now generally called a *pediment*.   Pediment and bajada slopes
are relatively gentle, varying between ½ degree and about 7 degrees, whereas
the mountain fronts against which they abut are typically much steeper,
ranging from 15 degrees to nearly vertical.   The result is a sharp break in
slope or nick where a pediment meets a mountain front.

A true bajada consists of a series of coalescing alluvial fans built by
streams which debouch onto a piedmont slope and spread their detritus radially outward from the mouths of mountain valleys.   Materials of a bajada
consist of gravelly alluvium or even large boulders sometimes interbedded

with mudflow deposits. Large mountain streams build broad fans with low gradients, whereas shorter streams which do not head far back in the mountains build steeper alluvial cones. Frequent shiftings of stream channels

Fig. 11.5. Youthful pediment at base of Grapevine Range, northern Death Valley, California. Basal fans indent the mountain front. (Photo by Eliot Blackwelder.)

across fans are suggested by a sort of distributary system of dry channels. Present drainage lines are usually marked by washes which near the mountain are incised considerably below the tops of the fans. Such washes are often referred to as *fanhead trenches*.

Fig. 11.6. Pediment slope in front of Cady Mountains east of Barstow, California. A more advanced phase of pediment development than shown in Fig. 11.5. Long tentacles of the pediment now extend into the mountains, and small monticules have been isolated from the ends of the longer mountain spurs. (Photo by Eliot Blackwelder.)

The surface of a bajada is undulatory in character when traversed parallel to the mountain front. In contrast, a pediment surface is nearly flat or slightly concave when viewed at right angles to the bordering mountain front. A pediment may or may not have an alluvial veneer over it, but it is basically a bedrock surface, although it may have developed across older alluvial deposits. Pediments may have a thin veneer of gravel over them adjacent to the mountains, or this may be lacking, but away from the mountains they become more and more obscured by bajada deposits and thus become *concealed pediments;* then it becomes difficult to distinguish them from the bajadas unless well records are available. Even where concealed, a pediment is believed to be most typically a convex rock surface which "plunges" beneath the bajada that obscures it. Individual pediments at the point where mountain valleys discharge onto a piedmont slope may expand to form *coalescing pediments,* which may in time practically consume a mountain mass. Some pediments exhibit a considerable degree of dissection and are referred to as *dissected pediments.* The more common idea is that dissected pediments are a product of second-cycle erosion of originally nearly flat pediments, but Gilluly (1937) thought that those of the Ajo, Arizona, area may have been "born dissected," as did Sharp (1940) in reference to some of the pediments flanking the Ruby-East Humboldt Range of Nevada.

THE PEDIMENT PROBLEM

Two interrelated problems must be considered in any attempt to explain pediments: (1) what process or processes are responsible for their formation and (2) what is the explanation of the sharp break between the mountain front and the adjacent piedmont slope. At least three geomorphic processes have been suggested as operating to form pediments. They are sheetflood erosion or sheetwash, lateral planation by streams, and back-weathering.

Sheetflood theory. What was apparently the first recognition of the topographic form now generally called a pediment was made by McGee (1897) in the Sonoran desert of southwestern Arizona. At the base of mountains, he recognized plains of erosion which were lightly veneered with alluvial materials but were essentially rock plains and attributed them to sheetflood erosion. He believed that sheetfloods have great corrasive power because of their abundant tools and high velocities. The pediments thus formed tended to encroach upon the mountains by methods not made entirely clear by McGee.

Lateral planation theories. Paige, Blackwelder, and Douglas Johnson particularly have stressed the role of lateral erosion by streams in the formation of pediments. Paige (1912) concluded that the processes of interstream and lateral erosion at the edges of alluvial fans would produce a sloping planated surface cut on bedrock which would become buried toward the center of a basin under a cover of gravel which progressively grew mountainward as the rock-cut surface expanded at the expense of the mountains. He thought that the steep mountain front against which a pediment abuts was an expectable product of lateral erosion by streams. According to his thinking, sheetflood erosion was a *result* of the development of this planated surface rather than the *cause* of it.

Blackwelder (1931) concluded that pediments are essentially compound, graded plains produced by lateral erosion by ephemeral desert torrents. He considered a pediment, not a bajada, as the typical and normal desert land form and pictured it as expanding until mountains and hills had been consumed and the bajada regraded to form a smooth desert dome which was the desert-land equivalent of the humid-land peneplain. He considered bajadas characteristic features of regions where repeated uplift by faulting or warping interfered with development of normal graded slopes and doubted that a bajada a thousand feet or more in thickness could ever be formed except in areas of active diastrophism. In support of this last idea it is perhaps significant that extremely thick bajada deposits are found in southern California in the Los Angeles basin at the base of the San Gabriel mountain front, but there is little or no evidence of pediment formation. Apparently uplift by faulting has been so recurrent that there has been no opportunity for the development of a graded pediment surface.

In view of the importance that he attached to lateral erosion in humid lands, it is not surprising that Johnson (1932a, 1932b) considered lateral erosion the dominant process in pediment formation. He believed that in an arid region, where mountains lie adjacent to or surround intermontane basins, there exist three rather distinct zones. They are: (a) an inner zone consisting chiefly of mountains in which vertical downcutting by streams is dominant; (b) an outer zone corresponding to the area covered by a bajada in which aggradation prevails; and (c) an intermediate zone between the two and surrounding the mountain front, in which transportation and lateral erosion by streams are most significant. It is in the intermediate zone, according to him, that pedimentation takes place. He believed that the pediments evolve from features that he called *rock fans*. Johnson applied this name to fan-shaped rock surfaces which apex where mountain streams debouch upon a piedmont slope. Great significance was attached to the shifting back and forth of stream courses across fans where they leave the mountains. This shifting is not that of streams meandering across floodplains, but is more analogous to changes in courses which braided streams display. As a result of repeated shifting of their courses the streams at times occupy positions marginal to the mountain front in what were called *piedmont depressions* or *channels*. Johnson admitted that rock fans are commonly covered or obscured by alluvial deposits, but he cited some not too convincing examples of such rock fans along the eastern margin of the Sierrita Mountains and along the western side of the Dragoon Mountains in Arizona. Among the rather serious objections to Johnson's theory of lateral erosion are: the already-mentioned scarcity of exposed rock fans; secondly, there is considerable evidence to suggest that sheetfloods, which are, as Rich (1935) has pointed out, a sort of exaggerated form of sheetwash, are much more significant in desert erosion than streams; furthermore, it is difficult to explain the straightness so commonly exhibited by the mountain front if its recession is produced by lateral stream erosion. It would seem that the mountain front should be sinuous or scalloped with prominent embayments in it where major streams emerge from the mountains. If lateral erosion is significant, there should be evidence of what Davis (1938) called *basal trimming* of the steep mountain face where it has. been undercut by lateral erosion. Residual knobs or rock nubbins, which are often observed upon a pediment, hardly seem compatible with the idea of lateral planation. Most stream courses, where they emerge from mountains and become braided across fans, extend radially out from the apices of the fans and do not impinge upon the mountain front. Apical angles in the embayments of the mountain front should be much larger than is commonly observed if lateral erosion is significant here. These objections cause hesitancy in accepting lateral erosion as the dominant factor in pediment formation, although it can hardly be denied that it is a contributing factor. We are thus led to look with more favor upon what may be called the composite theory of pediment formation favored by such men as Bryan, Davis, Sharp, Rich, and Gilluly.

Composite theory. The composite theory of pediment formation attributes pediments to a combination of processes, chiefly three: backweathering, sheetwash or sheetflood erosion, and lateral planation. These three processes are combined in varying proportions by different individuals, but emphasis is upon a combination of processes rather than upon one dominant process. Bryan (1923) in his earliest discussion of pediments attributed them to: (*a*) lateral erosion by streams emerging from mountain canyons; (*b*) rill cutting at the foot of mountain slopes; and (*c*) weathering of outliers and unreduced mountain areas with ensuing transportation of debris by rills. He thought that lateral erosion became less important in the later stages as weathering and rill wash became dominant. In a later discussion, Bryan (1935) stated: "A considerable experience and many observations in the intervening years tend to confirm these conclusions." Davis (1938), instead, stressed the importance of backweathering of the mountain front with removal of its waste by sheetfloods and assigned to lateral erosion a minor or unimportant role. He was impressed by the importance of sheetfloods as described by McGee and believed the mountain front retreated parallel to itself after it had attained a degree of slope determined largely by the size of the waste shed from it, as was originally suggested by Lawson (1915). Sharp (1940), from a study of the pediments around the Ruby-East Humboldt Range in Nevada, concluded that varying geologic, climatic, and topographic conditions result in differing degrees of efficacy of the three processes of lateral erosion, backweathering, and rill wash. He found lateral erosion dominant in areas of permanent streams and on weak rock and rill wash, rainwash, and weathering dominant in areas of hard rock and ephemeral streams. Considering the pediments in the Ruby-East Humboldt area as a whole, he thought that lateral erosion was responsible for about 40 per cent of their development, and he attributed the remaining 60 per cent to weathering and various types of rainwash.

Gilluly (1937), from a study of the pediments of the Ajo, Arizona area, concluded: "Nothing seen in the Ajo quadrangle appears inconsistent with the conclusion of Bryan that pediments are slopes of transportation peculiar to arid regions and formed by cooperation of lateral erosion, rill wash, and weathering with subsequent removal of detritus by rills. One process may dominate in one place and another elsewhere but all are operative."

Rich (1935) thought that lateral erosion might contribute to the formation of pediments but did not believe that it is a necessary process. He concluded: "Rock fans and pediments are normal features of wasting, retreating escarpments, and mountain fronts, and are best developed in arid or semi-arid climates · · ·. Wasting and sheetwash, in conjunction with the blanketing effect of alluvial debris, are essential factors required for the production of rock fans and pediments. Lateral corrasion by streams is not necessary."

Bradley (1940) has described miniature pediments cut in the Eocene Bridger formation in the Washakie Basin of southwestern Wyoming, which seem to have much in common with pediments on harder rocks. He believed that they were formed largely by the combined action of rill wash and pelting rain, acting along a narrow zone near the base of the retreating escarpment of a badland area. He further concluded that, after initiation, a pediment surface is further lowered by sheetflood erosion. He considered a pediment surface to be a graded slope of transportation and thought that streams basinward take over the function of transport and cut their courses below the pediment surfaces and seem to be destroying rather than creating and extending the pediment surfaces.

Before leaving the subject of pediments, it is pertinent to point out that here we encounter the nearest approach to Penck's idea of parallel retreat of slopes. In the slope of the mountain front we have a good example of the steilwand of Penck or the gravity slope, as Meyerhoff (1940) interpreted it. It has a certain inclination, dependent chiefly upon the size of the materials being shed from it, being steeper where materials are large and less steep where they are small. Bryan (1940) has shown that movement of material down the mountain front is not entirely a direct effect of gravity. Although direct gravitational sliding and creep operate upon this slope, rain-wash seems to be the dominant process in the transport of material down it. A pediment surface would seem to be essentially the equivalent of the haldenhang of Penck or what Meyerhoff called the wash slope, and the sharp angle made by the two slopes would be a knickpunkt. In contrast to the ungraded gravity slope, the wash slope is essentially a graded slope of trans-portation over which a thin layer of detritus is moving downward to form the bajada with which it merges.

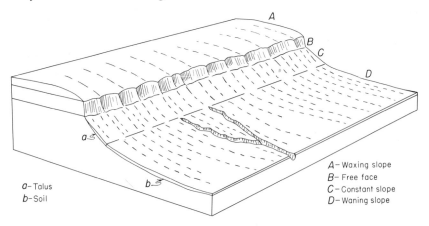

a–Talus
b–Soil

A–Waxing slope
B–Free face
C–Constant slope
D–Waning slope

FIG. 11.7.   Elements of a slope.   (After L. C. King and T. J. D. Fair.)

King and Fair (1944) and Fair (1947) in discussing slope development in Natal, South Africa, concluded that four elements comprise slopes: (*a*) an upper convex or waxing slope; (*b*) a free face or outcrop of bare rock; (*c*) a talus or detrital slope; and (*d*) a lower concave or waning slope, known also by such names as wash slope or pediment. Not all four elements are necessarily apparent at all times. Presence of all four elements seemed to them to depend upon the existence of a hard cap rock at the top of the slope. Where this is present they concluded that slopes retreat parallel to them-selves. They thought that there are two possible ways by which the scarp which rises above the pediment can be maintained, either by lateral plana-tion by streams at the foot of the scarp or by the removal of weathered waste by sheetwash.

THE ARID EROSION CYCLE

It may be questioned by some whether the geomorphic processes in arid regions are well enough understood to speak of an arid erosion cycle. Davis (1905) in his early writings treated the arid cycle as a modification imposed upon the humid cycle by a change to aridity, one of his so-called climatic accidents. Some of his ideas would hold true today, but with modifications to take into account such processes as pedimentation. The idealized cycle as suggested by Davis was one that would apply particularly to a desert area, such as exists in western United States, where block faulting has produced many enclosed basins. Not all desert regions present such conditions.

Some of the differences envisaged by Davis which would distinguish the cycle in arid regions from that of humid regions were: significant differences in the manner of runoff as previously described; maximum relief in youth instead of in maturity, as in the humid cycle; decrease instead of increase in relief as the cycle progresses; a prevalence of consequent drainage into enclosed basins with few antecedent streams; active dissection of the highlands in youth and accompanying aggradation of basins; lack of through streams with resulting predominance of local base levels of erosion; and continued rise of these local base levels as a result of basin aggradation. Maturity was presumed to be marked by reduction of relief and a certain degree of integration of the drainage of adjacent basins. Old-age conditions were largely conjectural, but it was believed that exportation by winds of finer waste would produce a gradual lowering of the area, resulting in a "leveling without base-leveling." Scattered areas of more resistant rocks might be left standing as monadnock-like features to which the name *inselberge* had been given in South Africa. By some such sequence of events it was believed that there might develop a "desert peneplain" analogous to the peneplain of humid regions.

Within more recent years, the tendency has been to emphasize the formation and extension of pediments as the major geomorphic processes in development of desert landscapes. Even Davis in his later papers stressed this aspect of desert erosion, and he recognized (1936) that the evolution of desert land forms might proceed with: (*a*) a slowly rising, local base level resulting from aggradation of an enclosed playa; (*b*) a slowly sinking base level while a playa in an enclosed basin was being lowered by deflation; or (*c*) normal base level control where the desert drainage flows to the sea.

At least four names have been proposed to designate the graded surface which results from advanced pedimentation. Lawson (1915) proposed the term *panfan* to designate "an end stage in the process of geomorphic development in an arid region in the same sense that the peneplain is an end stage of the general process of degradation in a humid climate." He also recognized that both peneplains and panfans represent penultimate rather than ultimate stages of degradation. Serious objection has arisen to the name panfan because, literally interpreted, it means "all fan," and, as fan is generally applied to alluvial deposits at the base of steep slopes, it is illogical to designate as a panfan an eroded bedrock surface consisting of a series of co-alescing pediments.

Davis (1933) designated as *granitic domes* certain surfaces in the Mojave Desert of southern California, which he believed represented the penultimate stage of desert erosion.   Later (1938), he called them more appropriately *desert domes,* because not all were developed across granites.   There are objections to using dome to describe a topographic feature when it is so intimately associated with a type of geologic structure.   Furthermore, it is questionable whether the penultimate desert erosion surface is always domal in profile.

FIG. 11.8. Cima dome northwest of Cima, southeastern California, a pediplain or desert dome above which rise a few bornhardts or inselbergs of metamorphic rock. (Photo by Eliot Blackwelder.)

The phrase *desert peneplain* has been used by various writers to describe erosional surfaces in Africa which have been attributed to eolian erosion and other processes.   Disregarding for the time the question whether wind ero-sion can produce plains of low relief which simulate peneplains, it hardly seems desirable to designate as a peneplain a surface produced under dif-ferent conditions and by different processes from a humid-land peneplain. Even the qualifying adjective, desert, will hardly prevent some from con-cluding that the two forms are of similar origin.

The term *pediplain* was proposed by Maxson and Anderson (1935) to describe "widely extending rock-cut and alluviated surfaces ··· formed by the coalesence of a number of pediments and occasional desert domes." *Pediplane* was proposed by Howard (1942) "as a general term for all de-gradational piedmont surfaces produced in arid climates which are either ex-posed or covered with a veneer of contemporary alluvium no thicker than that which can be moved during floods."   He encompassed under *pedipla-nation* all the processes by which pediplanes are formed.   Those who like to think of topographic surfaces in terms of mathematical planes may prefer the later spelling, but in general the less mathematical pediplain has come into wider usage.

Pediment slopes on opposite sides of a mountain mass may extend themselves through the mountains and join, although frequently the pediments on

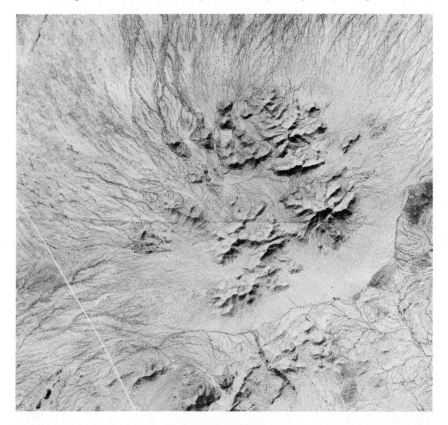

Fɪɢ. 11.9. Vertical photo of a group of bornhardts or inselbergs and surrounding pediment, Pima Papazo area, southern Arizona. Note the pediment gaps and passes. (Soil Conservation Service photo.)

the two sides are at different levels. Sauer (1930) applied the name *pediment passes* to "the narrow, flat, rock-floored tongues extending back from the general pediment, but still penetrating along the mountain sufficiently to meet another pediment slope extending into the mountain front from the other side." At a later stage, pediment passes may enlarge to such an extent that coalescing pediments are characteristic features of the topography, and pediments on opposing sides of the mountain range may join through broad openings which Sauer called *pediment gaps*. As portions of a mountain range become rimmed with pediments, a pediplain makes its appearance. Expansion of a pediplain may continue until all that remains of the original mountain mass are scattered knolls or hills which rise above its surface.

FIG. 11.10. Two bornhardts in South Africa. *Upper,* the "Lion" bornhardt, Chiweshe Reserve, Southern Rhodesia. *Lower,* unnamed bornhardt, Rarie Umvukwes, Southern Rhodesia. (Photos by L. C. King.)

Such residual hills are analogous to monadnocks on the peneplain and most commonly have been called inselbergs. These features are strikingly developed in the savanna areas on either side of the tropical rain forest areas in Africa, where numerous steep-sided residual hills and mountains rise prominently above the general level of the country. This type of topography has been made famous by Bornhardt, Passarge and others under the name of *inselberglandschaft* (inselberg landscape). Some have questioned whether pedimentation was responsible for these residual hills. Others have applied inselberg rather indiscriminately to any island-like hill which stands conspicuously above its surrounding, such as the so-called sugarloafs of tropical rainy climates. Willis (1936) attempted to clarify this situation by suggesting usage of the term *bornhardt* to designate the forms originally called inselbergs. Despite the possible ambiguity of the term inselberg, many writers continue to apply it to residual hills and mountains in arid landscapes.

Actually, there is considerable difference of opinion among those who have studied the inselberglandschaft of Africa as to what particular processes were responsible for its formation. Originally the landscape was attributed to eolian erosion, but there has been a marked reluctance among geologists to admit that wind erosion is capable of reducing extensive areas to conditions simulating a peneplain. Passarge in later years recognized the possibility that wind scour may not have been the chief agent in the creation of this landscape. Within more recent years, there has been a tendency to believe that the African landscapes are "relict landscapes" of more humid climatic conditions, when streams were more widespread and powerful than at present and carried on active lateral erosion to develop the plains above which the inselbergs stand. Whether the inselbergs of Africa are really surrounded by pediments similar to those in arid western United States is in dispute. Cotton (1942), in a thorough discussion of the inselberg problem, suggested that there may be two similar types of topography in Africa, the inselberg landscape of the savanna areas, which is probably largely the result of lateral erosion by streams, and the pedimented landscapes of the arid and semiarid regions in which residual hills or nubbins similar in general appearance to true inselbergs rise above coalescing pediments or pediplains.

With recognition of the pediplain as a distinctive old-age surface different in origin from a peneplain, there has been an inclination among some geomorphologists to reinterpret some erosional surfaces long-called peneplains. Howard (1941) suggested that the two rather generally recognized erosional surfaces of the Rocky Mountain region, the so-called Flattop and Rocky Mountain peneplains, may in reality be pediments (pediplanes, according to his later usage of the term). Rich (1938) had previously suggested that this possible origin for these two topographic surfaces be given serious consideration.

Mackin (1947) interpreted what has been called the Subsummit peneplain of the Bighorn Mountains as dissected pediments. He thought that the steep regional slopes of this erosional surface were more logically explained as inherited from pediments than as the result of upwarping of a former peneplain. According to this explanation only 2000 feet of Pliocene and Pleistocene tilting of the Bighorns need be postulated instead of 6000 feet or more that is required to explain the present slopes of the Subsummit surface as an uptilted peneplain. Mackin further thought that the sharp topographic break which exists between the axial peak area in the Bighorns and the adjacent planation surface is one that would be expected in a pediment landscape and need not be attributed to faulting.

Bradley (1936) considered both the Gilbert Peak and the Bear Mountain erosional surfaces on the north flank of the Uinta Mountains to be pediment surfaces. Both are covered with conglomerates rather than the deep residual soils expectable upon a peneplain. Their steep slopes, ranging from 400 feet to the mile near the crest of the range to 55 feet per mile near the adjacent basin, seemed more typical of pediplains than peneplains.

A most sweeping application of the pediplain concept has been made by King (1950), who has interpreted widely separated erosion surfaces in Africa, Asia, North America, Europe, South America, and Australia as ancient pediplains dating back as far as the Cretaceous. Although few geologists would draw such sweeping conclusions either as to the origin or age of the surfaces interpreted by him as pediplains, it is desirable that the erosional surfaces of arid and semiarid regions that have previously been called peneplains be reexamined in light of the growing recognition of the importance of pedimentation in such environments.

REFERENCES CITED IN TEXT

Blackwelder, Eliot (1928). Mudflow as a geologic agent in semiarid mountains, *Geol. Soc. Am., Bull. 39*, pp. 465–484.

Blackwelder, Eliot (1931). Rock-cut surfaces in desert ranges, *J. Geol., 20*, pp. 442–450.

Bradley, W. H. (1936). Geomorphology of the north flank of the Uinta Mountains, *U. S. Geol. Survey, Profess. Paper 185-I*, pp. 163–199.

Bradley, W. H. (1940). Pediments and pedestals in miniature, *J. Geomorph., 3*, pp. 244–255.

Bryan, Kirk (1923). Erosion and sedimentation in the Papago country, Arizona, *U. S. Geol. Survey, Bull. 730*, pp. 19–90.

Bryan, Kirk (1935). The formation of pediments, *Rept. 16th Intern. Geol. Cong.*, Pt. 2, pp. 765–775.

Bryan, Kirk (1940). The retreat of slopes, *Assoc. Am. Geog., Anns., 30*, pp. 254–268.

Chapman, R. W., and M. A. Greenfield (1949). Spheroidal weathering of igneous rocks, *Am. J. Sci., 247,* pp. 407–429.

Cotton, C. A. (1942). *Climatic Accidents,* Chapters 7 and 8, Whitcombe and Tombs, Ltd., Wellington.

Davis, W. M. (1905). The geographical cycle in an arid climate, *J. Geol., 13,* pp. 381–407; also in *Geographical Essays,* pp. 296–322, Ginn and Co., New York.

Davis, W. M. (1933). Granitic domes of the Mojave Desert, California, *Trans. San Diego Soc. Nat. Hist., 7,* pp. 211–258.

Davis, W. M. (1936). Geomorphology of mountainous deserts, *Rept. 16th Intern. Geol. Congr.,* Pt. 2, pp. 703–714.

Davis, W. M. (1938). Sheetfloods and streamfloods, *Geol. Soc. Am., Bull. 49,* pp. 1337–1416.

Fair, T. J. D. (1947). Slope form and development in the interior of Natal, South Africa, *Trans. Proc. Geol. Soc. S. Africa, 50,* pp. 105–119.

Gilluly, James (1937). Physiography of the Ajo region, Arizona, *Geol. Soc. Am., Bull. 48,* pp. 323–348.

Howard, A. D. (1941). Rocky Mountain peneplains or pediments, *J. Geomorph., 4,* pp. 138–141.

Howard, A. D. (1942). Pediment passes and the pediment problem, *J. Geomorph., 5,* pp. 2–31 and 95–136.

Johnson, D. W. (1932*a*). Rock fans of arid regions, *Am. J. Sci., 223,* pp. 389–416.

Johnson, D. W. (1932*b*). Rock planes of arid regions, *Geog. Rev., 22,* pp. 656–665.

King, L. C., and T. J. D. Fair (1944). Hillslopes and dongas, *Trans. Proc. Geol. Soc. S. Africa, 47,* pp. 1–4.

King, L. C. (1950). The study of the world's plainlands: a new approach in geomorphology, *Quart. J. Geol. Soc. London, 106,* pp. 101–127.

Lawson, A. C. (1915). The epigene profile of the desert, *Univ. Calif. Pub. Geol., 9,* pp. 23–48.

McGee, W. J. (1897). Sheetflood erosion, *Geol. Soc. Am., Bull. 8,* pp. 87–112.

Mackin, J. H. (1947). Altitude and local relief of the Bighorn area during the Cenozoic, *Wyo. Geol. Assoc., Field conference in Bighorn Basin, Guidebook,* pp. 103–120.

Maxson, J. H., and G. H. Anderson (1935). Terminology of surface forms of the erosion cycle, *J. Geol., 43,* pp. 88–96.

Meyerhoff, H. A. (1940). Migration of erosional surfaces, *Assoc. Am. Geog., Anns., 30,* pp. 247–254.

Paige, Sidney (1912). Rock-cut surfaces in the desert ranges, *J. Geol., 20,* pp. 442–450.

Rich, J. L. (1935). Origin and evolution of rock fans and pediments, *Geol. Soc. Am., Bull. 46,* pp. 999–1024.

Rich, J. L. (1938). Recognition and significance of multiple erosion surfaces, *Geol. Soc. Am., Bull. 49,* pp. 1695–1722.

Russell, R. J. (1936). The desert-rainfall factor in denudation, *Rept. 16th Intern. Geol. Congr.,* Pt. 2, pp. 753–763.

Sauer, Carl (1930). Basin and range forms in the Chiricahua area (Arizona and New Mexico), *Univ. Calif. Publs. Geog., 3,* pp. 339–414.

Sharp, R. P. (1940). Geomorphology of the Ruby-East Humboldt Range, Nevada, *Geol. Soc. Am., Bull. 51,* pp. 337–372.

Willis, Bailey (1936). East African Plateaus and Rift Valleys, *Carnegie Inst. Wash. Publ.. 470,* p. 121.

Additional References

Bryan, Kirk (1932). Pediments developed in basins with through drainage, *Geol. Soc. Am., Bull. 43,* pp. 128–129.

Bryan, Kirk, and McCann, F. T. (1936). Successive pediments of the Upper Rio Puerco in New Mexico, *J. Geol., 44,* pp. 145–172.

Davis, W. M. (1936). Geomorphic processes in arid regions and their resulting forms and products, *Rept. 16th Intern. Geol. Congr.,* Pt. 2, pp. 703–714.

Frye, J. C., and H. T. U. Smith (1942). Preliminary observations on pediment-like slopes in the central High Plains, *J. Geomorph., 5,* pp. 215–221.

Johnson, D. W. (1931). Planes of lateral corrasion, *Science, 73,* pp. 174–177.

King, L. C. (1947). Landscape study in Southern Africa, *Trans. Proc. Geol. Soc. S. Africa, 50,* pp. 23–52.

King, L. C. (1949). The pediment landform: some current problems, *Geol. Mag., 86,* pp. 245–250.

King, L. C. (1953). Canons of landscape evolution, *Geol. Soc. Am., Bull. 64,* pp. 721–752.

Mabbut, J. A. (1952). A study of granite relief from south-west Africa, *Geol. Mag., 89,* pp. 87–96.

Strahler, A. N. (1950). Equilibrium theory of erosional slopes approached by frequency distribution analysis, *Am. J. Sci., 248,* pp. 673–696 and 800–814.

12 · Eolian Land Forms

TOPOGRAPHIC EFFECTS OF WIND EROSION

As indicated in Chapter 3, wind erosion manifests itself in three forms: abrasion or corrasion, the natural sandblast action of wind-blown sand; deflation, the lifting and removal of loose material from the earth's surface; and attrition, the mutual wear of particles carried along by the wind. It was formerly common practice to attribute great erosional power to wind

FIG. 12.1. Yardangs produced by wind abrasion, Umtamvuna, southern Natal, South Africa. (Photo by L. C. King.)

and to explain most desert land forms in terms of wind erosion, but it has become increasingly apparent that few major topographic features of arid regions are so formed.

Wind abrasion may aid in the shaping of some of the details of major forms but is itself hardly capable of producing features of great areal extent. Possibly the only topographic form that can be attributed unequivocally to wind abrasion is the *yardang*. This name is applied (Blackwelder, 1934) to elongated grooves or furrows which were first described by Hedin in Turkestan. They usually are elongated in the direction of the prevailing winds and are nearly always carved from relatively weak materials.   Those in Turkestan have been cut in lacustrine silts.

Wind abrasion can operate only near the ground because of the inability of wind to lift sand more than a few feet.   Bagnold (1941) stated that sandblast action on posts is seldom appreciable above a height of 18 inches and concluded that sand is seldom lifted more than 2 meters above the ground.

Fɪɢ. 12.2.   A ventifact near Garnet in the Coachella Valley, southern California.
(Photo by R. P. Sharp.)

According to Blackwelder (1928) wind abrasion manifests itself through (*a*) polishing and pitting, (*b*) grooving, and (*c*) shaping and faceting.   He believed that effects of the first are rather widespread throughout arid regions, but that they are of little geologic significance.   It is important, however, that we recognize that polishing is a natural effect of abrasion or sandblast action, for many etched boulders and surfaces have been attributed to wind abrasion that lack polishing and are more likely the result of differential weathering than of eolian erosion.   Wind abrasion may produce *ventifacts* which exhibit one or more polished and faceted surfaces, but these are relatively rare because strong winds, abundance of sand, and absence of vegetation are essential to their best development.

Minor forms such as alcoves and niches in rock walls, sometimes called wind caves, may have in part resulted from wind abrasion, where they are at the base of a cliff, but more commonly they are products of differential weathering and rainwash or deflation. Stone lattice or honeycombed surfaces, names that have been given to intricately pitted rock surfaces, have often been attributed to wind abrasion, but it is likely that abrasion has played a minor role compared with differential weathering and solution.

FIG. 12.3. Small residual mesas resulting from deflation in Danby Playa in the southwest part of the Mojave Desert, California. The tabular masses are capped with gypsum. (Photo by Eliot Blackwelder.)

Certain topographic forms, commonly called *pedestal rocks,* which consist of residual masses of weak rock capped with harder rock, have been explained by wind abrasion. It is doubtful whether this process has contributed significantly to their shaping. They are more commonly the result of differential weathering aided by rainwash (Bryan, 1926). Deflation has possibly been responsible for pedestal rocks, especially those found in intermont basins in which detritus has accumulated. Blackwelder (1931) has described small tabular mounds or hills in the Danby playa of southeastern California which are composed of lacustrine materials and capped with a veneer of coarsely crystalline selenite which has proved more resistant to deflation than the underlying clays and silts.

That deflation is responsible for the formation of many depressions called *blowouts* is rather generally agreed. These are commonly found in areas of sand accumulation where they form small basins on or within dunes and other types of sand accumulation. Blowouts may also develop in areas where non-indurated or poorly indurated materials lie beneath the surface. The High Plains region of the United States extending from Texas to Montana is an area which has over it a cover of older alluvium of Tertiary age which weathers rather easily because of its poor induration. Here are found thousands of shallow basins, many of which are sites of permanent or intermittent lakes, that are believed to have been produced mainly by deflation (Judson, 1950). Whether large hollows such as have been described in Mongolia by Berkey and Morris (1927), and called by them P'ang Kiang hollows, can be produced by deflation or wind scour is questionable. The P'ang Kiang hollows are as much as 5 miles across and 200 to 400 feet deep. Berkey and Morris attributed them to wind scour because no other origin seemed plausible, as these hollows were developed upon granites as well as sedimentary rocks. Big Hollow in the Laramie Basin of Wyoming, which is 9 miles long, 3 miles wide, and 150 feet deep, has been considered a blowout.

Thousands of shallow basins exist in the western Great Plains region. Various origins have been suggested for them, such as: (1) subsidence caused by deep-seated solution; (2) differential compaction of poorly indurated Tertiary sediments; (3) removal of surface materials by animals (buffalo wallows) and (4) removal of materials by deflation, particularly during drier interglacial times. Doubtless examples of depressions having each of the above-suggested origins exist, but it seems more likely that deflation has been responsible for more basins than any other cause. Judson (1950) has explained those in New Mexico as the result of alternating periods of leaching and deflation coinciding with wet and dry periods of glacial and interglacial times. Even if it is admitted that deflation was a major cause of the basins, it is difficult to prove that the periods of deflation were largely restricted to more arid glacial times.

Opinions vary as to the relative effectiveness of deflation. Blackwelder (1928) took a conservative view of the importance of wind erosion in arid regions and credited running water with the development of the major relief forms of deserts. He recognized, however, that vast quantities of debris must have been washed down into desert basins and thought that present basin deposits represent only a part of what must have been eroded from the surrounding mountains. Thus, he concluded that a great deal of the debris carried down into the basins, after being reduced to dust size, is removed by deflation. Some geologists, however, believe that deflation is a relatively insignificant process, except locally, in the reduction of desert landscapes and that, in general, intermont basins are being filled rather than lowered.

Lag deposits. In the process of removal of sand and smaller-sized particles by deflation there is a sorting of materials according to size with the coarser materials being left behind. These concentrations of pebbles and boulders have been designated by the general name of *lag deposits*. Locally they are striking features of desert surfaces. *Desert pavement* and *desert armor* are terms often applied to them. Such surfaces characterize the *hammadas* and *regs* of the Sahara, which are areas of stony desert in contrast to the areas of sand accumulation or *ergs* (Gautier, 1935). The stones that constitute these lag deposits may be brightly polished. This may be in part the effect of wind abrasion but more commonly coatings of oxides of iron and manganese give to the surfaces of the stones an enamel-like appearance known as *desert varnish*.

EOLIAN DEPOSITS

Materials undergoing transport by the wind are subjected to a winnowing which gives rise to two distinct types of wind deposits, accumulations of sand and deposits of silt and clay known as loess. Variations in wind velocities may produce a certain amount of interbedding of the two, but in general they are fairly distinct with loess being found farther leeward from the source area than sand. This is well illustrated in Nebraska where loess lies to the east and leeward of the large area of sand hills.

Environments in which wind deposits are found. Wind deposits are by no means restricted either in origin or location to deserts. Not only is loess found in humid or subhumid regions but eolian sand deposits are common in several non-desert environments. The four non-desert environments in which sand deposits may be found are: (*a*) along shore lines, (*b*) along stream courses in semiarid regions, (*c*) in areas where loosely cemented sandstones have disintegrated to supply sand, and (*d*) in areas of glacial outwash. Along most shore lines, either ocean or lake, unless the shore be rocky and rugged, dune sands are likely to be found. They are present in discontinuous patches along the Atlantic coast of the United States from Massachusetts to Florida. They are somewhat less extensive along the Pacific coast because of its ruggedness, but locally they are conspicuous. Dunes are found locally around the Great Lakes and mark both present and former shore lines of these lakes. Dune sizes along shore lines are controlled by the extent of the source area, the prevalence of onshore winds and the amount of vegetation (in general, the size of shore-line dunes is limited by the amount of open space between vegetated patches). A stream in a semiarid region varies greatly in volume, and at times extensive stretches of the stream's bed are exposed to wind action. Thus, the leeward sides of such valleys as the Platte, Arkansas, and Missouri where they cross the semiarid Great Plains are blanketed with sand deposits which often have dunal form. The Sand Hills region of western Nebraska, an area covering some 24,000 square miles, is an area where the sand is derived primarily from the weathering of poorly indurated Tertiary sandstones.

Numerous areas in North America and Europe have extensive sand and loess deposits of Pleistocene age which were derived from glacial outwash or lacustrine materials. Sand and silt were blown from glacial sluiceways and associated valley trains (see p. 313), outwash plains, outwash deltas, and lake beaches and deposited leeward as sand sheets, dunes, or blankets of loess.

Transportation by wind. The mechanics of wind transportation and deposition have not been adequately studied. The most extended quantitative

FIG. 12.4. Excessive deflation and resulting dust storm near Springfield, Colorado. Total darkness prevailed for one-half hour. (Soil Conservation Service photo.)

treatment of these processes has been made by Bagnold (1941); his conclusions to a large degree will form the basis of the following discussion. Sand storms and dust storms are frequently confused. Wind has the ability to lift and transport dust particles for long distances, but the bulk of sand movement probably takes place within 2 meters of the ground. Bagnold recognized three types of movement involved in the transport of sand: suspension, saltation, and surface creep. *Suspension* is relatively unimportant in the transport of sand because the vertical velocities required to lift sand-size particles are rarely attained. It is important in the transport of dust and results chiefly from turbulent flow of air in contrast to stream-line or laminar flow. Wind velocity is practically never constant but characteristically exhibits short-period variations in velocity known as gustiness. Gustiness results in eddies of air which circulate in many directions. Upward currents make possible suspension of finer earth particles and may keep them aloft for great distances, but suspension is unimportant in the transport of grains of sand size. *Saltation* is a bounding movement, resulting from impact and rebound of wind-driven sand. Saltation is initiated directly by wind pressure upon sand grains; the moving particles describe trajectory-like paths. Most wind-blown sand moves this way. *Surface creep* is produced by the impact of sand grains moving by saltation. A part of the energy producing saltation is retained by the bounding sand grains and continues their movement, but some is dissipated in friction against the surface upon which the sand grains fall. However, this energy is not lost, for, as a result of the continued bombardment of the sand surface, there is a slow movement forward of it by surface creep.

Deposition by wind. Bagnold thought that deposition of wind-blown materials may take place in three ways, through sedimentation, accretion, and encroachment. *Sedimentation* takes place when grains which are being carried by slowly moving air fall with insufficient force to be carried forward themselves by saltation or move other grains forward by surface creep. Most of the material carried by suspension is deposited in this manner. Deposition by *accretion* results when grains being moved by saltation strike the surface with such force that some grains continue to move forward as surface creep but more come to rest than continue their forward motion. Accretion deposits are thus a result of a combination of saltation and surface creep. When sedimentation takes place the particles do not move forward after hitting the ground, but in deposition by accretion the particles may move along the surface until they find a resting place in some slight hollow. Deposition by *encroachment* takes place when the surface upon which deposition is taking place is not smooth but is marked by an obstruction such as an abrupt rise or drop. Particles being moved by surface creep are detained, but those being moved by saltation can move on. An example of deposition by encroachment is that which takes place on the front of a dune when the particles roll down its surface and come to rest.

Types of sand deposits. Several classifications of sand deposits have been proposed, but none is completely satisfactory. Bagnold's classification is probably most applicable to areas of freely moving sand and was based upon extended observations in the deserts of Africa and Asia as well as laboratory experiments in wind tunnels designed to establish the mechanics of wind transportation and deposition. It seems, however, to be somewhat inadequate for areas where vegetation has played an important role in the shaping of sand deposits. He recognized the following types of eolian sand deposits:

- *A.* Small-scale forms: small ripples and ridges on sand surfaces which have little geomorphic significance
- *B.* Large-scale forms
  1. Sand shadows and sand drifts
  2. True dunes
     - *a.* The barchan or crescentic dune
     - *b.* The seif or longitudinal dune
  3. Whalebacks or sand levees
  4. Undulations
  5. Sand sheets

A *sand shadow* is an accumulation of sand to the lee of and in the shelter of an obstruction, such as a boulder, bush, or cliff, which interferes with

FIG. 12.5. Gypsum dunes in White Sands National Monument, New Mexico. (Photo by National Park Service.)

stream-line air flow and checks the wind's velocity. Sand accumulates to the lee of an obstruction until its front slope has attained the limiting angle of repose, which is about 34 degrees. Sand will then slide down the front of the advancing deposit along what is called its *slip face*. The height of a sand shadow is limited by the size of the ground plan of the wind shadow in which it forms, for any sand which gets outside this shadow will be swept into the main sand stream by the stronger wind outside it. Sand shadows may also form where wind sweeps sand over a cliff or escarpment. Such deposits have also been called *sandfalls*.

Sand drifts are found to the lee of a gap between two obstructions. This gap acts as funnel through which sand trails out to the leeward. This usage

Fig. 12.6. Sand dunes in the vicinity of Delta, Utah. (Photo by Fairchild Aerial Surveys, Inc.)

by Bagnold of the term sand drift may be confused with the more common application of it to thin deposits of sand that are called by him sand sheets.

Bagnold (1933) defined a *dune* as a "mobile heap of sand whose existence is independent of either ground form or fixed wind obstruction." Groups of dunes are often referred to as *dune complexes, dune colonies,* or *dune chains.* Unlike sand shadows and drifts, their existence is not dependent upon an obstruction or topographic break. They usually attain their maximum development upon relatively flat terrain.

Bagnold considered the barchan and the seif as the only true dune forms; they, according to him, rarely exist together. Dunes are most typically found

in deserts, for along coasts and rivers moisture and vegetation usually interfere with their growth and prevent their orderly development as well as limit
their size. A *barchan* is a crescentic-shaped dune with tips extending to
the leeward, making this side concave in plan and the windward side convex.

Fig. 12.7. Vertical photo of barchans near Fort St. John, Peace River District, Canada.
The effective winds were from the east. (Royal Canadian Air Force photo.)

The slip face, if present, is transverse to the wind. Barchans tend to arrange
themselves in chains extending in the direction of the most effective winds.
Changes of wind directions may modify their form and even temporarily cause
a reversal in positions of the slip faces, or destroy them with resultant development of conical-shaped sand masses. In areas of abundant sand supply,
barchans are large and closely packed, whereas in areas of lesser sand supply
they are smaller and more dispersed.

A *seif* or *longitudinal dune* instead of being transverse to the prevailing
wind is parallel to it. Many seifs in the Egyptian Sand Sea attain heights of

100 meters, and some in Iran are as much as 210 meters high. According to Bagnold, the width of a seif is roughly 6 times its height. Seif chains have been described which are 300 kilometers long. Their crests form knife-like ridges with many peaks and sags. One side of the crest may be rounded, and the other may show a collapsing front (slip face) at right angles to the prevalent wind direction. There may be gaps or corridors between adjacent

Fig. 12.8. Seif ridges in the western Sahara Desert. (Air Force photo.)

seifs in which bare desert floor is exposed. Contrary to the more commonly held idea that longitudinal dunes or seifs are built mainly by winds blowing sand rather constantly from one direction, Bagnold maintained that they grow in height and width largely through the action of cross winds and grow in length during periods when the prevailing wind is parallel to the trend of the seif chain. He believed that the seif is a modification of the fundamental barchan form produced by strong cross winds transverse to the prevailing wind direction and that their knife-like crests represent barchan forms super-posed upon a ridge of sand extending parallel to the prevailing wind.

Whalebacks or *sand levees* are flat-topped sand ridges which extend parallel to the prevailing winds but lack the collapsing fronts which mark seifs. They also have much larger dimensions. A whaleback may be 100 miles long, 2 miles wide, and 150 feet high. Locally there may be seif forms upon their tops. They seem to be confined largely to the Egyptian Sand Sea. Bagnold suggested that they may constitute the residue from the march of a dune chain or series of dune chains downwind. *Undulations* are sand deposits somewhat similar in nature to whalebacks but shorter in length and lacking the definite form of the whaleback. They form billowy surfaces between whalebacks or seifs. Both whalebacks and undulations appear to be areas of "dead" or non-mobile sand in contrast to the mobile barchans and seifs.

Bagnold applied the name *sand sheet* (more commonly called sand drift) to a sand area marked by an extremely flat surface and absence of any topographic relief other than small ripples. The famous Selima sand sheet of the Libyan desert is an outstanding example. It covers an area of at least 3000 square miles and is practically flat, as far as the eye can detect, except for an occasional line of dunes. It consists of a few feet of sand resting upon bedrock. Its flatness and uniformity of surface were attested to by Peel (1941) when he stated that he was able to drive his car at a consistent speed of 60 to 70 kilometers for 3 hours without having to decrease his speed or deviate from his compass course. This is the type of surface that has been referred to by some as a desert peneplain.

Hack (1941) proposed an empirical classification of dunes which he thought fitted the Navajo country of Arizona. He recognized three types of dunes, transverse, parabolic, and longitudinal. *Transverse dunes,* including the common barchan, are nearly always free of vegetation; their form seemed to him not to be influenced by vegetation. The tips of a transverse dune extend to the leeward. *Parabolic dunes* were defined as "long, scoop-shaped hollows, or parabolas, of sand with points tapering to windward" and with a windward slope much more gentle than the leeward in contrast to the transverse dune. He thought that these dunes were formed where the wind removed sand from windward hollows and deposited it on leeward slopes. This type of dune seemed to be associated always with a cover of vegetation. It would appear that they are not original dune forms but rather forms produced by sand blowouts and subsequent redeposition. Hack described them as "parabolic dunes of deflation." Elongate ridges of sand extending parallel to the prevailing wind direction were designated as longitudinal dunes.

Madigan (1936) has described the striking longitudinal dunes in the Great Sandy, Simpson, and Victoria deserts of Australia and attributed their remarkable parallelism to the prevailing southeast trade winds. He found that the barchan or crescentic dune form is practically lacking there. The longitudinal dunes of Australia are fixed in position by a thin cover of vegetation. They appear to be the equivalent of the seif form of Bagnold, but, contrary to Bagnold's idea that they were formed by transverse winds, Madigan thought that they were produced by the continued drift of sand by strong winds dominantly from one direction.

Melton (1940) proposed a classification of dunes which recognized three groups of dunes: (*a*) simple dune forms produced by winds of constant direction, (*b*) dunes formed by wind in conflict with vegetation, and (*c*) complex dunes formed by cross winds. This classification emphasized more than any previously proposed the role that vegetation plays in the building and shaping of dunes and also pointed out the possibilities of inferring former wind directions from dune forms.

Dune cycles. Aufrere (1931) suggested a cycle of dune-field history which, although largely theoretical, is of interest because it suggests cyclical evolution in dune areas. His study of the great erg regions of the Sahara led him to emphasize the changing character of the area of sand accumulation, or chains of dunes as he called them, and the depressions between them, called *gassis*. He believed that chains of dunes grew largely by the transfer of sand from the gassis. The youthful stage of a dune field, according to him, is marked by more or less isolated dunes which develop transverse to the wind and take on the barchan form in desert regions or the parabolic form in areas where there is vegetation. Maturity is marked by a dune field in which dunes parallel to the wind exceed in height the transverse dunes of the intervening gassis. During old age, dune chains continue to grow in size at the expense of the gassis, but the dune field as a whole loses volume as the result of removal of material as dust. Senility is marked by reduction in size of dune chains and increase in area of the gassis until the underlying bedrock of the gassis covers more area than the dune fields. The final product of the cycle will be an area of deflation in which the bedrock comprises most of the area. This is the desert peneplain in which some students of arid land forms believe.

This proposed cycle may have some validity, but, as Bryan (1932) pointed out, it fails to take into consideration the complications arising from variations in wind directions, the tendency of vegetation to fix dunes, and the effects of climatic fluctuations in the Sahara region during glacial and postglacial times.

Smith (1939, 1940) has described a dune cycle that he thought applied to the dunes of western Kansas, an area where dunes are largely fixed in position by vegetation. Two phases are recognized, an eolian and an eluvial phase. The *eolian phase* is marked by diminished vegetal control and active dune growth. Blowouts are typical of this phase, and adjacent to them sand piles up under the control of vegetation to form sand sheets, ridges, or mounds. Dune growth is by accretion of either foreset, topset, or backset beds or by combinations of these. Steep foreset beds form only where sand is supplied more rapidly than the vegetation grows and as a result is carried over dune crests and deposited down their slip faces. Foreset bedding is rare in the dunes of western Kansas. Backset bedding is far more common; it forms where plant growth is more rapid than the influx of sand. Vegetation traps the sand and backs it up windward. Backset bedding is at a low angle in contrast to foreset bedding. Topset beds may be deposited over either foreset or backset beds. Most of the dunes in western Kansas grew backward and upward by accretion rather than forward and were fixed in position by vegetation from their beginning.

The *eluvial phase* of the dune cycle is a passive one and is marked by degradation of the dunes rather than by growth. It begins when vegetation becomes extensive enough to check deflation. During the eluvial phase soilforming processes, creep, and slope wash predominate. The result is gradual reduction of relief, stabilization of dunes, thickening of the soil profile and simplification of land forms ending in a dunal topography with a faint undulating surface. Partial reduction of vegetal cover may terminate the eluvial phase and renew the eolian phase. Thus multicyclic dunal forms are produced. The records of former partial cycles consist of stratigraphic unconformities and soil zones in the dunes.

Internal structure of sand deposits. There are three structural features that wind-blown sands commonly exhibit. They are cross-bedding, lamination, and abrupt changes from well compacted to poorly compacted or "quick sand" deposits. Cross-bedding is to be expected as a result of the many changes in wind direction and also because of the varying angles of deposition along the slip face of a fore dune as well as on its windward side. Lamination may not always be apparent to the eye but it will usually become so if water is permitted to seep through the sand. The individual laminae represent the successive layers which were laid down by accretion. Because of varying wind velocities at the time that the accretion layers were deposited, a slight variation in the texture of the individual laminae is common.

Bagnold has offered what seems to be a logical explanation for the commonly observed alternation of areas of firmly compacted sand and soft or loose sand, commonly alluded to as "dry quicksand." His explanation is that the areas of firm sand represent accretion deposits which were well compacted by the impact of particles moving by saltation, whereas the areas of loose sand are encroachment deposits laid down in front of dunal slip faces. Their looseness is the result of the way in which the sand came to rest, for, instead of having the sand grains fitted selectively into place by continued agitation and bombardment by saltation, as happens in accretion deposits, the deposits ahead of a slip face avalanche down its surface without subsequent compaction.

Loess. This name is applied to wind-blown silts which commonly are buffcolored, non-indurated, calcareous, permeable, particularly in the vertical direction, and consist of angular to subangular particles of quartz, feldspar, calcite, dolomite, and other minerals held together with a montmorillonite binder. Unweathered loess is usually gray in color, but because of its permeability exposures of unweathered loess are uncommon.

It has been argued at various times that loess was of alluvial, lacustrine, eolian, residual, fluvial-colluvial, or fluvial-eolian origin. Its eolian origin is now accepted by most geologists, but Russell (1940) and Fisk (1951) have questioned the eolian origin of the loess along the lower Mississippi Valley and maintained instead that it was largely colluvial materials derived from back-swamp alluvial deposits. This theory has not been widely accepted and Leighton and Willman (1950) and Smith (1942) have presented convincing arguments that the loess of the Mississippi Valley had an eolian origin, being derived in the main from glacial outwash. The loess thins away from the valleys that carried glacial outwash; has definite stratigraphic relationships with respect to particular drift sheets (actually several loesses exist in the Mississippi Valley); shows a decrease in the calcium carbonate content with increasing distance from the valley floors from which it was derived; and has faunal and floral assemblages which are in keeping with a theory of eolian origin.

Loess that was probably derived in the main from glacial outwash is found in North America along the Mississippi, Missouri, Ohio, and Wabash valleys and in eastern Washington and western Idaho. Loess of similar origin is found in Europe along the Rhine, Rhone, and Danube valleys, and most extensively in the Ukraine region of Russia.

Not all loess, however, was derived from glacial sources. In some areas the loess consists of materials of silt and clay size removed from arid regions by deflation. This type of loess has sometimes been called *desert loess* to distinguish it from that of glacial origin. The loess of northeastern China seems to have been derived in this way from the Gobi Desert. Loess deposits in the steppe region of Siberia and in Turkestan east of the Caspian Sea probably were of similar origin, as were more local accumulations in northern Africa.

Some have attributed a desert origin to the extensive deposits of loess in Oklahoma, Kansas, and Nebraska, but it seems unlikely that much of this body of loess could have been derived from deserts to the west. More likely it came chiefly from disintegrated Tertiary alluvium which mantles much of the Great Plains and in part from the valley floors of the rivers which cross these states.

A geomorphologist is interested in loess mainly as it affects the topography upon which it lies. Generally it is laid down as a mantle over preexisting topography and does not exhibit distinctive topographic forms. The so-called *paha hills* in Iowa have been considered loess ridges, but it seems more likely that they are veneers of loess over bedrock hills or drumlins. In general, a loess-mantled topography displays less relief and more subdued topography than adjacent areas lacking such a mantle. The most striking topographic effects are noted along rivers where loess may form remarkably steep bluffs.

FIG. 12.9.   Loess bluff north of Vincennes, Indiana.

This ability to maintain steep faces was responsible for the fact that the loess of the Mississippi Valley was called the "Bluff formation" by early geologists.

REFERENCES CITED IN TEXT

Aufrere, L. (1931).   Le cycle morphologique des dunes, *Ann. Geog., 41,* pp. 362–385.

Bagnold, R. A. (1933).   A further journey through the Libyan desert, *Geog. J., 82,* pp. 103–129.

Bagnold, R. A. (1941).   *The Physics of Blown Sand and Desert Dunes,* Chapters 12–17, William Morrow and Co., New York.

Berkey, C. P., and F. K. Morris (1927).   *Geology of Mongolia,* pp. 52–55 and 336–341, American Museum of Natural History, New York.

Blackwelder, Eliot (1928).   Origin of desert basins of southwest United States, *Geol. Soc. Am., Bull. 39,* pp. 262–263.

Blackwelder, Eliot (1931).   The lowering of playas by deflation, *Am. J. Sci., 221,* pp. 140–144.

Blackwelder, Eliot (1934).   Yardangs, *Geol. Soc. Am., Bull. 45,* pp. 159–166.

Bryan, Kirk (1926). Pedestal rocks formed by differential erosion, *U. S. Geol. Survey, Bull. 790-A*, pp. 1–15.

Bryan, Kirk (1932). Review of "Le cycle morphologique des dunes" by Aufrère, *Geog. Rev., 22*, pp. 325–327.

Fisk, H. N. (1951). Loess and Quaternary geology of the lower Mississippi Valley, *J. Geol., 59*, pp. 333–356.

Gautier, E. F. (1935). *Sahara: The Great Desert*, pp. 29–118, Columbia University Press, New York.

Hack, J. T. (1941). Dunes of the western Navajo country, *Geog. Rev., 31*, pp. 240–263.

Judson, Sheldon (1950). Depressions of the northern portion of the southern high plains of eastern New Mexico, *Geol. Soc. Am., Bull. 61*, pp. 253–274.

Leighton, M. M., and H. B. Willman (1950). Loess formations of the Mississippi Valley, *J. Geol., 58*, pp. 599–623.

Madigan, C. T. (1936). The Australian sand-ridge deserts, *Geog. Rev., 26*, pp. 205–227.

Melton, F. A. (1940). A tentative classification of sand dunes: its application to dune history in the southern high plains, *J. Geol., 48*, pp. 113–174.

Peel, R. F. (1941). Denudational landforms of the central Libyan desert, *J. Geomorph., 4*, pp. 3–23.

Russell, R. J. (1940). Lower Mississippi Valley loess, *Geol. Soc. Am., Bull. 55*, pp. 1–40.

Smith, G. D. (1942). Illinois loess—variations in its properties and distribution: a pedologic interpretation, *Univ. Illinois Agr. Expt. Sta., Bull. 490*, 184 pp.

Smith, H. T. U. (1939). Sand dune cycle in western Kansas, *Geol. Soc. Am., Bull. 50*, pp. 1934–1935.

Smith, H. T. U. (1940). Geologic studies in southwestern Kansas, *Kans. Geol. Survey, Bull. 34*, pp. 159–168.

Additional References

Bagnold, R. A. (1937). The transport of sand by wind, *Geog. J., 89*, pp. 409–438.

Blackwelder, Eliot (1929). Cavernous rock surfaces of the desert, *Am. J. Sci., 217*, pp. 393–399.

Bryan, Kirk (1923). Wind erosion near Lees Ferry, Arizona, *Am. J. Sci., 206*, pp. 291–307.

Gautier, E. F. (1926). The Ahaggar: heart of the Sahara, *Geog. Rev., 16*, pp. 378–394.

Price, W. A. (1950). Saharan sand dunes and the origin of the longitudinal dune; a review, *Geog. Rev., 40*, pp. 462–465.

Smith, H. T. U. (1940). Review of "A tentative classification of sand dunes, etc.," by F. A. Melton, *J. Geomorph., 3*, pp. 359–361.

13 · Karst Topography

The geomorphic significance of rock solution has been previously considered in connection with rock weathering and as it contributes to the reduction of land masses, particularly in the late stages of the fluvial cycle. In certain regions, however, solution becomes a dominant process in landform development with resulting production of a unique type of topography to which the name karst has been applied. The word *karst* is a comprehensive term applied to limestone or dolomite areas that possess a topography peculiar to and dependent upon underground solution and the diversion of surface waters to underground routes. The term comes from the narrow strip of limestone pl.. eau in Jugoslavia and adjacent portions of Italy bordering the Adriatic Sea, where there exists a remarkable assembly of features dependent upon subsurface solution (Wray, 1922). Although the area in Jugoslavia has given its name to the general assemblage of land forms, not all the names of individual karst features have come from there. Nearly every country has its own peculiar set of karst terms. A few of the foreign names have been taken into the English terminology, particularly when a good English equivalent was lacking, but in general it seems desirable to use available English forms.

IMPORTANT KARST AREAS

Most of the notable karst areas are in regions where limestones underlie the surface, although in some localities the rocks are dolomites or dolomitic limestones. Solutional features may develop upon other soluble rocks such as gypsum and rock salt, but in general they are not of major importance because of the limited areal extent of these rocks. Limestones are abundant in nature; hence it might be expected that karst topography would also be widespread. Actually, full development of karst features is restricted to a relatively small number of localities, in view of the extensive distribution of limestones.

Significant karst development is to be found, in addition to the type locality, in the Causse region of southern France, Spanish Andalusia, Greece, northern Yucatan, Jamaica, northern Puerto Rico, western Cuba, the coastal plain fringing the Great Australian Bight, central Florida, the Great Valley of Virginia and Tennessee, southern Indiana, west-central Kentucky, north-central Tennessee, and numerous other localities.

In any of the above-mentioned areas numerous karst features may be seen but in none are all the possible individual forms found. A comprehension of the varying aspects of karst topography can be obtained only by studying several regions because they exhibit varying stages of karst development and different types of geologic structures. As far as the United States is concerned, probably the Indiana-Kentucky-Tennessee area presents the best region for the study of the manifold manifestations of karst features.

LESSER KARST AREAS

In addition to the areas mentioned above there are innumerable areas where karst features are present but do not dominate the landscape. This may be a result of the topographic youthfulness of the karst features, but more commonly it is attributable to the absence of one or more of the essential conditions prerequisite to ideal karst development. Wherever there is found soluble rock such as limestone, dolomite, rock salt, or gypsum beneath the land surface, some solution may be expected. Almost every state in the United States has within it such soluble rocks and local features, such as caverns, but caves in themselves do not constitute karst topography. The area in which is found what is probably the world's largest and most beautiful cavern, Carlsbad, in southeastern New Mexico, is a case in point. It is doubtful if any geomorphologist would characterize the plateau beneath which this cavern lies as an area of typical karst topography. Certainly it exhibits few of the features found in the type region or in most of the other important karst areas. The chalk region of England and France is another example of an area which exhibits some karst features, but the assemblage is too incomplete for it to be classed as a major area.

CONDITIONS ESSENTIAL TO FULL DEVELOPMENT OF KARST

What then are the conditions which must be met for the development of karst par excellence? Four conditions which contribute to maximum development of karst are essential. First, there must be present at or near the surface a soluble rock, preferably limestone. Dolomite may suffice but it is not so readily soluble as limestone. Chalk is soluble, but it usually lacks one of the other prerequisites. Secondly, and this is one of the most important factors, this soluble rock should be dense, highly jointed, and preferably thinly bedded. This is a point frequently lost sight of by persons who lack first-hand familiarity with karst. It is sometimes stated in textbooks that the major prerequisite for karst development is the presence of a permeable or porous limestone. As a matter of fact, permeability in the sense of mass permeability is unfavorable. Permeability as permitted by numerous joints and bedding planes is favorable. If a rock is highly porous and permeable throughout, rainfall will be absorbed en masse and move through the whole body of the rock rather than be concentrated along restricted lines of movement. This is perhaps the major reason why karst features are weakly developed in the chalk region of England and France. The importance of this factor is well-illustrated in the karst region of southern Indiana. In the physiographic region known as the Mitchell Plain there are four limestone formations of Mississippian age known as the Harrodsburg, Salem, St. Louis, and Ste. Genevieve. Each is a relatively pure limestone, but they differ greatly in their structural characteristics. The Harrodsburg is a dense, crystalline limestone characterized by a moderate number of bedding planes and joints; the Salem is a massive, relatively permeable limestone with thick beds and widely spaced joints; the St. Louis limestone is a dense, thinly bedded and highly jointed limestone, as is the Ste. Genevieve. It is in these two latter limestones that we find karst features most excellently developed. Sinkholes and caves do form in the other two, but not to the degree that they do in the St. Louis and Ste. Genevieve.

A third condition essential to excellent karst development is that there exist entrenched major valleys below uplands underlain by soluble and well-jointed rock. It is essential that groundwater be able to descend through a limestone, carry on its solutional work, and emerge into surface streams. The importance of this factor will be elaborated upon later in the discussion of the Lost River region of southern Indiana.

Finally, such a region should be one with at least a moderate amount of rainfall. It is significant that nearly all the notable karst regions are in areas of moderate to abundant rainfall. One apparent exception is the area in Yucatan, but it is probable that during the pluvial epochs of the Pleistocene the rainfall there may have been considerably greater than it is now. In general, arid and semiarid regions with limestone terrains do not exhibit marked development of karst, although certain of the features may be found.

FEATURES CHARACTERISTIC OF KARST REGIONS

It is not intended that the student suppose that all the features discussed below will be found in any one area, but an attempt will be made to describe the more important forms that may be found at one place or another. Some of the features discussed are not strictly topographic in nature, but they are so intimately associated with topographic forms that they merit description. Others may be present only here and there and still be significant.

Terra rossa. Surface and near-surface solution by descending ground-water usually leaves a residue of a red, clayey soil mantling the surface and extending down into opened joints. This material may be lacking on steep slopes, but it is characteristically present on moderate to gentle slopes. It may vary in thickness from a few to many feet and may completely mask the rock surface. The name *terra rossa* has been aptly applied to it. Terra rossa resembles in appearance the lateritic soil of the tropics, if indeed it is

FIG. 13.1. Terra rossa resting upon limestone in which there are numerous solutionally enlarged joint openings. (Photo by C. A. Malott.)

not a species of it. However, it is by no means limited to tropical or sub-tropical locations, being found as far north as southern Europe and southern Indiana.

Lapiés. Locally, where relief is considerable, limestone surfaces are bare of terra rossa and there is exposed an etched, pitted, grooved, fluted and otherwise rugged surface to which the name *lapiés* is most commonly applied. Cvijić (1924) has described and pictured the amazing diversity of surface and form which lapiés exhibits in the Dalmatian karst region. He maintained that lapiés is found chiefly on outcrops of naked rock and pointed out the influences of such factors as rock composition, texture, structure, surface slope, and amount of vegetation upon its formation. He further claimed that it was rare on horizontal rocks, being replaced there by sink-holes. Others have argued that it develops under a soil and vegetation cover and is exposed later by erosion accelerated perhaps as a result of deforestation. Lapiés surfaces are rarely seen in the Indiana-Kentucky karst area, but where the terra rossa has been removed artificially we find a surface exhibiting most of the characteristics of true lapiés. Whether lapiés should be applied to any solutional grooving of bare rock surfaces may be open to question. Smith and Albritton (1941), in discussing solutional effects upon limestones in the Sierra Blanca region of Texas as related to variations in slope, pointed out a gradation from shallow solutional pits on gentle slopes through solution facets on moderate slopes to solution furrows on steep slopes which they designated as lapiés. Palmer (1927) has also described

FIG. 13.2. Lapiés near Mitchell, Indiana, exposed by removal of soil for use in road fill. (Photo by C. A. Malott.)

both small and large-scaled solution grooves in basalt in Hawaii to which he applied the term lapiés. It may be questionable whether these forms exhibit the complexity of form and surface which characterizes the lapiés of karst areas. Lapiés shows, perhaps better than any other karst feature, how minute differences in rock solubility, permeability, jointing, bedding, and other physical and chemical attributes influence the rate and direction of solution by descending meteoritic waters.

Sinkholes and associated forms. By far the most common and widespread topographic form in a karst terrain is the sinkhole. In any major karst area sinkholes are found by the hundreds of thousands. Some idea of their frequency is indicated by a detailed mapping of 1 square mile of sinkhole terrain that was done by Malott and Shrock (Malott, 1945) in Orange County, Indiana. In this square mile they counted and mapped 1022 sinkholes. Malott estimated that there may be as many as 300,000 sinkholes in the karst area of southern Indiana.

Topographically, a sinkhole is a depression that varies in depth from a mere indentation of a few feet to a maximum of 100 feet or even more. Most of them vary in depth from 10 to 30 feet. In area, they range from a few

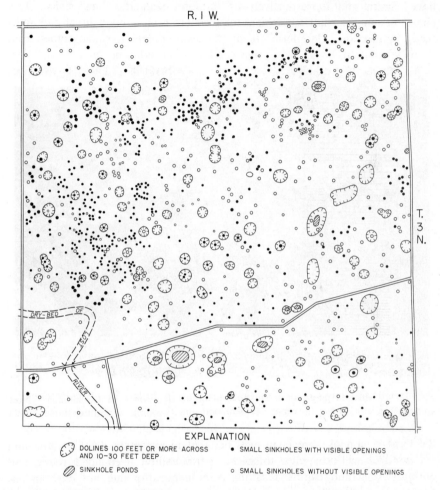

EXPLANATION

⬤ DOLINES 100 FEET OR MORE ACROSS AND 10-30 FEET DEEP   ● SMALL SINKHOLES WITH VISIBLE OPENINGS

⬤ SINKHOLE PONDS   ○ SMALL SINKHOLES WITHOUT VISIBLE OPENINGS

FIG. 13.3. Map of the sinkholes in 1 square mile southwest of Orleans, Indiana. There are 1022 sinkholes shown. (After C. A. Malott.)

square yards to an acre or more. The most common form is a funnel-shaped depression broadly open upward, but there are many variations from this.

Any attempt to classify sinkholes encounters difficulties because of the many variations that they exhibit and the varying local usage of terms applied to them. Fundamentally, they fall into two major classes, those that are developed slowly downward by solution beneath a soil mantle without physical disturbance of the rock in which they are developing and those that are produced by collapse of rock above an underground void. These two types have been referred to, respectively, as solution sinks and collapse sinks. This choice of names is not too logical for·they are both caused directly or indirectly by solution. The term *doline* (Serbian—*dolinas*), which comes from

Fig. 13.4. Vertical photo of a sinkhole plain in Harrison County, Indiana. Note complete lack of surface streams. (Production and Marketing Administration photo.)

the karst region of Jugoslavia, will be used to designate the first type and *collapse sink* will be applied to sinkholes that exhibit steep-sided, rocky, and abruptly descending forms, resulting from collapse of the roof over an underground solutional opening. Dolines are far more common than collapse sinks, even in the mature stage of karst development. Runoff, which drains into dolines, usually finds its way to underground routes by slow percolation through the soil in the bottom of the sinkhole, but surface waters enter some sinkholes through surface openings called *swallow holes*.

A variant of the doline is the *solution pan*. It differs from the typical doline in that it is much shallower and may embrace a much larger area. One, in the upper Lost River region in Washington County, Indiana, occupies more than 30 acres. Dolines may become clogged with inwashed clay to such an extent that they will hold water above the water table. These are called *sinkhole ponds* or *karst lakes,* if their size merits the latter designation. Sinkhole ponds and lakes whose surfaces coincide with the top of the water table are common features in the Floridian karst area.

In a region where sinkholes are numbered by the hundreds per square mile, individual sinkholes expand in diameter and coalesce to form *compound sinkholes.* These usually consist of a major solutional depression covering

Fig. 13.5. A small sinkhole pond. (Photo by P. B. Stockdale.)

several acres with numerous lesser dolines or swallow holes superposed upon it. A special type of collapse sinkhole is the *karst window.* This term has been applied (Malott, 1932) to an unroofed portion of an underground stream course through which may be seen a stream which flows out of a cavern at one side, across an open space, and into a cavern at the opposite side. The opening may vary from a mere peephole to one of considerable size. Some karst windows have become greatly enlarged and possess alluviated floors across which cavern streams briefly flow as surface streams. The Bosnian term *uvala* is most commonly applied to the larger depressions resulting from the collapse of extensive roof sections over underground watercourses. What have been designated above as compound sinkholes are sometimes called uvalas, but this usage of the term does not seem justified. The development of karst windows and uvalas represents a significant stage in the evolution of karst topography, as will be pointed out in the discussion of the karst cycle.

A somewhat special form of sinkhole is the *polje*. This name has been applied to depressions of various forms and origins, but the typical Bosnian polje is an elongated basin with a flat floor and steep enclosing walls which owes its existence to solutional modification of downfaulted or downfolded limestone blocks. A polje may somewhat resemble an uvala, but it differs from it in origin and extent. A typical polje is a sizable feature covering many square miles, whereas an uvala commonly covers only a few acres. The largest polje in the Western Balkans, the Livno polje, is 40 miles long and 3 to 7 miles wide. As far as possible the term polje should be restricted to these structure-controlled solutional forms.

Other features of a sinkhole or karst plain. In regions of nearly horizontal or gently dipping limestone strata, there develops what is most aptly designated as a *sinkhole* or *karst plain,* which over large areas acts as a veritable regional sieve with thousands of sinkholes acting as hoppers which convey meteoric waters to underground routes. The sinkholes are so effective in collecting surface runoff that few streams are able to cross a sinkhole plain. Larger streams which rise beyond it and are entrenched below it may do so, but lesser streams lose their waters to underground routes. They are then called *sinking creeks* and the point at which each terminates is its *sink*. The sink of a surface stream usually takes place in an observable swallow hole, but in some streams the water disappears through alluvial materials upon the stream bed so that the exact point of disappearance is uncertain. Some sinking creeks disappear in a single swallow hole, but others may have several and use one or all of them, depending upon the volume of water in the stream. Sinking creeks constitute an important aspect of karst terrains in relation both to their surface forms and to the development of caverns beneath a sinkhole plain. The underground course followed by a sinking creek before it reappears may be several miles long. Lost River in southern Indiana flows some 8 miles underground before it again becomes a surface stream. The Reka River, back of Trieste, is said (Wray, 1922) to have a subterranean section of its course that is over 18 miles long.

The little-used or unused continuation of a valley below the point where the stream in it sinks is called its *dry bed*. Most dry beds are used only during periods of heavy rainfall when swallow holes and connecting underground conduits are unable to accommodate all the storm waters. Then streams may flow for a period in their dry beds until the storm waters have subsided enough to allow the underground courses to accommodate them. It frequently happens that a sinking creek may have terminated at a swallow hole for so long that it has cut its valley upstream from its sink so much below the level of the karst plain that the valley ends at the swallow hole. Such a valley is called a *blind valley*. The surface continuation of the dry bed below the sink through long-continued disuse becomes obscured and gradually loses the characteristics of a surface valley. During periods of storm waters a blind valley may become the site of a temporary lake as a result of the inability of the swallow hole or holes to accommodate the increased volume of water. It may require several days or even weeks for these temporary lakes to drain. Ponded waters in broad, shallow blind valleys often present a real flood menace to towns located upon a karst plain and may interrupt traffic on highways across them where fills have not been made higher than the levels reached by the storm waters.

FIG. 13.6. Vortex produced as stormwaters of Lost River enter a swallow hole in its bed. Only after heavy rains does water follow this dry-bed route. (Photo by C. A. Malott.)

A special type of blind valley, sufficiently distinct from that just described to merit a special name, is the *solution valley* or *karst valley* (Malott, 1939). It differs from an ordinary blind valley in that it is not part of a typical karst plain but rather is completely or nearly enclosed by clastic rocks. It is a particularly significant karst feature because its development throws light upon the process by which a limestone terrain may expand at the expense of a region that originally was drained entirely by surface drainage and had the topographic characteristics produced by surface streams. A solution or karst valley represents a transitional stage between surface drainage and underground drainage. Sloans Valley, Elk Spring Valley, Logsdon Valley, and Waterloo Valley, shown on the Burnside, Monticello, Mumfordville, and Horse Cave, Kentucky, topographic sheets, respectively, are good examples of solution or karst valleys. In Indiana, they are extensively developed in

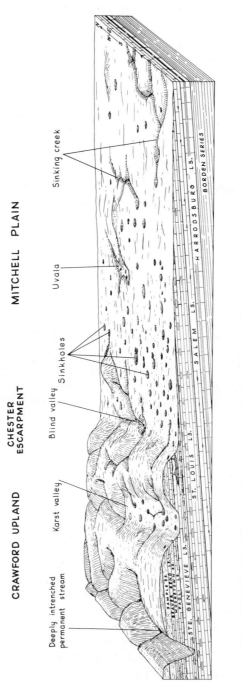

CRAWFORD UPLAND   CHESTER   MITCHELL PLAIN
                       ESCARPMENT

Deeply intrenched   Karst valley   Blind valley   Sinkholes   Uvala   Sinking creek
permanent stream

STE. GENEVIEVE LS.   ST. LOUIS LS.   SALEM LS.   HARRODSBURG LS.   BORDEN SERIES

FIG. 13.7.   Idealized diagram of a portion of the karst region of southern Indiana.   (Drawing by William J. Wayne.)

the Crawford Upland, immediately west of the Mitchell Sinkhole Plain. The geological conditions in Indiana and Kentucky are similar. In both areas there is a sinkhole plain developed upon limestones separated by an escarpment, the Chester escarpment in Indiana and the Dripping Springs escarpment in Kentucky, from a higher area of clastic rocks. The low plateaus adjacent to the karst plains in the main exhibit stream-formed topography, but at their edges adjacent to the karst plain there are numerous karst valleys. Rugged sandstone ridges separate deeply entrenched valleys which have cut their floors down to limestone. Surface drainage has begun to disintegrate through abandonment of surface routes for underground ones. Tributary valleys from the sandstone and shale hills descend into the limestone valleys and disappear in swallow holes. Dismemberment of a portion of the surface drainage results. In many karst valleys it is impossible to find evidence of the former surface trunk stream courses because their floors are dotted with sinkholes and other karst features. Some still exhibit dry-bed stretches which either end at swallow holes or carry storm waters to the lower ends of the karst valleys. Many of the dismembered side valleys terminate in swallow holes as blind valleys within the major karst valley. The underground drainage of karst valleys may be diverted completely to near-by, more deeply entrenched surface streams or it may reappear further down the same valley.

Return to the surface of the waters of a sinking creek is a common phenomenon in karst areas. Springs of varying sizes and volumes are thereby produced. The terms *rise* and *resurgence* have been applied to the reappearance of surface waters which have been diverted to underground routes. Some rises can be associated with specific sinking creeks but more often the source of the water is vague and undetermined. The larger rises are artesian in character and issue from subcircular *rise pits* of considerable size and depth. The numerous large springs of Missouri, Indiana, and Kentucky, in the main, belong in this class. Perhaps the most famous example is Silver Springs, Florida, which is noted equally for its volume and strikingly clear waters. Springs that mark the resurgences of sinking creeks vary greatly in volume and clearness of their waters. During dry spells their volumes decrease and their water may be rather clear, but during periods when heavy rains have increased the flow of the sinking creeks and added great quantities of mud and silt to them they increase greatly in size and become turbid. One can hardly see the muddy waters emerging from one of these rise pits without being convinced that there exists beneath the region a large water-filled cavern which terminates at the rise. Most writers on the origin of caverns have not given enough attention to sinking creeks and their rises. It may require considerable time and effort, but it is often possible to ascertain whence came the waters that emerge as these large artesian springs. By dumping such coloring matter as fluorescein, or even such materials as sawdust or oats, into near-by swallow holes it may be possible to ascertain where the waters that empty into them reappear.

Not all springs, however, in karst areas represent the rises of the waters of sinking creeks. Usually only the larger ones do. Most of the smaller

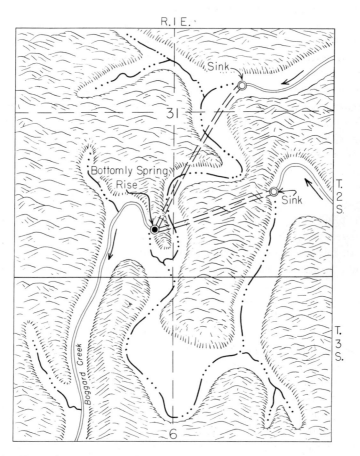

FIG. 13.8. Two subterranean cutoffs along Bogard Creek, Crawford County, Indiana. (Drawing by John Minton.)

springs found along hillsides are simple gravity springs in which the water collected in sinkholes and swallow holes moves down through the rocks until it comes in contact with an impervious bed and then is directed to an outlet where the impervious bed crops out along a valley side.

An interesting and sometimes unappreciated manifestation of karst development is the *subterranean cutoff,* which may be defined as an underground diversion of a part or the whole of a surface stream beneath a meander spur

along an entrenched valley.   A subterranean cutoff consists of three parts, a
swallow hole on the upstream part of the meander spur into which the water
sinks, an underground route or conduit through which the water runs, often
called a *natural tunnel,* and a rise or resurgence of the water on the downstream side of the meander spur.

F̦ɪɢ. 13.9.   Artesian spring marking the rise of the waters that flow through the subterranean cutoffs shown in Fig. 13.8.

Natural tunnels and bridges.   A spectacular example of a natural tunnel,
which is either the result of a subterranean cutoff or subterranean stream
piracy, is the Natural Tunnel of Virginia (Woodward, 1936).   Natural Tunnel is located in Scott County, Virginia, about 3 miles north of the town of
Clinchport and about 8 miles north of the Tennessee-Virginia state line.
This natural tunnel is unique because it serves as a railway tunnel for a
branch of the Southern Railway.   It is 900 feet long and is followed by
Stock Creek.   It averages 75 feet in height and 130 feet in width.   Woodward
favored the theory that it developed as a result of subterranean stream
piracy, resulting in the diversion of surface waters from a higher topographic
level on one side of Purchase Ridge to a lower topographic level on the other
side of the ridge.   He suggested that at the time of the development of the
Harrisburg peneplain the headwaters of present Stock Creek did not flow
into the Clinch River as at present, but rather followed the course of the
North Fork of the Clinch.   The portion of present Stock Creek below
Natural Tunnel was at that time the site of a short tributary to the Clinch
River.   The floor of the Clinch River was thus some 300 feet lower than the

floor of the valley of the North Fork north of Purchase Ridge. This relationship allowed water from the higher stream to seek an underground route, and thus the upper part of the North Fork was diverted to the short tributary of the Clinch, which flowed down the south side of Purchase Ridge and thus

FIG. 13.10. Natural Tunnel, Scott County, Virginia. The bed of Stock Creek shows to the left of the Southern Railway tracks.

became present Stock Creek which flows through Natural Tunnel. Natural Tunnel is only the remnant of a once much longer subterranean passageway. Another possible explanation of Natural Tunnel is that it was produced as a result of the development of a subterranean cutoff by Stock Creek through the ridge beneath which the Natural Tunnel extends.

A rather striking case of subterranean stream piracy, described by Beede (1911), involves an area shown on the Bloomington, Indiana, topographic sheet. Here some 15 square miles of former surface drainage of Indian Creek have been diverted by underground routes to adjacent Richland Creek on the west and to Clear Creek on the east, because the valleys of these streams had been cut 100 to 150 feet below the level of the sinkhole plain between them, across which Indian Creek at one time flowed. This piracy, however, was not through a single underground channel but through innumerable underground routes which collect the waters from hundreds of sink-

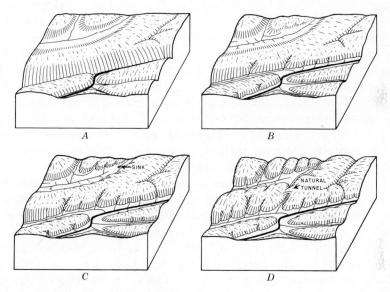

Fig. 13.11. Stages in the evolution of Natural Tunnel, Virginia, according to the theory that it formed as the result of subterranean stream piracy. (After H. P. Woodward.)

holes. The diverted surface waters reappear at lower levels as springs emerging along hillsides or at ravine heads. This seems to be the more common method by which subterranean stream piracy takes place. The presence of a single underground route, such as is exhibited at Natural Tunnel, seems more suggestive of a subterranean cutoff.

Collapse of the roof of a natural tunnel will in time reduce its length until it is better described as a *natural bridge*. Since natural bridges are developed in several ways, those resulting from solution by groundwater may well be designated as *karst bridges*. The most famous karst bridge in this country, if not in the world, is Natural Bridge, Virginia. It is located in Rockbridge County, 14 miles southwest of Lexington. Just as Natural Tunnel is unique because it serves as a railway tunnel, Natural Bridge is unique in that it serves as a highway bridge. U. S. Highway 11, a much-traveled federal highway, uses its span across the gorge of Cedar Creek. Natural

Bridge consists of a span of massive magnesium limestone 40 to 50 feet in thickness, 90 to 100 feet in length, and 50 to 150 feet in width. The bottom of its arch is about 150 feet above Cedar Creek. It is situated on the slope of a small hill, which extends about 100 feet above it. Obviously, any theory as to its origin must account for the gorge of Cedar Creek which it spans.

Thomas Jefferson, who owned Natural Bridge for a time, suggested in 1794 that it was the result of some great "convulsion of nature." This idea, of course, merely reflected the cataclysmic ideas regarding land-form development that prevailed at that time. About a quarter of a century later, Gilmer (1818) attributed the formation of Natural Bridge to the solutional work of groundwater and argued against a cataclysmic origin. He thought that the solution was carried on by some sinking creek, of which there are numerous examples in Virginia, and cited Natural Tunnel as being similar to Natural Bridge. Thus the solutional origin was established rather early. Gilmer was vague, however, in regard to the conditions that made possible the diversion of a surface stream to a subterranean route. The argument since then has been largely about the particular method by which diversion of surface waters to an underground route took place.

Three distinct ideas have been proposed to account for Natural Bridge. Walcott (1893) advanced an idea that found its way into many textbooks and is found even yet in some. According to his theory, Cedar Creek once flowed as a surface stream across what is now Natural Bridge. At the time it did so the portion of the gorge of Cedar Creek upstream from Natural Bridge was non-existent. The James River, to which Cedar Creek is tributary, underwent rejuvenation, and in turn Cedar Creek felt the effect of this rejuvenation and started cutting a gorge headward. A waterfall was believed to have existed in this gorge at a point not far below the site of Natural Bridge. Water from the stream found an underground route above the falls through which it flowed and discharged below the brink of the falls. As this passage was enlarged more and more water was diverted to the underground route until finally all the water followed this route, leaving a natural or karst tunnel which through time has dwindled in size to the present bridge.

A second theory (Woodward, 1936) is that Natural Bridge was produced by subterranean stream piracy. According to this theory, the upper part of present Cedar Creek was once a part of the drainage of eastward-flowing Poague Run and what is now the lower part of the valley of Cedar Creek was occupied by a short stream which was the predecessor of present Cascade Creek. Both streams were tributaries of the James River, but it was only 3 miles to the James River via the ancestral Cascade Creek, whereas it was 15 miles by way of Poague Run. There may have been a difference in elevation of as much as 200 feet in the two tributaries of the James. As a result, the headwaters of Poague Run were diverted through the underlying soluble limestone to Cascade Creek, and an underground tunnel was formed which was subsequently deroofed except for the part which remains as Natural Bridge.

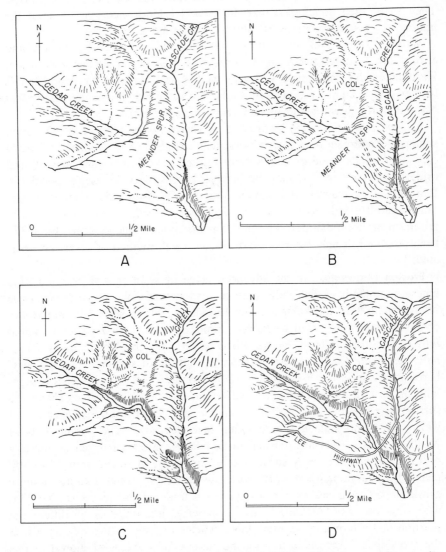

A   B

C   D

FIG. 13.12.   Stages in the development of Natural Bridge, Virginia, according to the theory that it is a roof remnant of a subterranean cutoff.   (After C. A. Malott and R. R. Shrock.)

A third theory, that of Malott and Shrock (1930), is that Natural Bridge is the result of a subterranean cutoff by Cedar Creek through a meander spur. This subterranean cutoff enlarged to the point where all the water of Cedar Creek went through it developing a karst tunnel which was gradually deroofed until only a small part of its roof was left as Natural Bridge. They based their conclusion that the meander loop once existed upon the finding of what they considered alluvial gravels along its course.

Wright (1936) did not believe that the evidence cited by Malott and Shrock for a subterranean cutoff through a meander spur was convincing. He doubted that the gravels noted by Malott and Shrock were proof of this former stream course followed by Cedar Creek and preferred the theory that Natural Bridge was a result of subterranean stream piracy.

It appears that natural bridges in limestone regions result chiefly from subterranean stream piracy or subterranean cutoffs. Only detailed field study can show which mode of origin more logically explains a particular natural bridge, and even then it may be difficult to establish the origin positively. If a high-level abandoned meander loop is still discernible, then a subterranean cutoff is indicated, but, if the diversion of surface waters to an underground route took place so long ago that the topographic evidence for it has been obscured, it may be impossible to say which method of diversion operated.

Erosion remnants. In an advanced stage of solutional destruction of a limestone upland, residual hills analogous to monadnocks in the fluvial cycle often remain. In the karst area of Jugoslavia, where such hills are rather common, they are known as *hums*. Similar features are known as *pepino hills* or *haystack hills* in Puerto Rico, as *mogotes* in Cuba, and as *buttes temoines* in the Causse region of France. The most striking examples, perhaps, are those in the West Indies. The mogotes in the Pinar del Rio highland of western Cuba (Bennett, 1928) stand 1000 to 1200 feet above the surrounding lowland and are full of caverns and lesser solutional openings in which are found stalactites and other cave deposits. The pepino hills of northern Puerto Rico are unusual in that they exhibit a marked asymmetry of form. They exist singly and in groups and range in height from 25 to 300 feet. Their asymmetrical sawtooth cross section typically exhibits a more gentle slope facing east or northeast into the trade winds. Hubbard (1923), who observed and described this asymmetry, attributed it to differential solution on eastern and western slopes related to temperature differences on the two slopes. According to him, the almost daily afternoon shower of this region is more nearly evaporated on the warmer western slope than on the cooler eastern slope; hence, there is less solution on the western slope than on the east with resulting asymmetry of form. Thorp (1934) has offered another explanation for the asymmetry. He attributed the greater solution on the eastern slope to the fact that the prevailingly strong, northeasterly trade winds drive more water against the windward slope than against the western leeward slope with greater solution resulting on the eastern side. Meyerhoff (1938) noted the absence of asymmetrical form in the mogotes of western Cuba and suggested that this was perhaps explained by the fact

Fig. 13.13. Pepino Hills near Arecibo, Puerto Rico. The local relief is about 200 feet. (Photo by R. N. Young.)

that here the rains are more convectional in origin and lack a notable lateral component of movement, along with the fact that the pepino hills of Puerto Rico are much smaller than the mogotes of Cuba and hence more commonly rise to peaks rather than have flat summits. Such residual hills have not been described to any extent in the karst regions of the United States. Perhaps their absence or scarcity is related to the nearness of the water table to the surface in most of the American karst regions.

Caverns and associated features. A *cavern* or *cave* may be defined as a natural subterranean runway void. It may be simple in plan or have complex ramifications. It may extend vertically or horizontally, and it may occupy one or more levels. It may or may not be presently occupied by a stream. Dry caverns usually have two or more levels and thus may be referred to as galleried caverns. As many as five cavern levels have been observed. Smaller caverns usually show rather clearly that their development took place along lines largely controlled by joints and bedding planes. These joints and bedding planes are systematic three-dimensional features which have been enlarged by selective solution by water moving along them. Frequently the control of joint systems is strikingly evident in a cave pattern. This is especially true of young caves, but in more complex caverns it may be obscure.

Fɪɢ. 13.14.   Former and present cave outlets at Sequiota Spring, Missouri.   (Photo by Dale Miller, Missouri State Highway Department.)

Caverns exhibit a variety of depositional features which add to their beauty or interest and throw light upon their mode of origin and history. The most striking features are usually the accumulations of calcium carbonate on the ceilings, walls, and floors. *Cave travertine* is an inclusive term generally applied to the many forms. Davis (1930) suggested that *dripstone* be used to include the forms developed by water dripping from the cavern ceiling. This would include the downward extending *stalactites,* the upward-growing *stalagmites,* and *columns* and *pillars* produced when the preceding forms grow together. Davis further suggested the term *flowstone* for the forms made by flowing water and *rimstone* for those found where waters overflow from basins. Neither of the two latter terms has come into wide use. In addition to stalactites, stalagmites, and columns, which exhibit an infinite variety of

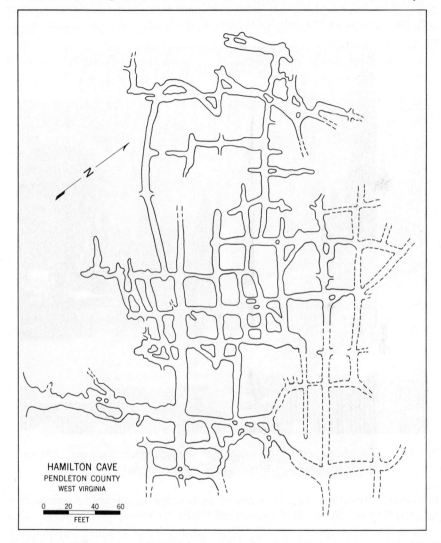

HAMILTON CAVE
PENDLETON COUNTY
WEST VIRGINIA

0   20   40   60
FEET

FIG. 13.15. A cavern plan which shows notable joint control. (After W. E. Davies, *West Virginia Geol. Survey, Bull. 19*.)

shape and size, there is found in portions of some caves a form known as a *helictite*. It is unusual in that its growth does not necessarily extend along vertical lines. Its individual parts may grow upward, horizontally, obliquely, or in curves, as well as downward. It is not certain why helictites are apparently able to defy gravity, but the most logical explanation seems to be that they develop where water is not entering the cave in sufficient quantity to give rise to drops that fall. If there is only enough water to keep the surface wet, then growth of the individual parts of the helictite depends upon chance

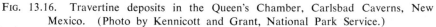

Fig. 13.16. Travertine deposits in the Queen's Chamber, Carlsbad Caverns, New Mexico. (Photo by Kennicott and Grant, National Park Service.)

orientation of the crystal axes of calcium carbonate. This may be in any direction.

Various attempts have been made to estimate the rate of formation of cave travertine, but so many variable factors affect the rate of deposition that it is doubtful if cavern ages arrived at by this method are accurate. It is possible to arrive at the geologic age of caverns if the time of cavern development can be related to datable topographic levels such as stream terraces and other cyclic land forms in adjacent areas. Even then we are arriving at comparative rather than actual age.

Less beautiful than the many forms of cave travertine are other features of caverns, which, however, often throw much light upon the origin and his-

Fig. 13.17. Monument Mountain, Wyandotte Cave, Indiana. The pile of fallen rock is 135 feet high and is capped with flowstone and stalagmites. (Photo by Perry Griffith.)

tory of the cavern. Few cavern routes exhibit uniformity of size for a great distance. Rather, they characteristically expand and contract in size. Large cavern rooms like the famous Big Room in Carlsbad Cavern, although most are of much smaller proportions, are striking features. In the semidarkness in which they are viewed their dimensions usually seem exaggerated. Their high, vaulted ceilings may be covered with stalactites, attesting to their stability. In many instances, however, there are large piles of rock debris on the floors of the rooms, as in Rothrocks Grand Cathedral in Wyandotte Cave,

Indiana. Here a mass of fallen rock forms what is known as Monument Mountain, which with its cap of flowstone and stalagmites rises 135 feet above the floor of the room to within 35 or 40 feet of the ceiling. Such features of rockfall attest to the existence of a cavern in partial ruins. Generally these rock mounts are more prominent in the parts of a cavern where the limestone is more thinly bedded and highly jointed. In the portion of Wyandotte Cave known as the Old Cave, in a distance of little more than one-half mile there are seventeen of these mounds of rock infall varying in height from 30 to 60 feet.

Large cavern rooms are usually connected by relatively simple passageways or conduits whose floors are nearly level and usually consist of deposits of clay and silt. These graded floors attest to the fact that a stream once flowed through the cavern. They may be of such proportions as to be designated as avenues, like Gothic Avenue and Blacksnake Avenue in Mammoth Cave or Creeping Avenue in Wyandotte Cave, which got its name from the fact that the passageway was filled with clay and silt to within 3 or 4 feet of the ceiling. Stream channels with associated mudbanks and bars may often be observed. Sometimes pebbles and boulders of sandstone or other non-calcareous rocks may be found in these runways. These entered the cavern through surface swallow holes connecting downward with the cavern stream. In some caverns, the opening into a room is so constricted that it merits such a name as "auger hole" or "fat man's misery." More frequently, perhaps, these constricted portions connect one level of a cavern with another. The Corkscrew in Mammoth Cave and Worm Alley in Wyandotte Cave are examples of such connections. Many other features add detail to the characteristics of individual portions of cavern routes. Grooves and flutings in the sides, floors, and ceilings attest to erosion, both by corrosion and corrasion, by the waters which flow or flowed through the cavern. Potholes, pockets, and tube-like openings are also commonly observed.

THE ORIGIN OF LIMESTONE CAVERNS

General statement of problem. In 1930 an extended paper by W. M. Davis entitled "Origin of Limestone Caverns" challenged some of the existing ideas of cavern formation and stimulated the thinking of geologists who were interested in limestone terrains regarding the mode of development of caverns. As a result ideas regarding caverns have been considerably clarified although there still remains divergence of opinions as to their origin. Actually, prior to 1930, ideas of how caves formed were rather vague. It was taken for granted that they were produced by solution by groundwater, and there the matter usually ended. The prevalent idea was somewhat as follows (Matson, 1909):

As the belt of rapid solution is restricted to the zone of the active water circulation, the formation of caverns takes place largely above the level of the surface streams that receive underground drainage, but this does not imply that there is no deep-seated solution or that active circulation may not extend slightly below the level of surface drainage · · ·. In the process of deepening its valley a stream lowers the base-level of its tributaries and thus affords them an opportunity to degrade their channels. When the tributaries flow in caverns they lower their channels along some joint plane, and the process goes forward most rapidly along some particular line in the same manner as during the formation of the original cavern · · ·. Doubtless the formation of caverns and especially pits and domes is aided by the mechanical action of the water, especially when it contains sediment · · ·. Migration from higher to lower levels may take place at successive periods, leaving a series of abandoned channels at different levels · · ·. After a section of the underground stream has been abandoned the amount of water entering the old channel becomes less and less until it is confined to small seeps. At this stage much of the water is removed by evaporation, so that solution gives place to deposition, the principal deposit being calcium carbonate.

In this statement we see several ideas either stated or implied. Among them are: (1) caverns develop chiefly above the water table through the action of diverted surface waters; (2) circulation may extend below the water table, but it is implied that this is not significant in cavern formation; (3) local base levels of surface streams control the downward development of caverns; (4) joint planes largely direct the lines of cavern extension; (5) mechanical erosion may contribute significantly to cavern enlargement; (6) one cavern level may be abandoned for a lower as downcutting of surface valleys permits and thus give rise to several cavern levels; and (7) after abandonment of a cavern level, formation of cave travertine becomes the dominant process.

It was particularly the ideas that caverns are developed above the water table by diverted surface waters, that circulation and solution below the water are insignificant, and that mechanical erosion (corrasion) is significant in cavern enlargement that Davis challenged. Briefly, his theory, which has come to be called the two-cycle theory, is that the major part of cavern development takes place by solution below the water table by phreatic water. Then follow a lowering of the water table and a draining of the caverns of their groundwater with a resulting occupation of the caverns by vadose water and air. It is, according to Davis, during this second cycle that the formation of cave travertine takes place. Although Grund (1903), in Europe, seems to have anticipated the ideas of Davis, lack of familiarity with his writings has resulted in Davis being looked upon in this country as the father of the two-cycle theory. Whether one accepts the tenets of Davis or not, it cannot be denied that his challenge to accepted ideas has stimulated greatly the interest in this phase of karst and thereby has ultimately contributed to a better understanding of caverns.

Water table in karst areas. Some of the difference of opinion as to the mode of formation of caverns results from varying ideas of the character of the water table in a karst region and the position of the zone of major erosion with respect to it. Some students of karst maintain that there is no water table in the sense of a continuous surface separating saturated and non-saturated zones. Rather, they picture the groundwater as being discontinuous and confined to joint and bedding plane openings rather than as a continuous mass throughout the rock. The difficulty of obtaining wells from limestones, except where they represent closed artesian systems between impermeable beds, and the frequency with which air-filled cavities are encountered in drilling are cited as evidence for this conclusion. Where limestone extends downward deeply enough it can hardly be doubted that the usual concept of a water table applies, even though the bulk of the water may be held in and move through joint and bedding plane networks.

Most students of cavern formation are willing to accept the basic concept that a water table level may exist in limestone regions and are also willing to agree that solution can take place below the water table. The main point of argument is whether or not the large amount of rock solution represented by caverns that are miles long and display well-integrated linear systems at one or several levels takes place above, at, or below the water table.

Relative importance of corrosion and corrasion. Geologists who have championed the idea that cavern courses are largely formed by flowing streams analogous in their essential characteristics to surface streams and actually in most instances surface streams that have sunk in swallow holes attach considerable importance to corrasion in cavern development. In this class belong the French geologists Lapparent, Martonne, and Martel and the American geologists Weller, Lobeck, and Malott. Davis and his followers have in general denied the importance of corrasion, although they do not deny that locally its effects may be noted. Those who believe that corrasion contributes significantly to cavern enlargement call attention to the great quantities of clay, silt, and even in some places sand and gravel, which may be seen in caves. Movement of this material through cavern routes must have caused corrasive action on the floors and sides of caves. Other cavern features that are cited as evidence of mechanical erosion are potholes, flutings, grooves, and miniature incised meanders. In view of the unquestioned existence of such features, it would be foolish to deny the possibility of corrasion in cavern routes, but it seems that in general it is of secondary importance. The fact, however, that it does take place is significant and needs to be kept in mind in evaluating the various theories of cavern origin.

Two-cycle theory of Davis. Because this theory has been the focal point about which much of the argument has revolved, we may begin our consideration of the various theories with a further consideration of it. The basic ideas of the theory have been stated previously. Space limitations will not permit a detailed discussion of all the reasons for this viewpoint advanced by Davis (1930) and by Bretz (1938, 1942, and 1953), who has strongly championed Davis's theory, but a few of the more pertinent arguments may be summarized. (1) There is evidence for deep-seated solution below the water table in the great freshwater springs which emerge off the coast of Florida, Cuba, and elsewhere; (2) caverns display three-dimensional network and spongework patterns rather than the branching patterns which streams normally have; (3) the network plans which caverns so often exhibit are difficult to associate with streams; (4) the many blind galleries found in caverns are also difficult to explain by stream erosion; (5) the common lack of graded longitudinal cavern profiles argues against their being formed by underground streams above the water table; and (6) the numerous wall pockets, ceiling pockets, ceiling tubes, and rock spans across cavern chambers, many of which are apparently fortuitous in location and unrelated to joints or bedding planes, could never have been made by the streams that now flow through caverns.

Bretz (1953) from extensive studies of the caverns in the Ozarks concluded that: (1) most of the caves were formed below the water table by waters circulating under hydrostatic head beneath the mature landscape which preceded peneplanation of the Ozark dome; (2) cave making largely ceased when the topography had attained old age because of reduction of hydrostatic head and because the caves were then filled with red clays derived from the deep soil upon the peneplain surface above them; (3) uplift of the region followed by dissection and accompanying lowering of the water table brought most of the caves above the level of the water table; and (4) vadose water circulating through the caverns has since removed most of the clay fills and to some extent modified the original caverns.

Bretz was so impressed by the nearly universal presence of deposits of red, unctuous clay in caverns that he thought that these deposits suggested a distinct episode in cavern history which came between the time of solution below the water table and replenishment with travertine above the water table. These deposits he called clay fills, and, from their lack of evidence of current action, lamination, and deposits of dripstone, flowstone and rimstone, he inferred that they were made while the caverns were still beneath the water table. Clay was thought to have been brought into the caverns through enlarged joints connected with dolines at the surface. That such clay deposits are often found in caverns can hardly be denied. Moneymaker (1941) has described similarly filled cavities 100 feet or more beneath the Tennessee River, and he cited them as evidence that solution could take place beneath the water table.

He noted, however, that cavities above the water table were more numerous and of greater size and concluded that, although cavern development may start below the water table, the major part of the enlargement of cavities takes place above the water table. It would hardly seem necessary to postulate a distinct stage of clay filling in cavern history to account for the clay and silt deposits that are often found in them. Cavern streams carry large quantities of silt and clay, and during heavy rains much terra rossa is carried down through swallow holes into caverns.

One type of cavern about whose origin there is little argument is the so-called *crystal cavern,* a type found in some limestone terrains, whose walls are usually more or less lined with crystals of calcite and other minerals. Crystal caverns are especially common in the lead and zinc area around Joplin, Missouri, but they are known elsewhere. They are in a sense giant geodes. That they developed below the water table seems likely. Whether distinct epochs of solution and fill are necessarily implied, as Davis believes, may be open to question.

Water-table theory of Swinnerton. This theory represents a modification of the prevalent idea which was held prior to Davis's proposal of the two-cycle theory that caverns were formed by vadose waters. Swinnerton (1929, 1932) placed great emphasis upon the controlling factor of the water table and attributed cavern solution essentially to laterally flowing water at the level of the water table. He stated (1929):

> In the simplest case precipitation passes more or less directly downward through openings in the rock to the water table and then moves laterally in the fluctuating top of the water table into the surface drainage channels. Insignificant caves may occur both above and below the water table as temporary phases of the adjustment of subsurface flow to the level of the surface streams, but it does not seem possible that continuous systems of caves can develop below the zone of actively circulating ground water—that is, below the water table. The water table is dependent on the level of the surface streams; these in turn depend upon the regional base level.

This theory attributes caverns mainly to solution by vadose waters, but, because it emphasizes lateral movement at the level of the water table, the author referred to it as the water table theory. This, according to Cotton (1948), represented essentially a revival of similar ideas held by Grund and Penck. The lowering of valleys by surface streams accelerates the lateral flow of the water at the top of the water table, and its movement is localized largely along joints and bedding planes. Lowering of surface valleys and concomitant lowering of the water table accounts for different levels in caverns. The water which dissolves cavern ways is largely supplied by infiltration from above through innumerable joint openings. At times of heavy rainfall all openings are filled and the water table rises notably, but this higher level is maintained for only a short time. During this brief period, however, the caverns are filled with water and network development takes place. Fluctuation of the water table is thus sufficient to explain the network passages so emphasized by Davis and Bretz. Swinnerton admitted that the crystal caverns of the Missouri region may have been formed by solution beneath the water table under an artesian head but did not believe that two cycles in their development were suggested. He believed that deposition of crystals could accompany solution.

Static water-zone theory of Gardner. Gardner (1935) proposed a theory of cavern origin which attributed the inauguration of caverns to the tapping of zones of static water followed by the diversion of surface waters to routes opened by the draining out of this static water. According to him, large caverns are found mainly in thick limestone terrains on the flanks of structural uplifts. They are extended mainly in the direction of the dip, more rarely along the strike, and never for any distance up the dip. Where stream valleys have been cut through strata, caverns are on the up-dip sides of valleys and not on the down-dip side. Thus, he stressed the importance of rock dip in controlling the movement of groundwater. According to him, there are two stages in the history of a cavern system: the precavern stage during which groundwater remains static because valleys have not yet been cut deeply enough to drain the beds of their water, and the stage of cavern formation that starts when surface valleys have cut into the terrain and drained the strata of their static waters, permitting vadose and meteoric waters to move downward along the joints and bedding planes. The groundwater in this latter stage moves slowly at first, and it emerges mainly as seepages, but as corrosion and corrasion enlarge joint openings a definite subterranean waterway or cavern is formed. Further lowering of adjacent surface valleys will tap and drain lower zones of static water and permit the development of cavern systems at lower levels. Gardner admitted the probability of some solution below the water table but did not consider it quantitatively sufficient to account for large caverns. Although he rather implied that underground streams diverted from surface courses play an important role in cavern formation, he did not attach paramount importance to them.

Invasion theory of Malott. From his 25 years of intermittent study and mapping of caverns in Indiana and Kentucky, Malott (1937) concluded that most large caverns in this area gave evidence of having been the runways for waters that were diverted from surface to underground routes. These subterranean streams have definite places where they enter underground routes through swallow holes and definite places where they reappear at lower levels as resurgences or karst springs. He thought that these diverted surface streams played the major role in the formation of the caverns through which they flow rather than that they occupied caverns which already had been formed below the water table, as postulated by Davis. He admitted that solution could take place below the water table, resulting in the formation of primitive, poorly integrated passages or networks, as Davis termed them,

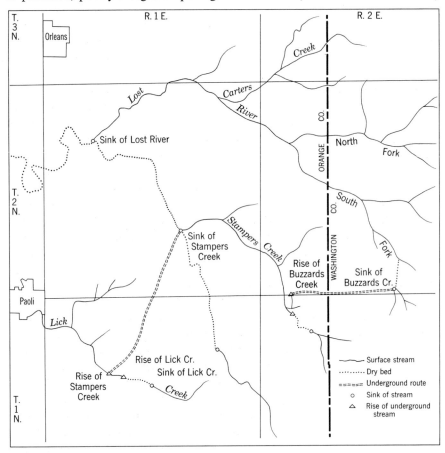

Fɪɢ. 13.18. Map showing the multiple subterranean stream piracies which have taken place in the headwater portions of Lost River, southern Indiana. (After C. A. Malott.)

but maintained that the waters of diverted surface streams occupied these poorly integrated passages and developed selected ones into well-integrated cavern routes at or near the water table.

The Lost River area of southern Indiana was cited (Malott, 1952) as a specific example of an area where a cavern was formed in this way. Lost River consists of three parts: a lower section which is a deeply entrenched surface stream, a middle dry-bed portion where the stream usually flows

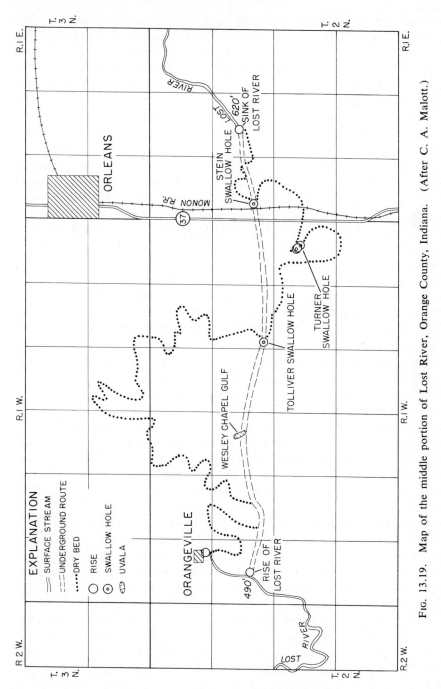

Fig. 13.19. Map of the middle portion of Lost River, Orange County, Indiana. (After C. A. Malott.)

underground, and an upper part with an unentrenched surface stream. The lower part of Lost River cuts through clastic rocks below which are thick beds of limestone. Along this portion there are numerous adjacent karst valleys which collect the surface drainage from sandstone-capped ridges. The middle portion of Lost River extends across a limestone plain developed upon the St. Louis and Ste. Genevieve limestone of Middle Mississippian age. This limestone plain is characterized by thousands of sinkholes and is almost entirely lacking in surface streams. It is in this section that the 22-mile meandering dry bed of Lost River and its related subterranean route are found. The upper part of Lost River extends across limestones just as soluble as those in which its underground route is developed, but karst features are either lacking or weakly developed. Surface drainage characterizes this stretch rather than the deep-seated phreatic type which would be expected under the two-cycle theory. The sinkhole plain on which the dry bed of Lost River is developed lies below an altitude of 700 feet and is thus below a late Tertiary erosion surface, which in this area is found at an altitude of about 750 feet. This suggests that Lost River's subterranean drainage developed within the present geomorphic cycle, following upon entrenchment of the lower course of Lost River below the late Tertiary erosion surface. Lost River sinks in five swallow holes which range in altitude from 625 feet to 560 feet. The main dry-weather sink is at 620 feet, but during periods of heavy rainfall water may follow a surface route to the fifth swallow hole at an elevation of 560 feet. The underground course is about 8 miles long and follows a fairly direct southwesterly route down the regional dip of the rock. Its resurgence as a large artesian spring is at an altitude of 490 feet. Its cavern, which is still undergoing expansion, can be entered at several places and stretches of it have been mapped. It exhibits none of the beauty that characterizes most of the caverns seen by the public. It is a cavern in the making with much mud in it from storm waters which periodically flow through it. Stalactites and stalagmites and other forms of cave travertine are relatively rare. Progressive entrenchment of lower Lost River to the present base level of erosion and accompanying lowering of the water table have made possible diversion of surface waters to subterranean routes with resulting cavern expansion and headward extension. In the upper course of Lost River, where stream entrenchment and accompanying lowering of the water table have not yet taken place, there is still surface drainage, and sinkholes are few in spite of the fact that a soluble limestone underlies this section of its course.

Dating caverns. The time of formation of a cavern can be determined by relating the formation of the cavern to local topographic surfaces or deposits of known age. Relative rather than absolute age is thus obtained. As noted above, Lost River cavern postdates a late Tertiary peneplain, for its formation was dependent upon uplift of this peneplain and entrenchment of Lost

River below it. The cavern probably started early in the Pleistocene, and it is still expanding.

Carlsbad Cavern has been dated by Bretz (1949) and Horberg (1949) by its relationship to Tertiary erosion surfaces in the area. They concluded that the cavern antedated three Pleistocene terraces along the Pecos depression and an older erosion surface in the Guadalupe Range. It was considered older than the erosion surface in the Guadalupe Range because this surface and present canyons transect Carlsbad and other caverns in the Guadalupe Range. A pre-Ogallala age is indicated by the presence of quartzose pebbles, thought to have been derived from the Ogallala (probably largely Pliocene), mixed with flowstone in the transected cavern passages.

Sweeting (1950), who favored the theory that caverns form at the level of the water table, has pointed out the coincidence between cavern levels and cyclical erosion surfaces in the Ingleborough district in England. Cavern levels are notably concentrated at altitudes of 1200 to 1300, 900 to 1000, and 800 to 900 feet. It was inferred that these levels mark former base levels and positions of the water table. A pre-Wisconsin age for the caverns is indicated by the presence in them of inwashed glacial materials of Wisconsin age. The 1300-foot level was tentatively assigned a Pliocene age, and the lower levels date somewhat between this and Wisconsin time.

Conclusions. What are we to conclude about how caverns develop when we find competent observers differing in their opinions in regard to the geologic conditions under which caverns are formed? As is often true when such divergence of opinion exists, there probably is some truth in each opinion. It certainly seems likely that some caverns form below the water table. Of the large caverns, Carlsbad is the best example. There is a strong probability that many of the Missouri caverns were similarly formed. Many small caves give evidence of having developed below the water table. However, any one who is familiar with the association between sinking creeks and caverns in the Indiana-Kentucky karst area is likely to have difficulty in escaping the conclusion that in this area at least there is a genetic relationship between the two rather than that the sinking creeks merely occupied ready-made caverns formed below the water table.

THE KARST GEOMORPHIC CYCLE

Whether there exists a distinct cycle of land-form evolution in limestone terrains which we may designate as a karst cycle or whether what has been so designated is better considered as the karst phase of a fluvial cycle is a disputed question. Davis, who gave only brief attention to the evolution of land forms in limestone terrains, was inclined to treat this as a special and transitory phase marking the mature stage of the "normal cycle" in special regions where "structure" was particularly favorable to solution. That there is some basis for such a contention cannot be denied, for in most, if not all, areas we start with surface drainage and end with it. Yet we probably do no harm if we think of a distinct karst cycle. In fact, there is probably no type of landscape to which the concept of cyclic evolution can be applied with less complication than a karst landscape. Here we usually find simple "structure" in the sense of relatively uniformly soluble rock. Faulting in the Jugoslavian area and folding in the Great Valley of the Appalachian region may represent slight departures from simplicity of "structure," but they alter the evolution of land forms only in minor details. One geomorphic process is dominant, that of solution. We lack many of the complications that have led some to question the concept of a geomorphic cycle. Thus it would seem permissible to apply such terms as youth, maturity, and old age to the stages of karst development, even though it be contended that the whole comprises only a special phase of a broader fluvial cycle.

We can conceive of two ways in which a karst cycle may be inaugurated: (1) through uplift above base level of a limestone terrain on which fluvial erosion had been in progress, or (2) through uplift of an area of clastic rocks beneath which are limestones above the new base level. In either event, the cycle begins with surface drainage lines but the transformation to underground drainage proceeds by somewhat different methods. Where limestone lies at and beneath the surface without a cap of insoluble material, initiation of karst features is rather simple. As the master streams, starting at their mouths, cut downward toward their new base level there will be a lowering of the water table immediately adjacent to them, and areas of limestone will become perched above surface drainage lines and the local water table. Dolines and swallow holes will form and divert surface drainage to underground courses. As the new base level of erosion is extended up major valleys and their tributaries there will be a progressive increase in the amount of terrain drained by underground routes.

Before carrying this further, let us consider how the procedure would differ in a region where the surface rocks are insoluble ones, such as shales and sandstone, but below which are limestones. The transformation in such a region from surface to underground drainage would not be quite so simple as previously outlined, particularly if we assume that a considerable thickness of clastic rocks overlies the limestone. First, the rejuvenated streams will need to cut through the clastic rocks into the limestone. Once this is done, underground drainage may start, but rather than starting as innumerable small dolines and swallow holes, usually the initial stage will be marked by the formation of solution or karst valleys. As these karst valleys expand at the expense of the areas of clastic rocks, the beginning of a sinkhole plain with the many dolines and swallow holes which characterize it may be seen.

This sequence can be readily seen on the Mammoth Cave, Kentucky, topographic sheet. Three distinct types of topography are apparent here: a typical sinkhole or karst plain at the south; north of it, an area with karst valleys between sandstone-capped hills; and, still farther north, an area of normal surface drainage upon clastic rocks in which karst valleys have yet to appear.

The concept of a karst cycle has perhaps been most intimately associated with Cvijić. He, according to Sanders (1921), recognized four stages in the evolution of karst, which he designated as youth, maturity, late maturity, and old age. Youth, according to Cvijić, begins with surface drainage on either an initial limestone surface or one that has been laid bare and is marked by a progressive expansion of underground drainage. Lapiés and scattered dolines are particularly characteristic of this stage. No large caverns exist and underground drainage is far from complete. During maturity there is maximum underground drainage. Surface drainage is limited to short sinking creeks ending in swallow holes or blind valleys. Cavern networks are characteristic of this stage. This is the time of maximum karst development. Late maturity marks the beginning of the decline of karst features. Portions of cavern streams are exposed through what we have called karst windows. These expand to form large uvalas, and detached areas of the original limestone upland have begun to stand out as hums. Old age is marked by a return to surface drainage with only a few isolated hums remaining as remnants of the original limestone terrain.

Cvijić first published a summary of his concept of the karst cycle in 1918 in an article entitled "Hydrographie souterraine et évolution morphologique du karst." Essentially the same ideas were expressed in this country by Beede in an article appearing in 1911, but, because of the obscurity of the publication in which it appeared, his ideas were overlooked by most geologists. Beede divided the karst cycle into the three conventional stages of youth, maturity, and old age, rather than the four as proposed by Cvijić. Youth is marked by the beginnings of diversion of surface drainage to subterranean routes with dolines as the characteristic land form of this stage. In maturity, there is a maximum of sinkholes and underground drainage with only major entrenched streams persisting as surface streams; sinking creeks, swallow holes, blind valleys, compound sinks, and dolines by the thousands typify this stage. Old age is initiated with the beginning of return to surface drainage; karst windows, karst tunnels, natural bridges, and hums are the most diagnostic features.

It should be recognized that, just as a region undergoing dissection by stream erosion may exhibit the characteristics of youth in one section, maturity in another, and old age in another, so in a karst region the various stages of the karst cycle may be present. Much as in the fluvial cycle, the various stages move progressively into a region accompanying and following entrenchment of major drainage lines across the terrain. Areas remote from entrenched streams are likely to be less advanced in the cycle than those adjacent to them.

REFERENCES CITED IN TEXT

Beede, J. W. (1911). The cycle of subterranean drainage as illustrated in the Bloomington, Indiana, quadrangle, *Proc. Indiana Acad. Sci., 20,* pp. 81–111.

Bennett, H. H. (1928). Some geographic aspects of Cuban soils, *Geog. Rev., 18,* pp. 62–82.

Bretz, J. H. (1938). Caves in the Galena formation, *J. Geol., 46,* pp. 828–841.

Bretz, J. H. (1942). Vadose and phreatic features of limestone caverns, *J. Geol., 50,* pp. 675–811.

Bretz, J. H. (1949). Carlsbad Caverns and other caves of the Guadalupe block, New Mexico, *J. Geol., 57,* pp. 447–463.

Bretz, J. H. (1953). Genetic relations of caves to peneplains and big springs in the Ozarks, *Am. J. Sci., 251,* pp. 1–24.

Cvijić, Jovan (1918). Hydrographie souterraine et évolution morphologique du karst, *Rec. trav. insts. géog. alpine* (Grenoble), *6 (4),* 56 pp.

Cvijić, Jovan (1924). The evolution of lapiés, *Geog. Rev., 14,* pp. 26–49.

Cotton, C. A. (1948). *Landscape,* 2nd ed., pp. 445–493, John Wiley and Sons, New York.

Davis, W. M. (1930). Origin of limestone caverns, *Geol. Soc. Am., Bull. 41,* pp. 475–628.

Gardner, J. H. (1935). Origin and development of limestone caverns, *Geol. Soc. Am., Bull. 46,* pp. 1255–1274.

Gilmer, F. W. (1818). On the geological formation of the Natural Bridge of Virginia, *Amer. Phil. Soc., Trans. 1,* pp. 187–192.

Grund, A. (1903). Die karsthydrographie: Studien aus Westbosnien, *Penck's Geog. Abhandl., 7,* pp. 103–200.

Horberg, Leland (1949). Geomorphic history of the Carlsbad Caverns area, New Mexico, *J. Geol., 57,* pp. 464–476.

Hubbard, Bela (1923). The geology of the Lares district, Puerto Rico, *N. Y. Acad. Sci., Survey of Puerto Rico and Virgin Islands, 2,* Pt. 1, pp. 83–93.

Malott, C. A. (1932). Lost River at Wesley Chapel Gulf, Orange County, Indiana, *Proc. Indiana Acad. Sci., 41,* pp. 285–316.

Malott, C. A. (1937). Invasion theory of cavern development, *Proc. Geol. Soc. Am. for 1937,* p. 323.

Malott, C. A. (1939). Karst valleys, *Geol. Soc. Am., Bull. 50,* p. 1984.

Malott, C. A. (1945). Significant features of the Indiana karst, *Proc. Indiana Acad. Sci., 54,* pp. 8–24.

Malott, C. A. (1952). The swallow-holes of Lost River, Orange County, Indiana, *Indiana Acad. Sci., Proc. 61,* pp. 187–231.

Malott, C. A., and R. R. Shrock (1930). Origin and development of Natural Bridge, Virginia, *Am. J. Sci., 219,* pp. 257–273.

Matson, G. C. (1909). Water resources of the Bluegrass region, Kentucky, *U. S. Geol. Survey, W.S.P., 233,* pp. 42–44.

Meyerhoff, H. A. (1938). The texture of karst topography in Cuba and Puerto Rico, *J. Geomorph., 1,* pp. 279–295.

Moneymaker, B. C. (1941). Subriver solution cavities in the Tennessee Valley, *J. Geol., 49,* pp. 74–86.

Palmer, H. S. (1927). Lapiés in Hawaiian basalts, *Geog. Rev., 17,* pp. 627–631.

Sanders, E. W. (1921). The cycle of erosion in a karst region (after Cvijić), *Geog. Rev., 11,* pp. 593–604.

Smith, J. F., Jr., and C. C. Albritton, Jr. (1941). Solution effects on limestone as a function of slope, *Geol. Soc. Am., Bull. 52,* pp. 61–78.

Sweeting, M. M. (1950). Erosion cycles and limestone caverns in the Ingleborough district, *Geog. J., 115,* pp. 63–78.

Swinnerton, A. C. (1929). Changes of baselevel indicated by caves in Kentucky and Bermuda, *Geol. Soc. Am., Bull. 40,* p. 194.

Swinnerton, A. C. (1932). Origin of limestone caverns, *Geol. Soc. Am., Bull. 43,* pp. 663–694.

Thorp, James (1934). The asymmetry of the "Pepino Hills" of Puerto Rico in relation to the trade winds, *J. Geol., 42,* pp. 537–545.

Walcott, C. D. (1893). The natural bridge of Virginia, *Nat. Geog. Mag., 5,* pp. 59–62.

Woodward, H. P. (1936). Natural Bridge and Natural Tunnel, Virginia, *J. Geol., 44,* pp. 604–616.

Wray, D. A. (1922). The karstlands of western Jugoslavia, *Geol. Mag., 59,* pp. 392–408.

Wright, F. J. (1936). The Natural Bridge of Virginia, *Virginia Geol. Survey, Bull. 46-G,* pp. 53–78.

Additional References

Beckman, H. C., and N. S. Hinchey (1944). The large springs of Missouri, *Missouri Geol. Survey and Water Res., Bull. 29,* 2nd ser., 141 pp.

Dicken, S. N. (1935). Kentucky karst landscapes, *J. Geol., 43,* pp. 708–728.

Frye, J. C., and S. L. Schoff (1942). Deep-seated solution in the Meade Basin and vicinity, Kansas and Oklahoma, *Am. Geophys. Union, Trans. for 1942,* pp. 35–39.

Hubbert, M. K. (1940). Theory of goundwater motion, *J. Geol., 48,* pp. 785–944.

Jordan, R. H. (1950). An interpretation of Floridian karst, *J. Geol., 58,* pp. 261–268.

Lobeck, A. K. (1928). The geology and physiography of Mammoth Cave national park, *Kentucky Geol. Survey,* Ser. 6, Pamphlet 21, pp. 1–69.

Malott, C. A. (1929). Three cavern pictures, *Proc. Indiana Acad. Sci., 38,* pp. 201–206.

Morgan, A. M. (1942). Solution-phenomena in the Pecos basin, *Am. Geophys. Union, Trans. for 1942,* pp. 27–35.

Thornbury, W. D. (1931). Two subterranean cut-offs in central Crawford County, Indiana, *Proc. Indiana Acad. Sci., 40,* pp. 237–242.

14 · Types and Characteristics of Glaciers

INTRODUCTION

Glaciers today, except in high latitudes and at high altitudes, are of minor importance in the present-day shaping of land forms, but those that existed during the Pleistocene left their imprint upon many millions of square miles of the earth's surface. Some 4,000,000 square miles of North America, 2,000,000 square miles or more of Europe, and an as yet little known but possibly comparable area in Siberia were glaciated. In addition, many lesser areas were covered by local ice caps. Thousands of valley glaciers existed in mountains where today there are either no glaciers or only small ones.

Some of our finest mountain scenery is to a large degree a product of glacial geomorphic processes, as are many interesting lowland land forms. The effects of glaciation are particularly significant in the thickly populated sections of east-central North America and northwest Europe. Furthermore, as suggested in Chapter 2, the influence of the Pleistocene ice age far exceeded the direct effects of glacial erosion and deposition.

Our ideas concerning Pleistocene glaciation have changed greatly since the days of Venetz, Charpentier, and Agassiz (see p. 8). After many years of bitter argument, there seems to be agreement that the Pleistocene epoch consisted of four glacial ages separated by interglacial ages of probably far greater duration than the glacial. The last glaciation, or perhaps more correctly the latest, if we are not to rule out the possibility that we are living in interglacial time, has left the most obvious imprint upon our topography, but the effects of the earlier glaciations are still apparent in many areas.

TYPES OF GLACIERS

Following Flint (1947), we may define a glacier as a mass of ice, lying entirely or largely on land, which was formed chiefly by compaction and recrystallization of snow and which flows or at some time has flowed. Glaciers exhibit many variations in form, size, and origin; as a result, many different

354

names have been applied to individual types. Ahlmann (1948) has given a satisfactory morphological classification of glaciers which is as follows:

A. Glaciers that are continuous sheets from which the ice may move in all directions.
  1. Continental glaciers or inland ice covering large areas.
  2. Glacier caps, which are smaller than continental glaciers.
  3. Highland glaciers, which cover the highest and central portion of a mountain area.
B. Glaciers confined to more or less marked courses which direct their main movement. Included in this group are both independent glaciers and outlets of ice from glaciers of group *A*.
  4. Valley glaciers of the alpine type.
  5. Transection glaciers, which more or less fill a whole valley system.
  6. Cirque glaciers, which occupy localized niches on sides of mountains.
  7. Wall-sided glaciers, which cover a side of a valley or part of it.
  8. Glacier tongues afloat.
C. Glacier ice which spreads in large or small cake-like sheets over level ground at the foot of glaciated areas. None in this class is independent, but all connect with some other type of glacier.
  9. Piedmont glaciers formed by fusion of the lower parts of 4, 5, or 7 above.
  10. Foot glaciers, which are the lower ends and more extended portions of types 4, 5, or 7 above.
  11. Shelf ice.

Glaciers may be classified according to their mechanics of motion into two groups, ice streams and ice caps. The members of group *A* above are ice caps, those of group *B* are ice streams, whereas group *C* includes transitional types. Motion in ice caps is produced largely by differential pressures within an ice mass, is multidirectional, and is not controlled by underlying topography, although it may be influenced by it. Ice streams flow largely under the direct influence of gravity, and their motion is essentially unidirectional and controlled by underlying topography.

GLACIAL MOTION

An extended discussion of glacial motion would take us beyond the scope of geomorphology and into the field of glaciology, but certain aspects of the subject need to be comprehended in order to appreciate glaciers properly as erosional and depositional agents. Glaciers originate in snowfields or névés through transformation of snow into ice. In many respects this transformation is analogous to the various changes which sediments undergo in their change into sedimentary rock. Stratification is especially evident in the intermediate stage between the granular snow of a snowfield and the compact ice of a glacier. This intermediate stage is often called *firn*. Compaction results in increased density largely through expulsion of air. Freshly fallen snow has a density ranging between 0.06 and 0.16, whereas firn has a density ranging between 0.72 and 0.84. Banding representing individual snowfalls is commonly evident in firn. The ice of a glacier proper is still more dense

Fig. 14.1. The birthplace of glaciers, Bernese Oberland between Meiringen and Wassen, Switzerland. (Photo by Swiss National Travel Office.)

than that of firn, being about 0.9 as compared with 0.918 for ice derived from water. Air entrapped in glacial ice causes this slight difference. This air may be under pressure of as much as 150 pounds per square inch and is responsible for the explosive effects often exhibited by icebergs. Banding is also exhibited by glacial ice, but it usually is not the original stratification lines marking successive snowfalls.

The transformation of névé into firn and glacial ice is the result of several processes, not all of which are clearly understood. Among them are sublimation, recrystallization with growth of large crystals at the expense of the smaller ones, melting, refreezing under pressure (regelation), and compaction under weight. A certain minimum thickness of névé seems necessary for this change to take place. This is thought to be somewhere between 100 and 150 feet. Failure to attain this critical thickness, whatever it may be, explains why some snowfields never develop into glaciers. Where conditions of topography and exposure permit the accumulation of a necessary minimum thickness of snow, glaciers may form even below the snow line. In the mountains of western United States and in Norway, cirque glaciers exist 1000 to 1500 feet below the snow line.

FIG. 14.2. The 15-mile long Aletsch glacier, Switzerland. (Photo by Swiss National Travel Office.)

There have been two schools of thought during the past hundred years in regard to the nature of glacial motion. One group, including such men as Hugi, Forbes, and Agassiz, held to the concept of flow by continuous plastic yield. According to this concept a glacier is a *fleuve de glace,* as it was termed by Bishop Rendu, which exhibits the characteristic flow of a highly viscous liquid. Proponents of the viscosity theory found support for their contentions in the many measurements made on Alpine glaciers which indicated differential velocities within a valley glacier essentially similar to those of a stream except that they are slower. Laboratory experiments which demonstrated that crushed ice placed in pressure chambers could be made to flow out through an orifice seemed to substantiate this theory.

Another group, which included Tyndall, T. C. Chamberlin, and R. T. Chamberlin, believed that fracture and shear accompanied by regelation were largely responsible for glacial motion. Crevasses in ice were cited as evidence of its rigidity and brittleness. Movement was thought to be by individual particles, while the glacier as a whole retained features of a solid and rigid mass. Proponents of this theory cited the existence of shear planes, overthrusts, and "blue bands," which are common in the lower parts of glaciers and were interpreted as slippage planes, as further evidence of flow by fracture and shearing.

Although "the riddle of glacial motion" is by no means solved, we are approaching an understanding of it. It has become increasingly evident that neither theory can adequately explain glacial motion, for there is evidence of both types of motion. Demorest (1942, 1943) thought that there was evidence of four types of flow, which he designated as gravity flow, extrusion flow, obstructed gravity flow, and obstructed extrusion flow. Gravity flow and obstructed gravity flow he thought are characteristic of ice streams and extrusion flow and obstructed extrusion flow of ice caps.

If the gradient of the valley floor beneath an ice stream is sufficient for the glacier to overcome the frictional retardation which the floor exerts but is not great enough to produce an ice cascade or ice falls, a valley glacier moves much like a stream of water with a line of maximum velocity at or near the surface. This type of movement Demorest designated as *gravity flow.* Wherever a glacier finds its progress retarded by some obstacle, such as slowly moving or stagnant ice or a topographic obstruction, it tends to override this slower-moving or stagnant mass. It will do this largely by shearing. Shear planes develop which, in general, curve upward and forward. This type of flow Demorest called *obstructed gravity flow,* and he thought it to be particularly characteristic of the lower portions of valley glaciers or ice streams. The blue bands so emphasized by R. T. Chamberlin (1928, 1936) are evidence of this type of flow. They represent narrow zones of localized movement and are the result of recrystallization of ice subsequent to shearing. Discrete fault planes and folds in glacial ice give further evidence of this type of flow. Although shear and fault planes are most common in the terminal zone of a glacier, they may be found elsewhere. They form especially where the down-valley slope of a glacier decreases, for there the movement of the glacier may slow down to the point at which obstructed gravity flow results.

Demorest (1938), from a study of a portion of the abandoned bed of the Clements Glacier in Glacier National Park, found what he thought was evidence of existence in ice streams of another type of flow, which he designated as *extrusion flow.* This takes place where a bedrock floor is basined and ice thickness is greater than usual. This type of movement is produced by differential pressures within an ice mass and partakes of plastic flow in the basal portion of the ice. The maximum velocity seems to be near the base of the glacier rather than near its top, as is true of simple gravity flow.

The exact mechanics of movement of ice caps is still somewhat conjectural because of the difficulty of making direct observations except over an extremely limited thickness of ice, but it seems likely that extrusion flow is of major importance in movements of ice caps. Although all glacial motion is ultimately dependent upon gravity, we may think of the motion of ice caps as being directly caused by pressure differences due to varying thickness of

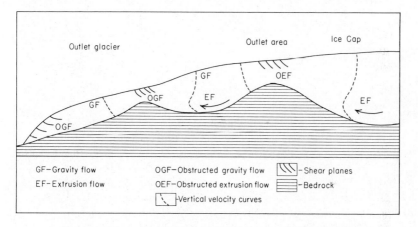

FIG. 14.3. Schematic diagram of the types of glacial motion.

the ice. The slopes or gradients of the ice surface determine the directions and rates of motion of an ice cap rather than the configuration of the topography upon which it rests. An ice cap can move uphill with respect to underlying topography, provided the slope of the ice surface is in this direction. That they have done so is evidenced by the many glacial boulders which have been lifted hundreds or even thousands of feet above their sources. Movement in an ice cap is likely to be greatest in or near its basal portions, and it is multidirectional, extending out from areas of greatest surface altitudes toward the thinning ice edge. The topography beneath an ice cap may localize movement and contribute to more rapid movement and greater erosion where the axes of valleys or basins beneath the ice coincide in direction with the surface gradients of an ice cap. This helps to explain why some areas have been greatly eroded and other areas have suffered little erosion. Greatest erosion is usually found where subglacial topography has favored ice movement rather than obstructed it.

Just as there may be obstructed gravity flow in ice streams where topographic obstacles or slowly moving or stagnant ice masses retard gravity flow, so there may be under similar conditions *obstructed extrusion flow* in ice caps. Ice caps may give rise to masses of ice that move primarily by gravity flow.

FIG. 14.4. Outlet glaciers extending into a fjord from the ice cap over Greenland. (Air Force photo.)

This is well illustrated in the outlet glaciers which extend out from the ice cap over Greenland. Movement in this ice cap is probably mainly by extrusion flow but, where the ice impinges upon the mountain rim which encloses it, obstructed extrusion flow would be expected. Ice moving out from the central basin and down mountain valleys becomes ice streams moving by gravity flow or obstructed gravity flow.

THE REGIMEN OF GLACIERS

The *regimen* of a glacier refers to its variations in nourishment, movement, and wastage. Changes in the regimen of a glacier determine whether it is in a state of advance, recession, or stagnation.

Nourishment of glaciers. Nourishment of ice streams, except perhaps for outlet glaciers, is dependent upon well-understood meteorological conditions. Ice streams form in mountainous areas and are nourished by orographic snowfall. The snow that falls directly on a glacier's surface represents but a part of that which feeds the glacier. Snow is shed onto the snowfield and glacier by avalanching from the enclosing cirque and valley walls, and wind may drift snow into a glacial basin from adjacent upland areas or may even drift it across divides into valleys to the leeward, as is evidenced by the fact that glaciers are sometimes more extensive on the leeward than on the windward side of a mountain range.

If a mountain range is unusually high, the heaviest snowfall may be considerably below its crest, in which case glaciers may head well below the mountain tops. Local topographic configuration and exposure to the sun to a large degree determine whether snowfields attain the depth required to form glaciers. Deep valleys which are shaded from the sun, particularly during midday and early afternoon, are most likely to accumulate snow in an amount sufficient to nourish glaciers. Lewis (1938) has pointed out that the line of most frequent ice-stream extension lies between northeast and east-northeast in the northern hemisphere.

Difference of opinion exists as to how ice caps are nourished. Hobbs (1943), impressed by the existence of strong outblowing winds over the Greenland and Antarctic ice caps, developed the idea of a glacial anticyclone as the chief mechanism by which ice caps are nourished. According to him, winds that blow outward from ice caps originate in descending air of permanent areas of high pressure which exist over large ice caps. Aloft, the air is presumably moving inward to replace the descending air. Water vapor condenses as the air ascends to form high cirrus clouds composed of small ice particles. On descending, these ice particles are vaporized as a result of adiabatic heating, but upon coming in contact with a cold ice surface condensation takes place in the form of rime and hoarfrost. It is this condensation which presumably feeds the glacier. According to Hobbs, the central part of an ice cap is essentially motionless with movement being restricted largely to its periphery. The ice that moves outward is replaced by snow which frequent blizzards carry out from the center.

Most students of glaciers are likely to look with skepticism upon this theory. Too much emphasis is placed upon supposed permanent areas of high pressure over Greenland and Antarctica. That average pressure is high cannot be denied, but to a certain degree the high-pressure areas are "statistical highs." Matthes (1946, 1950) has pointed out that the several expeditions to Greenland found that moving cyclones do penetrate its interior and bring snowstorms. The periods of quiet weather which Hobbs stressed so much are characteristic only of the intervals between these storms. Matthes further showed that the radially outflowing winds which Hobbs took as evidence of the existence of a permanent anticyclone over the interior of the Greenland ice cap are really katabatic winds or cold air draining downslope under the influence of gravity rather than the general outward radial winds of a high-pressure system.

Furthermore, it seems unlikely that the amount of condensation that could be produced by contact cooling could be adequate to maintain glaciers of such size as those on Greenland and Antarctica. Precipitation comes not from descending air but from ascending air. Contact cooling can take place in descending air, but the amount of water vapor so condensed is never great. It seems more logical that the edges of cold-air masses over the two ice caps act essentially as polar fronts which force maritime air masses to rise over them and cause cyclonic precipitation. Precipitation produced in this way can be adequate to nourish a large ice sheet.

Wastage of glaciers. Glaciers waste chiefly by melting and evaporation. Wastage by these methods is called *ablation*. Glaciers that send tongues of ice into standing bodies of water may waste by calving of ice into the water.

As Ahlmann (1948) and Sharp (1951) have stressed, a glacier may have a positive or negative balance sheet, depending upon whether nourishment exceeds or falls short of wastage. Ahlmann classified glaciers on the basis of their dynamics as active, inactive or passive, and dead. To which of these classes a particular glacier belongs largely depends upon whether its regimen has been for many years positive or negative or just balanced. The consequences of a protracted positive regimen will be an active glacier with an advancing front; an extended negative regimen will cause thinning of the ice, movement of less ice into the ablation area, and eventually a receding ice front. A negative regimen, however, does not necessarily result in immediate frontal recession, for ablation may be balanced for a time by ice transported from the accumulation area. A glacier may remain active for a time, even though it is undergoing frontal recession. Persistence of a negative regimen finally leads to an inactive glacier and eventually a dead one. Even dead ice may not be completely stagnant, but such movement as takes place is entirely consequent upon topographic slopes rather than a positive regimen. Locally, dead ice may exist around the margin of an otherwise active glacier.

Recession of an ice front is often referred to as *backwasting*. This term may lead, as Johnson (1941) has stressed, to an erroneous idea that ablation is taking place only at the front of the ice. Glacial thinning or *downwasting* is probably more important at all times in producing retreat of an ice front than is backwasting. Certainly ablation is not limited to the terminus of a glacier but extends over its whole length, although it may be considerably greater in its lower portion. Studies on the Nisqually Glacier, on Mt. Rainier, showed a mean annual recession of its front of 70 feet and a thinning of 6.6 feet. The Hintereisferner Glacier in the Alps receded 1006 feet in 24 years and thinned 340 feet in the same time. In both cases more ice was lost by downwasting than by backwasting. Thinning of a glacier may eventually lead to stagnation or lack of forward motion. Whether this can happen to an ice stream in mountainous country may be open to question, but it commonly happens with piedmont glaciers and ice caps. Certainly mere maintenance of a fixed ice front is no proof of stagnation. True stagnation probably requires detachment of an ice mass from the main body of actively moving ice.

At present, most glaciers are receding. Observations have not extended over a long enough period in the United States to indicate just when recession began, but in the Alps, where observations cover a much longer period, recession began in the last decade of the nineteenth century and has continued intermittently to the present. There have been minor oscillations during which glaciers temporarily advanced, but the general trend has been toward recession. Individual glaciers, however, may depart from the general trend because of the influence of local conditions. There is considerable evidence to suggest that the present recession may represent a temporary episode which will in time be succeeded by a period of expansion. Similar periods of recession and advance within historic time (Matthes, 1942) are known. Glaciers in the Alps during the Middle Ages were much less extended than now. It is even doubtful whether present-day glaciers in western United States are the direct lineal descendants of Pleistocene glaciers. There is much evidence to suggest that some 4000 to 6000 years ago there was a period of milder climate during which most middle-latitude valley glaciers disappeared. This has been called not too aptly the Climatic Optimum. It is now more commonly designated as the Thermal Maximum or Megathermal Phase.

Present-day glaciers formed after a shift toward cooler climates, resulting in what has been called by Matthes the Little Ice Age. The causes of these changes in the regimens of glaciers are not known. It is natural to associate them with climatic cycles, but so far attempts to correlate them with such suggested cycles as the Brückner 35-year cycle and the 11-year sunspot cycle have failed, or at least not been convincing.

EFFECTIVENESS OF GLACIAL EROSION

The particular processes by which glaciers erode were discussed in Chapter 3. Our purpose here is to present two contrasting ideas which have developed regarding the ability of glaciers to modify by erosion the landscapes over which they move. During the latter part of the nineteenth century and the first few decades of the present century, a decided difference of opinion existed in regard to the effectiveness of glaciers as erosional agents. Most students of glacial land forms were impressed by the amount of erosion that glaciers apparently had performed, but a small group, which came to be called the glacial protectionists, took an opposite viewpoint. It was argued that glaciers are not as effective erosive agents as streams and that a glacier lying over a region actually protects it from more rapid sculpturing by stream erosion and weathering.

Even those who looked upon glaciers as extremely effective erosional agents recognized that there are zones where erosion or deposition may predominate. Areas that have been covered by an ice sheet commonly display three fairly distinct zones: (1) a central protected area where both erosion and deposition have been relatively slight; (2) an intermediate zone where ice scouring is pronounced; and (3) a peripheral area in which depositional features are prominent. Abundant striated, polished, and grooved surfaces give evidence of effective erosion. Smith (1948) has described giant grooves in the MacKenzie Valley which are as much as 100 feet deep and upwards of a mile in length. Lack or scarcity of residual soils over much of the glaciated area has been used as an argument for effective glacial erosion. The composition of glacial deposits indicates that they were derived mainly from unweathered rather than weathered rock. The total volume of glacial deposits in most areas is greatly in excess of the volume of weathered rock that might reasonably be assumed to have been present in the glaciated areas prior to glaciation. Numerous rock basins in areas where it is difficult to otherwise explain them argue for effective erosion as does the depth of many glaciated valleys.

Those who doubted the effectiveness of glacial erosion cited in support of their view that striated, polished, and grooved surfaces are scarce or lacking in many areas; that in many places preglacially weathered rock underlies glacial deposits above it; that in many areas glacial deposits of one age are found overlying old soils developed upon older glacial materials; that many of the features along glaciated valleys which were attributed to glacial erosion were formed by streams prior to glaciation and only modified slightly by glacial erosion.

Most of these arguments can be equally well cited as evidence for differential erosion. When we attempt to visualize the movement of ice within and beneath an ice cap, it seems that we should expect localization of flow and erosion as determined by variations in ice pressure and influenced and directed by subglacial topography. It is then not incompatible to find evidence of effective erosion near areas showing little erosion, for glacial erosion like other types of erosion is differential.

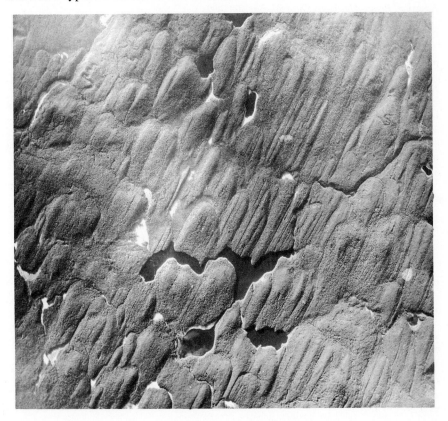

Fɪɢ. 14.5. Vertical photo of a drumlinized and grooved glacial plain. (Photo by Geol. Survey of Canada.)

REFERENCES CITED IN TEXT

Ahlmann, H. W. (1948). Glaciological research on the North Atlantic coasts, *Royal Geog. Soc., Res. ser., 1,* 80 pp.

Chamberlin, R. T. (1928). Instrumental work on the nature of glacial motion, *J. Geol., 36,* pp. 1–30.

Chamberlin, R. T. (1936). Glacier movement as typical rock deformation, *J. Geol., 44,* pp. 93–104.

Demorest, Max (1938). Ice flowage as revealed by glacial striae, *J. Geol., 46,* pp. 700–725.

Demorest, Max (1942). Glacial thinning during deglaciation: Part I, Glacier regimens and ice movement within glaciers, *Am. J. Sci., 240,* pp. 31–66.

Demorest, Max (1943). Ice sheets, *Geol. Soc. Am., Bull. 54,* pp. 363–400.

Flint, R. F. (1947). *Glacial Geology and the Pleistocene Epoch,* p. 15, John Wiley and Sons, New York.

Hobbs, W. H. (1943). The glacial anticyclone and the continental glaciers of North America, *Proc. Am. Phil. Soc., 86,* pp. 368–402.

Johnson, D. W. (1941). Normal ice retreat or downwasting?, *J. Geomorph., 4,* pp. 85–94.

Lewis, W. V. (1938). A meltwater hypothesis of cirque formation, *Geol. Mag., 75,* pp. 263–264.

Matthes, F. E. (1942). Glaciers, Chapter 5, in Physics of the Earth, Part IX, *Hydrology,* McGraw-Hill Book Co., New York.

Matthes, F. E. (1946). The glacial anticyclone theory examined in the light of recent meteorological data from Greenland, Part 1, *Trans. Am. Geophys. Union, 27,* pp. 324–341.

Matthes, F. E. (1950). The glacial anticyclone theory examined in the light of recent meteorological data from Greenland, Part 2, *Trans. Am. Geophys. Union, 31,* pp. 174–182.

Sharp, R. P. (1951). Accumulation and ablation on the Seward-Malaspina glacier system, Canada-Alaska, *Geol. Soc. Am., Bull. 62,* pp. 725–744.

Smith, H. T. U. (1948). Giant grooves in northwest Canada, *Am. J. Sci., 246,* pp. 503–514.

Additional References

Ahlmann, H. W. (1935). Contributions to the physics of glaciers, *Geog. J., 86,* pp. 97–113.

Garwood, E. J. (1910). Features of alpine scenery due to glacial protection, *Geog. Jour., 36,* pp. 310–339.

Gould, L. M. (1940). Glaciers of Antarctica, *Proc. Am. Phil. Soc., 82,* pp. 835–876.

Seligman, Gerald (1941). The structure of a temperate glacier, *Geog. J., 97,* pp. 295–317.

Taylor, Griffith (1914). Physiography and glacial geology of east Antarctica, *Geog. J., 44,* pp. 365–382, 452–467, and 553–571.

Von Engeln, O. D. (1938). Glacial geomorphology and glacier motion, *Am. J. Sci., 235,* pp. 426–440.

15 · Mountain Glaciation

In most of the world's high mountains ice streams have so profoundly modified valleys that their forms are distinctly different from those produced by fluvial erosion. Although the effects of glacial erosion are most strikingly displayed, depositional features are also present and add distinctiveness to the landscape.

MAJOR FEATURES RESULTING FROM GLACIAL EROSION

The cirque. The most common, and probably the most striking, land form in glaciated mountains is the *cirque*. This French term is applied to amphitheater-like basins (the term amphitheater is not strictly correct, for they are not completely enclosed by walls) which are found most frequently at valley heads, but which may not connect in their distal parts with valleys. The widespread distribution of cirques is attested by the many languages which have an equivalent word: the German *kar,* the Welsh *cwm,* the Scotch *corrie,* and the Scandinavian *botn* and *kjedel.*

That cirques exhibit many characteristics is suggested by the many qualifying terms which have been given to them. They are variously described as simple, compound, hanging, tandem or two-staired, intersecting and nivation cirques. All of these forms are associated with present-day or past glaciers except the nivation cirque. This name has been applied to cirque-like basins which mark the sites of snowbanks which never grow into glaciers. They are shallow basins found on nearly level upland tracts or in tundra areas of subpolar regions. They are produced by solifluction and rill wash beneath snowbanks. They lack the steep headwalls which distinguish true cirques and do not have moraines associated with them, although in their lower parts there may be heaps of solifluction materials which might be mistaken for moraines. The hollowed, worn floor of a typical cirque is usually lacking.

An empty cirque usually displays three distinguishable parts, a headwall, basin, and threshold. A cirque headwall may be as much as 2000 to 3000 feet high and is notably steep and free from talus at its base, even in empty cirques. Scarcity of talus suggests that ordinary weathering processes have not played an important role in cirque formation but rather that some subglacial process was in operation which permitted glaciers to remove material as rapidly as it was detached from the headwall. The most widely held theory of cirque formation was proposed by Johnson (1904). This is generally called the bergschrund theory because of the importance that he attached to basal sapping or plucking at the base of the crevasse, called a *bergschrund,* which commonly is found at the head of a valley glacier. Johnson had himself lowered into an open bergschrund and was much impressed by the evidence of frost riving on the cirque headwall.

Objections have arisen to Johnson's theory because cirque headwalls frequently far exceed in height any observed bergschrund depths and because many glaciers do not exhibit bergschrunds. Bowman (1916) emphasized the fact that bergschrunds are frequently lacking or are not open continuously and thought that cirque headwalls could develop without the existence of a bergschrund through plucking and abrasion beneath moving névé and glacial ice. More recently, Lewis (1938, 1940) has presented a modification of the bergschrund hypothesis in which he stressed the importance of meltwater which finds its way down the cirque headwall back of the névé and then freezes in the rocks and shatters them, after which the loosened rock fragments become incorporated in the moving ice. This process, he thought, helps to account for deeply incised cirques to which the bergschrund theory does not seem applicable.

A cirque usually shows a definite basining of its floor which extends forward from the headwall and terminates at a bedrock riser called its *threshold.* Consequently, cirque basins are commonly sites of small *cirque lakes* or *tarns* as they are called in the British Isles.

Cirques vary in plan from simple, subcircular outlines to compound ones which display many scallops. Compound cirques represent a more advanced stage of development and are products of headward cirque extension at varying rates in different parts of a cirque headwall. Hobbs (1910*a*, 1910*b*) considered differences in cirque characteristics as largely indicative of their stage of advancement in a cycle of mountain glaciation. This is doubtless true as a general rule but local variations in lithology and structure may also account in part for more expanded cirques in some areas than in others. Cirques are more conspicuous in some regions than in others, even where climatic and topographic conditions seem to have been equally favorable to the formation of glaciers. Conditions that seem to favor maximum cirque development are: (1) rather wide spacing of preglacial valleys so as to permit expansion without intersection of adjacent cirques at an early stage; (2) snowfall sufficient to form large snowfields and glaciers but not heavy enough to form ice caps; and (3) fairly homogeneous rocks which permit cirque extension equally well in any direction.

Glacial troughs. Next to a cirque, the most distinctive topographic feature in glaciated mountains is the *glacial trough.* Most glacial troughs were originally stream-cut valleys, but glaciers usually have so altered them that neither in cross profile nor in long profile do they resemble greatly stream-carved valleys. Trough characteristics may be scanty or lacking where glaciers were not large or where valleys were not sites of glaciers during Wisconsin glaciation.

A glacial trough heads not at a cirque headwall but at the lower edge of a cirque threshold. There is usually a conspicuous drop from a cirque threshold

FIG. 15.1. A glacial trough in the Bridge River District, British Columbia. (Photo by Geol. Survey of Canada.)

to the floor of a glacial trough, which has been called a *trough headwall*. Trough headwalls are likely to be most conspicuous where more than one cirque contributes ice to a glacial trough. Glacial troughs are notably irregular and ungraded in their long profiles. Few exhibit the relatively smooth, concave longitudinal profiles characteristic of stream-cut valleys. Descent of trough floors takes place in a series of *glacial steps* or what is sometimes called a *glacial stairway*. In general, these steps are more pronounced in the upper than in the lower part of a trough, probably because the glacier persisted there longer. Each step typically has three component parts: a riser, a riegel, and a tread. A *riser* marks the down-valley end of each step and the ascent to it from the step below. A *riegel* is a sort of rock bar at the top of and just back of a riser. A *tread* is the relatively flat floor or surface of a step. It frequently has a reverse up-valley slope with resulting development upon it of a basin which becomes the site of a rock-basin lake or a chain of lakes, which because of the similarity of their arrangement to beads on a rosary are called *paternoster lakes.*

Glacial steps present an intriguing and puzzling problem to geomorphologists. Many theories have been advanced to account for them, but none seems to be universally applicable. One theory would explain steps as the result of differential glacial abrasion in constricted and open sections of valleys. It is argued that where a trough is narrow the thickness of a glacier will increase, glacial abrasion will be intensified, and greater trough-deepening will take place than in the wider stretches. This theory may be applicable to some steps, but it will hardly apply to all for not all risers are associated with constrictions in glacial troughs.

Steps also have been attributed to the effects of varying rock hardness, the idea being that risers and associated riegels mark hard rock "bars" across a glacial trough. Examples can be found where lithology seems to control the location of steps, but steps are numerous in troughs cut in apparently homogeneous rocks. It has been suggested that steps represent preglacial irregularities or nickpoints. One view holds that the nickpoints resulted from more resistant rock and that glacial abrasion simply accentuated them while they maintained a fixed position. Another view is that the preglacial valley floor irregularity was sufficient to cause crevassing of ice with resulting sapping at the base of the crevasses, much as Johnson postulated to explain cirque headwall recession. Risers so developed would not be stationary but would migrate up a trough and perhaps disappear if the glacier in it persisted long enough. Matthes (1930) attributed the steps in Yosemite Valley to varying erosibility of massive and jointed rocks. Well-jointed or fractured rocks would favor glacial plucking and hence would be more rapidly eroded than rocks that are massive. He believed that glacial erosion was effected largely by plucking and only to a small degree by abrasion. Because the positions of the steps were believed to be related to varying rock structure, he thought that, once risers were initiated, they would remain more or less fixed in position. Steps have also been attributed to the effects of incoming tributary glaciers. Increase in volume and weight of ice below the junction of a tributary glacier presumably would result in increased erosion and deepening of the trough at the point of junction. One can hardly deny the logic of this, and, in fact, examples can be found where this origin is suggested, but not all steps are so situated. The divergences of opinion concerning their origin suggests that glacial steps are of diverse origin. Although there may be difference of opinion as to their origin, there is general agreement that they are expectable features along glacial troughs.

The cross profile of a glacial trough usually is significantly different from that of an unglaciated valley in a mountainous area. The difference is usually described as the difference between a U-shaped valley and a V-shaped valley. This comparison is not too appropriate, for not all glacial troughs are U-shaped and not all stream-formed valleys in the mountains are V-shaped. Probably the best description of the cross profile of a glacial trough is that it resembles a catenary curve, as was suggested by Davis (1916). If a piece of string or rope is held by each end it is possible to reproduce most of the cross profiles exhibited by glacial troughs by varying the distances that the ends of the string are held apart. An overdeepened and oversteepened U-shaped cross profile, such as that shown by Yosemite Valley, would be produced if the ends are held close together; shallower more open forms would be suggested if the ends of the string were held farther apart. Differences in cross profiles may be related to such factors as the thickness of the glacier, the lithology and structure of the rocks in which the trough is cut and the number of times a valley was glaciated.

Portions of glacial troughs may exhibit remarkably flat floors. More commonly these flat-floored sections are the result of deposition subsequent to trough development than of uniform glacial erosion. They may represent aggraded, postglacial, alluvial floors, outwash materials deposited accompanying glacial recession, or lacustrine plains produced by filling of lakes.

Hanging valleys. In contrast to river valleys, which usually have their tributary valleys joining them accordantly, glacial troughs so commonly have

Fig. 15.2. Staubbach Falls descending from a hanging valley into Lauterbrunnen Valley, Bernese Oberland, Switzerland. (Photo by Swiss National Travel Office.)

tributary troughs or valleys joining the main trough discordantly that this may be considered normal. At one time hanging valleys were almost considered prima facie evidence of glaciation. We understand now that hanging valleys may be produced in several ways. They may characterize stream valleys, particularly if the tributary streams are intermittent or much smaller than the main stream. Tilting of a region may steepen the gradient and accelerate the downcutting of a main valley without a corresponding effect upon its tributaries, as Matthes (1930) has pointed out in the Sierra Nevada area. Faulting may also produce hanging valleys. Usually it is not difficult to distinguish hanging valleys of non-glacial origin from those caused by glaciation, because the former is not associated with a glacial trough. Not all trough junctions are discordant. Glaciers of nearly equal size in tributary valleys may join accordantly.

Arêtes or serrate ridges. As cirques enlarge by the sapping process mentioned above, preglacial uplands are gradually consumed by headwall reces-

Fig. 15.3. Mount Assiniboine, a glacial horn, and related cirques of the Canadian Rockies. (Royal Canadian Air Force photo.)

sion. Whether this exhibits a cyclical progression as postulated by Hobbs (1910*a*, 1910*b*) may be open to some question. Certainly preglacial uplands exhibit varying degrees of destruction. The somewhat extreme view

(Wooldridge and Morgan, 1937) that cirques are rather static in position and that sharpening and demolition of peaks takes place through ordinary weathering processes is not widely held. At an advanced stage of cirque recession, mountain divides may be nearly consumed. All that will be left will be a sharp sawtooth-like ridge, which the French have called an *arête* and to which the English term *serrate ridge* is often applied. An arête or serrate ridge consists essentially of alternating sags or *glacial cols* produced by intersection of opposed cirques and pointed peaks or *horns* representing unreduced portions of the original mountain range. Well-known examples of horns are the Matterhorn and Weisshorn in the Alps. Somewhat related in origin to horns but detached from the main mountain range are *monuments* or *tinds,* as they are called in Scandinavia. These are formed where lateral cirque recession cuts through an upland spur between two glacial troughs.

Truncated spurs. In general, glacial troughs are straighter than unglaciated valleys. Glacial troughs do conform to the original valley course, but ice streams may straighten their troughs by abrasion of spur ends and thereby produce *truncated* or *faceted spurs*. Different degrees of spur truncation may be seen, varying from spur ends that have been cut off completely to partially trimmed spurs still recognizable from rocky knobs or nubbins.

Fjords and piedmont lakes. Most students of glacial land forms agree that *fjords* are glacial troughs eroded by ice below sea level, but a few attribute the great depths of water in them to submergence of troughs formed above sea level. Fjords are characteristic features of shore lines in high latitudes and are well-developed along the coasts of Norway, Greenland, British Columbia, Alaska, Chile, and New Zealand. It was once held by some that fjords were largely tectonic in origin, but that idea has largely disappeared. It is possible that the plans of some fjords reflect joint or fault control, but it seems unlikely that many of them were initially grabens.

A feature of a fjord that has evoked considerable speculation is the threshold or sill which is often found at its terminus. The depth of water here is typically much less than farther seaward or headward in the fjord. Thus a fjord is really an elongate basin which would become a lake, if sea level were lowered below the top of its threshold. It has been suggested that fjord thresholds are submerged terminal moraines, and in some instances this may be partly true, but it seems more likely that they result from greater glacial erosion up fjords, where the ice was thick and actively eroding, and less erosion in terminal zones, where the ice was too thin for effective erosion.

In the same class with fjords, but differing from them in that they are above sea level, are numerous elongated *trough lakes*. The Italian lakes Como, Lugano, and Maggiore and Lake McDonald in Glacier National Park are examples of trough lakes. They exhibit the basining which characterizes

fjords as well as an elongate form and great depth of water. The Finger
Lakes of western New York display similar features but they are not included
in this class because they owe their existence to localized erosion by ice caps
rather than by ice streams in mountains.

Fig. 15.4. Milford Sound, New Zealand, a typical fjord. The Tasman Sea shows in
the background. (Photo by Whites Aviation, Ltd.)

DEPOSITIONAL LAND FORMS

Glacial forms. Glacial deposits, in contrast to those of glacially fed
streams and glacial lakes, are marked by heterogeneity of materials and lack
of stratification. Three types of glacial deposits—end moraine, lateral
moraine, and ground moraine—may be distinguished, depending upon
whether deposition took place at the end of, at the side of, or beneath an ice
stream. The term *recessional moraine* has been used in the past to designate
end moraines back of the outermost one, the implication being that a series
of end moraines marks successive pauses in the position of a retreating ice
front. Actually, some may be retreatal moraines, and others may mark readvances. Determining which interpretation is correct is one of the difficult
problems encountered in working out the history of a glaciated valley.

Slight oscillations of an ice front as it recedes may result in an irregular
belt of knolls and basins, usually described as *knob and basin topography*.

This type of end moraine is more commonly formed by ice caps than by ice streams. Not all glaciers build conspicuous end moraines. This depends upon such factors as whether an ice front maintains itself in one position long enough; whether ice-fed streams emerging from glaciers are capable of removing material as rapidly as it is dumped; and whether the glaciers are carrying large loads.

FIG. 15.5. Lateral moraines (foreground) along Green Valley, northwest of Mono Lake, California. (Photo by J. S. Shelton and R. C. Frampton.)

Lateral moraines form along the sides of an ice stream chiefly from materials which are contributed from the valley sides above the glacier by weathering, snowslides, avalanches, and other types of mass movement. Two lateral moraines may join to form a medial moraine, but a medial moraine is more a feature of a glacier's surface than a land form, because it will not persist long after a stream occupies the trough. Lateral moraines are frequently patchy and may or may not be present on both sides of a trough, for usually portions have been removed by postglacial stream erosion. Despite this, lateral moraines are usually among the most impressive features found along glacial troughs. Small lakes are sometimes found perched above the floor of a glacial trough between a lateral moraine and the trough wall.

Ground moraine is not nearly so abundantly associated with ice streams as with ice caps. It is scarce because an ice stream is a particularly effective erosional agent which vigorously abrades and plucks its bedrock floor. Hence, ground moraine is likely to be at best thin and patchy in glacial troughs.

Glacio-fluviatile forms.   Much of the material acquired and transported by an ice stream is ultimately got hold of, transported, and finally deposited

FIG. 15.6. The terminus of a valley glacier. (Air Force photo.)

by streams flowing on, within, beneath, and beyond the glacier. Deposits of such origin are classed as *glacio-fluviatile*. They may retain some of the characteristics of glacial debris, but they show a degree of assortment and stratification roughly proportional to the distance that they were carried by streams. The most common land forms belonging to this class found in areas of mountain glaciation are valley trains, eskers, kame terraces, kames, and outwash fans or deltas.

Valley trains consist of outwash sand and gravel heading usually at an end moraine and extending down-valley from it. Except for those still in the process of formation, they are likely to be marked by terraces above present

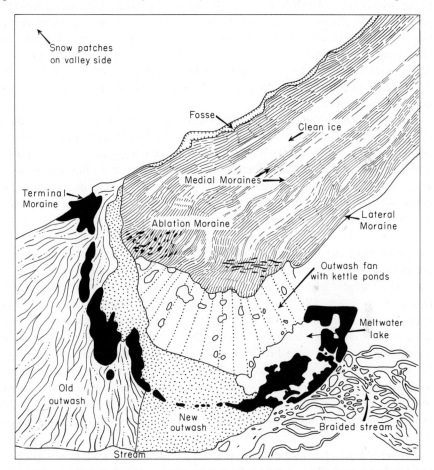

FIG. 15.7. Explanation of Fig. 15.6. (Modified after Gilluly, Waters, and Woodford *Principles of Geology*, W. H. Freeman and Co.)

valley floors. Study of the valley trains extending down from the Alps led Penck and Brückner to a recognition of four stages of glaciation. The valley trains of the Alps, from highest and oldest to lowest and youngest, were named the older deckenschotter (gravel sheet), the younger deckenschotter, the high-terrace gravel, and the low-terrace gravel. It is extending the name valley train somewhat to apply it to the older deckenschotter, for it appears to have been laid down on an old erosion surface, possibly a peneplain, more as great sheets of outwash (so-called outwash plains) than as valley trains confined within valley walls.

Eskers are sinuous ridges of assorted and somewhat stratified sand and gravel which are believed to represent fillings of superglacial, englacial, or sub-glacial stream channels. They are far more common in areas of continental glaciation than in glacial troughs, for eskers are features formed when ice stagnates, and stagnant ice is rare in valley glaciers. They might be confused with lateral moraines, for any persisting in a glacial trough are likely to be found at or near its sides. In contrast to eskers, however, lateral moraines consist of unassorted materials.

Kame terraces may form in glacial troughs, but not too commonly. They, along with eskers and kames, belong to a class known as ice-contact features, so-called because the materials of which they are composed were laid down against an ice surface. Kame terraces are fillings or partial fillings of depressions between a glacier and the sides of its trough. These depressions are called *fosses,* and they owe their existence to the more rapid rate of melting which takes place here because of the added effect of heat absorbed or reflected from the valley sides. Fosses may be the sites of short lakes, or they may have stream courses through them. Materials of a kame terrace are more poorly assorted than those in an esker because they did not move far. The surfaces of most kame terraces are irregular, but some may be flat and have a down-valley slope, much like a valley-train terrace, and may even merge down-valley into a valley train. Kame terraces can be distinguished from valley-train terraces by the poorer degree of assortment of their materials and by the fact that indentations in their fronts, which mark the ice-contact surface, bear no relationship in size to cusps which a stream would develop in cutting out a valley train. Most kame terraces are rather short and discontinuous. They may be mistaken for lateral moraines, but examination of their materials will usually make possible a distinction between the two.

The term *kame* was introduced into geologic literature by Jamieson in 1874 and has been applied to so many different types of glacial and glacio-fluviatile deposits that it has been argued by some that it is so ambiguous that it should be abandoned. Cook (1946) thought that the term should be discontinued and suggested the term *perforation deposit* for individual mounds of sand and gravel and *kame complex* for areas of sag and swell topography. Holmes (1947) defended the usefulness of the term, although he admitted that features called kames have diverse origins. Much like glacial drift, which no longer has its original connotation but is still a handy term, kame has undergone change from its original meaning, but it is still useful when we do not want to be or cannot be specific about the origin of a particular deposit of sand and gravel. It is difficult to define a kame, but almost everyone agrees that it is a mound or hummock composed usually of poorly assorted water-laid materials. Most of them are probably ice-contact forms whose materials were laid down in intimate contact with an ice surface, although they may have undergone modification by slumping subsequent to melting of the ice. Some represent moulin (circular or subcircular shafts extending down into a glacier) and crevasse fillings. Some may be the type of deposit called a perforation deposit by Cook. He thought that such a deposit accumulated in a shallow well or cistern in ice which gradually melted its way downward into the glacier without connecting with the bed of the glacier as a moulin is supposed to do. Many kames formed as small delta cones or fan deposits in small reentrants along an ice front or against ice faces of a mass of stagnant ice and later slumped down when the ice melted. Others, as suggested by Holmes (1947), may represent fillings of fosses around ice-freed hilltops during final downwasting of a stagnant ice mass. Kames of this origin may grade into kame terraces or eskers.

The term kame complex is a useful one if we restrict it to an assemblage of kames and do not apply it to any area of sag and swell topography. Kames are sometimes so numerous in end moraines as to cause moraines to be designated as *kame moraines,* but even here till is likely to be more abundant than water-laid materials.

Glacio-lacustrine features. Lakes, present or extinct, are common features of glaciated valleys. The more common types are: rock-basin lakes in cirques and upon the treads of glacial steps; lakes back of terminal and lateral moraines; lakes formed by the damming of tributary valleys by a valley train in a main valley; and those in the basins of knob and basin topography. Lakes are ephemeral land forms, and their sites soon become *lacustrine plains* by filling with inwash or lowering of their outlets by erosion. The small lacustrine plains which are often found in glaciated mountain valleys are often striking because their flatness contrasts sharply with their surroundings.

The land forms that have been described above as being characteristic of glaciated mountain valleys arrange themselves in a somewhat systematic sequence. In the Colorado Front Range the following down-valley sequence is evident (Jones and Quam, 1944): a series of cirques near the crest of the range; ice-scoured glacial troughs extending from these cirques for a considerable distance down valleys; below these, broad, flat-floored, alluviated basins or "parks" (lacustrine plains) bordered by lateral moraines; terminal moraine zones; and, beyond these, terraced remnants of valley trains which extend beyond the limits of glaciation and often reach beyond the mountain front onto the adjacent piedmont. Similar sequences are encountered in most areas of mountain glaciation.

MULTIPLE MOUNTAIN GLACIATION

It is not yet definitely established that episodes of mountain glaciation were exactly the same in number and duration as in the areas of continental glaciation, but there is abundant evidence of multiple glaciation in most high mountains. It is only reasonable to assume that the major expansions of ice streams in mountain valleys coincided roughly with those of ice caps in lowland areas, and most attempts at correlations of stages of mountain glaciation have been based upon this assumption. The problem of correlating stages of mountain glaciation with those of continental glaciation or with stages of mountain glaciation in separate areas is a difficult one. Even in the Alps, where the more intensive studies have been made, opinions differ as to whether there is evidence of four or as many as six stages of mountain glaciation.

Following are some of the reasons why it is more difficult to distinguish and date stages of mountain glaciation than of continental glaciation. (1) Soil profiles on glacial deposits in mountain regions are commonly immature or incomplete, and glacial materials are generally non-calcareous; hence comparative depths of weathering and leaching cannot be used as effectively in arriving at ages as in many areas of continental glaciation. (2) Rarely found are interbedded soil zones, interglacial deposits, loess, or other stratigraphic markers, which often separate the drift sheets of continental glaciers. (3) Usually there is lack of variation in texture, lithology, and color of the deposits of different stages, for a glacier in a mountain valley erodes the same rocks during each glaciation. (4) Ice streams are more likely than ice caps to destroy the evidence of previous glaciations, because they are confined to one route of movement. Ice caps exhibit fluctuations in directions of maximum movement so that some areas are not glaciated more than once. Only if the first advance of mountain glaciers was the farthest and during each succeeding age the ice failed to reach as far down a valley as in the preceding one, would we be apt to find deposits of each glacial stage preserved.

In spite of the inherent difficulties encountered, there are several methods that can be used to distinguish stages of mountain glaciation. Blackwelder (1931) and Sharp (1938) have discussed this problem in detail and listed the following criteria. The topographic relationships of moraines are suggestive, particularly moraines marking stages of glaciation. Earlier moraines are higher than later ones, although this is not true so much of moraines marking substages. Older moraines may be farther down a valley, but this is not necessarily true.

The topography of moraines also is suggestive of their ages. Moraines of later stages exhibit more distinctly glacial characteristics, are more hummocky, less dissected, and less flattened by the effects of mass-wasting. Older moraines frequently lack the well-defined features of glacial topography. There may be difficulty in distinguishing old moraines from solifluction deposits and landslide jumbles unless exposures extending down into the materials several feet are available. Lateral moraines are often all that is left of older glacial deposits. Deposits of older stages are much more patchy in distribution than those of later stages. A valley or portion of a valley not glaciated during Wisconsin time but glaciated during a pre-Wisconsin glaciation will show a greater degree of postglacial modification than one glaciated in Wisconsin time. Use of varying depths of postglacial stream incision as a measure of the time since glaciation is hazardous, for varying rock hardness and geological structure may account for the differences. Just as moraines of a later glaciation are fresher than older ones so are the cirques. The amount of vegetation in a cirque may be suggestive of its age, but this criterion must be used with caution and applied only to cirques cut in similar rock types.

The extent to which a transverse terminal moraine has been destroyed by a stream crossing it is suggestive of its age. Young end moraines may be only notched or partially destroyed, whereas older ones will be more nearly or completely destroyed. The same principle applies to the degree of destruction of lacustrine flats marking lake sites of different ages. The degree of preservation of striated and polished surfaces may have some value in indicating relative ages, providing they are compared for similar rock types. Existence of such features in granitoid rocks or marbles usually suggests a late age, for they are not preserved long in these rocks.

The ages of moraines can sometimes be determined by relating them to other topographic or geologic features that can be dated, such as faults, lake terraces, beaches, deltas, and stream-cut terraces beyond the moraines.

Careful study of the types of boulders on moraines and their degree of weathering may prove useful. Boulders are less common on old moraines and consist chiefly of resistant-rock types. Young moraines show more boulders at the surface, and boulders of non-resistant rocks will be about as common as more resistant types. Young moraines have about the same percentages of different rock types at the surface as at depth, whereas older moraines show a higher percentage of resistant rocks at the surface than at depth. The thickness of weathering rinds on boulders may be used to advantage, provided similar rock types are compared. The effects of rock weathering extend to a greater depth in older moraines. Granitoid rocks in old moraines are often easily broken with a hammer but not those in younger ones.

Despite the fact that soil profiles are less well-developed on deposits of valley glaciers than those of ice sheets, they may be useful in dating materials of different ages. Richmond (1950) has pointed out the possibility of distinguishing deposits of the Wisconsin substages by comparison of the thickness of the oxidized and colored horizon in profiles developed in materials of different ages.

A careful application of these criteria will usually permit the geologist to recognize evidence for more than one stage or substage of glaciation within a particular area, but it may still be difficult to correlate them with those of other areas. Most interarea correlations reflect consciously or unconsciously an attempt to find evidence of the same number of stages of glaciation in mountains as in lowlands, particularly with attempts at correlation of substages of the Wisconsin. It is further assumed that glacial maxima and minima were contemporaneous in mountains and lowlands. There may be some question whether there was as close correlation between substages in mountains and lowlands as is usually assumed, for mountain glaciers may have maintained themselves relatively undiminished in size while a major ice recession took place in the lowlands.

REFERENCES CITED IN TEXT

Blackwelder, Eliot (1931). Pleistocene glaciation in the Sierra Nevada and Basin ranges, *Geol. Soc. Am., Bull. 42,* pp. 865–922.

Bowman, Isaiah (1916). The Andes of southern Peru, *Am. Geog. Soc., Publ. 2,* pp. 274–313.

Cook, J. H. (1946). Kame-complexes and perforation deposits, *Am. J. Sci., 244,* pp. 573–583.

Davis, W. M. (1916). The Mission Range, Montana, *Geog. Rev., 2,* pp. 267–288.

Hobbs, W. H. (1910a). The cycle of mountain glaciation, *Geog. J., 35,* pp. 146–163 and 268–284.

Hobbs, W. H. (1910b). Studies of the cycle of glaciation, *J. Geol., 29,* pp. 370–386.

Holmes, C. D. (1947). Kames, *Am. J. Sci., 245,* pp. 240–249.

Johnson, W. D. (1904). The profile of maturity in Alpine glacial erosion, *J. Geol., 2,* pp. 569–578.

Jones, W. D., and L. O. Quam (1944). Glacial landforms in Rocky Mountain National Park, Colorado, *J. Geol., 52,* pp. 217–234.

Lewis, W. V. (1938). A meltwater hypothesis of cirque formation, *Geol. Mag., 75,* pp. 249–265.

Lewis, W. V. (1940). The function of meltwater in cirque formation, *Geog. Rev., 30,* pp. 64–83.

Matthes, F. E. (1930). Geologic history of Yosemite Valley, *U. S. Geol. Survey, Profess. Paper 160,* pp. 94–98.

Richmond, G. M. (1950). Interstadial soils as possible stratigraphic horizons in Wisconsin chronology, *Geol. Soc. Am., Bull. 61,* p. 1497.

Sharp, R. P. (1938). Pleistocene glaciation in the Ruby-East Humboldt Range, northeastern Nevada, *J. Geomorph., 1,* pp. 296–323.

Wooldridge, S. W., and R. S. Morgan (1937). *The Physical Basis of Geography,* pp. 377–378, Longmans, Green and Co., London.

Additional References

Atwood, W. W., and K. F. Mather (1932). Physiography and Quaternary geology of the San Juan Mountains, Colorado, *U. S. Geol. Survey, Profess. Paper 166*, 171 pp.

Cotton, C. A. (1941). The longitudinal profiles of glaciated valleys, *J. Geol., 49*, pp. 113–128.

Cotton, C. A. (1942). *Climatic Accidents,* Chapters 13–21, Whitcombe and Tombs, Ltd., Wellington.

Kerr, F. A. (1936). Glaciation in northern British Columbia and Alaska, *J. Geol., 44*, pp. 681–700.

Matthes, F. E. (1900). Glacial sculpture of the Bighorn Mountains, Wyoming, *U. S. Geol. Survey, 21st Ann. Rept.*, Pt. 2, pp. 167–190.

Russell, R. J. (1933). Alpine land forms of western United States, *Geol. Soc. Am., Bull. 44*, pp. 927–950.

Sharp, R. P. (1949). Studies of superglacial debris on valley glaciers, *Am. J. Sci., 247*, pp. 289–315.

Tuck, Ralph (1935). Asymmetrical topography in high altitudes resulting from glacial erosion, *J. Geol., 43*, pp. 530–538.

Wentworth, C. K., and D. M. Delo (1931). Dinwoody glaciers, Wind River Mountains, Wyoming: with a brief survey of existing glaciers in the United States, *Geol. Soc. Am., Bull. 42*, pp. 605–620.

16 · Ice Caps and Their Topographic Effects

As previously stated, the most fundamental differences between ice caps and ice streams are their place of development, mode of nourishment, and mechanics of motion. Other differences are those of size and thickness of ice. No accurate data are available on the thickness of Pleistocene ice sheets but at their centers they must have been many thousands of feet thick. Ice caps certainly move less rapidly than ice streams. The center of an ice cap may appear to be motionless, but this does not preclude motion at depth. Tongues from ice caps extending down valleys may attain rates of motion comparable with those of ice streams. The most rapid movements which have been measured are those of some outlet glaciers in Greenland, where rates up to 100 feet per day have been observed.

One feature of the marginal zones of the ice sheets which existed over North America and Europe that should be emphasized is the notable degree of lobation that they displayed. The edges of the ice caps were probably never straight for any great distance, but in addition to many minor reentrants and projections along their margins there were numerous larger protrusions or *ice lobes* down lowlands. Examples of such lobes in central United States were the Michigan, Erie, and Superior lobes, which existed in pre-glacial lowlands now occupied by members of the Great Lakes. As a result of lobation, the terminal moraines which mark former positions of the ice fronts are commonly arcuate in pattern. An appreciation of lobation helps account for the apparently anomalous Driftless Area in parts of Wisconsin, Minnesota, and Iowa.

Three facts help explain why this area escaped glaciation. First, there is a highland area, the Superior Highland, to the north of it, which interfered with southward advance of the ice; secondly, lowlands to the west and east of this highland area in the basins now occupied by Lake Superior, Green Bay, and Lake Michigan made it easier for the ice to deploy down them than to continue straight southward over the Superior Highland; and shifting of centers from which the ice advanced resulted in the area never being invaded from several directions at the same time.

FEATURES PRODUCED BY EROSION BY ICE CAPS

An *ice-scoured plain,* if the word plain be used with some latitude, is per-haps the most distinctive result of ice-cap erosion. It is well-illustrated in the Canadian Shield area and in parts of Scandinavia and Finland. It is not so much a land form as an assemblage of land forms, most of which are erosional

FIG. 16.1. A portion of the glaciated Canadian shield, Northwest Territories, Canada. (Royal Canadian Air Force photo.)

in origin. Striated, grooved, and polished surfaces, rock basins, and rounded rock knobs, called roche moutonnées, are interspersed with patches of glacial drift.

In areas where the surface over which the ice caps moved was mountain-ous, as in the Adirondacks, Green Mountains, and White Mountains of east-ern United States, the result was not an ice-scoured plain, but a general smoothing off or streamlining of the topography. The many granite bosses of New England exhibit this to a notable degree and form a special type of topog-raphy that is well described as a *mammillated surface.*

On the whole, the effects of glacial erosion were deleterious. Much of the preglacial soil was removed and extensive areas have little agricultural value today. Except for some extensive areas of fertile lacustrine soils and scattered mining camps and resorts for fishing in summer and skiing in winter, the vast Canadian Shield presents a great population void in Canada between the St. Lawrence lowland and the Prairie provinces. To a considerable degree this is a result of glaciation, although it may be questionable whether this area of igneous and metamorphic rocks, even if unglaciated, could have supported a population comparable with that of the sedimentary rock belts in other parts of Canada.

FEATURES PRODUCED BY GLACIAL DEPOSITION

The regional result of deposition by ice caps, where complete burial of preglacial topography took place, is a *till plain*. Where burial was incomplete we find glacially modified topography which is largely bedrock-controlled. The assemblage of topographic forms that make up a till plain consists of glacial deposits and associated water-laid materials. In some places the two types of deposits can be easily distinguished; elsewhere they may be so intimately associated with each other that it is difficult to draw a boundary between them.

Glacial till and its characteristics. An understanding of some of the features that distinguish till plains entails an appreciation of the physical characteristics of till. The terms till and drift should be carefully used and distinguished in discussions of glacial deposits. The name *drift* has come down from the days when what we now know to be glacial deposits were thought to represent materials rafted by icebergs during the Noachian deluge. The word survives even though it has lost its original connotation. In fact, it is sometimes a convenient term to have at hand, particularly if we want to be noncommittal and merely indicate that materials of glacial origin are present in an area without saying whether they were deposited by ice or water. We should, however, restrict the term *till* to those deposits laid down by ice. The terms glacio-fluviatile, glacio-eolian, glacio-lacustrine, and glacio-marine should be applied to the deposits of glacially derived materials made by the indicated agencies.

One outstanding feature of till is its physical heterogeneity. There is no size assortment and no evidence of stratification. The bulk of the material usually is of clay, silt, or sand sizes, but pebbles and huge boulders may be present. Tills high in clay may show a certain amount of lamination or fissility caused by compaction under ice pressure that may be mistaken for stratification. Alternating layers of slightly different texture and composition may also give an effect of pseudostratification.

Lithologic and mineralogic heterogeneity characterize tills, for most tills were derived from several rock types. Despite this, tills usually do exhibit a certain consistency in lithology which makes it possible to describe them as clay, sandy, gravelly, or stony tills. Till lithology is commonly related to local bedrock, and in most places the greater part of the material comprising till was derived from near-by bedrock and not from rocks hundreds of miles away. Till in areas of granitoid rocks, as in New England, is characteristically pebbly and bouldery; till in areas of sandstone is sandy, and till overlying shales, dolomites, and limestones is typically clay till, although where there are limestones and dolomites much of the material of clay size is pulverized limestone or dolomite rock flour.

Intercalated lenses of sand and gravel are common, and 'their presence emphasizes that there is no sharp distinction between ice-laid and water-laid materials. Most sand and gravel lenses are of local extent, and many connect with the surface, as is indicated by the ease with which water wells are obtained from them, even in areas of clay till. Some subsurface beds of sand and gravel have great horizontal extent, and most of these represent interglacial outwash deposits that were subsequently overridden and buried.

Frequently a till section displays two distinctly different types of material, a lower part which is compact and fine-textured, and an upper part from which most of the clay and silt fractions are lacking. These are sometimes interpreted as tills of different ages, when actually they are the same age. The compact lower portion is *basal till* and is believed to have been deposited largely by lodgment beneath the ice. The less compact and more permeable material above is called *superglacial till* or *ablation moraine* and is thought to have been let down during final downwasting of the ice. The large amount of water that resulted from downwasting flushed out most of the clay and silt fractions, leaving a coarse-textured till.

TILL PLAINS AND ASSOCIATED FEATURES

The term moraine as originally used by the French peasants was applied to ridge-like embankments of glacial deposits in alpine valleys or what are now called end moraines. Its usage has since been extended to glacial materials lacking a ridge-like form. Basically there are two classes of moraines. That which is deposited in sheets over the landscape and lacks conspicuous ridge-like form is called *ground moraine* and that which does show ridge-like form is *end moraine*.

Ground moraine. Ground moraine may consist of both basal and superglacial till. If thick, its typical topographic expression is a till plain. Over several of the midwestern states ground moraine is thick enough to completely bury preexisting topography. In Ohio, Indiana, Illinois, and Iowa its average thickness probably exceeds 100 feet, and locally over preglacial valleys it may be several hundred feet thick.

Lithologic differences in tills are reflected in the topographic characteristics of a till plain. Clay tills give rise to flat till plains or undulating surfaces which have been described as *swell and swale topography*. A till plain may be so flat as to possess no relief detectable by eye. It is difficult to understand how glaciers could leave deposits with so level a surface. A flat till plain is not merely a preexisting flat bedrock surface which has been

Fig. 16.2. Flat till plain near Clark Hill, Indiana.   (Photo by Chicago Aerial Survey Co.)

veneered with till. Well records frequently show the existence of several hundred feet of bedrock relief beneath the most level till plain. Flat till plains are usually in areas of clay till and multiple glaciation. The earlier glaciations probably obscured most of the preglacial relief. The final deposits of clay till may have been so water-saturated that they could flow and spread out evenly over the landscape. Postglacial mass-wasting has probably helped to erase many minor topographic irregularities by moving material from the higher to the lower areas. If gravel, pebbles, and boulders are abundant in ground moraine, a till plain is not likely to develop but instead knob and basin type of topography will develop, because the angle of repose for these materials is much higher than that for clay till.

End moraines. End moraines frequently add diversity and relief to a till plain. They vary in character from conspicuous ridges to faint swells.

The materials in end moraines may be deposited in three ways, by lodgment, pushing, or dumping (Flint, 1947) and on this basis may be designated as *lodge, dump,* and *push moraines.* End moraines composed of stony till are usually more conspicuous than those consisting of clay till. End moraines composed of clay till are likely to be broad and flat rather than ridge-like. Broad moraines composed of stony till display a knob and basin type of topography with numerous lakes or swamps in the closed basins. Broad, clay till moraines may exhibit slopes so gentle that without topographic maps one is hardly aware that they stand above the general level of the adjacent ground moraine. End moraines are seldom continuous for great distances but usually are segmented because of breaks made through them by drainage outlets, lack of fixity of the ice front, or local scarcity of debris in the glacier. The crescentic or arcuate plan of end moraines related to ice lobation is not always apparent in the field, but it usually shows up well on aerial photographs and topographic maps.

Many end moraines have been classed as recessional moraines because of their positions back of another end moraine. This term may be misleading, for some mark pauses of a receding ice front and others mark readvances. Only careful field study will permit the differentiation of the two types but probably most so-called recessional moraines really mark ice readvances. If exposures can be found showing till over outwash or lacustrine deposits of an earlier age or if a weathered zone between two tills is present, a readvance is indicated. Moraines with smooth outlines are suggestive of deposition during a readvance, for an ice front is likely to be irregular during backwasting, but this is by no means conclusive evidence.

Interlobate moraines form in a reentrant between two ice lobes. The famous Kettle moraine of Wisconsin, which formed between the Green Bay and Lake Michigan lobes, is a well-known example. The Packerton moraine of Indiana, which formed between the Saginaw and Erie lobes, is another. Most end moraines have varying amounts of water-laid material associated with them, but interlobate moraines are particularly likely to contain large amounts of water-deposited materials. This is understandable when we realize that drainage from two ice fronts is directed toward the area occupied by an interlobate moraine.

Another type of end moraine is that often designated as *kame moraine* because of the association with it of numerous kames. It is believed by some that this type of moraine is deposited in water produced by downwasting of a stagnant ice mass. Kame-like hummocks of water-laid gravels are so common in end moraines that it may be questionable whether numerous kames indicate stagnant ice.

Areal mapping of glaciated areas usually entails delineation of end moraines and their demarcation from ground moraine and other ridge-like glacial forms.

Certain persistent characteristics aid in their recognition.   The ridge form of a moraine is transverse to the direction of ice movement.   In areas of stony till, end moraines usually consist of a complex of minor relief forms, consisting of "short hills" or knobs and irregularly shaped, closed basins or kettles.   This is not so characteristic of clay till moraines.   Moraine summits usually lack uniformity of level, and where stratified materials are present they are usually coarse and poorly assorted and are intimately associated with till.

Fig. 16.3.   An interlobate moraine, north of Lake Ontario, with two moulin kames in the foreground.   (Photo by Conrad Gravenor, Geol. Survey of Canada.)

Drumlins.   Strikingly streamlined hills called *drumlins* are found on some till plains.   Why drumlins are found in some areas and not in others is not understood.   Most drumlins are composed of till, but some have stratified materials as a surficial coating or lens beneath till.   A typical drumlin is half ellipsoid in shape like an inverted spoon, but many variations are found. Double and triple forms may be found arranged en echelon.   Generally, drumlins display a striking parallelism of arrangement.   Drumlins vary in height from 20 feet to 200 feet or more and in length from a fraction of a mile to several miles.   They are rarely found singly but exist in great fields or swarms.

There are four major drumlin belts in North America: an area in New England, in southern New Hampshire and eastern Massachusetts; a belt in New York and Ontario south and north of Lake Ontario; another in Nova Scotia; and a fourth belt including parts of Wisconsin, Iowa, and Minnesota. Drumlins are also present in Ireland, England, Germany, and Switzerland. Probably the most famous drumlin in the world is Bunker Hill.   It and Breed's Hill, another drumlin, played historic roles in the Revolutionary War.

Drumlins characteristically lie several miles back of end moraines. Some consider them to be erosional features, and others attribute them to glacial deposition. Those who favor an erosional origin argue that they are the result of the reshaping of previously deposited end moraines. Three things seem to argue against such an origin. They lack the roche moutonnée profile asso-

FIG. 16.4. A partially submerged drumlin, Chester, Nova Scotia. (Royal Canadian Air Force photo.)

ciated with glacial erosion; the widths of drumlin belts are too wide for them to represent former end moraines; and they usually are composed of the youngest drift in the area. Their streamlined forms suggest lodgment of till, with subsequent overriding and reshaping by ice. Their typical development back of end moraines may possibly be related to rapid thinning of the ice in this zone. The glacier may have been clogged with drift and forced to override some of its load and thereby shape it into drumlin forms. Elongate forms paralleling the direction of ice movement certainly suggest an erosional origin.

GLACIO-FLUVIATILE FEATURES

Depositional Forms

Proglacial forms. Proglacial land forms are those that were built by streams extending beyond an ice front. Included in this group are outwash fans, deltas and aprons, valley trains and both pitted and non-pitted outwash plains. Innumerable small meltwater streams built local outwash fans or aprons beyond an ice front or end moraine. If a proglacial lake existed immediately beyond the ice margin, streams deposited their loads in it as outwash deltas.

Down major drainage lines or *sluiceways* more extensive and continued outwash produced valley trains extending many miles into unglaciated areas. As the ice receded the valley trains were extended headward as long as the sluiceways continued to receive glacial meltwaters. Thus some valley trains are several hundred miles long. As in areas of mountain glaciation, most valley trains are preserved today as terrace remnants along former sluiceways. Such terraces are common features along the Mississippi, Missouri, Illinois, Wabash, Ohio, and many lesser valleys in North America and are also found along many of the valleys of the North European plain. The headward portions of valley trains consist largely of sand and gravel, but the outwash becomes progressively finer down valley and grades into silts and clays in the lower valley courses. Along major sluiceways, however, silts and clays may have been largely carried to the sea.

Outwash plains are produced by the merging of a series of outwash fans or aprons. Those built into non-glaciated areas or areas not recently glaciated are non-pitted. Their surface may exhibit shallow depressions produced by melting of small ice blocks incorporated in the outwash or by deflation before establishment of vegetation upon the outwash plain. A well-known example of a non-pitted outwash plain is the one which extends across the southern part of Long Island south of the Ronkonkoma moraine.

Pitted outwash plains are found where ice blocks have been buried beneath outwash. The ice subsequently melts, resulting in subsidence of overlying material to form *kettles*. Extreme pitting implies that the outwash sands and gravels were laid down over a nearly continuous mass of stagnant ice of irregular thickness. Sometimes kettles align themselves so as to constitute kettle chains, which may mark preglacial or interglacial valleys in which ice blocks persisted.

A pitted outwash plain differs from other types of topography with which it may be confused, such as kame moraines and hummocky end moraines, in that its material is better assorted than that of kame moraines and its surface exhibits a more level skyline than that displayed by end moraines. Pitted out-

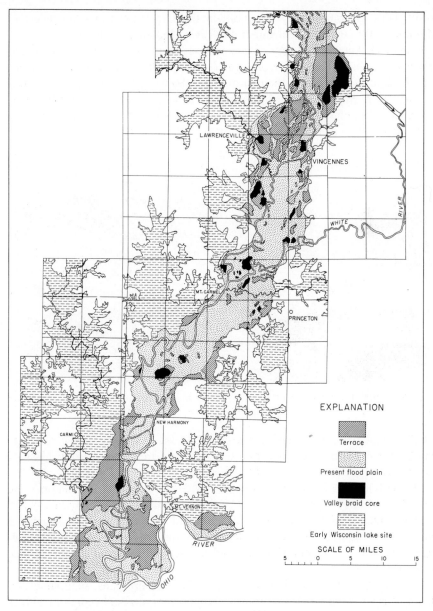

Fig. 16.5. Map of lacustrine areas in the lower Wabash Valley, which mark the sites of former lakes formed in the tributary valleys of the Wabash as the result of their damming by Wisconsin valley train materials. The altitudes of the lacustrine plains decrease down valley in accordance with the gradient of the valley train. (After M. M. Fidlar, *Indiana State Geol. Survey, Bull. 2.*)

wash plains are particularly common on the late Wisconsin drift sheets in Minnesota, Wisconsin, Michigan, and northern Indiana and Illinois.

Eskers. Eskers locally may be conspicuous features in areas of continental glaciation. Numerous and prominent eskers probably indicate that the ice was stagnant at the time of their formation. Some of the finest eskers in this country are in Maine. One of these eskers can be traced for over 100 miles. Many show tributary eskers joining a main esker. Most eskers are definitely associated with end moraines and lie within or tributary to them. They are

FIG. 16.6. Pitted outwash plain on Cape Cod, Massachusetts. (Photo by Fairchild Aerial Surveys, Inc.)

usually segmented, and it has been argued that the individual segments represent yearly deposits made in connection with a retreating ice front. Eskers may merge at their lower ends with deltas, kames, or outwash plains.

Mannerfelt (1945) considered eskers evidence of glacial stagnation. He distinguished three types of eskers: (1) those deposited completely in blocked tunnels under stagnant ice, (2) eskers formed at the mouths of subglacial tunnels where glacial meltwater under hydrostatic pressure flowed into standing water, and (3) eskers which formed in subaerial trenches on the surface of the ice.

Some eskers extend uphill and across divides and drumlins, suggesting that these, at least, were let down onto the ground moraine and represent fillings in englacial or supraglacial stream channels.

Kames and kame terraces. Kames and kame terraces are expectable features in areas of continental glaciation. They were discussed in the preceding chapter and need only to be mentioned here. One special variety of kame which does merit attention is that commonly designated as a *crevasse* *filling.* This variety of kame is ridge-like in contrast to other types of kames. In some respects crevasse fillings resemble eskers except that they are usually smaller and, where they are in groups, may extend in any direction rather than parallel to the direction of ice movement, as do most eskers. Their relationships to each other are not such as to suggest that they are parts of a single linear system like eskers. Flint (1928) believed that they represent fillings made in crevasses at or near the margin of a stagnant ice mass.

FIG. 16.7. An esker near Fort Ripley, Minnesota. (Photo by W. S. Cooper.)

Glacio-fluviatile features have distinctive forms which usually cause them to stand out on topographic maps and aerial photographs. Typically, drainage texture is coarser upon them than upon topographic forms composed of till. This usually will permit accurate mapping of boundaries between them and the ground moraine with which they are associated.

FEATURES PRODUCED BY WASTAGE OF STAGNANT ICE

It is now recognized that in several areas, particularly in New England (Flint, 1929) and the mountains of Norway and Sweden (Mannerfelt, 1945), the ice was stagnant during final wastage.   The topographic features and types of deposits formed under these conditions are notably different from those found where ice was actively moving during its wastage.   Till comprises a smaller part of the deposits as compared with water-laid materials.   End

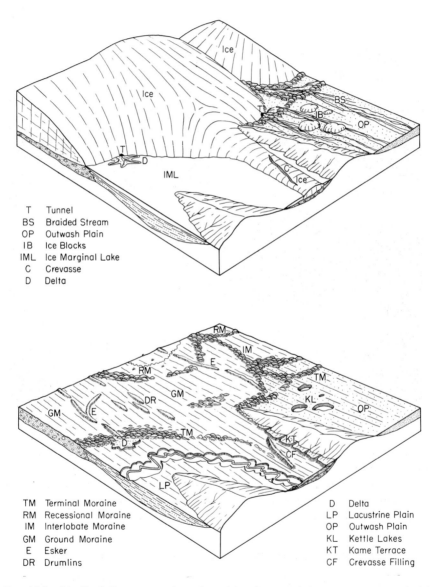

T	Tunnel
BS	Braided Stream
OP	Outwash Plain
IB	Ice Blocks
IML	Ice Marginal Lake
C	Crevasse
D	Delta

TM	Terminal Moraine		D	Delta
RM	Recessional Moraine		LP	Lacustrine Plain
IM	Interlobate Moraine		OP	Outwash Plain
GM	Ground Moraine		KL	Kettle Lakes
E	Esker		KT	Kame Terrace
DR	Drumlins		CF	Crevasse Filling

FIG. 16.8.   Idealized diagrams to show the origin of some of the more common glacial land forms.   (Drawings by W. C. Heisterkamp.)

moraines are faint or entirely lacking. No true end moraines of the push type exist, but faint dump moraines formed by the sliding of ablation moraine down an ice surface may be present. They will be found, however, as scattered hummocks rather than as linear ridges. Ablation moraine, ice contact deposits such as kames, kame terraces, crevasse fillings, eskers and esker systems, extramarginal deltas and outwash plains, lacustrine deposits, ice-marginal and lateral drainage lines and spillways are among the more common features that are found where ice wasted while stagnant.

Mannerfelt (1945) has given the accompanying summary of forms produced by the melting of stagnant ice.

Environment of Formation	Characteristic Features
Superglacial	Crevasse fillings Subaerial eskers Superglacial deltas Ablation moraine
Lateral	Lateral drainage channels Lateral serpentines (winding interconnecting drainage channels) Lateral terraces
Subglacial	Subglacial drainage channels Subglacially engorged eskers Subglacial collecting eskers Esker networks Subglacial dead ice moraines
Frontal and extramarginal	Col gullies or saddle channels Ice-lake drainage channels (spillways) Erosional shore lines Accumulation shore lines Ice-lake sediments Deltas Sanders (outwash plains) Submarginal or marginal dead ice moraine Terminal moraines

ICE-MARGINAL AND PROGLACIAL STREAMS

Glaciation was responsible for many drainage modifications in Europe and North America. Meltwaters from the ice caps usually flowed away as proglacial streams, but some meltwater streams were largely ice-marginal. As the ice fronts receded old routes were abandoned for new ones and in some areas a plexus of drainage channels developed. Many of these former drainage lines have little or no connection with existing stream courses. They may even cross present drainage divides through notches or grooves, as do the so-called Rinnentäler of Europe. Some of these abandoned stream courses may have been cut by subglacial streams, but more often they represent successive drainage lines roughly paralleling a receding ice front. Such are the Urstromtäler of the North European Plain. The "through valleys" south of the Finger Lake region in New York State are examples of abandoned glacial drainageways which carried glacial meltwater when the ice had crossed the divide between the drainage to the Atlantic and the Gulf of St. Lawrence (Holmes, 1952). Striking examples of "one-sided valleys" are found around the western margin of the Adirondacks, which are understandable only when we recognize that ice served as one side of these drainage lines. Bretz (1943) has described an extensive system of drainage trenches (coulees) in the plains of southern Alberta. On almost any glacial plain there are likely to be found abandoned sluiceways which temporarily carried glacial meltwaters.

Four major river systems in the United States show markedly the effects of glaciation upon their courses and the topographic features along them. These are the Mississippi, Missouri, Ohio, and Columbia. The Ohio and Missouri rivers are so closely related in position to the glacial boundary as to make them essentially ice-marginal streams. The Mississippi and Columbia are better thought of as proglacial streams which have undergone important modifications as a result of glaciation.

History of the Missouri River. The Missouri River is the best example in North America of what was an ice-marginal stream. It is generally agreed that in preglacial time the Upper Missouri River, along with the Yellowstone and Little Missouri, drained northward into a river system (probably the Souris-Assiniboine) which emptied into Hudson Bay (Todd, 1914). According to Flint (1949a), the preglacial Grand, Moreau, Cheyenne, Bad, and White rivers of South Dakota, which join the Missouri from the west, extended farther east than at present to join a northward-flowing river in the James River lowland. The segment of the Missouri River from Kansas City on to its junction with the Mississippi probably follows the preglacial course of the Kansas River.

Not enough details are known at present to make clear all the diversions of drainage that were involved in establishing the present course of the Missouri, nor are the dates of the various changes known. Warren (1952) has presented evidence that suggests that the segment of the Missouri River across South Dakota acquired its present course as a result of Illinoian glaciation rather than Kansan, as suggested by Flint. When the complete history is worked out, it may well show that the present Missouri course is the result of numerous drainage changes effected during several if not all of the glacial ages.

History of the Ohio River. The present Ohio River came into existence as the result of the derangement of several drainage lines by southward-moving ice. Its lower part is the preglacial Ohio; its middle portion consists of part

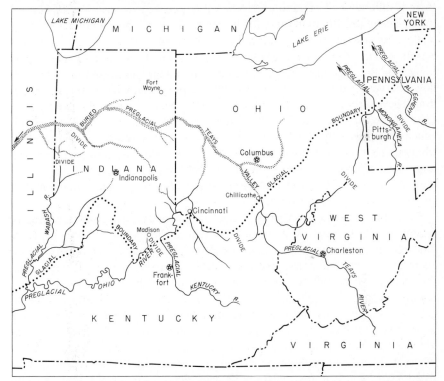

FIG. 16.9. Map showing the preglacial drainage lines from which the Ohio River evolved. (After Horberg, Leverett, Malott, Tight, et al.)

of the drainage of a preglacial river which Tight (1903) called the Teays, and its upper part is a portion of the Allegheny-Monongahela drainage, which in preglacial time flowed into the Gulf of St. Lawrence drainage basin.

The preglacial Ohio headed somewhere in southeastern Indiana (Wayne, 1952) or southwestern Ohio. The preglacial Teays headed in the Blue Ridge or Piedmont of North Carolina and Virginia and flowed northwestward across West Virginia and Ohio and thence slightly southwestward across Indiana into central Illinois (Horberg, 1945, 1950), where it followed approximately the present course of the Illinois River to the Mississippi. The buried course of the Teays Valley across Ohio, Indiana, and Illinois is indicated by many wells which penetrated unusual thicknesses of drift. The preglacial Allegheny-Monongahela drained northward into a lowland now occupied by Lake Erie and thence to the Gulf of St. Lawrence. Some of the preglacial valleys of this system are reflected through the glacial drift. One such can be seen on the Meadville, New York, topographic sheet. There is general agreement as to the diversion of former northward drainage into the preglacial Ohio, but there is difference of opinion as to when it happened. It seems certain that it took place in pre-Illinoian time, for Illinoian outwash is present down the Ohio Valley, but whether it was produced by Nebraskan or Kansan ice is as yet uncertain. One of these ice sheets moved across the Teays and Allegheny-Monongahela drainage lines, blocked and ponded them, and caused them to overflow intervening divides and connect with the preglacial Ohio to form a master drainage line roughly parallel to the ice front. Many deeply alluviated valleys exist in the Appalachian Plateau region of West Virginia and Pennsylvania which were partially filled with silts and clays as a result of this ponding. Many present-day streams in this area bear little relationship to preglacial valleys. Teays Valley, which is shown on the Milton and Saint Albans, West Virginia, topographic sheets, is one example of a preglacial valley which is largely unused by present streams.

Columbia Plateau scablands. One of the most interesting topographic regions in North America, if not in the world, is the so-called channeled scabland region southwest of Spokane in eastern Washington. Here, in an area enclosed by the Columbia, Spokane, and Snake rivers and exceeding in size the State of Maryland, are to be found some of the most unique topographic features produced by glacial meltwaters. The term scabland comes from the great scars which mar the face of the basaltic plateau of eastern Washington and break it up into a maze of buttes, mesas, and canyons. Intercanyon tracts are usually mantled with deposits of eolian sand or loess. Bretz (1923, 1928), who has published numerous papers on this area, listed at least twenty distinctive topographic features that may be seen here. Among the more striking are: interlocking and reticulated channels; abandoned canyons or coulees, as they are called in that region; dry waterfalls and cataracts; steep-walled midchannel mesas and buttes; rock-rimmed basins; hanging valleys; bar-like deposits of gravel; gravel terraces; slackwater deposits in tributary valleys; spillway plexi; and trenched spurs and divides. All observers agree on the bizarreness of topographic forms, but there is lack of agreement on their origins. The only point on which there is rather general agreement is that the scablands were produced by glacial meltwaters. There is even lack of unanimity on this, for Hobbs (1943) argued that the scablands were produced by abrasion by an ice lobe, which he called the Scabland lobe.

Bretz concluded that glacial meltwater occupied the preexisting valleys north of the scabland tract in such quantity that it overflowed the divide and constituted a vast, though short-lived, flood of catastrophic character, which he called the Spokane flood. He envisaged the "flood" as having been as much as 800 feet deep in places. The great streams of water which constituted the "flood" presumably built large constructional bars, which projected above the bottoms of the coulees, and then subsided rather quickly, and while so doing cut the coulees and other striking erosional features. He has been un-

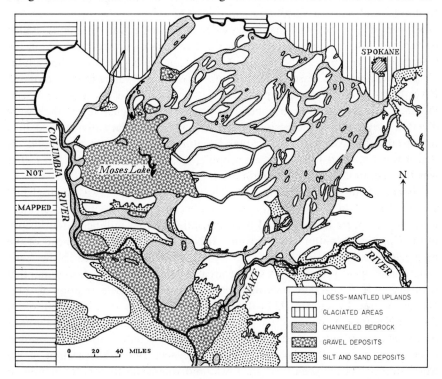

FIG. 16.10. The scabland area of eastern Washington. (After J. H. Bretz, courtesy American Geographical Society.)

able to account for such a flood but maintained that field evidence indicated its reality. This theory represents a return to catastrophism which many geologists have been reluctant to accept.

Flint's (1938) explanation of the scabland tract was that advance of ice lobes across the drainage divide at the northern edge of the plateau resulted in large volumes of glacial meltwater being poured southward across the plateau. This meltwater, following to a large degree preglacial drainage lines, scoured out canyons or coulees in the basalts. Glacial streams then gradually filled their channels with outwash silt, sand, and gravel, as the level of Lake Lewis in the Pasco Basin to the southwest rose, to form a thick fill which grew vertically and headward. As the level of this fill rose many streams flowed across former divides to develop divide crossings and cut scarps in the loess-mantled uplands. Draining of Lake Lewis was followed by progressive dissection of the fill, leaving portions as terraces (the so-called bars of Bretz). When by backwasting the ice front receded north of the divide, the supply of meltwater was cut off and the scablands ran dry except for the through drainage of the Columbia and Snake rivers.

Allison (1933) believed that field evidence pointed toward a "flood," which he believed was produced by ponding of meltwaters by an ice blockade in the Columbia River gorge at the Wallula Gateway, at the beginning of the Columbia gorge through the Cascade Mountains. Rise of the waters in the Columbia River began in the gorge and presumably grew headward into eastern Washington. As the waters were ponded to progressively higher levels they were diverted by ice into a succession of routes across secondary divides and thus produced the gravel bars and scabland features. His idea of the "flood" is that it was not short-lived and catastrophic, as suggested by Bretz, but rather was long-continued and of moderate flow.

History of the Mississippi River. The Mississippi River was a proglacial stream rather than an ice-marginal one, but because several significant changes in its upper course have resulted from glaciation a brief outline of the history of the Upper Mississippi will be given. It is not certain where the head of the preglacial Mississippi River was but it probably was somewhere in the same general area as at present. Its source area in Minnesota is so deeply mantled with glacial deposits that present stream courses bear little relation to preglacial routes. Leverett (1921) thought that from some 15 miles south of St. Paul to Hastings, Minnesota, the present Mississippi Valley follows the course of a preglacial tributary and that from Hastings to Clinton, Iowa, the present river course essentially coincides with its preglacial one. The section of the Mississippi River between Clinton, Iowa, and where the Illinois River joins it above St. Louis appears to have been established as a result of blocking of an earlier route from Clinton, Iowa, across Illinois to near Hennepin, Illinois, and thence southward by the route of the present Illinois River. Shaffer (1952) has presented evidence to indicate that this diversion of the Mississippi from its route through Illinois was effected during the Tazewell subage of the Wisconsin. It is possible that this portion of the Mississippi River has undergone several shiftings back and forth from easterly to more westerly routes as ice lobes alternately advanced from the northwest and northeast.

The Mississippi River near where the Ohio River joins it has undergone several shifts accompanying aggradation of its valley with glacial outwash. Its course was originally considerably farther west than at present in what is called the Drum Lowland (Fisk, 1944). This route was aggraded until it was higher than the Advance Lowland to the east, and the river spilled out of the Drum Lowland into the Advance Lowland; continued aggradation later caused it to spill out of the Advance Lowland into the Morehouse Lowland still farther east. Further aggradation of its valley led to abandonment of the Morehouse Lowland and establishment of its present route through Thebes Gap. The effects on the lower Mississippi of alternation between Pleistocene glacial and interglacial conditions are discussed later.

GLACIO-LACUSTRINE FEATURES

It is undoubtedly true that glaciation has been responsible for the formation of more lakes than all other geomorphic processes combined. Minnesota advertises itself as the "state of 10,000 lakes." Other thousands exist in Wisconsin, Michigan, and Canada, and comparable numbers are found in northwest Europe.

Origins of glacial lakes. Glacial lakes not only exist in great numbers but also have numerous origins. The following classification of lakes and lake basins is based primarily upon one given by Zumberge (1952) for the lakes of Minnesota, but is amended so as to have more general application.

A. Lakes resulting from glacial or glacio-fluviatile deposition.
    1. Lakes in basins resulting from irregular deposition of till.
    2. Lakes back of morainal dams.
    3. Kettle lakes or lakes in ice-block basins.
        a. Ice-block basins along preglacial or interglacial valleys.
            1. Ice-block basins in outwash.
            2. Ice-block basins in till.
        b. Kettles or ice-block basins having no relation to valleys.
            1. Basins in outwash.
                a. On pitted outwash plains.
                b. In troughs adjacent to eskers.
            2. Ice-block basins in till.
            3. Ice-block basins in till and outwash.
    4. Lakes back of valley trains.
        a. In areas previously glaciated.
        b. In unglaciated areas.
B. Lakes in basins produced by glacial erosion.
    1. Lakes in bedrock basins.
        a. Localized by preglacial or interglacial valleys.
        b. Having no apparent relation to preglacial or interglacial valleys.
        c. Along glacial troughs.
        d. In cirques.
C. Lakes in basins produced by glacial erosion and deposition.
    1. Bedrock basins partly dammed by drift.
D. Ice-marginal lakes.
    1. Where the topography slopes toward an ice sheet.
    2. In valleys tributary to a main valley down which a tongue of ice extends.

Lacustrine features.   Lakes are short-lived, and many of the smaller and shallower glacial lakes have already disappeared.   Their sites are now marked by lacustrine plains or areas of mucky soils or bogs, if not quite extinct.   Shore lines of the larger extinct lakes are marked by beaches, bars, and sand dunes. A lacustrine plain is probably the flattest topographic land form that exists, and usually it is readily recognized.   Lacustrine plains are underlain mainly by clays and silts, which may display well-developed lamination.   On aerial photographs they are marked by regularity of road patterns across them and uniform color tone, although silty areas, bars, and beaches may stand out because of their lighter color.

The deposits beneath some lacustrine plains consist of alternating pairs of light and dark layers called *varves*.   Each varve pair typically consists of a layer of fine, light-colored sand or silt overlain by a darker clay layer.   It is commonly believed that each varve represents a year's deposition, the lighter-colored silty layer being the summer layer and the darker clay layer the winter layer.   On the assumption that varves accumulated in successive ice-marginal lakes formed along a receding ice front, varve counts have been used as bases for estimating the time required for recession of the ice sheets from North America and Europe.   By counting and measuring the thickness of the varves deposited in each lake and plotting the results graphically, varve patterns are obtained for individual lakes.   From similarity in patterns obtained in adjacent lakes it may be possible to recognize portions of a varve sequence common to two lakes, and by extending this process to successive lakes it is possible to arrive at an estimate of the total length of time recorded by the varves. It was by this method that Antevs (1928, 1945) in America arrived at the estimate that 20,000 to 25,000 years were required for retreat of the ice from its maximum advance during the Mankato subage.   There has been criticism of this method of arriving at estimates of Pleistocene chronology.   In the first place, it involves a great deal of interpolation and extrapolation, which introduce possible errors.   Secondly, there is some question as to whether varves actually are annual deposits.   Deane (1950) from his study of the varves in the Lake Simcoe region of Ontario was led to doubt seriously that varves represent yearly deposits and was more inclined to think that they represent deposits of shorter lengths of time.   Results from radiocarbon dating of late Wisconsin deposits are not in complete agreement with ages arrived at by varve counting but are similar enough to suggest that varves are probably annual deposits (De Geer, 1951).

SOME MAJOR PLEISTOCENE LAKES

The Great Lakes.   The unraveling of the history of the Great Lakes is an outstanding example of deciphering an extremely complex sequence of events through painstaking field work by a large number of individuals. Not all the details have been worked out, but enough is known to give the basic picture of their sequential phases (Leverett and Taylor, 1915). The Great Lakes had their evolution during Cary and Mankato times. So far as known, there were no pre-Pleistocene lakes, although it is probable that lakes existed during one or more of the interglacial ages. In preglacial time, the sites of the present lakes were valleys or lowlands developed to a large degree upon belts of weak rock. The basins of Lake Ontario, Georgian Bay, the North Channel of Lake Huron, and Green Bay are upon rocks of Ordovician age. The main basin of Lake Huron and that of Michigan are upon Devonian rocks. The Lake Superior Basin seems to owe its existence in part to downfolding and downfaulting.

It is questionable how much glaciers deepened these preglacial lowlands. Shepard (1937) thought that glacial erosion had contributed significantly to the formation of the basins, but this view is not widely held. The bottoms of all the lakes except Erie are below sea level, but that may be the result of depression of the land under an ice load rather than of glacial scouring.

The Great Lakes began their evolution as a series of small ice-marginal lakes after the ice front had receded north of the drainage divide between the Gulf of St. Lawrence and the Gulf of Mexico. Complexities in their evolution are to be explained in terms of four variable factors: an oscillating ice front; irregularities in the uncovered topography; variations in directions of retreat and readvance of the ice front; and differential uplift during and after deglaciation.

Their history has been worked out by tracing topographic features that mark positions of former lake levels and outlets. Such features include: wave-cut cliffs and associated features such as arches and caves; beaches and associated bars; lacustrine deposits; dunes back of former shore lines; and spillways or outlets cut across bedrock or glacial deposits, which are today occupied by underfit streams and exhibit accumulations of peat or muck in abandoned channels. The beaches vary greatly in distinctness, in relation in part to the length of time that a particular beach was occupied and in part to the strength of waves as determined by lake size. Some beaches have been partly or wholly effaced because in later phases lake levels rose above them. In general, beaches terminated at the east, north, or west against an ice front or end moraine. Only during the relatively late Nipissing phase were beaches developed on all sides of the lakes. Beach elevations can be correlated with those of outlets or spillways to determine which outlet or outlets were in use during a particular phase. Unraveling the history of the lakes is complicated by the fact that beaches are horizontal up to a certain point, called the hinge line, but beyond this they are tilted. This tilting resulted from rise of the land accompanying unloading during shrinkage of the ice sheet. Furthermore, the beaches do not have an order of topographic succession which is the order of their development.

Names have been given to the various small lakes that represented the initial phases of what eventually evolved into the present-day water bodies. Lake

FIG. 16.11. Dunes and beach ridges along Lake Michigan at Gary, Indiana. The dunes near the upper left corner were derived from the beach of present Lake Michigan; those at the right mark the shore line of one phase of Pleistocene Lake Chicago. The beach ridges between the two dunal strips were formed during the lowering of Lake Chicago. (Photo by Chicago Aerial Survey Co.)

Maumee was the forerunner of the present Lake Erie; Lake Chicago, of Lake Michigan; Lake Duluth, of Lake Superior; Lake Iroquois, of Lake Ontario; Lake Saginaw, of Lake Huron; and Lake Nicolet, of Green Bay.

As a consequence of their changing outlines and altitudes, the lakes emptied through numerous outlets. Some of the more important of these outlets were: the Fort Wayne outlet down the Wabash Valley; the Lake Chicago outlet to the Des Plaines River and thence to the Illinois River; the St. Croix River outlet of Lake Duluth to the Mississippi River; the Huron Mountain outlet across upper Michigan; the Imlay outlet of Lake Maumee across Michigan to Lake Chicago; the Ubly outlet across Michigan; the Grand River outlet across Michigan to Lake Chicago; the Syracuse outlets to the Hudson River via the Mohawk Valley; the Rome outlet to the Hudson; the Trent Valley outlet across Ontario; the Kirkfield outlet across Ontario; and the North Bay-Ottawa River outlet to the St. Lawrence River (see Fig. 16.12).

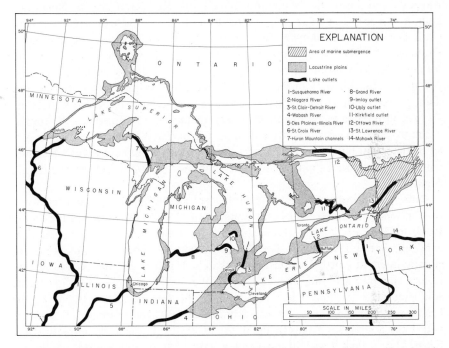

FIG. 16.12. Map of the lacustrine areas of the Pleistocene Great Lakes and the main outlets used by these lakes during their evolution into the present Great Lakes. (After Frank Leverett and F. B. Taylor, *U. S. Geol. Survey, Mon. 53.*)

The Finger Lakes. In western New York is a group of eleven lakes which, because of their elongate shapes and digitate arrangement, is known as the Finger Lakes. They occupy deep, elongate, steep-walled troughs which drain to the north. Lake Cayuga has a surface altitude of 381 feet but its bottom is 54 feet below sea level. Lake Seneca has a surface altitude of 444 feet with bottom 174 feet below sea level, but drilling went 600 feet below this without reaching bedrock. It seems most likely that these great depths are a result of glacial erosion concentrated in preglacial valleys, but there is no general agreement upon this point. Nor is there agreement on the direction of the preglacial drainage. The evolution of the Finger Lakes paralleled and was contemporaneous with that of the Great Lakes but was on a much smaller scale. As topography was uncovered by the retreating ice front a series of ice-marginal lakes came into existence. At first there was an independent lake in each valley but later these coalesced into a broad expanse of water held by the ice front at the north. With removal of the ice dam the present lakes came into existence.

Lake Agassiz. The largest of all ice-marginal lakes was Lake Agassiz, which existed during the Mankato subage in the Red River area of North Dakota, Manitoba, Saskatchewan, and western Ontario. An extensive lacustrine plain marks its site, along with as many as 50 beaches (Johnston, 1946). Drainage of Lake Agassiz during much of its existence was southeastward through the Minnesota River, or River Warren as it was called by Upham (1896), to the Mississippi River. The lake's history seems to have had two main phases related to a marked recession and readvance of the ice (Johnston, 1916), as indicated by an unconformity in the lake sediments. Where the outlet of Lake Agassiz was after its level fell below the Minnesota River outlet is unknown, but Johnston (1946), from a study of Lake Agassiz's beaches, concluded that various outlets to the east into northwestern Ontario could have been used.

Other ice-marginal lakes. Several other ice-marginal lakes existed for a time in western United States and Canada. Lake Dakota occupied the James River lowland of South Dakota and drained southward over a threshold at Mitchell, South Dakota. Lake Souris formed at about the same time, or slightly later, in North Dakota and Saskatchewan between the Altamont moraine (Mankato age) and the ice front to the north. It drained at first to the south into the Missouri River and later to the east into Lake Agassiz. Its beaches extend through a vertical distance of 500 feet or more. Coexistent with Lake Souris was Lake Regina to the northwest in the Saskatchewan River basin. For a time it was an independent lake with an outlet into Lake Souris, but later it too merged with Lake Agassiz. Contemporaneous with the Nipissing phase of the Great Lakes were Lakes Ojibway and Barlow in northern Ontario and northern Quebec. The clay belt south of the end of James Bay marks their sites. Their history has not been worked out in detail, but it is possible that they may have been connected with Lake Agassiz during part of their existence. Present-day fish in the lakes of western Ontario show a closer affinity to those of the Mississippi River than to those in streams flowing into the Atlantic (Deevey, 1949). The most logical explanation of this is that fish entered Lake Agassiz through its outlet to the Mississippi (River Warren) and then entered western Ontario via a connection of Lake Agassiz with Lakes Ojibway and Barlow.

The Baltic water bodies. In northwest Europe there developed a series of ice-marginal water bodies similar in many respects to the Pleistocene Great Lakes (Daly, 1934). However, an oscillating ice front in Scandinavia resulted in alternation between fresh and marine waters, depending upon whether the connection between the Baltic basin and the North Sea was open or closed.   Warped strand lines indicate that there were four major phases. The first of these was a freshwater lake known as the *Baltic Ice Lake*.   It was followed by the *Yoldia Sea*, so named from its characteristic mollusk *Yoldia arctica*.   A connection with the North Sea had been opened across southern Sweden which permitted salt waters to invade the Baltic basin.   Uplift then closed the connection with the North Sea and formed *Ancylus Lake*, so named from the characteristic freshwater mollusk *Ancylus fluviatilis*.   Another marine invasion followed to produce what has been called the *Littorina Sea* by some and the *Tapes Sea* by others.   This oversimplified discussion does little more than suggest the major changes that took place.   There were oscillations in the outlines and altitudes of the water bodies during each phase, which are reflected in multiple strand lines, but in general the conditions were analogous to those in the Great Lakes basins except for the alternation between freshwater and saltwater bodies.

TOPOGRAPHIC EFFECTS OF PLEISTOCENE CHANGES OF SEA LEVEL

Topographic effects of changing sea levels during late Pleistocene and postglacial times are world-wide in extent.   Varying estimates have been made of how much sea level has risen since deglaciation, but the most commonly

Fig. 16.13.   Emerged strand lines on Mount Pelley, Victoria Island, Northwest Territories, Canada, resulting from rise of land accompanying and following deglaciation. (Photo by A. L. Washburn.)

quoted figures are approximately 300 feet. Evidence for formerly higher sea levels is in part topographic and in part biologic in nature. In various parts of the world elevated strand lines and terraces exist which are believed to have had a marine origin. If these were local phenomena, their positions above sea level could be explained as the result of local diastrophism, but they are so world-wide in extent that they seem to be related to eustatic rise in sea level rather than to local uplift. Various attempts have been made to correlate the various terraces with interglacial ages or, in the case of some of the lower terraces, with high sea levels between subages of the Wisconsin, but there is lack of agreement on their number and ages.

Numerous terraces have been described along the Atlantic coast of the United States from New Jersey to the Gulf of Mexico. Cooke (1945), who has maintained that at least seven terraces are recognizable, has classified them as follows:

Brandywine terrace	270 feet	Aftonian
Coharie terrace	215 feet	
Sunderland terrace	170 feet	Yarmouth
Wicomico terrace	100 feet	
Penholoway terrace	70 feet	Sangamon
Talbot terrace	42 feet	
Pamlico terrace	25 feet	Mid-Wisconsin recession

Carlston (1950), working in the coastal plain of Alabama, found no evidence of the Brandywine terrace but did recognize the Coharie at 190 to 210 feet, the Sunderland at 150 to 160 feet, the Wicomico at 90 to 110 feet, the Penholoway at 60 to 70 feet, and the Pamlico at 20 to 30 feet. He attempted no age correlation of them. MacNeil (1950), working in Florida and Georgia, recognized four marine terraces which were tentatively dated as follows:

Okefenokee	150 feet	Yarmouth
Wicomico	100 feet	Sangamon
Pamlico	25–35 feet	Mid-Wisconsin glacial recession
Silver Bluff	8–10 feet	Post-Wisconsin

Flint (1940) was skeptical of so many former sea levels being recognizable along the Atlantic coast. He believed that south of the James River the field evidence indicated two principal marine scarps. These he called the Suffolk and Surry scarps. The younger and lower Suffolk scarp has its toe at 20 to 30 feet above sea level and extends up to 60 feet. He thought that it could be traced as far south as Florida. The higher and older Surry scarp has its toe at 90 to 100 feet above sea level. In general, it lies 10 to 50 miles back of the Suffolk scarp and is less distinct and continuous.

Topographic evidence of former low sea levels is more difficult to obtain because the evidence has either been destroyed or is beneath the sea. Submarine scarps, which have been interpreted as submerged strand lines, have been described. One, called the Franklin Shore, exists off eastern North America from the latitude of Philadelphia to lower Chesapeake Bay. A similar scarp in the Rhone delta has been noted, and a submerged strand line has been described off the coast of Alaska near Nome. It is not certain that these interpretations are correct, for there are other possible origins, such as faulting or submarine erosion.

Drowned valleys on the continental terrace offer convincing evidence of lower sea levels. These are not to be confused with the so-called submarine canyons to be discussed in a later chapter. Well-known examples are the Hudson, Rhine, and Sunda valleys (see p. 465). Buried peat, logs, and mammoth tusks which have been brought up from the floor of the North Sea strongly suggest that it was once land. A submerged valley system in the China Sea is well-established. In Bermuda (Sayles, 1931), vertical shafts 60 meters deep in limestone, which are now below sea level, attest to solution when the area was above sea level. The large amount of calcareous wind-blown sands or eolianites present in Bermuda have an inadequate present source of supply and suggest that these deposits were formed when a now submerged broad shelf was exposed.

Considerable biologic evidence points toward lower sea levels in the past. The faunas and floras of Tasmania and Australia are so similar as to suggest that these two land areas were formerly joined. A lowering of sea level of about 75 meters would connect them. Freshwater fish in Borneo and Sumatra are so alike as to indicate a former land connection between these two islands. The faunas of Barbuda and Antigua in the West Indies, which are 50 kilometers apart, are similar, and water only 20 meters deep separates them.

EFFECTS OF GLACIATION UPON AREAS MARGINAL TO AND BEYOND THE ICE CAPS

Aside from changes in world sea levels, alternation between glacial and non-glacial conditions had numerous far-reaching effects upon areas beyond the ice sheets. These effects will be discussed under the following headings: periglacial phenomena, effects upon stream regimens, and lake fluctuations in non-glaciated areas.

Periglacial phenomena. The term *periglacial* was introduced by Lozinski in 1909 to refer to: areas adjacent to the borders of the Pleistocene ice sheets; the climatic characteristics of these areas; and, by extension, the phenomena induced by this type of environment. A periglacial climate is characterized by low temperatures, many fluctuations above and below the freezing point, and strong wind action, at certain seasons at least. The so-called soil structures or structure soils discussed in Chapter 4 are particularly diagnostic of this environment. Topographic features and geomorphic processes associated with present periglacial climates are found in subpolar regions and at high altitudes, but numerous features related to former permafrost conditions and intensified frost action have been described in areas that were periglacial during the Pleistocene.

Realization that the dominant geomorphic processes are not the same in periglacial and humid temperate regions has come slowly. Matthes (1900) recognized the significance of nivation as a geomorphic process, and Andersson (1906), as stated in Chapter 4, called attention to the importance of solifluction in high latitudes. Cairnes (1912), from studies in Alaska, recognized the importance of frost riving in the shaping of landscapes in subpolar regions. He proposed the term *equiplanation* to include those processes that operate at high latitudes and tend toward leveling of the land without reference to a base level control. Eakin (1916), from observations in Alaska, was impressed with the importance there of solifluction and solifluction slopes, and he applied the name *altiplanation terrace* to "certain terrace forms and flattened summits and passes that are essentially accumulations of loose rock materials." By inference he defined the processes involved in their formation as *altiplanation*. Many of the features described by the above-mentioned individuals are soil structures as we now know them. De Terra (1940) in discussing the landscape of the Tibetan Plateau, an area where frost conditions prevail throughout most of the year, noted the abundance of soil structures, mass movement, and angular debris produced by physical weathering; he concluded that the dominant process involved in planation of this area was physical weathering.

Bryan (1946) suggested the term *cryoplanation* to describe land reduction by the processes of intense frost action. At the same time he proposed a rather extensive terminology for the various processes and features associated with intense frost action and areas of permanently frozen ground. Most of these terms have not caught on, but a few are being used increasingly and seem likely to become a permanent part of the terminology applied to periglacial processes and their products. The term *congeliturbate* was proposed by Bryan to describe "a body of material disturbed by frost action." It would include materials variously known as warp, trail, head, coombe, and churned earth. The surfaces of congeliturbates take the various forms described by Sharp (1942) as soil structures. *Congelifraction* was used by Bryan to designate the process of rock fragmentation by frost splitting or frost riving, and the mass of rubble so produced was called a *congelifractate*.

Peltier (1950) thought that there is sufficient evidence to indicate that there is a periglacial cycle in which cryoplanation is dominant and which is to the subpolar regions what peneplanation is to humid temperate regions. He further suggested the following characteristics for youth, maturity, and old age in the periglacial cycle.

Youth

1. Gently rounded, congeliturbate-covered, undissected upland remnants.
2. Jagged, frost-riven cliffs at the base of which a talus of angular, frost-shattered fragments has formed.
3. Gently sloping, congeliturbate-mantled, steep-sided reentrant valleys in the cliffs whose slopes are continuous with the top of the talus.
4. Isolated steep-sided, frost-riven remnants surrounded by a gently sloping congeliturbate-mantled surface of cryoplanation.

Maturity

1. Long, smooth, gently sloping, undissected, congeliturbate-mantled slopes.
2. Broadly rounded hill-tops and hill crests.
3. Broad, gently sloping, congeliturbate-covered valleys or broad, flat valleys filled with congeliturbate and coarse alluvium derived therefrom.
4. An absence of cliffs and remnants of the previous cycle.

Old age

1. Continued mechanical groundwater weathering.
2. Continued downhill movement of congeliturbate which has led to destruction of hills to slopes of less than 5 degrees.
3. Thoroughly comminuted congeliturbate produced by congelifraction.
4. Wind action may be important on the silt and sand sizes, resulting in loess and sand deposits and wind-swept pebble pavements.

Peltier concluded that the periglacial cycle merits recognition along with the fluvial, arid, and other cycles as a distinct type of denudation which is conditioned by the particular climatic conditions of subpolar and high-altitude regions. Three types of topography seem to distinguish periglacial landscape: surfaces of downwastage or denudation produced by congeliturbation; surfaces of lateral planation resulting from concentrated congelifraction and congeliturbation and stream-eroded surfaces which commonly are aggraded or littered with the waste produced by the preceding processes.

Relict features attributable to former existence of periglacial conditions have been described at many places in Europe and North America. Examples of stabilized block fields or felsenmeere have been described by Smith (1949) in Wisconsin and by Smith (1948) and Denny (1951) in Pennsylvania. Peltier (1950) has recognized what he considered relict periglacial features in the St. Francois Mountains of Missouri. Before such features as block fields, rock streams, and talus slopes are interpreted as evidence of former periglacial conditions, there should be evidence that they are stable at present. Evidence of stabilization consists of: growth of vegetation, particularly trees, upon them; secondary weathering effects upon rock blocks so as to indicate breakdown in place; the beginning of a soil profile; and presence of other types of overlying deposits in an undisturbed condition.

The assumption sometimes made that the periglacial zone is everywhere characterized by permanently frozen ground is, as Flint (1947) has pointed

FIG. 16.14. The Devil's Racecourse, west of Gladhill, Pennsylvania, a block stream probably produced under periglacial conditions. (Photo by G. W. Stose, U. S. Geol. Survey.)

out, not necessarily correct. Undoubtedly many areas peripheral to ice sheets had temperatures many degrees lower than those that exist today in the same areas. However, remarkably few features which can be associated with former permanently frozen ground or intensified frost riving have been described in such states as Ohio, Indiana, and Illinois. The evidence seems to suggest rather that climatic conditions in these states were not much more severe during glacial times than at present. Such pollen analyses as have been made of bogs in this part of the United States (Potzger and Wilson, 1941) indicate that there was no tundra belt peripheral to the ice sheets as seems to have been true in central Europe. The evidence suggests that conifers grew right up to the ice sheets. Probably the presence of periglacial features in Wisconsin, Pennsylvania, and Missouri, if they be such, may be accounted for by higher altitudes of the areas in which the periglacial features exist.

Stream regimens. That alternation between glacial and interglacial conditions had pronounced effects upon the regimens of streams leading from ice sheets is generally recognized. The Mississippi and its tributaries, the Missouri and Ohio, drained areas covered by ice sheets and their valleys bear evidence of alternating periods of aggradation and degradation. There seems to be little doubt that in areas adjacent to the ice sheets valley aggradation was dominant during glacial maxima and that valley trenching prevailed during glacial minima. Most students of midwestern physiography do not believe that eustatic changes of sea level had any appreciable effect upon stream regimens in the upper Mississippi Valley. As stated in Chapter 6, Fisk (1944) and his associates have interpreted the physiography of the lower Mississippi Valley largely in terms of the effects of eustatic rise and fall of sea level. They have described four terraces in the lower Mississippi Valley (see Fig. 6.3) which they explained as a result of alternation between periods of valley filling during interglacial times and valley cutting during glacial times. According to them, lowered sea levels during glacial times induced valley cutting and rise of sea level during glacial times caused valley filling to take place. Continued uplift of the lower Mississippi Valley was thought to account for the terraces being lower for each interglacial stage.

There has been vigorous objection to this interpretation of the terraces of the lower Mississippi Valley by Leighton and Willman (1950), but Fisk (1951) has stoutly defended his interpretation. It is conceivable that the effects of glaciation and deglaciation might be different in the inland portion of a large valley from what they would be in the seaward part. The upper part of a proglacial river might be subject to the climatic rhythm of interglacial erosion and glacial deposition with change from lesser load to greater load, whereas the lower part of the river is subject to the eustatic rhythm of interglacial deposition accompanying rise of sea level and erosion during glacial times consequent upon a lowered sea level.

Peltier (1949) has described six terraces along the Susquehanna Valley, which he interpreted as remnants of former valley trains built down this valley during glacial times. Four of the terraces were considered to be of Wisconsin age and were correlated with the four substages of the Wisconsin, and the other two were thought to be of early and late Illinoian age. Associated with these terraces are rubble-covered slopes, which he considered were produced by congeliturbation in a periglacial zone. He concluded that there was evidence to indicate that each glacial age was marked by valley aggradation and each interglacial age or subage was characterized by valley trenching.

Lake fluctuations in non-glaciated areas. The Basin and Range province of North America contains many closed basins which during the Pleistocene experienced alternating pluvial and arid climates coinciding with glacial and

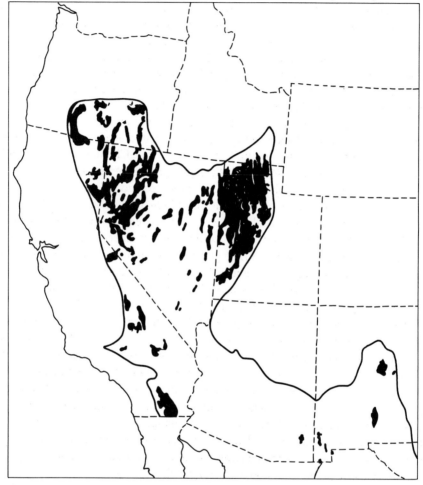

FIG. 16.15. Sites of Pleistocene lakes in the Great Basin region. (After O. E. Meinzer.)

interglacial ages. During glacial times lakes formed in the enclosed basins or existing ones expanded, and during the arid interglacial ages shrinkage or desiccation of lakes took place. A few lake histories have been worked out in detail, but most of them have yet to be studied. The largest and best-known lake was Lake Bonneville, whose history was described in a classic report by Gilbert (1890). It occupied a number of coalescent basins in Utah, Nevada, and Idaho and at its maximum extent covered nearly 20,000 square miles and was over 1000 feet deep. Present Great Salt Lake, Utah Lake, and Sevier Lake are remnants of it. Its former shore lines and associated bars and deltas are striking features of the present-day landscape. Four dis-

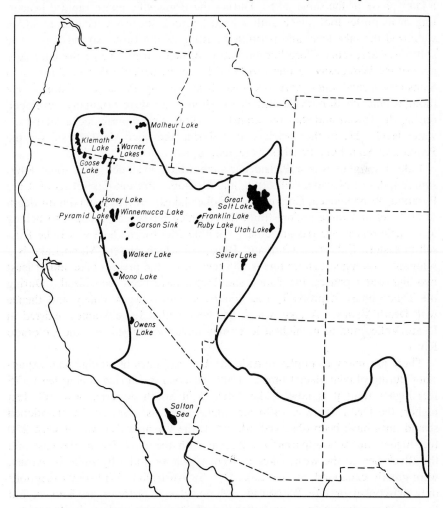

FIG. 16.16. Present lakes of the Great Basin region. (After O. E. Meinzer.)

tinct strand lines have been recognized: the Bonneville at 1000 feet, the pre-Bonneville at 910 feet, the Provo at 625 feet, and the Stansbury at 330 feet above present Salt Lake. The first high-water stage is marked by the pre-Bonneville strand line which is believed to have formed during Illinoian time. During Sangamon interglacial time, desiccation seems to have obliterated the lake. In early Wisconsin time (presumably Iowan), a new lake developed which rose above the level of the previous lake. It had an outlet through Red Rock Pass to the north into the Snake River. Although the Provo strand line is some 285 feet below the Bonneville beach, it is believed that it represents a later phase of the same lake. During the Bonneville phase rapid downcutting of its outlet took place until a hard bedrock sill was encountered which stabilized the lake level and permitted cutting of the Provo strand line. The still lower Stansbury shore line probably formed during one of the later subages of the Wisconsin. Antevs (1945), however, believed that the Provo and Stansbury strand lines were occupied twice during the lake's history. He thought that the Bonneville and Provo shore lines were originally developed during the Iowan and that the Stansbury shore line represented a later lowwater level. He further believed that during the later Mankato subage the Provo and Stansbury beaches were reoccupied.

Lake Lahontan was a Pleistocene lake which extended over some 8400 square miles of Nevada, California, and Oregon. Present-day relics of it are Pyramid, Winnemucca, Carson, and Walker lakes. It had a maximum depth of 500 feet and three major strand lines, which have been thought to belong to a single phase, are recognizable. Another Pleistocene lake was Lake Russell in eastern California (Putnam, 1950), of which present Mono Lake is a remnant. Its highest shore line is 655 feet above the present lake and at least two high-water phases, the Tahoe and Tioga, have been recognized. During the Tahoe phase it probably was connected with Owens Valley and thence with Death Valley. A lake, which has been called Lake Manley, existed in Death Valley and at its highest level may have overflowed into the Colorado River.

There are many examples outside the United States of similar lake expansions during pluvial glacial times. Lake Texcoco in Mexico was at least 175 feet higher than it is now; Lake Titicaca in South America was 300 feet higher; the Dead Sea was 1400 feet higher, and as many as 15 abandoned strand lines have been observed around it; the Caspian Sea was at least 250 feet higher and was apparently confluent with the Aral Sea to the east and the Black Sea to the west; lakes in Kenya Colony and Abyssinia, in Africa, were greatly expanded, as was Lake Eyre in Australia. With few exceptions only Wisconsin or late Wisconsin and postglacial histories are known, but there is every reason to assume that the effects during earlier glacial and interglacial ages were similar to those that have been recognized.

CRITERIA FOR DISTINGUISHING STAGES AND SUBSTAGES OF CONTINENTAL GLACIATION

Not all the methods used in distinguishing stages and substages of continental glaciation involve geomorphic evidence, but several do, and, inasmuch as a geomorphologist should be able to distinguish deposits and topographic features of one glacial stage from those of another, the methods by which this may be done are briefly discussed.

Topographic evidence. There is usually a marked topographic difference between glacial deposits of Wisconsin age and those of an earlier stage. Wisconsin glacial topography has a freshness of form which older features have lost through erosion and mass-wasting. End moraines are weak or obscure on pre-Wisconsin drift sheets. Pre-Wisconsin deposits are generally more dissected by stream erosion, although early Wisconsin materials may display

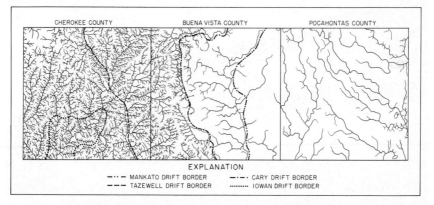

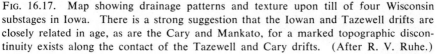

EXPLANATION

—··— MANKATO DRIFT BORDER   —·—· CARY DRIFT BORDER
——— TAZEWELL DRIFT BORDER   ·········· IOWAN DRIFT BORDER

FIG. 16.17. Map showing drainage patterns and texture upon till of four Wisconsin substages in Iowa. There is a strong suggestion that the Iowan and Tazewell drifts are closely related in age, as are the Cary and Mankato, for a marked topographic discontinuity exists along the contact of the Tazewell and Cary drifts. (After R. V. Ruhe.)

considerable dissection adjacent to major stream valleys. Lakes and closed basins are almost always restricted to the later subages of the Wisconsin.

Although they can usually be distinguished from Wisconsin materials, it is generally difficult to distinguish Nebraskan, Kansan, and Illinoian deposits from each other, by topographic differences alone. The more patchy glacial materials are, the older they are likely to be, but this criterion will seldom permit specific age determination. Substages of the Wisconsin can to some degree be distinguished on the basis of the freshness of the glacial topography. Abundant lakes suggest a Cary or later age as do numerous closed basins. Drainage lines are much more extensive and better integrated on drifts of Iowan and Tazewell age than on the younger Cary and Mankato drifts (Ruhe, 1952).

Weathering phenomena. Three methods of dating glacial materials which make use of the comparative degree of weathering have been used. One of these is by comparison of the depths of leaching. This is effective and useful where tills are calcareous but does not apply to non-calcareous tills. Depth of leaching is a function of several factors (Flint, 1949*b*), among which the most important are: composition and texture of the parent material; climate, particularly the factor of amount of rainfall; topography as related to the amount of slope and the position of the water table; amount and kind of vegetation; and the length of time that leaching has been going on. Under similar conditions of topography, climate, parent material, and vegetation, the time

Fig. 16.18. A roadcut through loess showing the line of contact between leached and unleached loess. This can rarely be seen. Special circumstances are responsible for its showing so well in this picture.

factor becomes most significant. Sites for determining comparative depths of carbonates should be selected judiciously. They should be on flat upland areas where soil erosion is at a minimum. Swales and basins should be avoided, for there inwash has thickened the soil profile and too great a depth to carbonates will be obtained. Early Wisconsin till, in the midwestern states, is typically leached to a depth of 4 to 5 feet as compared with 10 to 15 feet for Illinoian. Kansan and Nebraskan tills are leached to much greater depths. Depths of carbonates cannot be safely compared for regions remote from each other because climatic differences and varying till lithologies may have resulted in different depths of leaching.

Although there is a progressive shallowing of the leached zone in deposits of the various substages of the Wisconsin, usually the changes are gradual

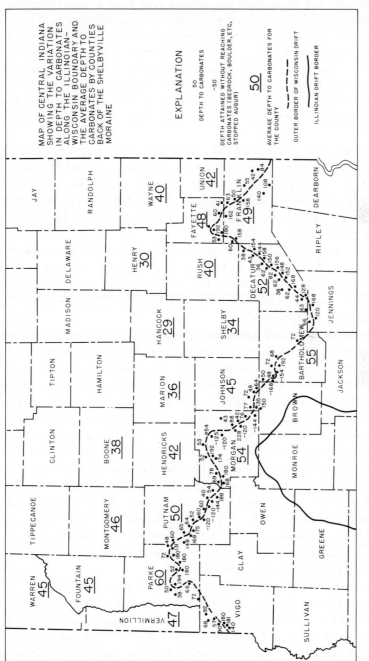

Fig. 16.19. Map of central Indiana showing the boundary between Wisconsin and Illinoian drifts as indicated by the difference in the depths of leaching.

and not abrupt, and rarely does this method lend itself to exact marking of substage boundaries. Typical depths of carbonates of the various Wisconsin substages in the midwestern states are approximately as follows: Iowan, 5 to 6 feet; Tazewell, 4 to 5 feet; Cary, 3 to 4 feet; and Mankato, 2 to 3 feet. Different representative values have been obtained in New England and in the northern Great Plains. The time intervals between the subages of the Wisconsin were relatively short; hence other factors that influence the depth of leaching become as significant as the time factor in determining the depth of carbonates.

All till sheets, except the Wisconsin, have reached such an advanced stage of weathering that there has developed in their *B* horizons a deeply weathered material which is called gumbotil, mesotil, or silttil, depending upon its texture (Leighton and MacClintock, 1930). Comparative thicknesses of this horizon may be used to ascertain the age of a till, for it will be thicker on older than younger tills.

In areas of non-calcareous drift it may be necessary to use such methods as measurement of thicknesses of weathering rinds on boulders, along with the general degree of weathering exhibited by the whole mass of till, to determine its age. Care must be taken to compare boulders of similar rock types when weathering rinds are measured.

Interglacial deposits and buried soil horizons. A buried soil or weathered zone between two tills is one of the most effective means of differentiating two till sheets. It may not always be possible to determine with certainty the age of a buried till, unless a complete soil profile is preserved. Where such profiles are found the depth of the profile or the thickness of a gumbotil or correlative horizon in the profile may permit age determination of the buried till by comparison with known typical soil profiles in tills whose ages are known.

Intertill deposits, such as buried loess, lacustrine materials, wind-blown sands and outwash, may be helpful in differentiating two or more tills. These materials are particularly useful when they have soil profiles developed upon them that were developed during interglacial times. Care should be taken not to confuse the numerous local lenses of sand and gravel which are found in most tills with extensive sheets of glacial outwash. Overridden outwash is especially useful in indicating tills deposited during readvances. Lithologic differences in till sheets may also be diagnostic.

Radiocarbon dating. More accurate dating of some glacial deposits has become possible through age determinations arrived at from radiocarbon (Flint and Deevey, 1951). Carbonized wood, bones, peat, and other types of materials may be used for dating the deposits in which they are found. At present this method is applicable only to the later substages of the Wisconsin, but improvements in techniques are being made which may eventually permit its application to any materials of Wisconsin age.

REFERENCES CITED IN TEXT

Allison, I. S. (1933). New version of the Spokane flood, *Geol. Soc. Am., Bull. 44,* pp. 675–722.

Andersson, J. G. (1906). Solifluction, a component of subaerial denudation, *J. Geol., 14,* pp. 91–112.

Antevs, Ernst (1928). The last glaciation, *Am. Geog. Soc., Research series, No. 17,* pp. 105–171.

Antevs, Ernst (1945). Correlation of Wisconsin glacial maxima, *Am. J. Sci., 243-A,* pp. 1–39.

Bretz, J. H. (1923). The channeled scablands of the Columbia River plateau, *J. Geol., 31,* pp. 617–649.

Bretz, J. H. (1928). The channeled scablands of eastern Washington, *Geog. Rev., 18,* pp. 446–477.

Bretz, J. H. (1943). Keewatin end moraines in Alberta, Canada, *Geol. Soc. Am., Bull. 54,* pp. 31–52.

Bryan, Kirk (1946). Cryopedology—The study of frozen ground and intensive frost-action with suggestions of nomenclature, *Am. J. Sci., 244,* pp. 622–642.

Cairnes, D. D. (1912). Differential erosion and equiplanation in portions of Yukon and Alaska, *Geol. Soc. Am., Bull. 23,* pp. 333–348.

Carlston, C. W. (1950). Pleistocene history of coastal Alabama, *Geol. Soc. Am., Bull. 61,* pp. 1119–1130.

Cooke, C. W. (1945). Geology of Florida, *Florida Geol. Survey, Bull. 29,* pp. 12–13 and pp. 245–248.

Daly, R. A. (1934). *The Changing World of the Ice Age,* pp. 51–80, Yale University Press.

Deane, R. E. (1950). Pleistocene geology of the Lake Simcoe district, Ontario, *Geol. Survey, Canada, Mem. 256,* pp. 36–41.

Deevey, E. S., Jr. (1949). Biogeography of the Pleistocene, *Geol. Soc. Am., Bull. 60,* p. 1393.

De Geer, E. H. (1951). De Geer's chronology confirmed by radioactive carbon, C 14, *Geol. Fören. i Stockholm Förhandl., 73,* pp. 517–518 and 557–570.

Denny, C. S. (1951). Pleistocene frost action near the border of the Wisconsin drift in Pennsylvania, *Ohio J. Sci., 51,* pp. 116–125.

de Terra, Hellmut (1940). Some critical remarks concerning W. Penck's theory of piedmont benchlands in mobile mountain belts, *Assoc. Am. Geog., Anns., 30,* pp. 241–246.

Eakin, H. M. (1916). The Yukon-Koyukuk region, Alaska, *U. S. Geol. Survey, Bull. 631,* pp. 67–82.

Fisk, H. N. (1944). *Geological Investigation of the Alluvial Valley of the Lower Mississippi River,* Mississippi River Commission, Vicksburg, 78 pp.

Fisk, H. N. (1951). Loess and Quaternary geology of the lower Mississippi Valley, *J. Geol., 59,* pp. 333–356.

Flint, R. F. (1928). Eskers and crevasse fillings, *Am. J. Sci., 215,* pp. 410–416.

Flint, R. F. (1929). Stagnation and dissipation of the last ice sheet, *Geog. Rev., 19*, pp. 256–289.

Flint, R. F. (1938). Origin of the Cheney-Palouse scabland tract, Washington, *Geol. Soc. Am., Bull. 49*, pp. 461–524.

Flint, R. F. (1940). Pleistocene features of the Atlantic coastal plain, *Am. J. Sci., 238*, pp. 757–787.

Flint, R. F. (1947). *Glacial Geology and the Pleistocene Epoch*, John Wiley and Sons, New York, 589 pp.

Flint, R. F. (1949*a*). Pleistocene drainage diversions in South Dakota, *Geograf. Annaler, 31*, pp. 56–74.

Flint, R. F. (1949*b*). Leaching of carbonates in glacial drift and loess as a basis for age determination, *J. Geol., 57*, pp. 297–303.

Flint, R. F., and E. S. Deevey, Jr. (1951). Radiocarbon dating of late-Pleistocene events, *Am. J. Sci., 249*, pp. 257–300.

Gilbert, G. K. (1890). Lake Bonneville, *U. S. Geol. Survey, Mon. 1*, 438 pp.

Hobbs, W. H. (1943). Discovery in eastern Washington of a new lobe of the Pleistocene glacier, *Science, 98*, pp. 227–230.

Holmes, C. D. (1952). Drift dispersion in west-central New York, *Geol. Soc. Am., Bull. 63*, pp. 993–1010.

Horberg, Leland (1945). A major buried valley in east-central Illinois and its regional significance, *J. Geol., 53*, pp. 349–359.

Horberg, Leland (1950). Bedrock topography of Illinois, *Illinois Geol. Survey, Bull. 73*, pp. 67–72.

Johnston, W. A. (1916). The genesis of Lake Agassiz: a confirmation, *J. Geol., 24*, pp. 625–638.

Johnston, W. A. (1946). Glacial Lake Agassiz, with special reference to the mode of deformation of the beaches, *Can. Dept. Mines and Resources, Geol. Survey, Bull. 7*, pp. 1–20.

Leighton, M. M., and Paul MacClintock (1930). Weathered zones of the drift sheets of Illinois, *J. Geol., 38*, pp. 28–53.

Leighton, M. M., and H. B. Willman (1950). Loess formations of the Mississippi Valley, *J. Geol., 58*, pp. 599–623.

Leverett, Frank (1921). Outline of Pleistocene history of Mississippi Valley, *J. Geol., 29*, pp. 615–626.

Leverett, Frank, and F. B. Taylor (1915). The Pleistocene of Indiana and Michigan and the history of the Great Lakes, *U. S. Geol. Survey, Mon. 53*, pp. 316–518.

MacNeil, F. S. (1950). Pleistocene shore lines in Florida and Georgia, *U. S. Geol. Survey, Profess. Paper 221-F*, pp. 95–106.

Mannerfelt, C. M. (1945). Några glacialmorfologiska formelement, *Geograf. Annaler, 27*, pp. 3–239.

Matthes, F. E. (1900). Glacial sculpture of the Bighorn Mountains, Wyoming, *U. S. Geol. Survey, 21st Ann. Rept.*, Pt. 2, p. 183.

Peltier, Louis (1949). Pleistocene terraces of the Susquehanna River, Pennsylvania, *Penn. Geol. Survey, Bull. G-23*, 4th ser., 147 pp.

Peltier, Louis (1950). The geographic cycle in periglacial regions as it is related to climatic geomorphology, *Assoc. Am. Geog., Anns., 40*, pp. 214–236.

Potzger, J. E., and I. T. Wilson (1941). Post-Pleistocene forest migration as indicated by sediments from three deep inland lakes, *Am. Midland Naturalist, 25*, pp. 270–289.

Putnam, W. C. (1950). Moraine and shoreline relationships at Mono Lake, California, *Geol. Soc. Am., Bull. 61*, pp. 115–122.

Ruhe, R. V. (1952). Topographic discontinuities of the Des Moines lobe, *Am. J. Sci., 250*, pp. 46–56.

Sayles, R. W. (1931). Bermuda during the ice age, *Proc. Am. Acad. Arts Sci., 66*, pp. 381–468.

Shaffer, P. R. (1952). Tazewell glacial substage of western Illinois and eastern Iowa, *Geol. Soc. Am., Bull. 63*, p. 1296.

Sharp, R. P. (1942). Soil structures in the St. Elias Range, Yukon Territory, *J. Geo morph., 5*, pp. 274–301.

Shepard, F. P. (1937). Origin of the Great Lakes basins, *J. Geol., 45*, pp. 76–88.

Smith, H. T. U. (1948). Periglacial boulder field in northeastern Pennsylvania, *Geol. Soc. Am., Bull. 59*, pp. 1352–1353.

Smith, H. T. U. (1949). Periglacial features in the driftless area of southern Wisconsin, *J. Geol., 57*, pp. 196–215.

Tight, W. G. (1903). Drainage modifications in southeastern Ohio and adjacent parts of West Virginia and Kentucky, *U. S. Geol. Survey, Profess. Paper 13*, 108 pp.

Todd, J. E. (1914). The Pleistocene history of the Missouri River, *Science, 39*, pp. 263–274.

Upham, Warren (1896). The glacial Lake Agassiz, *U. S. Geol. Survey, Mon. 25*, 658 pp.

Warren, C. R. (1952). Probable Illinoian age of part of the Missouri River, South Dakota, *Geol. Soc. Am., Bull. 63*, pp. 1143–1156.

Wayne, W. J. (1952). Pleistocene evolution of the Ohio and Wabash valleys, *J. Geol., 60*, pp. 575–585.

Zumberge, J. H. (1952). The lakes of Minnesota—their origin and classification, *Minn. Geol. Survey, Bull. 35*, pp. 10–48.

Additional References

Andersen, S. A. (1931). The waning of the last continental glacier in Denmark as illustrated by varved clays and eskers, *J. Geol., 39*, pp. 609–624.

Black, R. F. (1950). Permafrost, pp. 247–275, in Trask's *Applied Sedimentation*, John Wiley and Sons, New York.

Cooke, C. W. (1930). Correlation of coastal terraces, *J. Geol., 38*, pp. 577–589.

De Geer, G. J. (1940). *Geochronologia Suecica Principles*, Almqvist and Wiksells Boktryckeri A.-B., Stockholm, 367 pp. (in English).

Dines, H. G., et al. (1940). The mapping of head deposits, *Geol. Mag., 77*, pp. 198–226.

Goldthwait, R. P. (1951). Development of end moraines in east-central Baffin Island, *J. Geol., 59*, pp. 567–577.

Gould, L. M. (1940). Glaciers of Antarctica, *Proc. Am. Phil. Soc., 82*, pp. 835–877.

Gwynne, C. S. (1942). Swell and swale pattern in the Mankato lobe of the Wisconsin drift plain in Iowa, *J. Geol., 50*, pp. 200–208.

Holmes, C. D. (1941). Till fabric, *Geol. Soc. Am., Bull. 52*, pp. 1299–1354.

Krumbein, W. L. (1933). Textural and lithological variations in glacial till, *J. Geol., 41*, pp. 383–408.

Leighton, M. M., and W. E. Powers (1934). Evaluation of boundaries in the mapping of glaciated areas, *J. Geol., 42*, pp. 77–87.

Lougee, R. J. (1940). Deglaciation of New England, *J. Geomorph., 3*, pp. 189–217.

Speight, Robert (1940).   Ice wasting and glacier retreat in New Zealand, *J. Geomorph.*, *3*, pp. 131–143.

Taber, Stephen (1943).   Perennially frozen ground in Alaska, *Geol. Soc. Am., Bull. 54,* pp. 1433–1548.

Thornbury, W. D. (1940).   Weathered zones and glacial chronology in southern Indiana, *J. Geol., 48,* pp. 449–475.

Thwaites, F. T. (1926).   The origin and significance of pitted outwash, *J. Geol., 34,* pp. 308–319.

Thwaites, F. T. (1946).   *Outline of Glacial Geology,* Edwards Brothers, Ann Arbor, 129 pp.

17 · Geomorphology of Coasts

MOVEMENTS OF WATER IN OCEANS AND LAKES

The following discussion deals chiefly with ocean shore lines, but much of it applies equally well to the shore lines of lakes. Lakes and oceans are bodies of standing water only in the sense that their margins and positions are fixed, subject in the case of the ocean to minor oscillations produced by the ebb and flow of tides. Movement of water goes on at nearly all times both at the surface and at depth. We are not concerned here with large-scale circulations whereby there is interchange of water between the surface and depths or slow latitudinal movement of oceanic waters. The three types of movement that carry on gradational work are waves, currents, and tides, and only rarely are tides of geomorphic significance.

Wind is not the only cause of wave generation but it is by far the most important. Wind generates and affects wave motion through friction upon a water surface, through "push" against the rear of a wave and "pull" at its front, and by suction over the crest and compression in the trough of a wave. *Wave length* is the horizontal distance between adjacent crests or troughs, and *wave height* is the vertical distance between them. *Wave period* is the time between the passage of two consecutive crests or troughs past a given point, and *wave velocity* is the speed at which the wave form advances. The size of a wave is related to the velocity of the generating wind, but maximum size is attained only in water that is deep enough for the ocean bottom not to interfere with the undulatory movement of the water and then only after the wind has been blowing for a considerable time. The greatest wave heights so far accurately measured were about 16 meters. Greater heights have been reported but their reality is doubtful. Wave height is determined not only by wind velocity but also by the extent of water over which the wind can blow. This is called the *fetch*. Extremely large waves can develop only where the fetch is great, which explains why such waves cannot develop on lakes or enclosed arms of the sea. Kuenen (1950) stated that a fetch of at least 1000 kilometers would be required to produce the largest waves that have been observed.

Another factor, besides wind velocity and fetch, that affects wave height is the duration of the wind. Studies made at the Scripps Institute of Oceanography at La Jolla, California, during World War II indicated that, with a wind velocity of 105 kilometers per hour and a fetch of 1500 kilometers, a wave height of 20 meters was theoretically possible, but the wind would need to blow continuously for 50 hours before it would be attained.

The type of wave that we have been discussing is that which develops in water deep enough to allow free orbital movement. This is called a *wave of oscillation*. In it the main movement of water is roughly circular with the water moving forward on the crest, upward on the front, backward in the trough, and downward on the back. There is also a slow mass movement of water forward in the direction of wave propagation, chiefly because water particles are moving forward at a greater velocity on the crest than in the trough. After having completed their orbits, water particles find themselves somewhat ahead of previous positions, and thus without influence of wind they achieve forward motion. This forward mass movement of the water is greater in high than in low waves.

In contrast to waves of oscillation are *solitary waves* or *waves of translation* in which the water moves in the direction of wave propagation without compensating backward motion. Unlike oscillatory waves, which develop in chains, this type of wave is a single and independent unity. Movement of water particles is in a parabolic path at the surface but flattens downward until at the bottom it is in straight lines. Solitary waves may be generated at sea or in shallow water through the breaking of oscillatory waves. They do not display the noticeable crests and troughs of oscillatory waves but rather appear like welts separated by practically flat water surfaces. Transitional forms between the oscillatory and solitary waves exist, but they are not well-understood. Solitary waves are more effective than oscillatory waves in moving forward material on the sea bottom and in doing geologic work, for all their energy is carried forward.

As Shepard (1948) pointed out, oscillatory waves may exhibit two rather distinct phases of development. In one phase their characteristics are directly controlled by the wind. Waves in this stage of development are variously known as *wind waves, forced waves,* or *sea waves.* None of these terms is especially appropriate, but they are intended to indicate waves that are actively growing in height under direct influence of wind in contrast to *swell* or *free waves,* which were wind-generated but have traveled beyond a storm area and are decreasing in height. It may be impossible to distinguish the two, but swell or free waves are less complex than forced waves and consist of a kind of heaving motion of water with lower wave heights, longer wave lengths, and fewer breaking wave crests. Swell accounts for strong wave action along coasts on days when the air is relatively calm.

As waves move into shallow water there is a decrease in velocity with a consequent crowding together of the waves and steepening of their fronts. Where the orbital velocity in the crest of the wave exceeds the rate at which the wave is moving forward there develops a curl in the top of the wave front which, lacking sufficient water to fill the cavity beneath it, lunges forward or breaks to form *surf*. In breaking, waves of oscillation change to waves of

FIG. 17.1. A plunging breaker at La Jolla, California. (Photo by F. P. Shepard.)

translation, which dash forward onto the shore as *swash*. After the swash reaches its maximum forward position the water runs back down the seaward slope as *backwash*.

Tsunamis are waves that are produced by submarine earthquakes, volcanic eruptions, or landsliding and slumping. They are often called tidal waves but are in no sense related to tides. They have wave lengths of the order of 100 miles and travel at velocities of 400 miles or more per hour. In open sea, their heights may be only a foot or two but when they come onto shore they may develop into waves having extraordinary heights. According to Kuenen (1950), tsunamis along the Japanese and Chilean coasts have attained heights of 30 to 40 meters. Shepard (1948) reported the destruction by tsunamis of a lighthouse at Dutch Cape, Alaska, 100 feet above sea level. Tsunamis may effect greater erosion along a shore line during a brief period than ordinary waves will in years. They are, however, infrequent even in those areas where geologic conditions permit their formation.

Currents differ from waves in that there is continued and progressive forward movement of water. Currents may be created in several ways, but we shall discuss only those that have geomorphic significance.

Winds are responsible directly or indirectly for the major surface ocean currents; hence they have a definite relation in direction to planetary wind belts, modified, however, by the deflective effect of the earth's rotation and the configuration of the continents. Where currents impinge upon land they are forced to deviate in direction and hence the directions of some currents are determined to a large degree by the outlines of the continents. This is one cause of *longshore* or *littoral currents,* a type of current which may have important geomorphic effects. Longshore currents are also generated when waves break obliquely against a shore. Wind drift currents seldom affect water to depths in excess of 300 feet. Their velocities are rarely great enough to produce significant erosive action, but they can move material along the sea bottom and thereby contribute directly or indirectly to the development of depositional features along shore lines. *Longshore drifting* as effected by currents should not be confused with a different type of longshore transportation called by Johnson (1919) *beach drifting.* If a wave approaches a shore obliquely, its swash will run up the shore in the direction of wave propogation, but its backwash will move down the steepest slope under the influence of gravity. This is usually at right angles to the shore, and consequently particles being moved by the swash and backwash pursue parabolic paths which gradually move the particles along shore in a direction parallel to the front of the breaking waves. Where beach drifting coincides in direction with a longshore current, pronounced movement of shore materials may result; if, however, beach drifting is in a direction opposite to a longshore current, the result will be that the finer beach material will travel in the direction of the longshore current and the coarser material will travel in the direction of the beach drifting. Probably as much material is moved along shore by beach drifting as by shore currents. If sediments are coarse, beach drifting may be more important in their transport.

Tidal currents locally may attain velocities sufficient to transport material, but generally they are of minor geomorphic significance. Tides are hardly recognizable in lakes. Lake Erie has a tidal range of about 8 centimeters, and in the Baltic sea the tidal range is barely 2 centimeters (Kuenen, 1950). In funnel-shaped bays, tidal range may be large because of piling up of water. In the Bay of Fundy, between Nova Scotia and New Brunswick, the tidal range varies from 30 to 50 feet, and flow of the tide is commonly marked by a front of advancing water known as a *tidal bore,* which may be as much as 6 feet high. Similar tidal bores form in the estuaries of the Severn and Trent in southern England, and tidal bores 5 meters in height have been described in the Amazon; even higher ones have been reported in the Tsientan River in China (Kuenen, 1950). It is doubtful, however, if tidal bores perform much erosive work. Johnson (1925) concluded that in the Bay of Fundy there was more evidence of deposition than of erosion by tides. Tidal cur-

Fig. 17.2. Vertical photo, Marino County, California, showing rip currents extending seaward. (Production and Marketing Administration photo.)

rents can locally transport considerable debris and even carry on a scouring action. Water movement extends downward to greater depths than in wind-induced currents and may be markedly concentrated by topographic features. According to Shepard (1948), tides are responsible for the most rapid movement of sea water known. In Seymour Narrows, between Vancouver Island and the coast of British Columbia, tidal currents measuring 12 knots (13.8 miles) an hour have been recorded.

It generally is assumed that the water which is carried landward by waves returns seaward mainly by a bottom flow called *undertow*. Some (Shepard, Emery, and LaFond, 1941) have claimed that rip currents are more important in returning water seaward than undertow. *Rip currents* consist of localized lanes or streaks of water moving seaward in contrast to the non-localized flow of undertow. Rip tide has also been applied to them, but it is a misnomer, for they are in no way related to tides. The presence of rip currents along a coast is indicated by such features as (*a*) brown streaks of sediment-laden water, (*b*) streaks of green water resulting from the greater depth of water where a rip current forms, (*c*) streaks of agitated water, (*d*) gaps in lines of breakers, (*e*) foam belts, especially where the bottom is rocky, and (*f*) seaward movement of floating objects. Shallow channels on the sea floor often attest to their presence. It is likely that rip currents represent a more significant method of carrying seaward suspended sediments than is generally realized.

MARINE EROSION

Waves, particularly storm waves and tsunamis, are the most important agents of marine erosion. Smaller waves, such as those associated with surf, may carry on attrition of material and minor amounts of abrasion, but, just as a stream during a single flood may do more geologic work than it will for months or years at low-water stage, so storm waves during a short period may effect more change than ordinary waves will in months. With minor local exceptions, as in constricted passageways, currents are probably relatively unimportant erosional agents, but they are an important means of transportation.

Conditioning factors. Aside from variations in wave strength, there are several factors that influence the rate at which marine erosion proceeds. Among the more important controlling factors are:

a. The kind and durability of the rock along the shore line.

b. Structural features of the rock, particularly its attitude and degree of jointing and fracturing.

c. Stability of shore-line position.

d. Openness of a coast to attack.

e. Depth of water offshore.

f. Abundance and size of tools.

Processes of marine erosion. Marine erosion is most effective where abundant tools are available, but wave impact itself, particularly upon highly jointed and fractured rocks, may exert great pressure. Johnson (1919) reported pressures measured by dynamometers along the Scottish coast of over 6000 pounds per square foot. The enormous force exerted by breaking waves is attested by recorded movements of masses weighing many thousands of pounds. Air in joints and cracks is suddenly compressed and acts as if a wedge were suddenly driven into them. Recession of the water is accompanied by a sudden expansion of air with explosive force. This driving of water into cracks not only exerts great mechanical stress but in soluble rocks may greatly accelerate solution.

The most effective process in marine erosion is the corrasive or abrasive action of sand, gravel, and pebbles moved by waves against the shore. This includes what has been called the "artillery action" of tools hurled against solid rock and the more common but less spectacular corrasive action as rock particles are moved back and forth over bedrock. In the absence of tools, marine abrasion may be ineffective, as is shown by the preservation of glacial striae along some coasts. Attrition contributes indirectly to marine abrasion by reducing rock particles to sizes that can be carried seaward by undertow and rip currents.

Opinion differs on the question to how great a depth effective wave attack may extend. The term *wave base* is used by some to designate this lower limit of wave action. Johnson (1919) concluded that oscillatory wave action may extend down to depths as great as 600 feet with sufficient force to agitate fine particles on the sea floor and that effective wave erosion may reach to depths as great as 200 feet. Others, like Shepard (1948), are of the opinion that erosive action by waves, as indicated by movements of sand during storms, does not extend below depths of 30 or 40 feet.

THE SHORE PROFILE

Thus far we have used the terms shore, shore line, and coast without definition. Further discussion requires clarification of these terms. The zone extending from the low-tide to the landward limit of effective wave action may be called the *shore*. A technical distinction may be made between shore and shore line in that the *shore line* marks the position of the water level at any given time and varies between the low-tide shore line and the high-tide shore line positions, but it is usually difficult to make this distinction without undue verbiage; hence, in the following discussion shore and shore line will be used as essentially synonymous terms. To many people coast has the same connotation as shore or shore line, but there is a real geologic distinction between them. The *coast* is a zone of somewhat indeterminate width which extends landward from a shore or shore line. The boundary between it and the shore is the *coast line*. In many places the seaward limit of a coast is marked by a nick or scarp resulting from wave erosion called a *sea cliff*. Usually, extending seaward from the base of a sea cliff, is a bench produced by wave erosion called a *wave-cut bench*. This bench may be bare rock or it may be covered with a temporary deposit of sand, gravel, and pebbles, called a *beach*. A beach is a depositional feature upon a wave-cut bench and is not permanent, but is composed of materials more or less in transit. A wave-cut bench may terminate abruptly or it may grade into a somewhat flatter surface resulting from long-continued abrasion by waves and currents, called an *abrasion platform*. There is no sharp demarcation between the wave-cut bench and abrasion platform, and the two together constitute a *marine-cut terrace*. Seaward from the marine-cut terrace there is usually a *marine-built terrace*, or what Johnson (1919) called the *continental terrace*, which consists of materials removed in the cutting of the marine-cut terrace. The

FIG. 17.3. Exposure of a wave-cut bench across a syncline at Green's Haven, near Berwick, England. (Photo by British Information Services.)

abrasion platform and marine-built terrace together constitute what is commonly called the *continental shelf*.

With these definitions in mind, we may proceed to a discussion of the conventional ideas of shore-profile development. To simplify our discussion we shall assume that the initial shore line is one that has been produced by recent submergence of a coastal region having moderate slopes and further that sea level remains fixed for a long period of geologic time. Under these conditions the initial coastal profile will be one with moderate slopes descending beneath the water but not with depths of water so great that waves come onto the shore and are reflected back without breaking. Marine erosion soon starts to cut a notch in the land and initiates a sea cliff and thereby starts development of a wave-cut bench. Material derived from the cutting of this bench will accumulate at its seaward edge as a sort of submarine talus, and thus initiate a marine-built terrace. This represents the youthful stage of shore-profile development. In many respects it presents conditions analogous to those of youth in the fluvial cycle. Wave action is vigorous; both erosive and transporting powers are great, and the slope over which the eroded materials are being moved seaward is still ungraded. As waves cut farther inland their erosive attack upon the sea cliff decreases because they cross increasing expanses of shallow water over a widening wave-cut terrace. This loss of erosive power is analogous to loss of corrasional power by a stream as it deepens its valley and reduces its gradient.

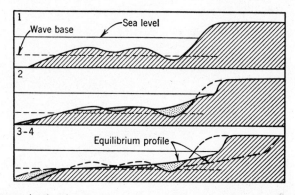

Fig. 17.4. Stages in the development of a profile of equilibrium along a steep coast. (After Ph. H. Kuenen, *Marine Geology,* John Wiley & Sons.)

Continued wave abrasion will in time reduce the wave-cut terrace to a slope which is just sufficient for movement seaward to the forward edge of the marine-built terrace of material derived from the sea cliff and shore. This marks attainment of a profile of equilibrium. The profile thus attained is concave in the part that is still actively undergoing erosion, nearly flat seaward from the wave-cut bench, and convex where active deposition is taking place at the forward edge of the marine-built terrace. Except for the convex portion at its distal end, it is similar in many respects to the profile of equilibrium of streams. Maturity may be said to have been attained when this stage is reached. Creation of a beach also characterizes maturity. Increasing distance from the sea cliff to deep water and decreasing effectiveness of wave action in shallowing water result in material remaining temporarily along the shore before being carried out into deeper water. The beach thickens during times of less active wave attack but may be completely destroyed during times of heavy storm waves. Whether a profile of equilibrium as sketched above is ever attained may be open to question. Because of complications introduced by diastrophic and eustatic changes in sea level, longshore currents, and deposition of sediments from rivers and glaciers, some

geologists, notably Shepard, have been inclined to think that it represents a
theoretical condition rather than a reality.

Certainly the concept of old age as represented by a sea cliff with a faint
slope, a wide wave-cut terrace adjoined by a wide wave-built terrace built
from waste supplied so slowly from the land that the wave-cut terrace largely
remains denuded is a theoretical abstraction as far as present-day shore lines

FIG. 17.5.   Marine terrace on Aguijan Island, in the Marianas.   The lower bench is
probably a high-water platform cut by storm waves.   (Air Force photo.)

are concerned.   There are those, however, who still believe in the ability
of the ocean to cut marine plains of denudation comparable in extent with
peneplains and other extensive erosional surfaces.   Numerous examples of
elevated marine terraces are to be found in different parts of the world, but
usually they are at best a few miles wide.   Marine terraces exist along the
coast of North Africa as much as 12 miles wide, and the Strandflat along the
west coast of Norway, which is thought by some to be of marine origin, has
a maximum width of 40 miles.   However, the frequent changes of sea level
during the Pleistocene have prevented development within recent geologic time
of any extensive marine plains.

Marine erosion is so localized along a narrow strip of land that cutting of
an extensive plain of marine abrasion would require such a vast length of
stillstand of sea level that it seems even less likely of attainment than the

peneplain stage in a fluvial cycle. Wooldridge and Morgan (1937) emphasized the difference between marine erosion and other types by comparing marine erosion with horizontal saw cutting in contrast with sand papering and filing by wind and running water over an entire land surface. Uplift of land of only 10 to 20 feet may start the whole marine cycle over, whereas with running water it merely accelerates the rate of erosion. The condition that would be most likely to favor the formation of an extensive marine plain

FIG. 17.6.   Sea cliff and elevated marine terrace with a former stack upon it, Port Harford, California.   (Photo by G. W. Stose, U. S. Geol. Survey.)

would be that of a slowly subsiding landmass or slowly rising sea level such as marked the great marine transgressions during the Cretaceous, but even then it is likely that waves merely added the final leveling to areas which had been already nearly destroyed by subaerial erosion.

TOPOGRAPHIC FEATURES RESULTING FROM MARINE EROSION

Sea cliffs and wave-cut terraces are the two most common features created by marine erosion. Details of sea cliffs vary greatly, depending upon the kind, structure, and attitude of the rocks in which they are being cut. The configuration of a cliff cut in rocks dipping seaward will be different from one in rocks that dip landward, and still different from one cut in horizontally bedded rocks; cliffs cut in granite differ in appearance from those

FIG. 17.7. *A*. Shore line on rocks dipping seaward, Langland Bay, Glamorgan, Wales. (British Official photo.)   *B*. Cormorant Rock, Aberystwyth, Wales, a stack on a shore platform cut on landward-dipping rocks.   (Photo by British Information Services.)

FIG. 17.8. *A*. Chalk cliffs at Birling Gap, Sussex, England, a shore line on nearly horizontal rocks. (British Official photo.) *B*. Land's End, Cornwall, England, a granite shore line. (British Official photo.)

FIG. 17.9. *A*. The West Cliffs, Sheringham, Norfolk, England, a shore line upon glacial
deposits. (British Official photo.)   *B*. Keanae coast, Island of Maui, Hawaii, a shore
line upon volcanic rocks. (Photo by Hawaii Visitor's Bureau.)

cut in basalts and cliffs in easily erodible material like glacial till or non-indurated Tertiary deposits will be bold and marked by much slumping and landsliding.

Above the base of a sea cliff there may be found a bench or *high-water platform* which owes its existence largely to the spray or splash of storm waves, aided by subaerial weathering and rainwash. Whether a high-water

Fig. 17.10. Fingal's Cave, Isle of Staffa, Scotland, a sea cave in columnar basalt. (Photo by British Information Services.)

platform is present seems to depend upon whether the processes of high-water leveling keep ahead of the more general process of marine terracing in the zone of breakers. Cotton (1945) has pointed out that optimum conditions for high-water platform development are attained in enclosed bays where normal wave action is not too effective, but these platforms are also found along open coasts, and so it seems that they are normal features of shore lines. Care should be exercised not to interpret high-water platforms as marine terraces exposed by recent shore-line emergence.

A shore line which is being extended landward by wave attack is said to be *retrograding*. During retrogradation, recession of the sea cliff will proceed at different rates along a coast, depending upon the durability of the rocks and the openness of the shore to wave attack. Thus the lesser indentations of a coast may result from differential marine erosion. These are known as *coves, bays,* or *bights.* It is doubtful, however, if wave attack is ever responsible for large bays and estuaries. These have resulted from submergence of topography produced in other ways. *Headlands* will be left in areas of more resistant rocks. Headlands subjected to wave attack from two sides

Fig. 17.11. Ecclesbourne Glen, Hastings, England, a hanging valley produced by marine erosion. (Photo by British Information Services.)

may have *sea arches* or *caves* cut in them. Portions of headland detached from the shore line are known by such names as *stacks, chimneys, skerries,* or *islands.*

Retrogradation of a shore line may go on so rapidly that small streams are unable to keep pace in downcutting with the rate of sea-cliff recession. As a result, these streams enter the sea from hanging valleys. These are common along the chalk cliffs on either side of the strait of Dover.

A shore line that is retrograding across a veneer of sedimentary materials over an undermass of older rocks may exhume or resurrect part of the buried topography. The shore line is then said to be *contraposed* (Clapp, 1913). Its characteristics may undergo notable change when waves encounter the resistant undermass. Contraposition of a shore line is analogous to superposition of valleys. Clapp cited examples of various stages of contraposition near

Victoria, British Columbia, where glacial drift covered a preglacial topography developed upon crystalline rocks.   At various places where wave erosion has cut through the glacial drift, there is a change from mature shoreline characteristics to youthful conditions on the exhumed crystalline rocks.

TOPOGRAPHIC FEATURES RESULTING FROM MARINE DEPOSITION

Beaches.   As suggested above, the term beach should be restricted to the temporary veneer of rock debris which accumulates along and on a wave-cut bench.   Its impermanence needs to be stressed, because the casual visitor is likely to consider it a permanent feature.   Beaches may extend continuously for hundreds of miles along a shore, as along the southeast coast of the United States, or they may be patchy.   Along rugged shore lines, beaches are mainly limited to strips at the heads of bays or coves, variously known as *bayhead beaches, pocket beaches,* or *crescent beaches,* but there may also be strips of beach materials on the tips of headlands, which are known as *headland beaches*.   Greater frequency of beaches at the heads of bays than at the tips of headlands results from the fact that waves converge upon headlands and diverge in bays, causing an intensification of wave erosion on headlands and a lessening of it in bays.   Longshore currents and beach drifting also tend to move sand into the bays to either side of a headland.

Movement of beach material goes on at all times, but reduction in beach thickness is most significant during periods of extreme storm waves, whereas beach accretion takes place chiefly during periods of quiet water.   Shepard and LaFond (1940) noted that at La Jolla, California, beach destruction or diminution characterized winter and spring months, and beach growth the calmer summer and fall months.   Waves produced by tsunamis and hurricanes are especially destructive of beaches and may result in either pronounced landward movement of beach material or complete removal of a beach.   The term *storm beach* is often applied to beach material piled up by waves when they are highest.   Beach features are particularly ephemeral forms along a retrograding shore line, but along a shore line that is advancing seaward or is *prograding* they may be semipermanent in nature.

The materials of a beach come both from land and sea.   Probably the greater part comes from the land, being contributed by streams, landslides, weathering of the sea cliff, marine erosion of the sea cliff, and slope wash. It appears, however, that some beach material comes from the sea.   Certainly solitary waves or waves of translation can move material landward, and Grant (1943) concluded that oscillatory waves can transport materials on a sea floor in the direction of their propagation, because there is a differential in their forward and backward velocities in favor of movement landward.   Johnson (1919) cited a striking example of this landward movement of material in

FIG. 17.12.   A, summer and B, winter conditions of beach at La Jolla, California.
(Photos by F. P. Shepard.)

which shingle and chalk ballast dropped from 7 to 10 miles off the coast of Sunderland, England, in water as much as 20 fathoms deep had been brought on shore by storm waves.

Bars. The term *bar* may be used in a generic sense to include the various types of submerged or emergent embankments of sand and gravel built on

FIG. 17.13. Fire Island spit, Long Island, New York. (Photo by Fairchild Aerial Surveys, Inc.)

the sea floor by waves and currents. Specific names are given to individual forms, depending upon their form and position. One of the most common types of bar is that known as a *spit*. It was defined by Evans (1942) as "a ridge or embankment of sediment attached to the land at one end and terminating in open water at the other." Most commonly the axis of a spit will extend in a straight line parallel to the coast, but, where currents are deflected landward or unusually strong tides exist, growth of a spit may be deflected landward, with the resulting creation of a *recurved spit* or *hook*. Several stages of hook development may produce a *compound recurved spit* or *com-*

pound hook. Spits may be attached to both sides of headlands to form what is called a *winged headland.*

If a spit has become tied or nearly tied at both ends to land or to another spit tied to land, it is likely to be known simply as a bar. The term bar may even be used when there are small gaps kept open by tidal currents between the distal end of a spit and the adjacent land or between the distal end of two spits which have nearly joined together. Bars formed in or across bays may be termed *bayhead, mid-bay,* or *baymouth bars,* depending upon their position. Islands sometimes have spits on their landward ends or sides which in time join to form *looped bars.* Convergence of two spits offshore or recurving of a simple or compound spit until it becomes attached to shore produces a *cuspate bar.* Growth of successive cuspate bars farther and farther seaward may cause prograding of a shore line, and the extended area of land thus formed is a *cuspate foreland.* Islands may become tied together or to land by growth of one or more spits. Such bars are called *tombolos.* Bars that are completely detached from the coast are known as *offshore bars.*

Origin of spits and bars. Spits have been commonly attributed to movement and deposition of materials by longshore currents, particularly of the type that Johnson (1919) called *wave currents.* These are currents produced by waves, both oscillatory and translatory, meeting a shore obliquely. Johnson referred to the movement of materials in this way as *shore drift* and included beach drifting under it. Movement of material in shallow water seaward from the shore commonly is called longshore drift. Locally, currents which are a part of the planetary circulation of the ocean or eddies in it, as well as tidal currents, may aid longshore drift. According to one view, a spit will be formed where a current passing a headland maintains a straight course rather than conforming to the irregularities of the coast line. An embankment will be built gradually in the direction in which the current is moving. Changes of current directions account for recurved spits and other curved forms. Tidal currents seem to be of minor importance in the production of bars as is indicated by the existence of bars which extend in a direction opposite to the tidal currents and the presence of spits in lakes where tidal currents are lacking.

Lewis (1931, 1932) believed that spits are built in a different way from that just described. According to him, longshore currents may supply materials from which a spit is built, but its construction is a result of spasmodic progradation by obliquely impinging waves at times of major storms. He thought that it is mainly major storm waves that contribute to the permanent part of a bar and that a bar extends itself in a direction at right angles to the line of dominant approach by these waves. Waves that are most effective in building bars are those that come from the direction of maximum fetch.

In explaining growth of spits as the result of wave rather than current action, stress was placed upon the difference between the effects of waves with high frequency and short wave length and those with lower frequency and longer wave lengths. It was claimed that with waves of high frequency there is weak development of swash, and backwashing of material from the beach is important, resulting in removal of much material from the shore, whereas with waves of low frequency there is a marked development of the swash with resulting movement of shingle onto the beach. Thus storm waves with their high frequency and short wave lengths dominate the configuration of a spit.

Steers (1948) believed that beach drifting, more than any other process, contributed to the movement of beach materials and the growth of spits. Beach drifting transports material to the end of a spit and dumps it there. In time, this material is built up into ridges by wave action, and the end of the spit is extended farther in the direction of dominant beach drifting.

Origin of offshore bars. Difference of opinion exists as to whether offshore bars are built chiefly by waves or currents. Gilbert (1890), in discussing the features associated with Lake Bonneville, said: "Where the sublittoral bottom of the lake has an exceedingly gentle inclination the waves break at a considerable distance from the water margin. The most violent agitation of the water is along the line of breakers; and the shore drift, depending upon agitation for its transportation, follows the line of breakers instead of the water margin. It is thus built into a continuous outlying ridge at some distance from the water's edge. It will be convenient to speak of this ridge as a barrier [offshore bar]." He thus attributed offshore bars to materials supplied by longshore currents. Johnson (1919) accepted a theory originally proposed by De Beaumont in 1845 and later elaborated by Davis (1909) that materials of an offshore bar are derived from the sea bottom by wave attack. Johnson arrived at this conclusion through study of the profiles extending seaward from offshore bars. It appeared to him that there was a new profile seaward from an offshore bar which had resulted from wave abrasion, and which, if projected landward, intersected the level of the sea landward from the *lagoon* or body of water between the offshore bar and the coast. He argued that, if the hypothesis of Gilbert that offshore bars are built on the sea bottom through deposition by currents is correct, projection of the profile of the sea bottom in front of an offshore bar should result in intersection of sea level seaward from the inner margin of the lagoon, since the profile back in the lagoon would be a landward continuation of the sea-bottom profile upon which the bar was built. According to Johnson's ideas, a bar is built above sea level through increase in the amount of material in the bar until it begins to be intermittently above sea level, appearing first, perhaps, as a chain of islands. These islands increase in size, length, and number until eventually a complete barrier is formed ·between the open sea and the lagoon back of the bar.

CLASSIFICATIONS OF COASTS AND SHORE LINES

Considerable difference of opinion, as well as confusion, characterizes present-day classifications of coasts and shore lines. Part of the confusion arises from the fact that one person has classified coasts and another shore lines. Most of the difficulty encountered, however, in arriving at a satisfactory classification stems from the fact that relatively few present-day shore lines are simple. Compound or multicyclic shore lines predominate, largely as a result of the oscillations of sea level which have marked the last few hundred thousand years of geologic time. Features associated with both low and high sea levels related to the waxing and waning of great ice sheets are to be found along most shore lines. Present coastal geomorphology, except in areas of active diastrophism, is determined largely by postglacial rise of sea level. No satisfactory classification is likely to be formulated which does not take into account *dominant relative emergence* or *dominant relative submergence,* and even then it may be difficult to decide which has been most significant along a specific coast. As Cotton (1951) has stated, "classification of coasts is in the melting pot." What eventually will prove the most satisfactory basis of classification remains to be seen.

Johnson's (1919) classification of shore lines, which has been widely accepted and used in elementary textbooks, recognizes four classes of shore lines: shore lines of emergence, shore lines of submergence, neutral shore lines and compound shore lines. *Shore lines of emergence* are those whose features are the result of dominant relative emergence of an ocean or lake floor; *shore lines of submergence* are those whose features are the result of dominant relative submergence of a land mass; the features of *neutral shore lines* are dependent upon neither submergence nor emergence; *compound shore lines* are those whose features present a combination of two or more of the preceding types. Under shore lines of submergence, Johnson recognized two types, *ria shore lines,* formed by the partial submergence of an area which was dissected by subaerial erosion, and *fjord shore lines,* produced from partial submergence of glacial troughs. Under *neutral shore lines* six types were recognized: delta shore lines, alluvial plain shore lines, outwash plain shore lines, volcano shore lines, coral reef shore lines, and fault shore lines. This classification is essentially a genetic classification taking into consideration the two main factors that influence shore-line configuration, the initial character or form of the land against which the sea came to rest, and whether the dominant change of sea level was such as to produce emergence or submergence.

Johnson's classification has much to recommend it. Lucke (1938) in defending it listed the following items in its favor: (*a*) it gives order to the former chaotic classifications; (*b*) it has simplicity; (*c*) it can be easily applied (a point which may be open to question); (*d*) it is complete; and (*e*) it is genetic in nature. Certainly a genetic classification is desirable. Objections have arisen to Johnson's classification of shore lines, particularly from Shepard

FIG. 17.14. An emergent shore line with offshore bar and lagoon back of it, Daytona Beach, Florida. (Photo by Daytona Beach Chamber of Commerce.)

(1937, 1938), chiefly because he thought that Johnson's classification implied that offshore bars are diagnostic of shore lines of emergence. This inference apparently was not intended by Johnson, but some persons using his classification have mistakenly assumed that the presence of an offshore bar along with straightness of shore line are diagnostic features of shore lines of emergence. Shepard has maintained that practically all shore lines exhibit evidence of both emergence and submergence and hence are compound shore lines under Johnson's classification. He further contended that delta shore lines, which were classed as neutral shore lines by Johnson, have in many localities proved to be areas where submergence is dominant. The Mississippi delta area is a case in point. Shepard also claimed that Johnson failed to recognize the significance of the eustatic changes in sea level which accompanied glaciation and deglaciation and that, as a result of these eustatic changes, practically all shore lines and coasts give evidence of both emergence and submergence. He admitted that it was easier to find fault with existing classifications than to devise a classification that would get away from their shortcomings. Shepard's latest proposed classification (1948), which fol-

FIG. 17.15. Whangaroa Harbour, northeast of Auckland, New Zealand, a submerged coast. (Photo by Whites Aviation, Ltd.)

lows, is one of both coasts and shore lines and represents a revision of a classification previously proposed in 1937.

I. Primary or youthful coasts and shore lines, whose configuration was produced chiefly by non-marine agencies.
    A. Those shaped by erosion on land and subsequently drowned as a result of rise of sea level because of deglaciation or downwarping.
        1. Drowned river coasts (Ria coasts).
        2. Drowned glaciated coasts.
    B. Those shaped by deposits made on land.
        1. River deposition coasts.
           a. Deltaic coasts.
           b. Drowned alluvial plains.

2. Glacial deposition coasts.
   a. Partially submerged moraines.
   b. Partially submerged drumlins.
3. Wind deposition coasts.
4. Coasts extended by vegetation.
C. Coasts shaped by volcanic activity.
  1. Coasts on recent lava flows.
  2. Shore lines caused by volcanic collapse or explosion.
D. Coasts shaped by diastrophism.
  1. Fault scarp coasts.
  2. Coasts on folded rocks.
II. Secondary or mature coasts and shore lines whose configuration is largely the result of marine agencies.
  A. Shore lines shaped by marine erosion.
    1. Shore lines straightened by marine erosion.
    2. Shore lines made irregular by marine erosion.
  B. Coasts and shore lines shaped by marine deposition.
    1. Straightened shore lines.
    2. Prograded shore lines.
    3. Shore lines with offshore bars and longshore spits.
    4. Coral reef coasts.

Shepard admitted that his classification may be incomplete, but he contended that it avoided the confusion which frequently results from attempting to designate a shore line as one of emergence or submergence. Lucke (1938) thought that there are four main objections to Shepard's classification: (1) it is a classification of coasts rather than shore lines; (2) it was devised largely from a study of coastal charts which give insufficient information for proper classification; (3) it is incomplete; and (4) it does not give any idea of the evolutionary changes which shore lines undergo; hence it would give no indication of the stage of development of a coast.

Johnson (1938*a*) in alluding to criticisms of his classification stated that its misinterpretations arose from a failure to distinguish between *characteristics* and *criteria* in evaluating the features that determine the type of a particular shore line. A feature such as an offshore bar may characterize different types of coasts, but it may not be a criterion of such diagnostic value as to determine alone the type of shore line. In another reply (1938*b*) to the criticisms of his classification he stated: * ··· "neither the writer nor others who employ this classification have ever taught, much less established, the theory that 'offshore bars can be found only on a shoreline of emergence' ···. On the contrary, ··· offshore bars have been cited by them as characteristic

* Quoted by permission of Columbia University Press from Douglas Johnson, Offshore bars and eustatic changes of sea level, *J. Geomorph., 1*, pp. 273–274.

although not inevitable features of certain types of compound shorelines; and as possible though not necessary features of several types of neutral shore-lines." He further stated that "the ··· classification was developed with full knowledge of ··· repeated and frequent oscillations of sea level, and ··· that glacial control was a major factor in these oscillations. In order that the classification might be applied equally well (*a*) to shorelines resulting from elevation and subsidence of the land, (*b*) to shorelines resulting from the rise and fall of sea level, and (*c*) to shorelines resulting from the rise and fall of lake levels, the words 'elevation,' 'subsidence,' 'rising,' 'falling,' 'positive,' 'negative,' and all similar terms implying any suggestion as to whether move-ments of land or water were responsible for observed shorelines, were spe-cifically excluded from the scheme of classification developed. In their place the terms 'submergence,' 'emergence,' 'neutral' and 'compound' were intro-duced, precisely because these terms involved no implication as to whether it was the land or water that had changed level or remained stationary."

Although the terms emergence and submergence may encompass Pleisto-cene oscillations of sea level, one searches in vain through Johnson's *Shore Processes and Shoreline Development* for any discussion of their effects upon shore line development. In *New England-Acadian Shoreline,* by the same author, some four pages are devoted to the effects of glaciation upon shore-line development. It is true, of course, that the effects of the Pleistocene upon world sea levels were much less understood at the time these books were written than today. It has become evident that Pleistocene changes in sea level were numerous and complex. Realization of this emphasizes all the more that the distinctions between shore lines of emergence and submergence can only be made by determining whether shore features reflect the dominant effect of submergence or emergence.

Although Cotton (1945) in his *Geomorphology* referred to coasts of sub-mergence and coasts of emergence, he barely mentioned neutral and com-pound coasts and throughout his discussion seemed to prefer coastal over shore-line classification. He followed the accompanying classification of coasts proposed previously by himself (1942).

It is apparent that the attempts at coastal and shore-line classification over-lap to some degree, and it may be that, as Wooldridge and Morgan (1937) have emphasized, it is inexpedient or impracticable to keep coast-line and shore-line features entirely separate. Along shore lines that are marked by submergence, geomorphic forms and geologic structures of the coast are bound to be dominant over lesser shore-line features in determining the configuration of the shore line, whereas along emergent shore lines the dominant features are those that result from marine processes. It may never be possible to de-velop a classification that satisfactorily separates features of coasts from those of shore lines.

CLASSIFICATION OF INITIAL COASTS

Initial coasts resulting from

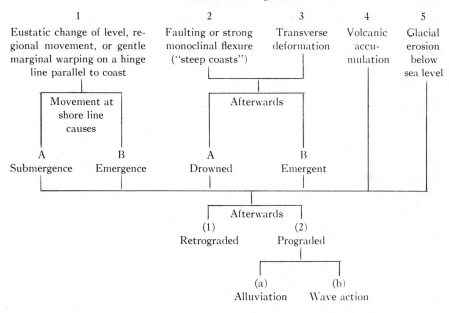

1	2	3	4	5
Eustatic change of level, regional movement, or gentle marginal warping on a hinge line parallel to coast	Faulting or strong monoclinal flexure ("steep coasts")	Transverse deformation	Volcanic accumulation	Glacial erosion below sea level

Movement at shore line causes

A	B	A	B
Submergence	Emergence	Drowned	Emergent

Afterwards

(1)	(2)
Retrograded	Prograded

(a)	(b)
Alluviation	Wave action

SHORE-LINE DEVELOPMENT

Along submerged coasts. The initial shore line of a coast which has undergone recent submergence is likely to be marked by great irregularity. Exceptions would be where flat alluvial, deltaic, or glacial plains are submerged. Drowned valleys characterize a submerged coast, and the coast is either a ria or fjord type, depending upon whether submergence affected "normal" or glacial topography. Drowning of a coast may result in dismemberment of the lower portions of stream systems. This is well illustrated in the Chesapeake Bay region where such streams as the James, Rappahannock, and Potomac are now separate streams, whereas they were formerly tributaries of the Susquehanna.

Interstream ridges project seaward as headlands, or, if nearly submerged, their higher parts form a group of aligned islands. Wave refraction concentrates erosion upon headlands and islands. In general the plan of the coast may be described as crenulate. Sea cliffs and associated erosional forms are most characteristic of a youthful submerged shore line. The shore profile is ungraded to start with but, with the development of a wave-cut bench and marine-built terrace in front of it, a graded profile is attained. Straightening or rectification of the shore line now begins. This is effected by the cutting

back of headlands and islands and the construction of spits and bars across
reentrants in the coast.   In the early stages of bar development two shore
lines really exist, an outer one in front of the bar and an inner one back of
the bar along the initial coast, but, when a baymouth bar has closed a coastal

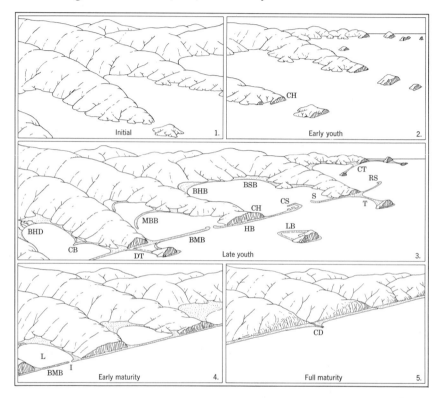

Fig. 17.16.   Stages in the development of a shore line of submergence.   T, tombolo;
S, spit; RS, recurved spit or hook; CS, complex spit or compound hook; Ct, complex
tombolo; LB, looped bar; CH, cliffed headland; DT, double tombolo; HB, headland
beach; BMB, baymouth bar; MBB, midbay bar; CB, cuspate bar; BHB, bayhead beach;
BSB, bayside beach; BHD, bayhead delta; L, lagoon; I, inlet; and CD, cuspate delta.
(After A. N. Strahler, *Physical Geography,* John Wiley & Sons.)

reentrant, the inner coast line ceases to function and the enclosed body of
water is converted into lagoon or marsh, which in time is filled with waste
from the land.   This stage which is marked by retrograded headlands and
nearly closed bays may be designated as submaturity.

Two things have usually been considered as marking the attainment of
maturity along a submerged coast line: first, attainment of a shore profile of
equilibrium, and secondly, retrogradation of the shore until it lies back of

the heads of the initially drowned valleys. Although a coast line may not be straight in maturity, it is marked by much greater regularity than in youth. A mature shore line is further distinguished from a youthful one in that the whole coast is now retrograding, although not necessarily at the same rate everywhere because of varying rock resistance. Structural and lithologic influences will be evident at this stage in the more rapid retrogradation of belts of weaker rock.

As previously suggested, there is some question whether an old-age shore line is more than a theoretical abstraction, for it is doubtful whether stability of sea level is maintained long enough for plains of marine denudation of sub-continental extent to be developed.

Along emergent coasts. It is most generally believed that the initial plan of an emergent coast would be simple and straight, but Shepard (1948) has maintained that in some areas there are sufficient irregularities on the continental shelf so that emergence would actually increase the irregularity of the coast lines of these regions. The geomorphic evolution of an emergent coast line to a large degree depends upon whether offshore slopes are gentle or steep. The geomorphic evolution of an emergent coast line as outlined by Johnson (1919) applies mainly to a coast line bordered by gentle offshore slopes. Under such conditions, waves will not be able to attack a shore line vigorously because the larger ones will break offshore. There will develop a submarine embankment at a distance from the shore determined by the rate at which the water shallows, which in time will emerge as an offshore bar or line of offshore bars with a lagoon back of it. Numerous breaches in the offshore bar will be maintained as tidal inlets, particularly opposite the mouths of major streams. Through these inlets tides will flow and ebb, and along rivers with low gradients their effects may be felt far enough upstream to temporarily pond the streams. Thus the lower portions of such streams are at times essentially tidal rivers. During low tide, extensive tidal flats may be exposed back of the offshore bars.

The most significant change along an emergent shore line with gently shelving waters involves the migration of its offshore bar. The bar may grow seaward temporarily, but its main movement is landward as it retreats under attack of the waves which have been partly, if not chiefly, responsible for its formation. Maturity is attained when the offshore bar has migrated landward until it reaches a position coinciding with the original coast line and when lagoons and marshes back of it have been eliminated. From this point on, the progress of shore-line development essentially parallels that sketched above for submerged coasts during maturity.

It is now recognized that there are regions where emergent coasts exist which do not in their initial conditions fit those described above. They are along the so-called steep coasts which are particularly prevalent around the

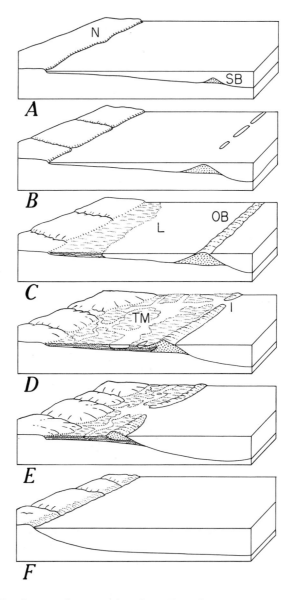

FIG. 17.17. Developmental stages of a shore line of emergence of the coastal-plain type involving the formation, migration, and destruction of an offshore bar. N, nip; SB, submarine bar; L, lagoon; OB, offshore bar; I, inlet; and TM, tidal marsh. (After A. N. Strahler, *Physical Geography,* John Wiley & Sons.)

Pacific Ocean. Putnam (1937) has described the probable shore-line cycle of such a coast as typified in the vicinity of Ventura, California. The coast there is an emergent one bordered landward by wave-cut terraces as high as 1400 feet above sea level and seaward by deep water. No offshore bar develops under these conditions. Early youth is marked by cutting of a sea cliff and entrenchment of consequent streams across an exposed sea floor. Late youth is characterized by a conspicuous sea cliff whose height is dependent upon the amount of original uplift. Larger streams have by now cut their courses nearly to sea level, and shorter streams have built alluvial fans at their mouths which have been truncated by wave erosion with resultant existence of a composite sea cliff partly in bedrock and partly in recent alluvium. At this stage, balance between prograding and retrograding has been reached. Putnam considered that maturity arrives when a shore line has retrograded to the position it occupied before uplift. At this stage, the terrace exposed by uplift will have been consumed and waves will again be attacking the original sea cliff. This stage is analogous to that along gently sloping shore lines when an offshore bar has migrated to the position of the original shore and when waves are again attacking that shore. From this point on, development will be similar along the two types of emergent shore lines.

REFERENCES CITED IN TEXT

Clapp, C. H. (1913). Contraposed shorelines, *J. Geol.*, *21*, pp. 537–540.

Cotton, C. A. (1942). Shorelines of transverse deformation, *J. Geomorph.*, *5*, pp. 45–58.

Cotton, C. A. (1945). *Geomorphology*, pp. 396–492, John Wiley and Sons, New York.

Cotton, C. A. (1951). Seacliffs of Banks peninsula and Wellington: some criteria for coastal classification (Part I), *New Zealand Geographer*, *7*, pp. 103–120.

Davis, W. M. (1909). The outline of Cape Cod, *Geographical Essays*, pp. 707–710, Ginn and Co., New York.

Evans, O. F. (1942). The origin of spits, bars, and related structures, *J. Geol.*, *50*, pp. 846–865.

Gilbert, G. K. (1890). Lake Bonneville, *U. S. Geol. Survey, Mon. 1*, p. 40.

Grant, U. S. (1943). Waves as a sand-transporting agent, *Am. J. Sci.*, *241*, pp. 117–123.

Johnson, D. W. (1919). *Shore Processes and Shoreline Development*, John Wiley and Sons, New York, 584 pp.

Johnson, D. W. (1925). *The New England-Acadian Shoreline*, pp. 507–510, John Wiley and Sons, New York.

Johnson, D. W. (1938*a*). Criterion or characteristic, *J. Geomorph.*, *1*, pp. 181–183.

Johnson, D. W. (1938*b*). Offshore bars and eustatic changes of sea level, *J. Geomorph.*, *1*, pp. 273–274.

Kuenen, Ph. H. (1950). *Marine Geology*, John Wiley and Sons, New York, 568 pp.

Lewis, W. V. (1931). The effect of wave incidence on the configuration of a shingle beach, *Geog. J.*, *78*, pp. 129–143.

Lewis, W. V. (1932). The formation of Dungeness foreland, *Geog. J., 80*, pp. 309–324.

Lucke, J. B. (1938). Marine shorelines reviewed, *J. Geol., 46*, pp. 985–995.

Putnam, W. C. (1937). The marine cycle of erosion for a steeply sloping shoreline of emergence, *J. Geol., 45*, pp. 844–850.

Shepard, F. P. (1937). Revised classification of marine shorelines, *J. Geol., 45*, pp. 602–624.

Shepard, F. P. (1938). Classification of marine shorelines: a reply, *J. Geol., 46*, pp. 996–1006.

Shepard, F. P. (1948). *Submarine Geology,* Harper and Brothers, New York, 348 pp.

Shepard, F. P., and E. C. LaFond (1940). Sand movements along the Scripps Institution pier, *Am. J. Sci., 238*, pp. 272–285.

Shepard, F. P., K. O. Emery, and E. C. LaFond (1941). Rip currents; a process of geological importance, *J. Geol., 49*, pp. 337–369.

Steers, J. A. (1948). *The Coastline of England and Wales,* pp. 44–70, Cambridge University Press, London.

Wooldridge, S. W., and R. S. Morgan (1937). *The Physical Basis of Geography,* p. 361, Longmans, Green and Co., London.

Additional References

Cotton, C. A. (1951*a*). Accidents and interruptions in the cycle of marine erosion, *Geog. J., 117*, pp. 343–349.

Cotton, C. A. (1951*b*). Atlantic gulfs, estuaries and cliffs, *Geol. Mag., 88*, pp. 113–128.

Cotton, C. A. (1952). The Wellington coast: an essay in coastal classification, *New Zealand Geographer, 8*, pp. 48–62.

Edwards, A. B. (1951). Wave action in shore platform formation, *Geol. Mag., 88*, pp. 41–49.

Jutson, J. T. (1939). Shore platforms near Sydney, New South Wales, *J. Geomorph., 2*, pp. 237–250.

Munk, W. H., and M. A. Traylor (1947). Refraction of ocean waves: a process linking underwater topography to beach erosion, *J. Geol., 40*, pp. 1–26.

Putnam, W. C., W. H. Munk, and M. A. Traylor (1949). Prediction of longshore currents, *Trans. Am. Geophys. Union, 30*, pp. 337–345.

In addition to the above references there are numerous publications of the Beach Erosion Board which will be of interest to persons who are particularly interested in shore processes.

18 · Topography of the Ocean Floors

INTRODUCTION

The present century has seen the rise of a new branch of geology, submarine geology, which, although still in its infancy, has revolutionized our views regarding the topography of the ocean floors. The old idea that the floors of the ocean basins consist mainly of vast monotonous plains with little relief has

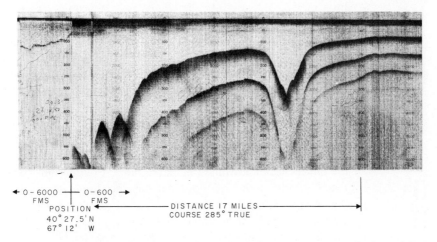

FIG. 18.1. Fathogram showing some small canyons off George's Bank at latitude 40° 27' north and longitude 67° 12' west. The depth scale is from 0 to 600 fathoms. (Courtesy Woods Hole Oceanographic Institution.)

had to be modified. Strangely enough, geologists long displayed little interest in the study of the ocean floors, and most of the pioneer work was done by oceanographers. That geologists are becoming interested in this branch of science is indicated by the appearance of two books on the subject written by geologists mainly for geologists, Shepard's *Submarine Geology* in 1948 and Kuenen's *Marine Geology* in 1950. The increasing interest of oil geologists in the possibilities of the continental shelves as sources of oil is certain to lead to a more thorough study and understanding of this portion of the sea floors.

459

As long as knowledge of the topography of the sea floors depended upon soundings made by letting out wire lines, progress was exceedingly slow and laborious. The historic voyage of H.M.S. *Challenger* between 1872 and 1876, which may well be considered as marking the beginning of submarine geology, succeeded, according to Veatch (1937), in obtaining a total of 504 deep-sea soundings; the Coast and Geodetic Survey ship, *Blake*, between 1874 and 1879 made only 3195 soundings. With development of sonic and supersonic sounding soon after the close of World War I, it became possible to obtain thousands of soundings in a matter of weeks instead of years. Water depths are now obtained by determining the time required for sonic or supersonic waves to travel from the ocean surface to the sea bottom and back to a recording device. The velocity of sound in water is nearly constant, being about 4800 feet per second. The continuous record of the depths to sea bottom obtained by a moving ship is called a *fathogram*. Difficulties are still encountered in determining accurately the successive positions of the ship and in the interpretation of data obtained where the slopes of the sea floors are so steep that the echoes come from cliffs rather than from flatter ocean floor, but these difficulties are being overcome and increasingly accurate results are being obtained.

Another significant advance has been the development by Ewing and his associates of a practical method of taking photographs of the sea bottom. The Ewing submarine camera can use either black and white or colored film. Colored film gives better details of the sea floor but has the disadvantage that the results cannot be obtained as quickly as for black and white film. Submarine photographs show most of the details of the sea floor, and from them it can be ascertained whether the floor is bedrock or covered with sediments. Ripple marks and other features such as pebbles, boulders, and sand deposits can be recognized. Submarine photography will probably prove increasingly helpful in further studies of the continental shelves.

There have been corresponding improvements in devices

Fig. 18.2. The Ewing underwater camera. The sail at its top is to retard rotational movement; below the sail is the camera and below it the lamp housing; below the lamp housing is a battery case and part of the triggering mechanism; a coring tube and the remainder of the triggering mechanism are at the bottom. (Courtesy of Maurice Ewing.)

used to obtain samples from the sea floor.   Cores as long as 70 feet have been obtained.   The Piggot gun, which uses gunpowder to shoot a tube into the ocean bottom, was used in 1939 to obtain a line of cores across the

FIG. 18.3.   Undersea photo taken from the research vessel *Atlantis,* at a depth of 18,000 feet, southeast of Bermuda, at latitude 30° 27' north and longitude 59° 07' west.   The largest objects shown are about 5 inches in diameter and are thought to be sponges. (Photo by D. M. Owen, Woods Hole Oceanographic Institution.)

North Atlantic.   Better devices are now available for dredging the ocean floor, and improved current meters can be used at great depths.   Ewing and others have pioneered in seismic-refraction determinations of the thickness of sediments on the sea floor.

FEATURES OF THE CONTINENTAL SHELVES AND SLOPES

The continental shelves.   Between the depths of the ocean basins proper and land there is a submarine terrace of varying width, commonly called the *continental terrace.* This terrace consists of two component parts, a relatively flat tread called the *continental shelf* and a descent from it to oceanic depths called the *continental slope.* Water is relatively shallow over the continental shelf. It frequently is stated that its forward edge lies approximately where the depth of water is about 100 fathoms (600 feet) but it now is known that this is far from uniform. According to Shepard (1948), the average depth of water at the edge of the continental shelf is about 72 fathoms but may be as great as 250 fathoms. The idea long held that the surface of the continental shelf is a relatively smooth slope which descends gradually to the upper edge of the continental slope applies to some areas, but it is not true often enough to be considered characteristic. Where accurate topographic maps have been made, as in the Gulf of Maine (Murray, 1947), detailed relief features have been shown to exist upon it. Shepard (1948) stated that eminences of 60 feet or more are found on 60 per cent of the profiles across the continental shelves and depressions of 60 feet or more appeared on 35 per cent of these same profiles. Irregularity of profile is particularly characteristic off coasts that have been glaciated. The shelves are deeper in general off coasts bordered by young mountain ranges. The continental shelf may vary in width from zero to as much as 750 miles. South of Korea, in the Yellow Sea, it attains a width of 750 miles, as it does in the Barents Sea off the Arctic coast of Europe. Shepard estimated its average width to be 42 miles.

Divergence of opinions exists in respect to the origin of the continental shelf. A view long held by many is that it is the result of the combined landward extension of the wave-cut terrace and the seaward growth of the wave-built terrace. The presumed uniform depth of its forward edge, about 600 feet, is said to represent the base of effective wave action. Shepard (1948) believed that the continental shelf was largely the result of marine erosion by waves and currents. Although admitting that some shelves consist of sediments, he was impressed by the many areas of non-deposition (Shepard, 1941) where bedrock rather than sediments constitutes the shelf surface. He further contended that there is little evidence to indicate that there is a grading from coarser to finer material seaward over the shelves as would be expected if the material of which they are composed was carried seaward from the land. Work along the Atlantic and Gulf coastal plains has indicated that the continental shelf here is underlain by many thousands of feet of sediments, which suggests that deposition took place in what was essentially a geosyncline. Off other coasts, however, there is evidence that the shelves are not comprised of sediments. Shepard and Emery (1941) proposed the term *continental borderland* to describe the type of shelf that exists off the coast of California, where there is a series of submarine basins and ridges similar in origin to the faulted structures landward from them. This is a different type of shelf from that off the Atlantic coast, but how common it is cannot be said with certainty.

We may conclude that the continental shelves are probably not so simple in topographic form or origin as was once thought. The shelves of various regions may differ notably in their characteristics relating to the geologic history of the coast which they border. Whether they are the result chiefly of erosion or deposition cannot be stated with finality.

The continental slope. The descent from the edge of the continental shelf to the floor of the deep ocean is the continental slope. The following significant facts about it were given by Shepard (1948). Its average height is about 12,000 feet but in some places it may be as great as 30,000 feet. Typically, its slope is steepest in the upper 6000 feet. Textbook diagrams usually exaggerate the difference in the slopes of the continental shelf and the continental slope, but there is a distinct difference of slope. The average gradient of the continental shelf was estimated to be 0° 07′ as compared with a gradient in the upper 6000 feet of the continental slope of 4° 17′. There is great variation in the steepness of the continental slope, which is well illustrated by the fact that off the northwest coast of Australia the slope is less than 1°, whereas off the southwestern coast of the same continent it is as great as 27°. According to Shepard, these variations seem to be related to the nature of the coast line; off rivers which have large deltas, it averages 1° 20′; off stable coasts lacking large rivers, it averages 3°; off coasts with young mountains, it averages 4° 40′; and off faulted coasts, it averages 5° 40′. The surface of the continental slope is by no means smooth. Numerous submarine canyons and furrows, to be discussed later, give great relief to it, and in many places it is marked by hills and ridges. In general, its trend is fairly straight or at most sinuous. The continental slope represents the line of juncture between the continental platforms and the ocean basins and hence is one of the most significant features of the earth's surface. This transition from one level to the other usually takes place over a horizontal distance ranging from 10 to 20 miles.

Much difference of opinion exists as to the origin of the continental slope, and to a considerable degree one's views will be influenced by his ideas regarding the origin of the continental shelf. Some consider the continental slope to be the forward edge of a wave-and-current-built terrace built seaward with debris brought into the sea by streams or obtained from the wave-cut portion of the continental shelf. Others explain it as the result of downwarping of ancient land masses which were perhaps peneplained before being downtilted. This idea has been applied particularly to the continental slope off the Atlantic coast of North America which is considered by some to be the edge of the ancient landmass of Appalachia. Shepard (1948), Holtedahl (1950), and others thought that the continental slope represents a major fault zone in

the earth's crust. It is somewhat beyond the field of the geomorphologist to
explain such a grand topographic feature as the continental slope, but it is

FIG. 18.4. Undersea photo of the escarpment at the edge of the continental shelf in the
Gulf of Mexico, west of Florida, at latitude 27° 18′ north and longitude 85° 30′ west.
The photo was taken from the vessel *Atlantis* at a depth of 1100 fathoms. The escarp-
ment has a slope as great as 35° and vertical relief as much as 4000 feet. The branched
object in the foreground is a crinoid about 1 foot long. (Photo by D. M. Owen, Woods
Hole Oceanographic Institution.)

important that he recognize its outstanding features and have in mind the
various theories that have been advanced to explain it, as well as the continental shelf.

GEOMORPHIC FEATURES ON THE CONTINENTAL SHELVES AND SLOPES

For obvious reasons, the detailed topography of the continental shelves and slopes is not as well-known as that of the continents but enough information is at hand to indicate that superposed upon these two major relief forms are numerous topographic features which in many places give to them a diversity comparable with that found on land.

Wherever detailed soundings of the continental shelves and slopes have been made it has been found that they have a surprising amount of relief. Detailed surveys of the Gulf of Maine (Murray, 1947), where over 155,000 soundings were obtained for an area covering 8900 square statute miles and contouring was done on a 30-foot interval, showed numerous banks, ridges, knolls, swells, and basins. Diversity of relief here may in part be attributed to morainal deposits and glacially scoured basins. It seems probable that many other areas of continental shelf, if surveyed in detail, would show similar if not greater diversity of surface. Not enough is known yet about this topography to be able to say how many of the land forms present above sea level exist upon the continental shelves. Certain ones, however, known to be present will be discussed. Kuenen (1950a) divided the negative features of the continental shelves and slopes into two major classes, shelf channels and submarine canyons, and he distinguished three types of shelf channels: drowned river valleys, channels resulting from tidal scour, and drowned glacial troughs.

Drowned valleys. The term drowned valley is here restricted to those valleys upon the continental shelf that can be taken without hesitancy as being submerged portions of valleys formed on land by the processes of subaerial erosion. Many are known to exist. Probably the best-known example is that of the Hudson River, which extends as a slightly winding channel from near Sandy Hook for a distance of 120 miles off the present coast. Lewis (1935) has described a drowned valley which extends northward beneath the North Sea to the latitude of the Orkney Islands. This valley represents the former course of the Rhine, and at the time that it was above water the Thames was a tributary to it. The submerged channels of the Elbe and Weser, which apparently were tributary to the submerged Rhine Valley, can be traced as far north as the latitude of Edinburgh. Kuenen (1950a) has described a system of submerged river channels on the Sunda shelf between Borneo and Sumatra, which has been called the Sunda River. Similar valleys submerged to a depth of about 280 feet are known in the Java and South China seas. These drowned river channels present no serious problem for their submergence is not too great to be explained by the rise of the sea level which accompanied deglaciation.

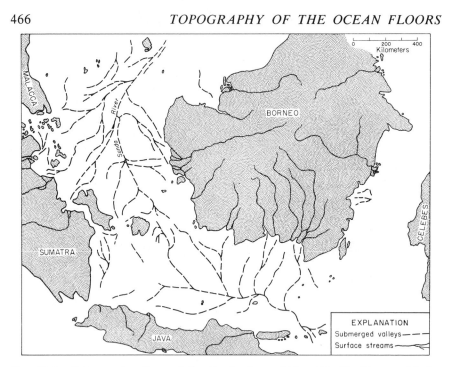

FIG. 18.5. Submerged valleys of the Sunda Sea. (After Ph. H. Kuenen, *Marine Geology,* John Wiley & Sons.)

Tidal channels. In numerous places, such as in the Sunda Sea, off the southeastern coast of the North Sea, and along the east coast of the United States, there are submerged channels between islands which apparently are not submerged river valleys but rather are channels cut by tidal scour. They usually have limited horizontal extent and are probably portions of former river valleys which have been deepened by tidal scour.

Drowned glacial troughs. In addition to the fjords which characterize many coast lines in high latitudes, and are generally admitted to be submerged glacial troughs, there exist certain trough-like depressions on the continental shelves about whose origin there is difference of opinion. Shepard, Holtedahl, Nansen, and others have maintained that they are glacial troughs which either were cut by ice below sea level or were cut in the shelves when the sea had withdrawn from the shelves and then later were submerged. Many of these supposed drowned glacial troughs connect with fjords but differ from them in being wider and less deep than typical fjords. Those who believe that they are the result of glacial erosion point out that: their longitudinal profiles are not graded but frequently show basining and reverse slopes; hanging valleys enter them from their sides; and their courses lack the sinuosity which characterizes drowned river valleys.

The different viewpoints as to the origin of these features may be illustrated by a brief discussion of the topography of the Gulf of St. Lawrence and the Bay of Fundy. Shepard (1931) maintained that, from about the point near the mouth of the Saguenay River out to the edge of the continental

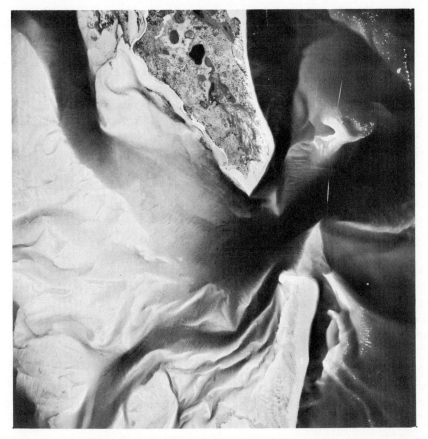

FIG. 18.6. Vertical photograph taken off the coast of Mozambique, Africa, showing troughs, bars, tidal channels, and other details of near-coastal submarine topography. (Courtesy Gulf Oil Corporation.)

shelf some 750 miles away, there is a trough in the Gulf of St. Lawrence which is the result primarily of glacial erosion. He admitted that its course may have been initiated by river erosion but contended that its final shaping was effected by glacial erosion. The tongue of ice which is supposed to have cut this trough presumably extended between Gaspé Peninsula and Anticosti Island and through Cabot Strait between Nova Scotia and Newfoundland out to the edge of the continental shelf. Shepard (1930, 1942) also contended that the Bay of Fundy was largely the result of the scooping action of a tongue of ice which extended down it.

Johnson (1925) believed that the Gulf of St. Lawrence was essentially a submerged lowland cut chiefly by stream erosion and modified only slightly by glaciation and that the feature that Shepard has called a glacial trough is nothing more than a drowned river valley. Johnson (1925) and Koons (1941, 1942) described the Bay of Fundy as a submerged lowland developed by stream erosion upon the weak Triassic rocks which underlie it and believed that it is bounded on the north by a fault-line scarp. Although they admitted that glaciation may have modified it incidentally, it was believed to be primarily the result of subaerial erosion and later submergence. Shepard denied the existence of a fault-line scarp along its northern border and ascribed what had been called a fault-line scarp to glacial erosion. Although it is not possible to definitely settle this argument, the evidence does appear to favor the subaerial origin of these two troughs. Shepard's view that the continental shelves in the higher latitudes have been modified greatly by glacial erosion is not generally accepted. Holtedahl (1950) believed in the existence of glacial troughs upon the continental shelf but thought that faults had played an important role in their location as well as in that of fjords.

Submarine canyons. Many deep gashes or valley-like trenches cut mainly in the continental slopes, but to some extent back into the continental shelves, have come to be known as *submarine canyons*. The term canyon is generally restricted to the larger and deeper ones. There exist many lesser grooves which have been called *furrows* by Daly (1936). Apparently the first recognition of the existence of these features was made by Lindenkohl (1891) in 1889, when, in tracing the drowned valley of the Hudson River across the continental shelf, he found that at a distance of 97 miles from land and at a water depth of only 240 feet it suddenly began to assume the form of a canyon, which he traced seaward for an additional distance of 23 miles to a depth of 2944 feet. Spencer (1903) in describing the numerous drowned valleys around the North Atlantic margins called attention to the existence of a number of these canyons and further pointed out that the Hudson Canyon extended at least 140 miles down the continental slope. More recent work seems to indicate that the drowned valley of the Hudson does not actually connect with Hudson Canyon. Tolstoy (1951) stated that recent soundings made by the vessel *Atlantis* indicated that Hudson Canyon extends at least 300 miles out on to the deep sea floor where the depth of water is 2500 fathoms.

Before undertaking a discussion of some of the theories that have been proposed to explain submarine canyons, we may well consider some of the general facts about them, as far as it is possible to be certain of facts, which any theory which seeks to account for them must be able to explain.

1. As more detailed soundings of submarine canyons have been made it has become apparent that many, if not most, submarine canyons consist of three rather distinct and interconnecting parts: (1) shallow valleys on the continental shelf, (2) deep, rock-walled canyons or furrows in the continental slope, and (3) broad shallow valleys on the floor of the ocean basins.

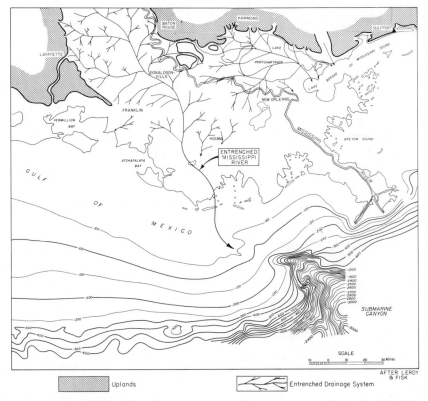

Fig. 18.7. Map showing possible relationship of the entrenched Mississippi Valley to the present river and submarine canyon in the Gulf of Mexico. (After T. E. Leroy and H. N. Fisk.)

2. They are apparently world-wide in distribution. They are known to exist off all the continents with the exception of Antarctica. The fact that they are not known off Antarctica is more likely the result of insufficient information than of their absence there.

3. Most of them lie seaward from the axial lines of streams on land, and some actually can be traced across the continental shelf into the estuaries of rivers, or to be more correct, into drowned valleys in these estuaries. There are, however, examples of canyons that seem to bear no genetic relation to streams on land.

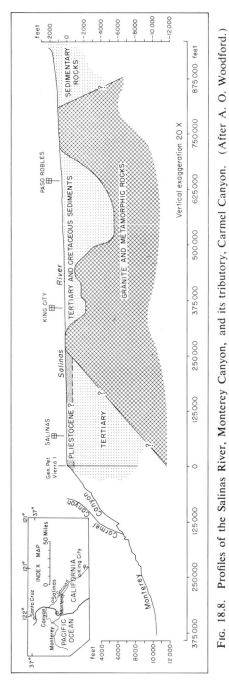

FIG. 18.8.   Profiles of the Salinas River, Monterey Canyon, and its tributary, Carmel Canyon.   (After A. O. Woodford.)

4. Deep canyons are interspersed with furrows and ridges and seem to be restricted mainly to the continental slopes.

5. Submarine canyons are of rather recent geologic age. Rocks of Pliocene age have been broken from the walls of some canyons, and the heads at least of some are cut in Pleistocene deposits.

6. Most canyons are cut in sedimentary rocks. Shepard (1948) stated that the southeast wall of Monterey Canyon, off the coast of California, at a depth of 500 fathoms yielded granite and that Carmel Canyon, a tributary to Monterey Canyon, is cut in granite, but Woodford (1951) has shown that it is unlikely that Monterey Canyon is cut in granite. The granite may have been obtained from a buried granite ridge encountered in the downcutting of the canyon. There seems, however, to be no doubt that Carmel Canyon is cut in granite.

7. Canyons are scarce or lacking where the continental slopes are less than 2 degrees.

8. Canyons apparently are distributed without any relation to the varying geologic histories of the coasts. They are found off submerged coasts, off emergent coasts, off mountainous coasts both young and old, off glaciated coasts, and off deltaic and alluvial coasts.

9. They seemingly have many striking similarities to canyons on land, such as V-shaped cross profiles, great depths, sinuous courses,

rock walls, and tributary canyons which commonly join the main canyons accordantly. It is well to be cautious, however, about concluding that they are in all respects similar to canyons on land. In the first place, contouring of most canyons is done from data that are far from adequate to give a detailed picture of the canyons. In the second place, the persons doing the contouring may have been influenced consciously or unconsciously in their drawing of contours by the fact that the only canyons with which they were familiar were those formed subaerially. Until more adequate surveys are made, we cannot be certain of all the topographic characteristics of the canyons. Except for some of the canyons off the coasts of the United States, few have been contoured in sufficient detail to give a satisfactory picture of what they are like.

10. Although the cross profiles of submarine canyons seem to resemble those of canyons on land, there is a significant difference in the longitudinal profiles of the two. The long profiles of canyons in the continental slope show gradients that average about ten times those of canyons on land.

11. Submarine canyons seem to exist off some oceanic islands and submarine banks, but information regarding these is meager, and it cannot be said with certainty that those reported are not drowned river valleys.

12. The canyons extend seaward to depths in excess of 2000 fathoms.

Theories to account for submarine canyons. No attempt will be made to consider all the theories that have been proposed to account for submarine canyons. Only those that have been given serious consideration will be discussed. Many of the earlier theories suggested that the canyons were of diastrophic origin and could be explained as the result of faulting and local subsidence of narrow blocks. It is possible that locally canyon sites may be related to faults or fault zones, but it seems unlikely that faulting could have been a major factor in their development. The sinuous courses of the canyons and the dendritic patterns that they and their tributaries apparently have, their graded longitudinal profiles, their presence in areas where faulting is not significant, as off the Atlantic coast, and trends in many areas across the structural grain, all seem to argue against such an origin.

It was only natural, in view of the many apparent similarities between submarine and land canyons and the common location of canyons opposite rivers on land, that a subaerial origin should be proposed. Explanation of how the canyons were submerged beneath thousands of feet of oceanic water has taken two forms. It was first suggested that they were formed on land and then drowned by downwarping or tilting. As long as only a few were known to exist, this theory was not too unreasonable, but now that they are known to exist off most coasts this explanation is hardly tenable. Since the canyons are cut in late Tertiary and Pleistocene sediments, it is apparent that the continental shelves would have to have been below sea level during these times in order to receive the sediments which underlie them, then universally uplifted enough to allow canyons thousands of feet deep to be cut in them, and then during late Pleistocene time submerged thousands of feet. Such behavior of the continental shelves does not seem compatible with our knowledge of diastrophism.

A second theory of subaerial origin attempts to get away from the diastrophic difficulties mentioned above by assuming that it was not the land that went up and down but the level of the sea, as a result of changes in volume produced by glaciation and deglaciation. Shepard was for many years a strong proponent of the theory that the canyons were cut subaerially during Pleistocene low sea levels, although he later abandoned it. It is most generally believed that the lowering of sea level produced by withdrawal of water from the oceans to form the Pleistocene ice sheets was of the order of 300 feet. This is at best a scientific guess, but it is probably of the right order of magnitude. At least such information as is available indicates that the lowering of sea level by the ice sheets was not of the magnitude of several thousand feet. Shepard and Emery (1941), by assuming a much greater thickness and extent of the ice sheets than is usually postulated, argued that there was the possibility of a lowering of sea level during earlier stages of the Pleistocene of as much as 3000 feet, but few geologists accept this estimate. Even this, however, is not enough. In his book, *Submarine Geology*, Shepard (1948) still argued that subaerial cutting during times of low sea level was still the more likely explanation, although he recognized the possibility that there were other contributing causes. More recently, Shepard (1952) has abandoned the idea of the canyons being strictly of subaerial origin and has proposed a theory of composite origin. The difficulties encountered in explaining the lowering of sea level necessary for the canyons to have been cut by streams seem insurmountable. Lack or scarcity of canyons across the continental shelf is hard to explain by this theory. If Tolstoy's conclusion that Hudson Canyon extends down to a depth of 15,000 feet is correct, the magnitude of lowering of sea level to permit subaerial canyon cutting seems beyond any possibility of realization.

Johnson (1939) suggested that the canyons might be the result of submarine spring sapping. His idea was that waters draining out of the Tertiary and Cretaceous formations beneath the coastal plains and continental shelves would have developed a great enough artesian head and pressure to emerge as submarine freshwater springs at the foot of or on the continental slope. Spring sapping, such as is responsible for the development of the steepheads of Florida and the spring alcoves of the Snake River Canyon, supposedly would have resulted in the headward growth of the canyons up the continental slope. There are so many fatal objections to this theory that few persons have given it serious consideration. In the first place, the type of shelf structure found along the Atlantic Coast is not found everywhere. Furthermore, it is necessary to assume that the sediments of the continental shelf once cropped out much farther inland and at much higher altitudes than they do now. It is impossible to explain by this theory the development of a canyon in granite, such as Carmel Canyon off the coast of California. Weathering is the most important process involved in the backward extension of spring alcoves, and it is practically inoperative under the ocean. The heads of the canyons do not seem to have the form that would be expected under Johnson's hypothesis. It is difficult to see how the steep gradients of the canyon floors can be explained by this hypothesis, for submarine springs would be held in position by the dip of the formations from which they emerged.

Bucher (1940) thought that the canyons might be the result of seismic sea waves or tsunamis and currents down the continental slope engendered by them. He believed that the late Tertiary and Pleistocene were times of pronounced crustal movements and that consequently tsunamis produced by submarine earthquakes were much more prevalent then than now. There are several serious objections to this theory. Much evidence seems to indicate that the canyons are still in an active stage of formation and are not filling up. Tsunamis are so infrequent and localized that they would not be operative along all coast lines. Major earthquakes are extremely rare in the North Atlantic where canyons seem to be most excellently developed, although their apparently greater number here may reflect better exploration of this area. Relatively few submarine canyons are known off the coasts of Japan and Peru, where tsunamis are particularly common. Tsunamis advance along a broad front; hence it is difficult to see how their energy could be localized enough to cut a canyon. They are oscillatory waves, so that there should be nearly as much material moved up the continental slope as down it. It is questionable whether currents produced by tsunamis exist at depths necessary to form canyons. Furthermore, it is difficult under this theory to explain the apparent dendritic patterns of the canyons and their greater frequency opposite rivers on land.

Daly (1936, 1942) has suggested that submarine canyons were cut by turbidity or suspension currents. Briefly, the theory is that trenching of the continental slopes was done by turbid sea water having mud and silt in suspension and thus constituting a *suspension* or *turbidity current*. Because such a current would have a greater density than normal sea water, it could flow down the continental shelf and slope to the floor of the ocean basin and in so doing erode sea bottom. Daly believed that turbidity currents were at a maximum during times of Pleistocene low sea levels when shores were farther seaward and the shallow waters over the continental shelves were carrying great quantities of glacially contributed mud and silt.

That turbidity currents can maintain their identity and flow as distinct currents is indicated by their existence in such freshwater bodies as Lake Geneva, Lake Constance, and Lake Mead. When the turbid waters of the Rhone enter Lake Geneva, they sink to the bottom of the lake and flow along as a distinct muddy current. The same thing is true where the Rhine enters Lake Constance and the muddy waters of the Colorado River flow into Lake Mead above Hoover Dam. Kuenen (1950*b*) cited evidence of the existence of turbidity currents in lakes and reservoirs and showed that they could be produced artificially in tanks, but their existence on the ocean floor is still largely a matter of inference. Submarine slumping, sliding, and mudflows are believed to aid in canyon cutting. It has been suggested that the common presence of canyons opposite streams on land results from the fact that the materials brought into the sea by streams straightway submerge to form turbidity currents. Canyons that are not in line with streams on land would be attributed to turbidity currents which originated on the continental shelf.

Shepard (1952), in the face of seemingly incontrovertible evidence that Pleistocene lowering of sea level was inadequate to permit subaerial cutting of the canyons, abandoned the theory of subaerial erosion and proposed in its stead a theory of composite origin. He maintained that submarine canyons have three distinct parts: inner valley heads in shallow water which give evidence of having been cut recently; intermediate portions of much more ancient origin with rocky walls which rise thousands of feet above canyon floors; and outer parts which are only slightly incised in unconsolidated sediments and which extend to the base of the continental slope or on to the ocean basin floors. He believed that the canyon heads were cut by streams during the times of glacial low sea levels. The intermediate and outer portions of the canyons were believed to be of much greater age. He thought that they dated back to times when the areas where the canyons are were geosynclinal tracts which experienced alternating uplift and subsidence and that these portions of the canyons were cut by streams during periods of emergence. The shallower distal parts of the canyons were attributed to the trenching of deltaic deposits by submarine landslides and engendered turbidity currents. Although he admitted the possibility that turbidity currents may have contributed to the cutting of the outer portions of the canyons, he still maintained that there is no evidence that they excavated the intermediate rock-walled sections of the canyons.

Landes (1952) has revived the generally discarded theory of a shrinking globe to account for cutting of submarine canyons by subaerial erosion. He argued that reduction in volume of the earth's interior has resulted in sporadic but continued sinking of crustal segments with the heavier ocean segments being the first to drop. By this method, he claimed, ocean basins might be differentially lowered as much as 20,000 or 30,000 feet and thereby permit cutting above sea level of canyons as deep as any so far measured. He did not believe that turbidity currents in deep sea waters had been demonstrated, despite the fact that many large rivers empty great quantities of sediment into the ocean near heads of submarine canyons. Instead, where rivers heavily laden with sediment enter the sea, fresh water rides over denser salt water carrying its sediment until decreasing velocity causes the sediment to settle downward through the salt water beneath.

The origin of submarine canyons remains a perplexing problem. The theory that they were cut by turbidity currents aided by submarine landsliding, slumping, and creep holds a slightly favored position, not so much because it answers all the questions connected with them but because it encounters fewer difficulties than any other theory. There is still some question as to the existence of turbidity currents at great depths and doubt whether such currents would have the power to cut canyons in bedrock, particularly in granite as in Carmel Canyon. It has been pointed out that in Lake Mead a delta is being formed where according to theory erosion should be taking place. Ericson et al. (1951) have presented what was considered evidence for the transport of material at great depth by turbidity currents. Several cores taken from a delta-like plain at the end of Hudson Canyon at depths between 2390 and 2700 fathoms showed sand layers ranging from thin films to 6 meters in thickness interbedded with clays of abyssal facies. Graded bedding characterized several of the sand layers, and it was thought that turbidity currents seemed the most likely means of transport of the sand.

Dietz (1953) has described five channels on the floor of the Indian Ocean, southeast of Ceylon, which he thought might be attributed to erosion by low-density turbidity currents of great volume. These channels vary in depth from 30 to 240 feet and in width from less than a mile to over 4 miles. Leveed banks that rise as much as 90 feet above the adjacent sea floor border them. Dietz thought it possible that the largest one might connect with the enormous Ganges Canyon some 1100 nautical miles to the north. Similar channels on the sea floor have been described elsewhere but it is as yet uncertain whether they have been eroded by turbidity currents.

TOPOGRAPHY OF THE DEEP-SEA FLOORS

Before the development of sonic sounding, the prevalent idea was that the floors of the ocean basins were largely vast, flat to undulating plains with relatively little relief. Although only a small fraction of the ocean floors has been surveyed in detail, enough information is at hand to indicate that this idea is far from correct. Extensive areas may lack major relief features, but it is probably the exception rather than the rule for the sea floor to be a monotonously level plain. This is surprising when we consider that sedimentation is dominant here and that the erosional processes which give diversity to land surfaces are inoperative. Apparently vulcanism and diastrophism are largely responsible for the major relief features on the floors of the ocean basins. The fact that such processes as weathering and mass-wasting are either lacking or act at reduced rates may help to explain the apparent sharpness of some of the topographic forms found. On land, a fault scarp is attacked by erosion, weathering, and mass-wasting and rapidly takes on a subdued form and in time is obliterated. Beneath the ocean, such a feature is preserved indefinitely. It is likely that, if suboceanic topography could be seen, it would exhibit a sharpness and angularity exceeding that on land. Probably all the major

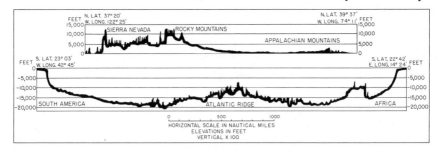

FIG. 18.9. Comparison of topographic profiles across North America and the South Atlantic basin. (After F. P. Shepard, *Submarine Geology,* Harper and Brothers, Publishers.)

structural and topographic forms found on land have their counterparts beneath the oceans.

Negative forms. The major negative features of the deep-sea floors are basins, trenches, and troughs. The term *basin* is applied to a large depression in the deep-sea floor which is more or less circular, oval, or elliptical in plan. Examples of basins in the North Atlantic are the West European, Canary, Cape Verde, Newfoundland, and North American basins. The Caribbean, the Mediterranean, Celebes, and Gulf of Mexico basins are partially enclosed by land.

A *trench* is an elongated narrow depression on the deep-sea floor which normally has steep sides, whereas a *trough* is a long, broad depression with more gently sloping sides. Actually, it may be difficult to make a sharp distinction between trenches and troughs, and application of the terms varies with individuals. Trenches seem to be invariably associated with tectonically active continental borders such as those around the Pacific. They are especially associated with island arcs like those off southeast Asia and in the Caribbean region. Examples of trenches are: the Mindanao Trench off the Philippines, which has a depth of water of 5740 fathoms and is the greatest known ocean

deep; the Java Trench with a maximum depth
of 4074 fathoms; the Aleutian Trench with a
depth of 4199 fathoms; the Japan Trench with
a depth of 5360 fathoms; and the Puerto Rico
Trench with a depth of 5041 fathoms. Examples of depressions in the deep-sea floor which
are better called troughs are the Bartlett
Trough (3958 fathoms deep) in the Caribbean
Sea south of Cuba, and the Weber Trough in
the Molucca Sea. *Deep* in common usage is
applied to ocean depths in excess of 3000
fathoms. Unfortunately it has no specific morphological significance, but it is so firmly
rooted in the literature that it is not likely to
be dropped.

Positive features. The major positive topographic features which rise above the deep-sea
floor are rises or swells, ridges, and plateaus.
Rise and *swell* have both been used to designate extensive, long, broad elevations which
rise gently from the deep-sea floor. A good
example of a rise or swell is that in the Pacific
Ocean known as the Hawaiian swell or rise.
It is a comparatively gentle rise some 600 miles
wide and about 1900 miles long, above which
have been built the volcanic domes which constitute the Hawaiian Islands. A *ridge* is a
long, narrow elevation above the deep-sea floor
which has steeper sides and somewhat rougher
topography than a rise or swell. By far the
largest and best known of the ridges is the Mid-
Atlantic Ridge, which extends from Iceland at
the north to a point in the South Atlantic southwest of the Cape of Good Hope. Hydro-
graphic work in the North Atlantic (Tolstoy
and Ewing, 1949, and Tolstoy, 1951) has
thrown much light upon the topography of this
ridge. This work indicates that it consists of
three rather distinct types of topography.
There is a high central zone, or what was
called the Main Range, which consists of

Fɪɢ. 18.10. Fathogram across the Mid-Atlantic Ridge at latitude 31° north. The horizontal distance is approximately 130 miles, and the depth range is between 0 and 2000 fathoms. (Courtesy Maurice Ewing.)

several parallel ridges extending in a general northeast-southwest direction. In many places the depth of water over these ridges is less than 800 fathoms. At depths between 1600 and 2500 fathoms on the flanks of the Main Range there is a series of flats, which were designated as the Terraced Zone. Individual terraces vary in width from 1 to 50 miles and collectively are between 200 and 300 miles in width. A third zone, at a depth of about 2900 fathoms lies between the terraced zone and the sea-floor plain. It is

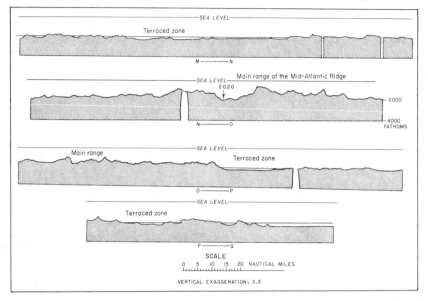

FIG. 18.11. Profiles across the Mid-Atlantic Ridge and adjacent areas close to latitude 31° north. (After Ivan Tolstoy.)

mountainous and rather distinct from the other two zones. It extends in a north-northeast-south-southwest direction and has individual peaks more than 3000 feet high. This zone was designated the Foothills of the Mid-Atlantic Ridge.

Several theories have been proposed to account for the Mid-Atlantic Ridge. Among them are: (1) that it is a horst; (2) that it is an anticlinal fold; (3) that it is the bottom of a rift which opened as Gondwana land began to fragment; (4) that it was produced largely by extrusion of volcanic rocks along great northeast-southeast fractures; (5) that it is an orogenic mountain belt similar to those on land; and (6) that it is a belt of blockfaulting. Tolstoy and Ewing concluded that they could not tell from fathograms whether the ridge owed its existence to folding or faulting. Their work seemed to demonstrate the existence in the ridge of consolidated limestone probably of Cenozoic age and also indicated extensive areas of sedimentary rocks. Even though their information about the ridge is largely topographic rather than geologic it presents the basis for interesting speculations as to the origins of the different types of topography.

Submarine elevations of considerable extent with relatively flat tops are called *plateaus*. Examples of plateaus are the Albatross Plateau in the Pacific Ocean off South and Central America, the Seychelles Plateau of the Indian Ocean, and the Azores Plateau in the North Atlantic. Tolstoy and Ewing have shown that the Azores Plateau has upon it numerous ridges, trending at right angles to the Mid-Atlantic Ridge, which are different in character from the ridges of the Main Range. The ridges on the Azores Plateau are separated by broad valleys instead of closely spaced, steep and narrow ones, as in the Main Range. Detailed soundings of the so-called plateaus may show that some have too much relief to merit this designation.

Portions of certain arcuate submarine ridges extend above sea level to form what are commonly called *island arcs*. These are particularly well-developed in the western Pacific and West Indies areas. On the convex sides of the arcs are deep trenches or troughs, often called *foredeeps*, which are the greatest ocean deeps. Usually three distinct belts are associated with island arcs— an outer belt of deep trenches or foredeeps where negative gravity anomalies commonly exist, the tectonic island arcs with positive gravity anomalies, and an inner belt of volcanoes which also exhibits positive gravity anomalies. Differences of opinion exist as to the exact mechanics involved in the formation of an island arc, but it is generally agreed that it represents a young and active orogenic section of the earth's crust.

What may be designated as lesser topographic features on the deep-sea floor are volcanic islands, seamounts, and guyots. Volcanic islands are particularly common in the Pacific Ocean. One of the interesting developments of sonic sounding has been the discovery of large numbers of submarine islands, to which the name *seamounts* has been given. Numerous seamounts in the Gulf of Alaska have been mapped (Murray, 1941; Menard and Dietz, 1951), which rise from 3500 to 12,400 feet above the floor of the gulf. Some are volcanic peaks and others are mountains of orogenic origin. Hess (1946) described some 160 seamounts in the Pacific Basin between Hawaii and the Mariana Islands. In the main, they seem to be truncated volcanic islands that rise 9000 to 12,000 feet above the sea floor with summits 3000 to 6000 feet below sea level. Hess gave the name *guyot* to the larger flat-topped forms. One guyot, south of Eniwetok atoll, has a basal circumference of 35 miles, and its remarkably flat top has a diameter of 9 miles. Another, northeast of Eniwetok, is 60 miles wide at its base and 35 miles at its top.

Shepard (1941) reported the existence of 77 seamounts off the coast of southern California, the largest of which, the San Juan seamount, rises 10,000 feet above the deep-sea floor. Tolstoy (1951) showed that numerous seamounts are present in the North Atlantic, and one particularly conspicuous group southeast of Cape Cod rises to about 11,000 feet above its surroundings. From the number which have been described it seems reasonable to assume that seamounts are common features on the ocean floors. Their truncated tops present a puzzling problem, particularly when it is recognized that the elevations of their summits vary greatly. Just how this truncation took place is not clear nor is it certain whether it was the result of one or more periods of erosion. Hess (1946) suggested that the summit surfaces of those in the

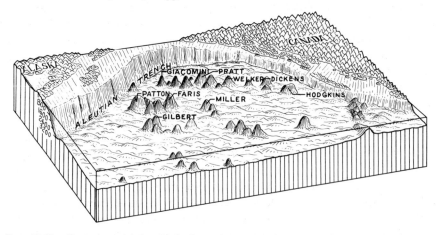

Fig. 18.12. Seamounts in the Gulf of Alaska. Drawn from a bathometric map by H. W. Menard and R. S. Dietz. (Drawing by W. C. Heisterkamp.)

Pacific Basin were very old and may represent the results of marine planation by Pre-Cambrian seas when sea level was much lower than it is now. Shepard has cited truncation of seamounts in support of his idea of a former great lowering of world sea level, although he recognized that the cause of truncation may be complex. As yet not enough information is available to draw any safe conclusion as to the cause or causes of the truncation.

REEFS AND ATOLLS

Discovery of oil in numerous reef structures, such as that at Leduc, Alberta, in 1947, led geologists to become interested in reef structures as geologic structures as well as topographic features. Actually, what is commonly called a coral reef was not built solely by corals. Numerous other organisms contribute to the growth of a reef, among them being numerous forms of calcareous algae, stromatoporoids, gastropods, echinoderms, foraminifera, and mollusca. The term *bioherm* (organic mound) is a more appropriate name for such structures, but they have been called reefs so long that the name persists despite its inadequacy. Colonial corals to a large degree do build the super-structure, but the voids between the corals are filled with the skeletons of other organisms as well as organic and inorganic detritus.

Types of reefs. Three main types of reefs are recognized. A *fringing reef* grows directly against the bedrock of a coast and actually constitutes its shore line. Such reefs are common along many tropical coasts. A *barrier reef* lies off the coast and is separated from it by a lagoon usually too deep to permit coral growth. The lagoon may vary in width from a narrow channel to many miles. The great barrier reef off the coast of northeast Australia extends for about 1000 miles with only minor breaks in it. Barrier reefs may change laterally into fringing reefs by joining a coast. Barrier reefs encircle many tropical islands in the Pacific, such as Tahiti, and form excellent enclosed harbors. A reef ring enclosing a lagoon lacking an island of non-reef origin is called an *atoll*. Shepard (1948) claimed that there is a fourth basic type of coral growth which he called a *pinnacle*. This name was applied to certain sharply upward-projecting forms found in the lagoon back of the Bikini atoll during the 1946 atom bomb test, which were called by Emery (1948) *coral knolls*. Some are nearly vertical; others have slopes of about 45 degrees. More detailed mapping of atoll lagoons will be necessary before it can be decided whether this is a distinctive reef form. Many variations of the funda-mental types are found. Some are nearly atolls but have a small island of non-reef origin, usually volcanic, within a lagoon; some are large reef masses without lagoons; some are reefs submerged below depths which permit the growth of corals; and small round or oblong reefs with enclosed lagoons called *faros* may form part of a larger barrier reef or atoll rim.

Ecological conditions for reef growth. Any consideration of the origin of reefs and atolls should take into account the ecological conditions which allow and condition the growth of the reef-forming organisms. Two groups of organisms seem to be most significant in the formation of reefs, corals and various types of calcareous algae which are attached to them in symbiotic relationship. The algae receive food and carbon dioxide from the coral polyps and in turn supply them with oxygen and possibly carbohydrates. Re-striction of active coral growth to shallow depths of water is apparently related to a need that the symbiotic algae have for light. The limiting depth of water in which the corals will grow varies according to the turbidity of the water and abundance of floating plankton and other material that affects the depth of light penetration. In exceptionally clear water, they may grow at depths as great as 300 feet but they are seldom found below depths of 150 to 200 feet.

Although water temperatures of 64 to 68 degrees Fahrenheit can be tol-erated, corals thrive most abundantly at temperatures between 77 and 86 degrees. Normal salinity ranging between 27 and 38 parts per thousand is most conducive to growth. Lack of sufficient salinity may account in part for the absence of reefs opposite the mouths of large rivers, or their absence here may result from greater turbidity of the waters. A large amount of silt is detrimental, not only because of its interference with light penetration but also because in settling on the organisms it interferes with their growth. Active circulation and dash of the waves are also essential. Corals are sedentary organisms and depend upon the dash of waves for their food supply of plankton. They also depend in part upon waves for their oxygen supply. Algae may supply them with sufficient oxygen during the day, but at night they get oxygen from aerated sea water. Reef-building corals prefer a firm foundation and do not thrive on muddy bottoms. Kuenen (1950a) attributed the scarcity of reefs in the Sunda Sea to the unfavorable type of bottom found there. It does seem, though, that corals are able to grow upon a variety of surfaces, and their scarcity in areas of muddy bottoms may be the result of the deleterious effects of silt in suspension rather than of muddy sea floors.

THEORIES OF ORIGIN OF BARRIER REEFS AND ATOLLS

A full discussion of this complex problem is beyond the scope of this book, but brief statements of some of the theories that have been proposed to explain atolls and barrier reefs will be given. For a more detailed discussion, the reader is referred particularly to Davis's *The Coral Reef Problem*. Probably no one theory is adequate to account for the many varying conditions under which atolls have formed. The main difference of opinion has been regarding the origin of the lagoons behind barrier reefs and atolls. Theories to explain atolls and barrier reefs fall into three main groups: (1) those which postulate that change of sea level is not necessary to their formation but that they were formed upon stillstanding platforms which existed beneath the sea antecedent to the attachment of the corals; (2) those which explain them as the results of the effects of changes in sea level during Pleistocene and post-Pleistocene times; and (3) those which maintain that they are the result of continued sinking of the foundations upon which the corals had established themselves.

Antecedent platform theories. Rein (1870) and Murray (1880) independently suggested that stillstanding submarine platforms could provide foundations upon which corals could establish themselves and develop into atolls or barrier reefs. Murray explained the lagoon back of the reef largely as the result of the solvent action of water thrown over the reef at high tide. The lagoon was extended, presumably, by the reef growing seaward on its front with accompanying enlargement of the lagoon back of it by solution. Most geologists do not consider solution by marine waters important enough to account for the formation of a lagoon. Moreover, most of the evidence points toward lagoons being areas of deposition rather than of erosion.

A more likely antecedent platform theory is that which was originally proposed by Alexander Agassiz (1903). According to this theory, whether a barrier reef or an atoll forms depends upon whether marine abrasion had partly or wholly destroyed the island before reef-building organisms established themselves. More recently Ladd and Hoffmeister (1936) have supported this theory in a modified form. They believed that any submarine platform, whether the result of long-continued wave erosion or the product of long-continued upbuilding through the slow rain of pelagic organisms or upbuilding by volcanic eruptions, that is at the proper depth and is located in the tropical coral reef zone is potentially a reef foundation, and that, if ecological conditions are right, a reef may grow upward to the ocean surface without progressive sea level changes. This theory neither precludes the effects of glacial lowering and rise of sea levels nor considers them necessary to account for atolls and barrier reefs.

Some of the difficulties of the antecedent platform theory are: it fails to explain the commonly embayed islands which lie back of barrier reefs; the thicknesses of coralline rock in the lagoons, as shown by recent borings and seismic tests, seem to be too great to have been produced by deposition of pelagic debris and organisms; the depths of many of the lagoons are too great to have permitted corals to have established themselves on submarine platforms.

The glacial control theory. The idea that atolls are the result of the influences of the changing sea levels of the Pleistocene is most generally associated with Daly (1934, 1942), although his theory as originally proposed in 1910 and elaborated in a number of later papers was based upon an idea originally put forth by A. Penck in 1894. Penck concluded from the surprisingly uniform depth of many lagoons back of atolls and the world-wide embayment of continental coasts that the ocean must have formerly stood for a considerable time at a level some 100 to 200 meters below its present level and that present high sea level is a result of the rise of sea level accompanying deglaciation. Penck did not go beyond this and it was Daly who elaborated upon the effects of such changes in sea level upon corals and other reef-building organisms. The essential points of Daly's hypothesis are: (1) the depth of lagoons back of atolls and barrier reefs is remarkably uniform and rarely exceeds 80 to 90 meters, which implies a cause world-wide in nature; (2) glacial conditions would result in a world-wide chilling of the seas and increased turbidity of oceanic waters as a result of the churning up of muds formerly below the reach of waves and thus would kill off reef-building organisms; (3) destruction of these organisms permitted marine abrasion to attack the islands and banks upon which they had grown and produce at a lowered sea level a great number of truncated islands or benches around islands; (4) return of interglacial high sea levels with their warmer and clearer waters favored reestablishment of corals and associated organisms upon submerged platforms with resulting development of an atoll or barrier reef, depending upon whether an island had been completely truncated or had only had a marine bench cut around it.   Corals established themselves around the periphery of a platform and gradually grew upward and forward as sea level rose, until their bases were ultimately submerged about 250 to 300 feet.   Aggradation of lagoons accompanied upgrowth of the reefs, but the rate of filling depended upon the size of lagoon, thus accounting for the variations found in their depths.   Some of the arguments against Daly's theory are: (1) many of the platforms are too broad to have been cut by marine abrasion during a glacial age; (2) lagoon depths are hardly as uniform as he claimed; (3) doubt exists as to the efficacy of low-level marine abrasion to form submarine platforms; and (4) it is questionable whether the conditions of low temperatures and turbidity postulated by Daly could have extended so far from the glaciated areas.

The subsidence theory.   The subsidence theory was originally proposed by Charles Darwin as one of the results of his famous voyage on the *Beagle*. It is remarkable that Darwin should have evolved what is now the more-favored theory after having visited only one good barrier reef, that of Tahiti, and one atoll, that of Keeling.   The subsidence theory is remarkably simple. It requires only sinking of a volcanic island foundation and upgrowth of a reef at a pace commensurate with the rate of subsidence.   According to this theory, there are three stages in the development of an atoll.   A reef starts as a fringing reef and through subsidence is transformed into a barrier reef and still later into an atoll.

Dana and Davis both thought that the embayed coasts of islands back of barrier reefs strongly supported the subsidence theory.   Davis (1928) pointed out that islands back of barrier reefs rarely have cliffed shore lines, as would be expected under Daly's theory if marine abrasion had not completely destroyed the islands.   He further thought that the geologic evidence argued against island stability in the regions where barrier reefs and atolls exist and called attention to the fact that coral rock had been obtained from cores near Honolulu at depths as great as 1178 feet.

Borings and geophysical tests made on Bikini seem to add support to the subsidence theory (Dobrin, Perkins, and Snavely, 1949).   They indicated that the upper 2500 feet of material consists of calcareous deposits such as are found on the sea floor within and around atolls and that below this there is a zone 5000 to 10,000 feet thick which may be either calcareous or volcanic material.   The evidence points toward several thousand feet of subsidence of this atoll.   Thus, Darwin's theory has returned to favor with the added recognition of the possibility of multiple factors involved in the formation of barrier reefs and atolls. However, it is too much to assume as yet that it is proved. Kuenen (1950) favored what he called "the glacially controlled subsidence theory," a combination of the ideas of Darwin and Daly. Although he believed that there is undeniable evidence of subsidence, he felt that existing reefs to a considerable degree owe their characteristics to the effects of glacially controlled sea levels. Stearns (1946) has advocated an integration of the antecedent platform, subsidence, and glacial control theories into a composite theory which he thought would explain the formation of barrier reefs and atolls during any geologic period.

REFERENCES CITED IN TEXT

Agassiz, Alexander (1903). On the formation of barrier reefs and of the different types of atolls, *Proc. Royal Soc., 71,* pp. 412–414.

Bucher, W. H. (1940). Submarine valleys and related geologic problems of the North Atlantic, *Geol. Soc. Am., Bull. 51,* pp. 489–512.

Daly, R. A. (1934). *The Changing World of the Ice Age,* Chapter 7, Yale University Press, New Haven.

Daly, R. A. (1936). Origin of submarine canyons, *Am. J. Sci., 231,* pp. 401–420.

Daly, R. A. (1942). *The Floor of the Ocean,* University of North Carolina Press, 157 pp.

Davis, W. M. (1928). The coral reef problem, *Am. Geog. Soc., Spec. Publ. 9,* 596 pp.

Dietz, R. S. (1953). Possible deep-sea turbidity current channels in the Indian Ocean, *Geol. Soc. Am., Bull. 64,* pp. 375–377.

Dobrin, M. B., B. Perkins, Jr., and B. L. Snavely (1949). Subsurface constitution of Bikini atoll as indicated by a seismic-refraction survey, *Geol. Soc. Am., Bull. 60,* pp. 807–828.

Emery, K. O. (1948). Submarine geology of Bikini atoll, *Geol. Soc. Am., Bull. 59,* pp. 855–860.

Ericson, D. B., Maurice Ewing, and B. C. Heezen (1951). Deep-sea sands and submarine canyons, *Geol. Soc. Am., Bull. 62,* pp. 961–966.

Hess, H. H. (1946). Drowned ancient islands of the Pacific basin, *Am. J. Sci., 244,* pp. 772–791.

Holtedahl, Olaf (1950). Supposed marginal fault lines in the shelf area off some high northern lands, *Geol. Soc. Am., Bull. 61,* pp. 493–500.

Johnson, D. W. (1925). *The New England-Acadian Shoreline,* pp. 63–66 and 286–294, John Wiley and Sons, New York.

Johnson, D. W. (1939). *The Origin of Submarine Canyons,* Columbia University Press, 126 pp.; also in *J. Geomorph., 1,* pp. 111–129, 230–244, and 324–340; and *J. Geomorph., 2,* pp. 42–60, 133–158, and 213–236.

Koons, E. D. (1941). The origin of the Bay of Fundy and associated submarine scarps, *J. Geomorph., 4,* pp. 237–249.

Koons, E. D. (1942). The origin of the Bay of Fundy: a discussion, *J. Geomorph., 5,* pp. 143–150.

Kuenen, Ph. H. (1950a). *Marine Geology,* 568 pp., John Wiley and Sons, New York.

Kuenen, Ph. H. (1950b). Turbidity currents of high density, *Rept. 18th Intern. Geol. Congr., 8,* pp. 44–52.

Ladd, H. S., and J. E. Hoffmeister (1936). A criticism of the glacial-control theory, *J. Geol., 44*, pp. 74–92.

Landes, K. K. (1952). Our shrinking globe, *Geol. Soc. Am., Bull. 63*, pp. 225–240.

Lewis, R. G. (1935). The orography of the North Sea bed, *Geog. J., 86*, pp. 334–342.

Lindenkohl, A. (1891). Notes of the sub-marine channel of the Hudson River and other evidences of Post-glacial subsidence of the Middle Atlantic Coast Region, *Am. J. Sci., 41*, pp. 489–499.

Menard, H. W., and R. S. Dietz (1951). Submarine geology of the Gulf of Alaska, *Geol. Soc. Am., Bull. 62*, pp. 1263–1286.

Murray, John (1880). On the structure and origin of coral reefs and islands, *Proc. Royal Soc. Edinburgh, 10*, pp. 505–518.

Murray, H. W. (1941). Submarine mountains in the Gulf of Alaska, *Geol. Soc. Am., Bull. 52*, pp. 333–362.

Murray, H. W. (1947). Topography of the Gulf of Maine, *Geol. Soc. Am., Bull. 58*, pp. 153–196.

Rein, J. J. (1870). Beiträge zur physikalischen Geographie der Bermuda-Inseln, *Bericht. Senckenb. Naturf. Gesell.*, Frankfurt am Main, pp. 140–158.

Shepard, F. P. (1930). Fundian faults or Fundian glaciers, *Geol. Soc. Am., Bull. 41*, pp. 659–674.

Shepard, F. P. (1931). Saint Lawrence (Cabot Strait) submarine trough, *Geol. Soc. Am., Bull. 42*, pp. 853–864.

Shepard, F. P. (1941). Nondepositional physiographic environments off the California coast, *Geol. Soc. Am., Bull. 52*, pp. 1869–1886.

Shepard, F. P. (1948). *Submarine Geology,* 348 pp., Harper and Brothers, New York.

Shepard, F. P. (1952). Composite origin of submarine canyons, *J. Geol., 60*, pp. 84–96.

Shepard, F. P., and K. O. Emery (1941). Submarine topography off the California coast, *Geol. Soc. Am., Spec. Paper 31*, 171 pp.

Spencer, J. W. (1903). Submarine valleys off the American coasts and in the North Atlantic, *Geol. Soc. Am., Bull. 14*, pp. 207–226.

Stearns, H. T. (1946). An integration of coral-reef hypotheses, *Am. J. Sci., 244*, pp. 772–791.

Tolstoy, Ivan (1951). Submarine topography in the North Atlantic, *Geol. Soc. Am., Bull. 62*, pp. 441–450.

Tolstoy, Ivan, and Maurice Ewing (1949). North Atlantic hydrography and the Mid-Atlantic ridge, *Geol. Soc. Am., Bull. 60*, pp. 1527–1540.

·Veatch, A. C. (1937). Recent advances in marine surveying, *Geog. Rev., 27*, pp. 625–629.

Woodford, A. O. (1951). Stream gradients and Monterey Sea Valley, *Geol. Soc Am., Bull. 62*, pp. 799–852.

Additional References

Betz, Frederick, Jr., and H. H. Hess (1942). The floor of the North Pacific ocean, *Geog. Rev., 32*, pp. 99–116.

Crowell, J. C. (1952). Submarine canyons bordering central and southern California, *J. Geol., 60*, pp. 58–83.

Dana, J. D. (1879). *Corals and Coral Islands,* 406 pp., Dodd, Mead and Co., New York.

Darwin, Charles (1898). *The Structure and Distribution of Coral Reefs,* 3rd ed., 344 pp., D. Appleton and Co., New York.

Dietz, R. S., and H. W. Menard (1951). Origin of abrupt change in slope in continental shelf margin, *Am. Assoc. Petroleum Geol., Bull. 35,* pp. 1994–2016.

Dietz, R. S. (1952). Geomorphic evolution of continental terrace (continental shelf and slope), *Am. Assoc. Petroleum Geol., Bull. 36,* pp. 1802–1819.

Du Toit, A. L. (1940). An hypothesis of submarine canyons, *Geol. Mag., 77,* pp. 395–404.

Emery, K. O. (1949). Topography and sediments of the Arctic basin, *J. Geol., 57,* pp. 512–521.

Emery, K. O. (1950). A suggested origin of continental slopes and of submarine canyons, *Geol. Mag., 87,* pp. 102–104.

Emery, K. O., and F. P. Shepard (1945). Lithology of the sea floor off southern California, *Geol. Soc. Am., Bull. 56,* pp. 431–478.

Jordan, G. F. (1951). Continental slope off Apalachicola River, Florida, *Am. Assoc. Petroleum Geol., Bull. 35,* pp. 1978–1991.

Menard, H. W., and R. S. Dietz (1952). Mendocino submarine escarpment, *J. Geol., 60,* pp. 266–278.

Murray, H. W. (1945). Profiles of the Aleutian trench, *Geol. Soc. Am., Bull. 56,* pp. 757–782.

Shepard, F. P. (1931). Glacial troughs on the continental shelves, *J. Geol., 39,* pp. 345–360.

Shepard, F. P., and K. O. Emery (1946). Submarine photography off the California coast, *J. Geol., 54,* pp. 306–321.

Smith, Paul (1937). The submarine topography of Bogoslof, *Geog. Rev., 27,* pp. 630–636.

Stetson, H. C., and J. F. Smith (1938). Behavior of suspension currents and mud slides on the continental slope, *Am. J. Sci., 235,* pp. 1–13.

Taber, Stephen (1927). Fault troughs, *J. Geol., 35,* pp. 577–606.

Veatch, A. C., and P. A. Smith (1939). Atlantic submarine valleys of the United States and the Congo submarine valley, *Geol. Soc. Am., Spec. Paper 7,* 101 pp.

Wiseman, J. D. H., and C. D. Ovey (1950). Recent investigations on the deep-sea floor, *Proc. Geol. Assoc., 61,* pp. 28–84.

19 · Land Forms Resulting from Volcanism

INTRODUCTION

Volcanism arises from forces which are endogenous in nature and are produced by physical and chemical changes taking place in the earth's interior. Extrusion of lava may produce distinctive land forms which do not owe their characteristics to gradational processes, although they do not exist long before they are modified by these processes. Intrusions are rarely directly responsible for topographic features, but their presence in the earth's upper crust may influence significantly the shaping of topographic forms by gradation.

To a large degree extrusions and intrusions may be considered special types of geologic structure upon which the gradational processes have to work. Localized volcanic activity may impose special features upon a landscape or may interfere with the progress of erosion. Volcanism on a regional scale may completely bury an existing landscape and present a totally different type of initial surface to the gradational agents.

PRESENT-DAY DISTRIBUTION OF VOLCANOES

As a result of their striking manifestations we often attribute to volcanoes an importance greater than they merit. Locally volcanism may be topographically significant but for the world as a whole it is far less important than several other less spectacular processes. One can travel widely over the earth's surface and never see an active volcano. There is only one such in the United States and that is Lassen Peak, in northern California. The most important belt of volcanoes is the so-called Ring of Fire or the circum-Pacific belt of active, dormant, or extinct volcanoes, which extends through the Andes of South America, Central America, Mexico, the Cascade Mountains of western United States, the Aleutian Islands, Kamchatka, the Kurile Islands, Japan, the Philippines, Celebes, New Guinea, the Solomon Islands, New Caledonia, and New Zealand. Other volcanic areas include: (1) scattered areas in the Pacific, including the Hawaiian Islands, the Galá-

488

pagos Islands, and the Juan Fernandez Isles; (2) the Indian Ocean girdle, including Timor, Java, Bali, and Sumatra; (3) a belt including Arabia, Madagascar, and the volcanoes of the Rift Valleys of Africa; (4) the so-called Mediterranean belt, including Mount Ararat in Asia, the Azores, and Canary Islands; (5) the volcanoes of the West Indies; and (6) other scattered areas such as Iceland. In addition, there are numerous areas of extinct volcanoes

FIG. 19.1.   Acatenango (foreground) and Fuego (background) volcanoes, some 30 miles south of Guatemala City. A side crater shows in Acatenango. (Photo by Fairchild Aerial Surveys, Inc.)

which are still topographically conspicuous. Extinct volcanoes are found in Arizona, New Mexico, Nevada, and Utah, the Auvergne region of France, the Eifel region of Germany, and the Faroe Islands.

There are probably fewer than 500 active volcanoes in the world today, but it is hazardous to classify definitely a volcano as active, dormant, or extinct. Many that are today dormant may well become active. Monte Somma was probably considered an extinct volcano by the inhabitants of Pompeii and Herculaneum in 79 A.D., for it had not been active for some 700 years, yet in that year there was a devastating eruption which marked the beginning of Mt. Vesuvius. Any volcano that has been active as recently as the Pleistocene is potentially an active volcano. A dozen or so volcanoes are known to have come into existence within historic time. The best-observed recent birth of a volcano was that of Paracutin, some 125 miles west of Mexico City.

It made its appearance on February 20, 1943, a large number of earthquakes having occurred on the previous day, and in a year's time had attained a height of 1410 feet.

TYPES OF ERUPTIONS

Numerous classifications of volcanoes according to their mode of eruption have been proposed. The most commonly used classification is that originally proposed by Lacroix in 1908. According to it, there are four principal types of eruptions—the Hawaiian, Strombolian, Vulcanian, and Peléan types. Cotton (1944) divided volcanoes, according to the type of material erupted, into lava volcanoes and pumice volcanoes, following somewhat Jaggar's grouping into lava and explosive volcanoes. Any classification is arbitrary to some degree, for there are gradations between types, and individual volcanoes may change their mode of eruption or alternate between one type of eruption and another. It is often stated that differences in volcanic activity are related to the type of magma emitted, non-explosive eruptions being characteristic of volcanoes with basaltic or basic lavas and explosive eruptions being associated with the emission of acid magmas. To some degree this is true, but it can hardly be considered a general rule, for not all the phenomena connected with volcanic eruption are determined by the character of the lava. It may be argued that Lacroix's classification of modes of eruption is purely descriptive and fails to take into account fundamental differences in magmatic conditions, but it probably suffices as well as any for a geomorphic discussion.

The Hawaiian type of eruption is marked by much fluid, effusive, basaltic lava. Explosive activity is relatively rare, but scoria mounds may be built around lava vents. Lava rarely pours out of the crater but more commonly issues through fissures around the sides of the volcanic pile as *flank eruptions*. Macdonald (1943) in describing the 1942 eruption of Mauna Loa, which was typical of most of its eruptions, recognized three phases in the flank eruption. The first phase covered a period of a few hours and was marked by the squirting of hot lava from a fissure as more or less continuous lava jets. The lava thus extruded formed thin flows or low mounds of agglutinated lava called *spatter cones*. The second phase was characterized by restriction of lava fountaining and the building of cinder and spatter cones. One or more major flows issued continuously from the cone, as well as several minor ones. During the final phase, there was a sharp decline in the amount of gas with consequent decrease in gas fountaining.

The Strombolian type of eruption emits basaltic lava which is less fluid than that of the Hawaiian type. Consequently, explosions are more common and more fragmental material is ejected. In neither type of eruption are dark smoke-like clouds formed.

Vulcanian eruptions eject viscous lavas which do not remain liquid long after coming in contact with air. The lava crusts over between eruptions, and each new explosion causes fragmentation of this frozen crust. Much ash is emitted, and ash-laden gases rise to form large dark cauliflower-like clouds. The ash may be distributed widely by the wind.

In a Peléan eruption the lava is extremely viscous and violent explosions are characteristic. One of its distinguishing features is the formation of *nuées ardentes*. These "glowing clouds" consist of a mixture of extremely hot, incandescent fine ash and coarser rock fragments permeated with hot gases to form a sort of emulsion. This material is extremely mobile yet is dense enough to rush with great velocity down the slopes of a volcanic cone. The great loss of life during the eruption of Mount Pelée in 1902, when all but one or two of the inhabitants of the city of St. Pierre at its base were killed, was caused by nuées ardentes.

Eruptions which take place through circumscribed vents are called *central eruptions*. They build volcanic cones of various types. Some eruptions, however, instead of being restricted to central vents may take place along a fissure or series of fissures as *fissure eruptions*. They usually do not build large volcanic cones but are more likely to produce lava plains or plateaus. Small cones may form along fissures, but large ones are rare. Fissure eruptions have not been common within recent times, if we exclude the flank eruptions of the Hawaiian type of volcanoes, but in 1783 there was an eruption from the Laki fissure in Iceland along some 20 miles, during which approximately 3 cubic miles of lava flowed out.

FEATURES OF LAVA FIELDS

The surface of a lava flow may be nearly flat, as in much of the Snake River lava plain of Idaho, or it may be so rough that travel across it is difficult, as in the "malpais country" of New Mexico. Diversity of surface may result from rock fragmentation during explosive eruptions, or it may be related to the mode of solidification.

The rock fragments ejected during explosive activity are known as pyroclastic materials. *Volcanic dust, volcanic ash, lapilli, scoriae, bombs,* and *blocks* are terms indicating increasing size which are applied to such materials. Not all pyroclastic material is produced by shattering of solid rock. Much of it forms from rapid cooling and solidification of molten lava upon coming in contact with air. Pyroclastic deposits are often very pervious, and springs commonly emerge at their forward edges.

Two distinct modes of solidification are exhibited by flowing lava. These are most commonly referred to as the *pahoehoe* and *aa* types of solidification. Jagger has termed these *dermolithic* and *clastolithic* solidification and has explained them in terms of variation in the gas content of the lavas, an explanation not accepted by all volcanologists.   Pahoehoe or dermolithic solidification takes place when lava containing much entrapped gas spreads out in thin sheets.   Such lava may be described as live lava.   It typically exhibits a wrinkled, twisted, ropy, and tapestry-like surface.   Its most distinguishing feature is a smooth, glistening skin.   Aa or clastolithic lava typically dis-

FIG. 19.2.   Vulcan's Throne and lava cascade above the Grand Canyon, Colorado. (Photo by J. S. Shelton and R. C. Frampton.)

plays a scoriaceous, clinkery, jagged, or blocky surface.   The lava is dead in the sense that most of its gases have escaped and the vesicles are filled with air.   This loss of gases is responsible for the more rapid cooling and greater viscosity of this type of lava.   Actually, the distinction between the two types of solidification is not so sharp as is sometimes suggested, and one type may change to the other within a single lava flow.

Numerous minor topographic features give diversity to a lava flow.   What Daly (1914) has called *tumuli* are bulging mounts on the surface of a lava flow with cracks in their crests through which lava may have been extruded. Tumuli do not seem to be gas blisters but rather are domings of the lava crust produced by the resistance which the lava surface has offered to the spreading of more fluid lava below and are thus much like laccoliths in origin. *Squeeze-ups* are small mounds or ridges which have resulted from the extrusion of viscous lava through a crack in the solidified crust. Nichols (1946) has described several on the McCartys lava flow in New Mexico, some of which are bulbous and others linear in form. *Pressure ridges* are somewhat similar in appearance but develop on a larger scale. In discussing the McCartys lava flow, Nichols (1939) stated that: "in tranverse cross-section

F<small>IG</small>. 19.3. Spatter cones and lapilli mounds, Craters of the Moon National Monument. (Photo by C. L. Heald.)

the sides of the pressure-ridges are steep, reminding one of the gable of a house or the cross-section of a broken anticline. They have a medial crack running along the crest of the ridge which may be as much as 15 feet in width but which is usually less. They are almost without exception lined parallel with the flow and they are in general close to its margin." That they are the result of lateral pressure is agreed, but it is not certain what the source of pressure is. Russell explained similar forms in the Snake River region as the result of lateral pressure originating from viscous drag of slowly moving subcrustal lava. Nichols believed that those of the McCartys flow could better be explained as corrugations resulting from the collapse of an originally domed lava surface.

Lava blisters are produced by the explosive effect of steam generated below a lava surface. *Spatter cones* or *scoriae mounds* are small mounds which form where gas or fire fountaining has thrown out liquid clots of lava. They sometimes attain heights sufficient to be called volcanic cones. Cotton (1944) cited Izalco, in San Salvador, which has a height of 3000 feet, as an example of a cone largely built in this manner. *Lava tunnels* and *lava caves,* which are common in basaltic lava fields, result from the draining out of lava from beneath a solidified crust. The so-called ice caves in Oregon and Idaho are lava caves in which ice formed during winter persists because the warm air of summer is unable to displace cold air.

TYPES OF VOLCANOES

Following Cotton (1944), we may recognize four types of volcanoes— *basalt cones, basalt domes* or *shield volcanoes, ash* or *cinder cones,* and *composite* or *strato-volcanoes.* This classification is based largely upon whether the volcanic pile was built up as the result of the outpouring of fluid or effusive lavas, as the product chiefly of ejected pyroclastic materials, or by a combination of the two. If this classification were applied rigorously most volcanoes would be classed as composite or strato-volcanoes, since there are few volcanoes where eruptions are continuously of one type. The dominant nature of the building of the volcanic pile must be considered in classifying volcanoes.

A central vent is the essential feature of a volcano, and the cone or dome associated with it is incidental as far as the history of the volcano is concerned. The shape and profile of the accumulated debris around the central vent are to a large degree influenced by the type of eruption. The angle of repose of the fluid or solid materials which are emitted from the central vent largely determine the steepness of a volcano's slopes.

Basalt cones are rare and are likely to be low rather than high cones because of the fluidity of basaltic lava. Cotton (1944) cited Rangitoto in New Zealand and Skjaldbreit in Iceland as examples.

The Hawaiian volcanoes are excellent examples of what have been called basalt domes or shield volcanoes, as are Mt. Etna and many of the volcanoes of Iceland. Shield volcanoes form where fluid basaltic lava is extruded, and, although they may attain great height (Mauna Loa has an altitude of 13,680 feet), they have such broad bases that they are not best described as cones. The great pile of volcanic material which rises 30,000 feet above the ocean floor to form the Hawaiian Islands is a complex of volcanic shields atop one another with Mauna Loa as the latest to form. Flank outflow of lava through radial fissures is typical of basalt domes or shield volcanoes, although it seems likely that in their early stages of development more of the outflow was from central orifices.

Ash or cinder cones are built where eruptions are of the explosive type with a predominance of pyroclastic materials. Growth of an ash or cinder cone begins around a crater with an encircling ring of pyroclastic debris con-

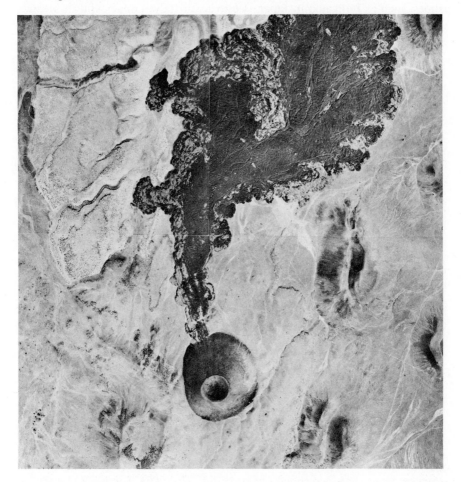

FIG. 19.4. Vertical photo of Sunset Crater, San Francisco Mountain area, Arizona, showing a cinder cone and basaltic lava flow extending from it. (Soil Conservation Service photo.)

sisting of ash, lapilli, and coarser materials. This is called a *tuff ring,* particularly when composed largely of the finer-sized materials. True ash or cinder cones seldom attain heights in excess of a few thousand feet.

A strato-volcano exhibits rough stratification produced by alternating sheets of lava and pyroclastic material. Its structure attests to alternating periods of explosive and quiet eruption. Lava intruded into fissures solidifies to form *dikes;* if injected between layers of fragmental ejecta, it forms *sills.* Individual lava flows from the crater or flank of a cone may form tongue-like extensions down the cone called *coulees.* Most of the world's larger volcanoes, such as Fujiyama in Japan, Vesuvius in Italy, Popocateptl in Mexico, Shasta in the United States, Cotopaxi in Ecuador, and Mayon in the Philippines are composite or strato-volcanoes.

FIG. 19.5.   Koko Head tuff ring, Island of Oahu, Hawaii.   (Photo by Hawaii Visitor's Bureau.)

OTHER FEATURES ASSOCIATED WITH VOLCANOES

Ash showers.   One of the chief products of Vulcanian and Peléan types of eruption is volcanic ash.   Ash showers within historic time have covered extensive areas with enormous quantities of ash.   The Tarawera shower, in New Zealand, in 1886, spread pumiceous lapilli over a 30-mile radius and beyond that, ash.   An explosion of Tamboro, in the East Indies, is estimated to have ejected some 150 cubic kilometers of rock debris.   Probably the most famous explosion of all was that of Krakatau, in 1883, when nearly 5 cubic miles of ejecta were produced, two-thirds of which fell within a radius of 9 miles.   The ejected materials accumulated to a maximum thickness of 200 feet.   It was estimated that as much as 95 per cent of the ejecta was formed from new magma and only 5 per cent from rock torn from the cone.   In 1912, an eruption of Katmai, in Alaska, (Fenner, 1920) blew out some 5 cubic miles of material. Fortunately, most great ash showers have fallen on sparsely populated areas with little loss of life, although the Tamboro shower in the East Indies is estimated to have cost a hundred thousand lives. Cotton (1944) estimated that the soils and subsoils of more than one-third of the North Island of New Zealand were derived from ash-shower materials. Ash showers have much the same effect upon preexisting topography as does loess. The material forms a mantle through which major, but no minor, relief forms are reflected. Where the ejected materials are permeable, surface streams are few, but fine ash may be impermeable and produce a fineness of drainage texture comparable to that of badlands.

Ash showers are commonly cold, but the nuées ardentes which formed during the eruption of Mt. Pelée in 1902 were hot enough to melt glass, which has a melting point between 650 to 700 degrees centigrade. The eruption of Katmai, in Alaska, in 1912, produced an ash and sand flow of nuée ardente origin which formed sandy tuff 100 or more feet thick over 53 square miles. The flow was first interpreted as a hot mud flow but later its nuée ardente origin was recognized. Cotton (1944) has described an area on North Island in New Zealand, where rock called ignimbrite is found, and suggested a similar origin for it. Ignimbrite is similar to what has been called welded tuff, and Cotton suggested that some of the "flow rhyolites" of Yellowstone Park, the Bishop "tuff" of southeastern California, and other similar rocks may have resulted from the agglutination of nuée ardente deposits.

Volcanic mudflows or lahars. Mudflows often accompany volcanic eruptions. To avoid confusion of them with those of semiarid regions, Cotton (1944) suggested that volcanic mudflows be designated *lahars,* a name of Javan origin. Lahars may be caused by: (1) heavy rain on nuée ardente deposits; (2) mingling of nuées ardentes with river waters; (3) emptying of lakes in volcanic craters onto ash; or (4) discharge onto ash of snow and ice melted during an eruption. Scrivenor (1929) has described a lahar from the volcano Gunong Keloet, in Java, which is 38 kilometers long and 4 kilometers wide. The material of the lahar resembles unstratified glacial deposits and even contains striated boulders. A lahar that accompanied an eruption of the volcano Galunggung, in Java, in 1822, spread over 114 villages, with a resulting loss of 4011 lives. A lahar commonly has a hummocky surface resembling that of landslide jumbles.

Plug domes. Some andesitic and rhyolitic lavas are so viscous when extruded that they do not flow or spread but form cylindrical or cumulo-form plugs in the crater of a volcano or cause doming of the rock above them. This form of extrusion is most commonly called a *plug dome,* but other names have been applied to it. Those whose tops spread in mushroom-like form have been called *cumulo-domes* or *tholoids,* with plug dome being restricted to the more cylinder-like protrusions. Care should be taken not to confuse a

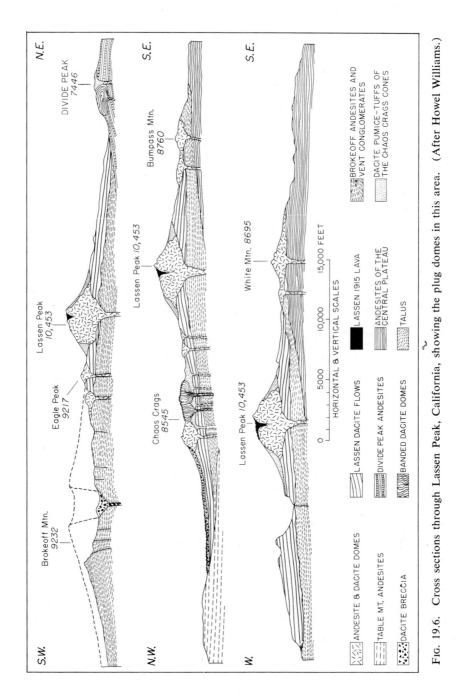

Fig. 19.6. Cross sections through Lassen Peak, California, showing the plug domes in this area. (After Howel Williams.)

plug dome with the volcanic plug formed when lava in the conduit of a volcano solidifies and which after extinction of the volcano may be exposed by erosion as a volcanic neck (see p. 510). A plug dome is formed by the active protrusion of lava in and above the crater of the volcano, not by erosion.

Plug domes are fairly common and often are striking topographic features. Lassen Peak, strictly speaking, is not a volcanic cone but a plug dome ele-

FIG. 19.7. The plug dome, which is Lassen Peak, viewed from Lake Helen. (Photo by National Park Service.)

vated in the crater of a greatly shattered former volcano known as Mt. Brokeoff (Williams, 1932*a*). It rises about 2500 feet above the original crater and has a diameter of 1½ miles. The Mono Craters (Putnam, 1938), at the eastern base of the Sierra Nevada, were formed by eruptions of rhyolitic pumice and lapilli, after which protrusions of obsidian formed plug domes in them. The Bogoslof Islands, in the Gulf of Alaska, are thought to be the summits of large andesitic plug domes. Several *puys* (volcanic hills) of the Auvergne region of France, such as Grand Sarcoui and Puy de Dome, are remnants of extinct plug domes. Most striking of all plug domes was that which formed on Mt. Pelée during its eruption in 1902. The spine that formed atop the plug dome was remarkable for its cylindrical form and great height. It attained a maximum height of 1000 feet and in one day grew 40 feet, but in a few months disintegrated into a heap of shattered rock. The plug dome from which the spine projected continued to grow after destruction of the spine. Within a year and a half it had attained a height of approximately 400 meters and a diameter of 1000 meters, in spite of repeated loss of mass by explosions. One day it increased 25 meters in height. The rapid growth of plug domes is further illustrated by one on Santa Maria volcano in Guatemala (Williams, 1932*b*) which in two years attained a maximum diameter of 1200 meters and a height of 500 meters. It grew in height as rapidly as 100 meters in a week. It had a spine which attained a height of 66 meters. Although some geologists have held that spines consist of rock that solidified in the neck of the volcano and was afterwards pushed up, it is more commonly believed that they are extruded as viscous lava.

DEPRESSION FORMS

The terminology for volcanic depressions is confusing, and any adopted here will be found at variance with other usages. This confusion in terminology is largely a result of varying opinions as to whether certain depression forms were produced by explosion or subsidence. The "caldera problem" is one of the perennial problems of volcanology. The terminology here adopted is somewhat arbitrary but space will not permit discussion of all the varying usages. Williams (1941) and Jaggar (1947) have reviewed them.

The following classification includes the more common type of volcanic depressions. It represents in the main a combination of classifications proposed by Jaggar and Williams.

Types of volcanic depressions
    I. Craters
        Explosion craters
        Craters within constructed rims
        Collapse craters
   II. Calderas
        Explosion calderas
        Subsidence calderas
        Composite calderas
  III. Volcanic-tectonic depressions

There are no completely satisfactory definitions of craters and calderas. Some class as craters all volcanic depressions, regardless of size and also include forms of non-volcanic origin, particularly those produced by meteorites, bombs, and mining. It is difficult to avoid application of the term crater to the two latter features, because it so well describes them. It seems, though, that as far as possible the term crater should be restricted to topographic depressions of volcanic origin. The distinction between craters and calderas must be arbitrary; craters are small, whereas calderas have large dimensions. We may define a *crater* as a bowl or funnel-shaped depression usually of volcanic origin which is more or less circular in plan and is rimmed by an infacing scarp commonly less than a mile in diameter. A *caldera* is a large volcanic depression more or less circular in plan with a diameter usually several times that of a crater.

Volcanic craters. If volcanic craters are associated with active volcanoes, their origin is clear; they may, however, mark sites of former volcanic activity. They may result from either explosive activity or from subsidence. Expulsion of volcanic ash, lapilli, and other forms of ejecta may build a ring about a volcanic vent and produce a crater. Cotton (1944) called craters of such origin *ubehebes*. The name comes from the Ubehebe Craters at the north end of Death Valley (Von Engeln, 1932), which mark the sites of several embryonic volcanoes. In the Eifel district of Germany, there are numerous lakes called *maare* or *maars* which occupy craters that seem to have been formed by volcanic activity that ended without the building of cones.

On the Hawaiian basalt domes or shield volcanoes are numerous small depressions called *pit craters* or *volcanic sinks,* which have resulted mainly from collapse following lowering of magma columns. Such large summit features as the Mokuaweoweo depression on Mauno Loa are better classed as calderas than pit craters, although it should be recognized that large depressions may originate through the intergrowth of several pit craters or volcanic sinks and thus make it difficult to draw a sharp line between pit craters and calderas. Sometimes a crater forms within another to form what are called *nested craters*. Nested craters may develop in two ways: explosion may enlarge an existing crater, after which another cone and crater are built within the older one; or, successive subsidences of parts of a crater may produce the nested form.

Calderas. Calderas typically are several miles across, as is indicated by the dimensions of the following calderas:

Widths of Calderas, in Miles

Mount Katmai, Alaska	3
Mokuaweoweo, Hawaii	3½
Krakatau, East Indies	4
Crater Lake, Oregon	5½
Aniakchak, Alaska	6½
Aira, Japan	15
Valles, New Mexico	18

Caldera has been applied by some, particularly in Germany, to any enclosed or walled basin, regardless of origin. It would simplify the problem of the origin of calderas if we restricted the name, as far as possible, to large circular basins of volcanic origin. Daly (1933) suggested that caldera be restricted to large basins produced by volcanic explosions and that basins formed by subsidence be called volcanic sinks, but caldera is still generally used without specific implication as to the origin of the depression. There seem to be three main types: (1) the explosion caldera, (2) the subsidence caldera, and (3) the caldera produced by the combined effects of explosion and subsidence. Stearns (1942) has recognized a fourth class, the erosional caldera, but it seems illogical to group it with forms produced by volcanic activity.

Formerly, the preponderance of opinion was that calderas originated from explosive paroxysms which accompany Vulcanian and Peléan types of eruptions, but today a collapse origin is more favored. Williams (1941), following such men as Van Bemmelen and Van den Bosch, has strongly advocated that collapse is more significant, although his theory of caldera formation may better be described as a composite theory, or the "explosion-collapse" theory as Cotton has called it. Williams recognized the following stages in caldera development, as previously outlined by Van Bemmelen:

1. Mild explosions of pumice. Magma stands high in the conduits.

2. Explosions increase in violence. Magma level falls into main chamber.

3. Culminating explosions. Part of the ejecta is hurled high above the cone, but most of it rushes down the flanks as nuées ardentes. Magma level is deep in the chamber. Roof begins to crack.

4. Lacking support, the top of the cone collapses into the magma chamber.

5. After an interval of quiescence and erosion, new cones appear on the caldera floor, especially near its rim.

The strongest arguments against an explosive origin of calderas are: first, around most calderas there is insufficient pyroclastic material to account for them, and, secondly, most of the ejecta consists of new lava rather than fragments of older rocks. Williams recognized that some calderas were formed by explosive action alone, but he claimed that this type is rare and usually small. Cited as examples were the calderas of Tarawera on the North Island of New Zealand, of Bandaisan on the main island of Japan, and the caldera at Chaos Crags on the north side of Lassen Peak, California. It is commonly stated that the great caldera of Krakatau was produced by the violent explosion which took place in 1883. Williams favored the view expressed by Verbeek and Stehn that it was as much the result of collapse which followed violent explosion as it was of removal of material by explosion. This view was held by Dana as early as 1890 and seems to be favored by most modern observers. In 1927, a basaltic cinder cone, Anak Krakatau, appeared on the floor of the caldera. This event appears to correlate with stage 5 in caldera development as outlined above.

The caldera on Mount Katmai has also been ascribed to an explosion there in 1912, but Fenner (1920) and Williams (1941) concluded that there was insufficient pyroclastic material around it to account for the volume of material removed to form it and that the scarcity in the pyroclastic ejecta of andesitic rock similar to that in the volcanic cone suggested that the pyroclastic material was derived from new magma rather than by fragmentation of old rock. The ejecta are mainly rhyolitic pumice or lapilli of probable nuées ardentes origin. Subsidence, along with explosion, seems to be necessary to account for the size of the caldera.

The caldera occupied by Crater Lake is about 5½ miles wide and is nearly circular in outline. Crater Lake itself covers over 20 square miles and is as much as 2000 feet deep. The encircling caldera rim rises from 500 to 2000 feet above the lake. An ancestral volcanic cone, whose summit has disappeared, has been called Mount Mazama. It was built during the Pleistocene and had numerous ice streams extending down its slopes, as is indicated by the presence at numerous places in the caldera wall of interbedded volcanic ejecta, fluvio-glacial sands and gravels, and glacial till. Diller, in 1902, attributed this caldera to subsidence because of the limited amount of pyroclastic debris around it. He believed that 17 cubic miles of material had been removed, whereas no more than 7 cubic miles could be accounted for in the ejecta and lavas around its rim. As at Katmai, nuées ardentes seem to have spread much of the material around the flanks of the cone.

An explosive origin was favored by Smith and Swartzlow (1936) for the following reasons:

1. The shape of the caldera is similar to that formed by known explosion craters, such as those produced by mines, and is strikingly different from that of known subsidence calderas as on Kilauea.

2. Much molten material may have issued from the vent, which is now buried beneath ejecta.

3. The distribution, character, and quantity of material around Crater Lake favors an explosive origin.

4. Other volcanoes which have partly destroyed themselves have done so by explosions.

5. The discrepancy between the amount of material necessary to fill the caldera and that present around Mt. Mazama may be accounted for by removal by wind and glaciation.

6. The apparent absence of coarser ejecta results from its being covered by pumice.

7. The absence of peripheral faults as at Kilauea and the growth of Wizard Island in Crater Lake do not harmonize with the idea of collapse.

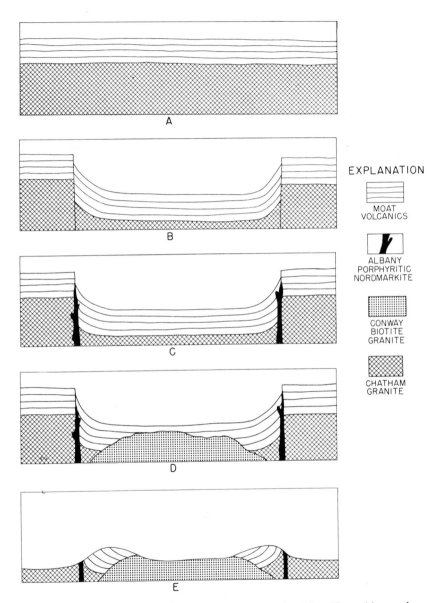

EXPLANATION

MOAT
VOLCANICS

ALBANY
PORPHYRITIC
NORDMARKITE

CONWAY
BIOTITE
GRANITE

CHATHAM
GRANITE

FIG. 19.8.   Stages in the genesis of the Ossipee Mountains, New Hampshire, and associated ring-dike.   (After Louise Kingsley.)

Williams (1941, 1942) has maintained that there is insufficient ejecta to account for the caldera and that therefore subsidence must have followed the

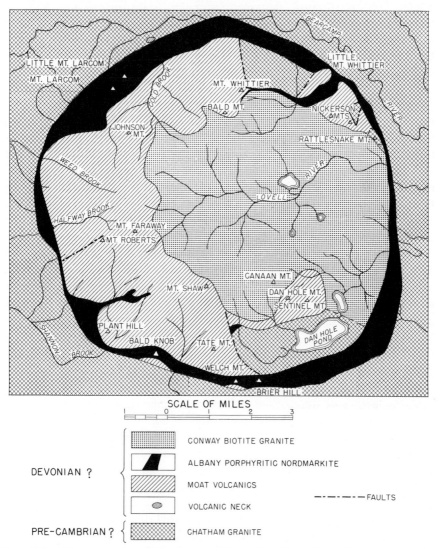

SCALE OF MILES

DEVONIAN ? {
- CONWAY BIOTITE GRANITE
- ALBANY PORPHYRITIC NORDMARKITE
- MOAT VOLCANICS
- VOLCANIC NECK

------- FAULTS

PRE-CAMBRIAN ? { CHATHAM GRANITE

FIG. 19.9. Geologic map of the Ossipee Mountains, New Hampshire, showing the nordmarkite ring-dike in these mountains. (After Louise Kingsley.)

terrific explosions with their accompanying nuées ardentes. Following a quiet period, the growth of Wizard Island took place through the collapsed caldera floor. This origin seems to be the more favored one at the present time.

That some calderas can be formed by subsidence without preceding or accompanying explosions seems certain. The large summit depressions on Mauna Loa and Kilauea, described as craters, sinks, or calderas, were so formed. Collapse followed lowering of the magma column either through withdrawal into the magma chamber or through intrusion of dikes at depth. Fault scarps are commonly associated with them and piecemeal caving in of their walls has contributed to their enlargement.

A special type of subsidence is that which has been called *cauldron-sub-sidence*. This was recognized originally by Suess and has been offered as an explanation of the igneous complex at Glen Coe, Scotland (Clough, Maufe, and Bailey, 1909). It involves a nearly circular block which subsides into a magma chamber and causes magma to well up along nearly vertical fractures to form what is called a *ring-dike*. Eroded ring-dike structures associated with cauldron-subsidence have been described in New Hampshire in the Ossipee Mountains (Kingsley, 1931), the Belknap Mountains (Modell, 1936), and in the Percy area in the White Mountains (Chapman, 1935). In the Ossipee Mountains there is a vertical ring-dike of nordmarkite 9 miles in diameter which encircles a block that is believed to have subsided at least 4500 feet and perhaps as much as 8000 feet. This ring-dike and associated volcanic rocks encircle a central lowland on the Conway granite in the zone of central subsidence. A total of 36 ring-dikes in New Hampshire have been described (Billings, 1945). Ring-dikes seldom encompass as much as 360 degrees. Those in New Hampshire average about one-half this. They vary in diameter from eight-tenths of a mile to over 9 miles. The Askja caldera, in Iceland, which is about 4½ miles across and has walls 1300 feet high, has been considered by some to be a result of cauldron-subsidence.

Volcanic-tectonic depressions. Certain volcanic depressions, presumably produced by collapse, lack the symmetry of craters and calderas. These have been called *volcanic-tectonic depressions*. Many deep trench-like depressions on the slopes of volcanic cones have been described as *sector grabens* and are thought to be the result of downfaulting of blocks, following withdrawal of magma at depth. Daly (1933) designated as *volcanic rents* those which seemed to be more the result of lateral than vertical movement. A large area on the North Island of New Zealand, including the Lake Taupo Basin, has been described by Williams (1941) as one of large-scale volcanic-tectonic collapse. Cotton (1944), however, was more inclined to ascribe the basin to the tectonic compression which he thought was responsible for many of the topographic forms in New Zealand, although he recognized the prob-able existence of volcanic-tectonic depressions. Williams (1941) considered depressions in Sumatra, originally described by van Bemmelen, to have formed by large-scale collapse, following eruptions of masses of rhyolitic pumice. The largest of these depressions is occupied by Lake Toba.

VOLCANIC PLATEAUS AND PLAINS

Many extensive *lava plateaus* have been built by great outpourings of lava. Some, as in the Snake River region of Idaho, have flat enough surfaces to be called *lava plains*. Examples of lava plateaus are the Columbia River Plateau of eastern Washington, Oregon, Nevada, and Idaho, the Deccan Plateau of west-central India, the Yellowstone Plateau of Wyoming, the Drakenberg Plateau of South Africa, the Paraná Plateau in southern Brazil, Uruguay, and Argentina, and the Ignimbrite Plateau in the central part of North Island, New Zealand. Most lava plateaus and plains are composed of basaltic lavas, but in the Yellowstone and Ignimbrite plateaus flow rhyolites and welded tuffs are abundant.

The Columbia Plateau represents one of the most voluminous outpourings of lava in geologic history. Basalt flows cover more than 100,000 square miles and have an average total thickness of 3000 to 4000 feet, with individual flows as much as 200 feet in thickness. The Snake River has cut a canyon into the basalts which locally is deeper than the Grand Canyon of the Colorado (Freeman, 1938). Along one stretch of 40 miles, its canyon averages 5500 feet in depth, and locally the lava flows total 5000 to 6000 feet in thickness. In places, the Snake River has cut as much as 1000 feet into the granite beneath the basalts. The Snake River lava plain of southern Idaho is geographically continuous with the Columbia Plateau but is usually considered a separate physiographic area because it is less dissected and is underlain by lavas of Pliocene and Pleistocene ages in contrast to the Miocene lavas of the Columbia Plateau.

In places, granite hills, called *steptoes* from Steptoe Butte north of Colfax, Washington, project through the lavas. Numerous explosion cones give diversity to what is otherwise a gently rolling surface. The Craters of the Moon (Stearns, 1924), in the Idaho section of the Plateau, were produced by recent fissure eruptions. Numerous crater pits, spatter cones, and ash and cinder cones mark the rift or fissure through which eruptions took place. The Thousand Springs along the Snake River Canyon (Stearns, 1936) include eleven of the largest springs in the United States. These springs discharge from porous pillow lavas in the basalt flows. Some issue from box-headed canyons, known as *spring alcoves,* which were formed by solutional spring sapping of the basalt.

Lava plateaus and plains are believed to have been formed by extrusions through fissures, such as the Laki fissure in Iceland and the "great rift" in the

Craters of the Moon area.   It seems that extrusion of the lavas took place
quietly, because rarely is pyroclastic material found interbedded with the
basalts.   The importance of fissures as lines of eruption is shown by the frequent alignment of spatter and cinder cones.   The volcanoes along the African
rift valleys and the Rhine graben, as well as the *puys* of central France, display
a similar linear arrangement.   When the great volume of lava involved in the

Fig. 19.10.   Steptoe Butte, north of Colfax, Washington, a quartzite hill that projects
above the lava flows of the Columbia Plateau.   (Official photograph U. S. Navy.)

building of a lava plateau and the remarkable uniformity in thickness of the
individual lava sheets over wide areas are considered, it seems that the lavas
must have been extruded in an extremely fluid condition from hundreds, if not
thousands, of fissures.

Such great regional outpourings of lava have been designated as *areal
eruptions*.   Daly (1933) offered the highly speculative theory that areal
eruptions result from the foundering of the roof of a batholith with resulting
upwelling of magma around its periphery.   Another theory is that of Reck,
who believed that lava plateaus and plains were formed from outpourings
through clusters of central vents rather than through many fissures.   Although
some of the lavas may have come from central vents, such extrusions do not

appear to have been numerous enough to account for the great quantities of lavas. Stearns and Clark (1930) have stressed the similarity between the origin of basalt domes or shield volcanoes and that of lava plateaus. Both are characterized by fissure eruptions and basaltic lavas. According to their ideas, whether a volcanic dome or lava plateau is produced depends upon the permanence of the fissure openings. If fissures are used continuously, the result is a volcanic dome; if new fissure openings repeatedly develop, a lava plateau or plain is formed. The great areal extent of some plateau basalts seems to require a multiplicity of openings not commonly associated with shield volcanoes, nor do the forms of basalt plateaus and plains suggest that they were built up as a succession of lava domes atop one another.

VOLCANIC SKELETONS

With cessation of volcanic activity erosion becomes dominant and dissection of a volcanic cone or dome ensues. A symmetrical, conical profile indi-

Fig. 19.11. Shiprock, New Mexico, a volcanic neck with radiating dikes. (Photo by Fairchild Aerial Surveys, Inc.)

cates that volcanic activity has only recently ended. Cones containing much coarse pyroclastic material may be relatively resistant to erosion because of their permeability. Radial drainage is characteristic of volcanic cones.

When streams have cut deep furrows in the slopes of the cones a mature stage of dissection has been reached.

Volcanic necks and *dike ridges* may be all that is left of former volcanic cones and domes in old age. They are preserved because of more resistant rock in the conduit of the volcano and the many fissures extending upward

Fɪɢ. 19.12. Devil's Tower, Wyoming. Columnar jointing is strikingly developed in what is probably a volcanic neck or small plug.

through the volcanic pile. They may persist as topographic features long after the volcanic form is obliterated. Familiar examples of volcanic necks are Shiprock in New Mexico and the oft-pictured Rocher Saint Michel at Le Puy, France. The Devil's Tower in Wyoming also is probably a volcanic neck. Dike ridges exist in amazing complexity in the Spanish Peaks region of southern Colorado and in the Crazy Mountains of Montana. Scores of volcanic necks and dikes are found throughout the high plateau country of Arizona, New Mexico, and Utah (Williams, 1936).

Dike ridges are sometimes called hogbacks, but this term should be restricted to sharp ridges upon steeply dipping sedimentary rocks. Dike ridges are likely to be more firmly rooted in position than hogbacks because of the steeper dips of the rocks responsible for them. Not all dikes give rise to

Fig. 19.13. Map of the major dikes and intrusive sheets of the Spanish Peaks area, Colorado. (After R. C. Hills, *U. S. Geol. Survey, Folio 71.*)

ridges. If the dike rock is less resistant to weathering and erosion than the surrounding rock, a topographic depression will mark its position. The famous Giant's Causeway in Ireland is a case in point. Dikes associated with large intrusive masses may have amazing size. The so-called Great Dike of Rhodesia, Africa, is some 37 miles long and varies in width from 2 to over 7 miles. Lightfoot (1940) concluded, however, that rather than being a dike it probably represented the filling of the space produced by downsinking of a block between two subparallel dikes. The Cleveland dike of northern England is over 100 miles long.

F IG. 19.14.   Intersecting dikes of the Spanish Peaks area, Colorado.   (Photo by J. S. Shelton and R. C. Frampton.)

REFERENCES CITED IN TEXT

Billings, M. P. (1945).   Mechanics of igneous intrusion in New Hampshire, *Am. J. Sci., 243-A*, pp. 40–68.

Chapman, R. W. (1935).   Percy ring-dike complex, *Am. J. Sci., 230*, pp. 401–431.

Clough, C. T., H. B. Maufe, and E. B. Bailey (1909).   The cauldron-subsidence of Glen Coe and the associated igneous phenomena, *Quart. J. Geol. Soc. London, 65*, pp. 611–678.

Cotton, C. A. (1944).   *Volcanoes as Landscape Forms,* Whitcombe and Tombs, Ltd., Wellington, 416 pp.

Daly, R. A. (1914).   *Igneous Rocks and their Origin,* pp. 133–134, McGraw-Hill Book Co., New York.

Daly, R. A. (1933).   *Igneous Rocks and the Depths of the Earth,* pp. 141–147 and 159–172, McGraw-Hill Book Co., New York.

Diller, J. S., and H. B. Patton (1902).   The geology and petrography of Crater Lake National Park, *U. S. Geol. Survey, Profess. Paper 3*, pp. 46–50.

Fenner, C. N. (1920).   The Katmai region, Alaska, and the great eruption of 1912, *J. Geol., 28*, pp. 569–606.

Freeman, O. W. (1938).   The Snake River Canyon, *Geog. Rev., 28*, pp. 597–608.

Jaggar, T. A. (1920).   Seismological investigation of the Hawaiian lava column, *Bull. Seismological Soc. Am., 10*, p. 162.

Jaggar, T. A. (1947).   Origin and development of craters, *Geol. Soc. Am., Mem. 21*, pp. 337–407.

REFERENCES

Kingsley, Louise (1931). Cauldron-subsidence of the Ossipee Mountains, *Am. J. Sci., 222,* pp. 139–168.

Lacroix, A. (1908). La montagne Pelée après ses éruptions, *Acad. Sci. Paris,* pp. 74–93.

Lightfoot, B. (1940). The great dike of southern Rhodesia, *Trans. Proc. Geol. Soc. S. Africa, 43,* pp. xxvii–xliii.

Macdonald, G. A. (1943). The 1942 eruption of Mauna Loa, Hawaii, *Am. J. Sci., 241,* pp. 241–256.

Modell, David (1936). Ring-dike complex of the Belknap Mountains, New Hampshire, *Geol. Soc. Am., Bull. 47,* pp. 1885–1932.

Nichols, R. L. (1939). Pressure-ridges and collapse-depressions on the McCartys basalt flow, New Mexico, *Trans. Am. Geophys. Union, 20,* pp. 432–433.

Nichols, R. L. (1946). McCartys basalt flow, Valencia County, New Mexico, *Geol. Soc. Am., Bull. 57,* pp. 1049–1086.

Putnam, W. C. (1938). The Mono Craters, California, *Geog. Rev., 28,* pp. 68–82.

Scrivenor, J. B. (1929). The mudstreams ("Lahars") of Gunong Keloet in Java, *Geol. Mag., 66,* pp. 433–434.

Smith, W. D., and C. R. Swartzlow (1936). Mount Mazama: explosion versus collapse, *Geol. Soc. Am., Bull. 47,* pp. 1809–1830.

Stearns, H. T. (1924). Craters of the Moon National Monument, *Geog. Rev., 14,* pp. 363–372.

Stearns, H. T. (1936). Origin of the large springs and their alcoves along the Snake River in southern Idaho, *J. Geol., 44,* pp. 429–450.

Stearns, H. T. (1942). Origin of Haleakala Crater, Island of Maui, Hawaii, *Geol. Soc. Am., Bull. 53,* pp. 1–14.

Stearns, H. T., and W. O. Clark (1930). Geology and water resources of the Kau district, Hawaii, *U. S. Geol. Survey, Water Supply Papers, 616,* pp. 138–140.

Von Engeln, O. D. (1932). The Ubehebe craters and explosion breccias in Death Valley, *J. Geol., 40,* pp. 726–734.

Williams, Howel (1932a). Geology of the Lassen Volcanic National Park, California, *Univ. Calif. Publs., Bull. Dept. Geol. Sci., 21,* pp. 190–385.

Williams, Howel (1932b). The history and characteristics of volcanic domes, *Univ. Calif. Publs., Bull. Dept. Geol. Sci., 21,* pp. 51–146.

Williams, Howel (1936). Pliocene volcanoes of the Navajo-Hopi country, *Geol. Soc. Am., Bull. 47,* pp. 111–172.

Williams, Howel (1941). Calderas and their origin, *Univ. Calif. Publs., Bull. Dept. Geol. Sci., 25,* pp. 235–346.

Williams, Howel (1942). The geology of Crater Lake National Park, Oregon, with a reconnaissance of the Cascade Range southwest to Mount Shasta, *Carnegie Inst. Wash. Publ., 540,* pp. 162.

Additional References

Atwood, W. W., Jr. (1935). The glacial history of an extinct volcano, Crater Lake National Park, *J. Geol., 43,* pp. 142–168.

Baker, C. L. (1923). The lava field of the Paraná Basin, South America, *J. Geol., 31,* pp. 66–79.

Coleman, Alice (1952). Selenomorphology, *J. Geol., 60,* pp. 451–460.

Cotton, C. A. (1941). Some volcanic landforms in New Zealand, *J. Geomorph., 4,* pp. 297–306.

Johnson, D. W. (1909).   Volcanic necks of the Mount Taylor region, New Mexico, *J. Geol., 18,* pp. 303–324.

Nichols, R. L. (1938).   Grooved lavas, *J. Geol., 46,* pp. 601–614.

Smith, W. R. (1925).   Aniakchak crater, Alaska peninsula, *U. S. Geol. Survey, Profess. Paper 132,* pp. 139–145.

Stearns, H. T. (1940).   Four-phase volcanism in Hawaii, *Geol. Soc. Am., Bull. 51,* pp. 1947–1948.

Williams, Howel (1941).   Volcanology, *Geol. Soc. Am., 50th Anniv. Vol.,* pp. 365–390.

20 · Pseudovolcanic Features

Certain topographic features resemble volcanic forms so much that, lacking a better name, we have designated them as *pseudovolcanic*.

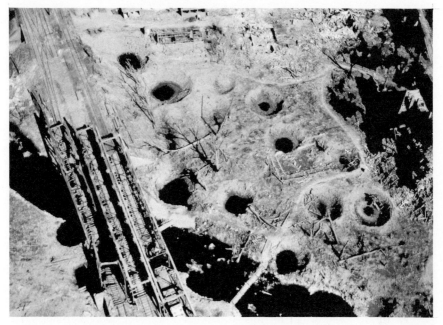

FIG. 20.1. Bomb craters between Herborn and Dillenburg, Germany, which resulted from bombing by 9th Air Force during World War II. (Air Force photo.)

Bomb and mine craters. Craters formed by bomb and mine blasts have several of the characteristics of volcanic explosion craters, including the encircling rim of ejected material.

Meteorite craters. A meteorite crater is produced by the impact and accompanying explosion of an object of extraterrestrial origin and is thus one of the most unusual land forms. For many years there was considerable skepticism as to the existence of meteorite craters but this can no longer be

515

questioned. The most famous meteorite crater is Meteor Crater, in Arizona.
Its meteoritic origin is now generally accepted, despite Darton's (1945) contention that it is more likely that it was produced by a gas explosion. It first
came to the public's attention with the discovery in 1891 in the adjacent
area of pieces of what has been called the Canon Diablo meteorite. Meteor
Crater is a rimmed basin 3950 feet wide and 570 feet deep. Its rim rises
130 to 160 feet above the surrounding desert and has in it blocks of limestone

FIG. 20.2.   Meteor Crater, near Winslow, Arizona.   (Photo by Jack Ammann Photogrammetric Engineers.)

weighing as much as 4000 tons. Permian sandstones and limestones form
the crater rim and dip radially outward at angles varying between 10 and 80
degrees.   Fossiliferous lacustrine deposits 70 to 90 feet thick lie at the
bottom of the crater. From this and other lines of evidence, Blackwelder
(1932) concluded that probably the crater was formed during late Wisconsin
time (possibly during the next to the last glacial subage). Two unsuccessful
searches have been made for iron believed to be buried below the bottom of
the crater, one of which involved an expenditure of $293,000. Failure to
find iron resulted in doubt as to its meteoritic origin until it was realized that
the evidence indicated that the meteorite had exploded rather than buried
itself.

Several other craters that seem to have been produced by meteorites have
been described (Spencer, 1933). The Henbury Craters in central Australia

consist of 13 craters within an area one-half mile square. The largest crater is 660 by 360 feet across and varies in depth from 3 to 25 feet. Bits of meteoritic iron have been found around them. Rims are not conspicuous, but their inner walls show powdered and shattered blocks of sandstone, quartzite, and slate of Ordovician age. Fused silica glass was found around the largest of the craters. The Wabar craters, which were discovered by Philby (1933) in the Rub 'al Khali in southern Arabia, consist of two distinct craters; there are suggestions of two others buried beneath sand. They have gentle outer slopes with steep inner slopes and are partly filled with drifted sand. Masses of silica glass are abundant in their rims, indicating that the meteorite created such intense heat as to fuse the wind-blown sand with which it came in contact. Pieces of meteoritic iron were also found adjacent to the craters. The Wolf Creek crater (Guppy and Matheson, 1950) in the Kimberly district of western Australia, is one of the more recently described craters. It is circular in form, is 2800 feet across, and has a maximum depth of 170 feet. Quartzites in the rim wall have outward dips up to 60 degrees, and shattered rock is much in evidence. Fragments of meteoritic iron have been found near it.

Another reported meteorite crater is Chubb Crater in Quebec, west of Ungava Bay (Meen, 1950); not enough information is available to conclude with certainty that it is a meteorite crater. This crater is about 10,000 feet in diameter, is in granite-gneiss, and is occupied by a lake. Its enclosing rim rises 300 to 500 feet above the surrounding plain, and in it are three sets of recognizable fractures which have been attributed to explosion. There is no sign of volcanic activity in the area. In many respects the description of Chubb Crater compares with that of the Ashanti or Lake Bosumtwi depression (Maclaren, 1931) in tropical West Africa, which is 6½ miles in diameter, is in Pre-Cambrian phyllites, and has been variously interpreted as a meteorite crater, a cryptovolcanic caldera, a result of faulting, and the product of gas explosion. Other probable meteorite craters are the Campo de Cielo craters in the Gran Chaco region of Argentina, the Odessa Crater in Ector County, Texas (Sellards, 1927), and a group of small craters in northern Siberia. Numerous other alleged meteorite craters have been described, but the evidence for this origin is not convincing.

We may well ask what the characteristics are that distinguish meteorite craters from similar depressions of other origins. We should recognize that meteorite craters are not mere dents in the ground made by percussion. The upward force resulting from explosion and probable partial vaporization of a meteorite upon striking the earth was more important in forming a crater than the downward percussive force. Four things in particular seem to point toward a meteoritic origin. The presence of meteoritic iron adjacent to a crater is strongly presumptive evidence of a meteoritic origin. Iron is rarely found in the crater for it is usually thrown some distance by the explosion. An abundance of silica glass around or adjacent to the rim so distributed and in such amount as not to be interpreted as fulgurites is further suggestive of meteoritic origin. This evidence can exist only where sands or sandy rocks are present. Thirdly, there should be shattered and pulverized rock in and about the enclosing rim. Particularly significant is evidence of disturbance of rim rock as shown by outward-dipping strata. This last feature, which should be evident before a meteoritic origin is assumed, may not be diagnostic of meteoritic impact because it can be produced in other ways, particularly by cryptovolcanic activity.

The Carolina "Bays." The term bay has long been applied to elliptical-shaped shallow depressions which are particularly numerous in the coastal plain of the Carolinas but extend from Florida to New Jersey. They were given little attention until Melton and Schriever (1933) suggested that they are scars produced by impact of a shower of meteorites. Since then there has been much argument about their origin. Their striking shapes and arrangements were poorly discernible until the days of aerial photography.

Prouty (1952) has given a detailed description of the bays. Some of the significant facts about them are:

1. The bays are restricted to an area between southern New Jersey and northeastern Florida, but are most notably developed in the two Carolinas.

2. They are very irregular in distribution and size.

3. There is no discernible relationship between the location of the bays and specific geologic formations or topography.

4. Bay groups are less numerous and smaller toward the northwest and southeast extremities of the area in which they are found but are as variable in distribution and size in these areas as in the Carolinas.

5. No bays are found outside the sand-covered coastal plain or its erosional remnants.

6. It was estimated that there are perhaps a half million of them.

7. Most typically they are elongated in a northwest-southeast direction.

8. Sand rims are characteristic and are best developed on the east or southeast sides and are larger around the larger bays.

9. Many bays have multiple rims.

10. The deepest part of a bay is generally found toward its southeast end and west of its axial line.

11. Many bays are partially filled with alluvial or eolian materials.

12. Practically every bay that has been surveyed has a well-defined magnetic high associated with it.

13. Metallic iron, fragments of basement rock, or fused sand have not been found in association with any of the bays.

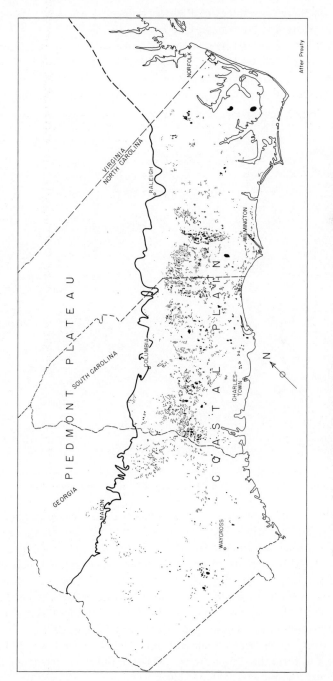

FIG. 20.3. Map showing distribution of the Carolina "Bays." (After W. F. Prouty.)

14. All the bays are on terraces of Pleistocene age.

15. A few bays are known to have lake-bottom springs.

Three hypotheses of origin have contended for priority. A meteoritic origin has been supported by Prouty (1935, 1952) and McCarthy (1937), who maintained that they were the result of "shock waves" associated with a cone of compressed air ahead of the meteorite rather than that they were

FIG. 20.4. Vertical photograph of some of the Carolina "Bays," Horry County, South Carolina. (Production and Marketing Administration photo.)

caused by the impact of meteorites. Cooke (1934, 1943) has contended that the bays are basins of extinct lakes or lagoons that existed on the coastal plain before its complete emergence from beneath the sea. He attributed their elliptical shapes to eddy currents produced by wind and considered their sandy rims to be beach ridges and bars of extinct lakes and lagoons. Johnson (1942) proposed what he called the "artesian-lacustrine-eolian" hypothesis of origin. He believed the bays were initiated by artesian springs rising through the coastal plain sediments. These springs enlarged their outlets by solution and formed basins which became sites of small lakes around which beach ridges were built to form the present rims. Many of the rims increased in size and height through the growth of dune ridges.

It has also been suggested informally that the bays may be sinkholes, blow-outs, or salt domes. Somewhat similar basins are found in other parts of the coastal plain as far south as Texas, although few display the striking symmetry of form and arrangement found in the Carolinas. Black and Barksdale (1949) have described numerous orientated lake basins along the Arctic coastal plain, but these basins seem to lack the sandy rims which characterize the Carolina bays; they are more likely permafrost phenomena.

Salt plugs. Salt plugs may produce topographic forms superficially similar to such volcanic forms as coulees, plug domes, and craters. Harrison (1930) has described some striking salt plugs in southern Persia which have many of the characteristics of plug domes. Salt has been extruded through the crests of anticlines, domes, and the ends of plunging folds. Salt extrusions take the form of salt hills, salt glaciers, salt corries, and salt marshes. Salt hills exhibit many of the features of plug domes, rising as abrupt hills with sinkholes and pinnacles on their summits. The so-called salt glaciers extend down slopes as tongues of salt similar in many respects to coulees produced by lava flows. Salt corries are cirque-like hollows resulting from solution; they somewhat resemble craters or calderas.

It is possible that some cryptovolcanic structures should be classed as pseudovolcanic features since their origin is still uncertain, but the present consensus is that they were produced by deep-seated igneous activity which fell short of the surface. Some alleged cryptovolcanic structures may prove to be meteorite craters.

REFERENCES CITED IN TEXT

Black, R. F., and W. L. Barksdale (1949). Oriented lakes of northern Alaska, *J. Geol., 57,* pp. 105–118.

Blackwelder, Eliot (1932). The age of Meteor Crater, *Science, 76,* pp. 557–560.

Cooke, C. W. (1934). Discussion of the origin of the supposed meteor-meteorite scars of South Carolina, *J. Geol., 42,* pp. 88–96.

Cooke, C. W. (1943). Elliptical bays, *J. Geol., 51,* pp. 419–427.

Darton, N. H. (1945). Crater mound, Arizona, *Geol. Soc. Am., Bull. 56,* p. 1154.

Guppy, D. J., and R. S. Matheson (1950). Wolf Creek meteorite crater, western Australia, *J. Geol., 58,* pp. 30–36.

Harrison, J. V. (1930). The geology of some salt-plugs in Laristan (Southern Persia), *Quart. J. Geol. Soc. London, 86,* pp. 463–522.

Johnson, D. W. (1942). *The Origin of the Carolina Bays,* Columbia University Press, New York, 341 pp.

McCarthy, G. R. (1937). The Carolina Bays, *Geol. Soc. Am., Bull. 48,* pp. 1211–1226.

Maclaren, Malcolm (1931). Lake Bosumtwi, *Geog. J., 78,* pp. 270–276.

Meen, V. B. (1950). Chubb Crater, Ungava, Quebec, *Geol. Soc. Am., Bull. 61,* p. 1485.

Melton, F. A., and William Schriever (1933). The Carolina "Bays"—Are they meteorite scars?, *J. Geol., 41,* pp. 52–66.

Melton, F. A. (1950). The Carolina "Bays," *J. Geol., 58,* pp. 128–134.

Philby, H. Ste. John (1933). Rub 'al Khali: An account of exploration in the Great South Desert of Arabia, *Geog. J., 81,* pp. 1–26.

Prouty, W. F. (1935). The Carolina Bays and elliptical lake basins, *J. Geol., 43,* pp. 200–207.

Prouty, W. F. (1952). Carolina Bays and their origin, *Geol. Soc. Am., Bull. 63,* pp. 167–224.

Sellards, E. H. (1927). Unusual structural feature in the plains region of Texas, *Geol. Soc. Am., Bull. 38,* p. 149.

Spencer, L. J. (1933). Meteorite craters as topographical features on the earth's surface, *Geog. J., 81,* pp. 227–248.

Additional References

Brown, I. C. (1951). Circular structures in the Arctic Islands, *Am. J. Sci., 249,* pp. 785–794.

Nininger, H. H. (1948). Geological significance of meteorites, *Am. J. Sci., 246,* pp. 101–108.

Rohleder, H. P. T. (1933). The Steinheim basin and the Pretoria salt pan, volcanic or meteoritic origin?, *Geol. Mag., 70,* pp. 489–498.

Schriever, William (1951). On the origin of the Carolina Bays, *Trans. Am. Geophys. Union, 32,* pp. 87–95.

21 · Tools of the Geomorphologist

Nothing can entirely replace careful field observations in the study of a geomorphic problem, but numerous aids can add to their effectiveness, reduce the amount of field work for many problems, make possible a more effective planning of the field program, and add support to the conclusions drawn. These aids are what may be called the "tools" of the geomorphologist, although their use is by no means restricted to him. Topographic maps, geologic maps, block diagrams, aerial photographs, soil maps, and climatic data are the most commonly used tools in geomorphic studies.

TOPOGRAPHIC MAPS

Unless it be the aerial photograph, there is no single tool that has as much usefulness to a geomorphologist as a topographic map. To a large degree topographic maps and aerial photographs are complementary to each other; for some purposes one may be more useful, and for other purposes the other. The great advantage of a topographic map is that, besides being three-dimensional, it gives quantitative information both as to form and altitude. It usually shows regional relationships better than any other method of representation.

It is important in using a topographic map that we early become aware of its quality. Many of the older maps represent topography poorly. They were made when the drawing of contours was a matter of personal judgment and interpretation and when topographic sketching was an art rather than a science. The application of aerial photography to topographic map making has removed a great deal of the personal equation from the making of contour maps. Comparison of old maps with new ones of the same areas made with the aid of aerial photographs brings out strikingly the increased accuracy of topographic mapping since about 1925. Many maps made prior to that date are of doubtful value in any problem that involves interpretation of topographic details.

Map scale and its significance. To a large degree the scale of a topographic map determines its usefulness. A small-scale map (a small map of a large area) may be useful in the study of problems involving regional relationships or in the study of larger land forms, but it is likely to depict poorly, or not at all, lesser topographic features. Generally, the large contour interval on a small-scale map further limits its value for study of smaller land forms. Many of the earlier topographic maps of western states were made on scales of $\frac{1}{250,000}$ or $\frac{1}{125,000}$ and with contour intervals of 50 or 100 feet. A large-scale map (a large map of a small area) brings out topographic details most effectively, but it may fail to show regional relationships as well as a small-scale map. These relationships, however, can be seen by using several maps, but this is generally a somewhat cumbersome procedure. In spite of its minor disadvantages, a large-scale map is to be preferred in geomorphic work. Several states have inaugurated topographic mapping on the scale of $\frac{1}{24,000}$. A map on this scale may have a high degree of accuracy and is also near enough to the scale of many aerial photographs to allow ready comparison of maps and photographs of the same area.

Contour interval. The contour interval is determined largely by the map scale, the amount of relief to be shown, and how many altitudes have been determined in the map area. A small contour interval permits the showing of great topographic detail. Most topographic maps are made with a 10-foot or larger contour interval; this means that some of the lesser features of the topography will not be shown. Such features as low sand dunes on an outwash plain or shallow depressions on a pitted outwash plain or sinkhole plain may be missed entirely if the contour interval is greater than the height of the dunes or the depth of the depressions. Many details of floodplains, deltas, and similar low-relief features can be shown effectively only if the contour interval is as little as 1 or 2 feet.

USES OF TOPOGRAPHIC MAPS

In the following discussion it is assumed that the student can read topographic maps and interpret the basic information on them. A keen appreciation of individual geomorphic forms and landscape assemblages is fundamental to a proper interpretation of topographic maps. Forms that appear alike superficially often may be differentiated if we take into consideration their relationships to other features on the map.

In addition to thorough study of a map for distinctive geomorphic forms, particular attention should be given to drainage patterns since they commonly give a clue to the geologic structure and geomorphic history of the area. Keeping in mind that dendritic drainage patterns are the "normal" ones, any marked departures from them should be carefully considered. Dendritic patterns reflect homogeneity of lithology or near horizontality of beds; trellis patterns suggest truncated, folded beds, steeply dipping beds, or possibly faulted structures where alternating bands of weak and strong rock are present; rectangular patterns may indicate joint or fault control upon drainage; radial and annular patterns develop on such features as volcanic cones, monadnocks, or some type of domal structure; a centripetal pattern may mark structural or topographic basins or domes with an inversion of topography; and a barbed pattern should at once suggest the possibility of stream piracy. Any abrupt changes in surface slopes, stream gradients or courses, as well as types of land forms, should be considered as possible reflections of varying lithology or structure. Differences in drainage texture may also provide clues to changes in types of underlying geologic materials.

Ability to interpret topographic maps accurately can be acquired only through painstaking attention to map details, combined with a thorough appreciation of the distinctive characteristics of the many types of terrains. It is not the primary purpose of this section, however, to explain how to interpret topographic maps, for usually this can be done better through laboratory or field study of maps. Rather, it is our purpose to discuss some uses that may be made of them and techniques that can be applied to their interpretation which may be unfamiliar to the reader.

Construction of topographic profiles. Some problems involving the use of topographic maps may be simplified through the construction of *topographic profiles*. Problems involving relationships of slopes, recognition of topographic levels such as remnants of former peneplains, correlation of terraces, and in general the correlation of topography with lithology and structure may be clarified by topographic profiles.

The simplest type of topographic profile is made by first drawing a straight line across the map along the course of the intended profile. A strip of cross-section paper is then laid parallel to the line of the profile, and some line on it is chosen as a datum line. The value attached to the datum line need not be sea level, but it must be less than the altitude of the lowest contour crossed by the line of profile. The horizontal scale of a profile will be the same as that of the map, but the vertical scale usually is exaggerated in order to show the relief more distinctly. If the cross-section paper is divided into 10 or 20 parts to the inch, for the vertical scale it will be convenient to let $\frac{1}{20}$ of an inch equal the contour interval. Thus if we are using a type of topographic map that is common today, one on a scale of $\frac{1}{24,000}$ with a contour interval of 10 feet, $\frac{1}{20}$ of an inch represents 10 feet vertically and an inch represents 200 feet or 2400 inches. Thus, the vertical scale is ten times the horizontal scale. If, with a map having a scale of $\frac{1}{62,500}$ and a contour interval of 20 feet, we let $\frac{1}{10}$ of an inch equal the contour interval of 20 feet, then 1 inch represents 2400 inches vertically and the vertical scale is exaggerated about 26 times. In general, the less the relief is, the greater the amount of vertical exaggeration needed to produce a useful topographic profile.

After a vertical scale for the profile has been selected, it is then a matter of marking on the cross-section paper at appropriate points above the datum line the value of each contour that is cut by the line of the profile. This can be done conveniently on a drawing table by laying the line of profile parallel to the lower edge of the table and using a T-square to project the points of intersection of the profile and contours onto the cross-section paper. After each contour value has been marked by a point, the successive points are connected by a smooth curve. Every topographic profile should be given a title such as "Topographic profile from Pilot Knob to Henley Schoolhouse, Bloomington, Indiana, quadrangle" or the line of profile shown on a reference map. The horizontal scale and the amount of vertical exaggeration must be indicated. If necessary information regarding dips and thicknesses of beds is available, a topographic profile may be converted into a combined *topographic profile and geologic structure section,* which will show such correlation as may exist between lithology, structure, and topography. To serve this purpose best it is usually advisable that the profile be made at right angles to the strike of the geologic formations.

Projected profiles. A simple topographic profile has the advantage of ease of construction and is valuable for showing the relation of topography to lithology and structure and for studying cross profiles of valleys for possible recognition of valley-in-valley features, structural benches and terraces, and asymmetry, but it does not lend itself as well to the study and correlation of peneplain remnants and similar topographic forms as does a *projected profile.* This type of profile was originated by Barrell (1920) and has been used extensively by geomorphologists. The chief disadvantage of a single vertical-plane profile in studies involving attempts at recognition and correlations of erosion surfaces is that the direction of the profile must be specifically chosen so as to pass through one or more supposed remnants of the erosion surfaces. Because these are usually restricted in extent, most of the profile will show surfaces of later age. Furthermore, selection of a line of profile introduces a subjective element and the accompanying danger of selecting features to support a particular viewpoint.

The construction of a projected profile, as most commonly done today, was described by Bates (1939) as follows:

The method of constructing the profiles consists of mounting the topographic maps of the area to be profiled on cloth in strips of convenient width, for example, the width of one 15-minute quadrangle. The long direction of the strips is made parallel to that in which the profiles are to be drawn. The direction usually chosen is at right angles to the trend of the structural and topographic features to be studied. These strips of mounted maps are then ruled lengthwise at intervals of two inches, one inch or one-half inch depending upon the detail desired. A profile of each of these narrow strips is then constructed on cross-section paper by

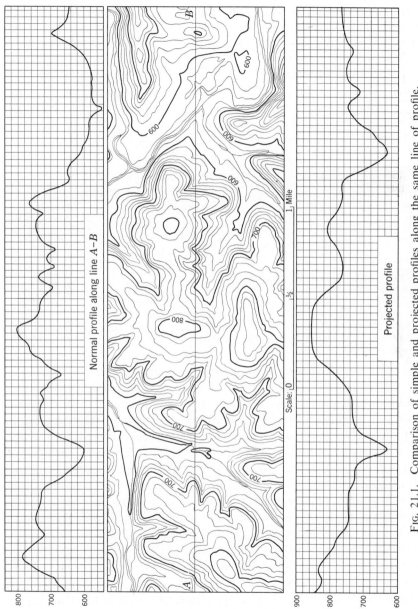

FIG. 21.1. Comparison of simple and projected profiles along the same line of profile.

plotting the highest contour within the strip at intervals of one-tenth of an inch along the strip and then joining the successive points with a line. The profile constructed in this way does not represent the profile along any single line, but rather that of the highest contours throughout the narrow strip. The profile is then traced on cardboard, cut out with a jig-saw, and finally is lined along both sides of the upper edge with a heavy black line. In order to study the profiles they are placed upright in their proper positions in slots cut in cardboard or wooden strips.

It is apparent that a projected profile thus made is a profile of the skyline and not that of the topography along any particular line. It is a useful tool,

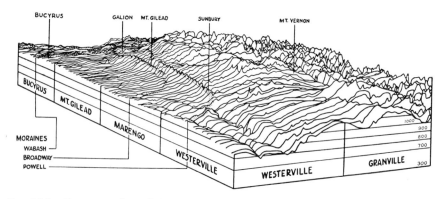

FIG. 21.2. Representation of the topography of several quadrangles in northern Ohio by projected profiles. (After George W. White.)

but, as Rich (1938) has indicated, should be used with care and discrimination. It is merely a device that is useful because it eliminates to some degree the subjective element of personal bias. It is certainly more trustworthy than impressions obtained by the eye, but by selecting only the highest contours within the profile strip the existence of extensive areas of near horizontality may be emphasized unduly. A subjective element still remains, for interpretation of a projected profile is largely a matter of personal judgment. Probably the chief advantage of projected profiles is that they make possible study of large areas on a much reduced horizontal scale and, with exaggeration of the vertical scale, many changes in slope not readily observable in the field are brought out. They further permit one, when groups are mounted together, to observe the topography in its relationships to geologic structure from all possible viewpoints. It should be obvious that the horizontal scale and amount of vertical exaggeration should be the same for all strips and great enough to bring out the details of the topography. Maps on a scale of less than $\frac{1}{62,500}$ are not too satisfactory for this method of study.

The projected profile as developed by Barrell differed from the type more commonly used today in one significant respect. Barrell started by drawing a continuous profile of the front of the area to be studied. For successive belts of topography back of the frontal strip only those areas were profiled which projected above the belt immediately in front of it. Thus each successive belt of topography shuts out all features behind it which are not higher than it is.

Since it has become more common practice to mount profiles in strips parallel to each other, the practice of showing only the highest topography is no longer followed. There has been a noticeable decline in the use of projected profiles, probably in part because too often they have not yielded results commensurate with the amount of work required for their preparation. White (1939) has shown, however, that projected profiles are particularly helpful in showing the position of end moraines that are inconspicuous in the field.

Mathematical analyses of topographic maps. The various methods of making mathematical analyses of topographic maps are quantitative and objective in nature, but they are tedious, time-consuming, and do not always yield results commensurate with the amount of time required for their preparation.

A simple approach to such a geomorphic problem as the identification and interpretation of topographic levels is by construction of *altimetric frequency curves.* Of the two methods that have been used in constructing these, one was used by Thompson (1936) in a study of the West Point and Schunemunk, New York, quadrangles. The contours that close and indicate summits were plotted according to their frequency at regular intervals, usually according to the value of the contour interval used. This method gives no indication of summit areas at various altitudes, but great frequency of certain altitudes may be suggestive at least of persistent topographic levels. Baulig (1935) using a variation of this method plotted the frequency of "spot elevations," by dividing a map into uniform small squares and using the elevation of the highest point within each square for plotting frequency of various altitudes.

Another type of altimetric curve is obtained by plotting the altitudinal distribution and areas of all summits (Thompson, 1941). This involves much tedious work, for the area enclosed by each summit contour must be ascertained. The measuring of summit areas where the areas are fairly large can be done with a planimeter; where they are small, it can be accomplished best by superposing the summit area over a grid of closely spaced lines made to the scale of the map. After this is done, the total area of summits at the altitude of each contour is plotted to scale. Maxima on the constructed curve then show areas where summits are either numerous or extensive and may thus possibly represent planation surfaces or reflect the influence of lithologic or structural control. The extent to which lithology does or does not influence the altitudes of summit areas may be suggested by the compilation in tabular form of the various geological formations upon which summits have developed.   This can be done readily if a geological map of the area is available, but otherwise it will entail considerable field work.

An even more tedious method of study of topographic surfaces involves the construction of a *hypsometric curve*, obtained by expressing the total area of land above each contour line as a percentage of the total area covered by the map.   Values for areas within each contour are usually obtained by means of a planimeter.   The hypsometric curve for an area that has been subjected to

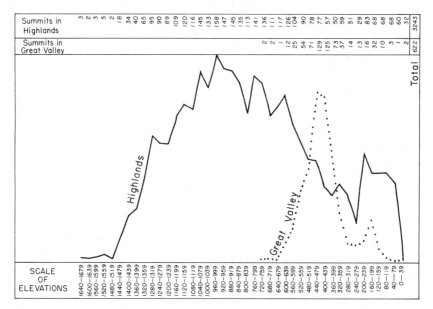

FIG. 21.3.   Altimetric frequency curves for the West Point and Schunemunk quadrangles, New York.   (After H. D. Thompson.)

continuous erosion and lacks any benches or planation surfaces is a curve concave upward, much like the curve of the long profile of a valley lacking any nickpoints along it.   If, however, there are extensive areas at persistent altitudes, convexities in the curve will appear at the altitudes at which persistent topographic surfaces exist.

Strahler (1952) has advocated for study of drainage basins a type of hypsometric analysis designated as a *percentage hypsometric method*.   In it two ratios $a:A$ and $h:H$ are involved, in which a equals the area enclosed between particular contours and the divide line marking the perimeter of a drainage basin, A equals the total drainage area, h equals the height of each contour above the basal plane of the drainage basin, and H equals the total vertical distance between the lowest point in the drainage basin and its summit point. A *hypsometric integral* is also computed which represents the ratio of the area below the hypsometric curve to the total area of the square in which it is plotted. A hypsometric curve obtained in this way permits comparison of forms of drainage basins of different sizes and altitudes. It expresses mainly

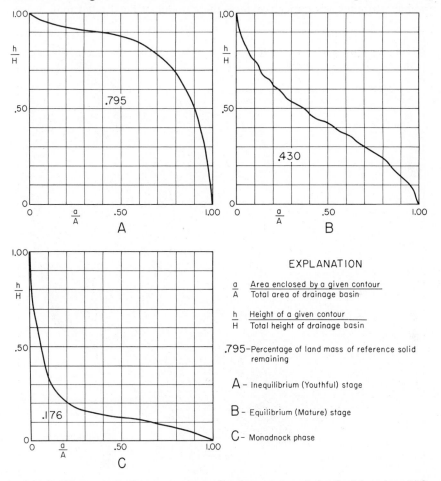

FIG. 21.4. Hypsometric frequency curves for three stages of the fluvial cycle. (After A. N. Strahler.)

the way in which the volume of earth beneath the ground surface is distributed from bottom to top of the drainage basin.

From a study by this method of five types of drainage basins, Strahler concluded that two stages seem to mark the evolution of drainage systems in a fluvial cycle: (1) an early inequilibrium stage, during which slope changes take place rapidly as drainage expands, and (2) an equilibrium stage, in which a stable hypsometric curve develops and persists as relief diminishes. A special monadnock phase may be recognized, but it is transitory and destruction of the monadnock is followed by restoration of the equilibrium form.

In the study of processes such as sheetwash, mass-wasting, and such topographic attributes as drainage texture, the amount of local relief may become

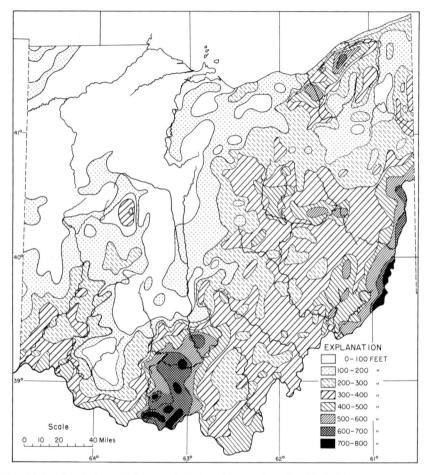

EXPLANATION

	0 – 100 FEET
	100 – 200 "
	200 – 300 "
	300 – 400 "
	400 – 500 "
	500 – 600 "
	600 – 700 "
	700 – 800 "

Scale

0   10   20   40 Miles

FIG. 21.5. A relative relief map of Ohio. (After Guy-Harold Smith, courtesy American Geographical Society.)

significant. Local relief has been variously expressed by such terms as *topographic relief, drainage relief,* and *relative relief.* What is involved is usually not the altitudes as shown by contour lines but the actual amount of relief within an area in contrast to absolute altitudes above sea level. Sometimes it may be desirable to prepare a map that emphasizes this factor. *Relative relief maps* have been rather widely used by European geographers, but they have not been used much in this country. Smith (1935) prepared a relative relief map of Ohio and his method can be applied to smaller areas. The procedure consists of dividing the area to be studied into a number of small blocks whose size will be determined by the amount of detail desired. The difference in altitude between the highest and lowest point within each block is noted and plotted at the proper point on the map. After this has been done, lines of equal relative relief, called *isopleths*, are drawn upon the map. The resulting map shows somewhat more strikingly the areas of greatest local relief than does the ordinary topographic map.

GEOLOGIC MAPS

The uses of geologic and structural maps in geomorphic studies are so self-evident that they require little discussion. A geologic map suggests the nature of underlying materials and is extremely useful in determining whether geomorphic features reflect control by lithology and structure or bear more the imprint of geomorphic processes and history. It is almost impossible to separate geomorphic from geologic studies, at least if they have regional significance. Unfortunately, geologic mapping has not kept pace with topographic mapping, and the chances are more than even that there will be no geologic map of any given area, for, as late as 1946, less than 10 per cent of the United States had been mapped geologically on a scale as large as 1 inch to the mile. Upon undertaking a geomorphic study of an area unfamiliar to the investigator, one of the first steps should be to find out whether the geology of the region has been mapped.*

The ideal situation is to have available a geologic map, a topographic map, and aerial photographs of an area, but this is rarely encountered. A geologic map, if available, is likely to be on a scale considerably less than that of the photos, but even under these conditions it may be of great value in interpreting the details of the topography as shown on the photographs and topographic map. With a knowledge of what the geology is, one is forewarned as to what to expect on the topographic map or photos and thus knows where to look for significant changes in topography as related to geologic variations; sometimes even minor tonal variations on the photographs come to have real significance.

* Information as to whether geologic information is available regarding a particular area and where to obtain geologic maps and other information regarding specific areas may be obtained from the two following sources: J. V. Howell and A. I. Levorsen (1946), Directory of geological material in North America, *Am. Assoc. Petroleum Geol., Bull. 30*, pp. 1321–1432, and Leona Boardman, et al., Index to geologic mapping in the United States, *U. S. Geol. Survey.* The latter consists of a series of state maps showing the portions of the states which have been mapped and a bibliography for each of the mapped areas.

BLOCK DIAGRAMS

A block diagram is not so much a tool which may be utilized in studying a geomorphic problem as it is a technique which may be utilized to show effectively results of a study. Their use in this country was introduced by G. K. Gilbert and was perfected by W. M. Davis. Lobeck (1924) and Raisz (1948) have been among the more successful users and teachers of block diagramming. Block diagrams are usually drawn in either one-point or two-point perspective, but sometimes they are drawn in more simplified forms, not in true perspective, which for some purposes are fairly satisfactory. During World War II significant progress was made by the Military Geology Unit of the United States Geological Survey through the development of two machines for the projecting of contours so that diagrams could be made showing how a terrain would appear from various angles. A block diagram is an exceedingly valuable illustrative tool when used with proper appreciation of its limitations. Its chief advantage is that it is diagrammatic and three-dimensional and hence can be made to emphasize pertinent features and omit irrelevant ones which may obscure the broader picture. A block diagram is easily understood and usually requires little explanation. Its chief disadvantages are that it is not quantitative and its construction requires a skill and an artistry that may be lacking, though they can be developed. Sometimes block diagrams are made so diagrammatic that they do not show features in their true relationships and thus give an impression of topography that does not coincide with what is seen in the field.

Somewhat similar in purpose to block diagrams are physiographic maps or what have also been called *morphographic* or *land form maps* (Raisz, 1948). They grew out of block diagrams but differ from them in that they are usually drawn on a small scale. Topography is shown by means of hachures and conventionalized symbols. The first major physiographic map was Lobeck's "Physiographic Diagram of the United States," which appeared in 1921. Raisz also published in 1939 a physiographic map of the United States entitled "Landforms of the United States." The terrain diagrams prepared for strategic uses by the Military Geology Unit of the U. S. Geological Survey during World War II were either block diagrams or physiographic maps. Physiographic maps are most useful in regional geomorphic studies.

AERIAL PHOTOGRAPHS

Probably the most significant geomorphic tool that has become available during the present century is the aerial photograph. A geomorphologist who does not understand at least the basic principles of airphoto interpretation is greatly handicapped. What has come to be called *photogeology* (Rea, 1941) deals particularly with the geologic interpretation of aerial photographs. The science of photogrammetry, which encompasses most of the manifold uses of aerial photographs, has already developed into a many-sided and highly technical subject, many aspects of which are highly mathematical. A geomorphologist, however, does not need to be familiar with all the technical aspects of photogrammetry in order to be able to use aerial photos to advantage. The literature on aerial photographs is already voluminous; a few of the more elementary and practical discussions from the viewpoint of the student of land forms are listed at the end of this chapter. It is hoped that the brief discussion that follows will serve as a foundation upon which the student may build and develop skill in the use of aerial photographs. Ability to interpret aerial photographs comes only through continued work with them in the laboratory and in the field. A proper interpretation of the geomorphic features which they show is first of all dependent upon a thorough background in the fundamentals of land-form development and land-form assemblages.

Types of aerial photographs. Aerial photographs are of three types: verticals, obliques, and composites. Simple *vertical photos* are taken with a single-lens camera with the optical axis in a vertical or nearly vertical position. They give a flat picture or plan of the area photographed. *Oblique photos* are taken with the optical axis of the camera inclined to the vertical. If the optical axis is more nearly in the horizontal than in the vertical, the photo is called a *high oblique;* if the optical axis is more nearly vertical than horizontal, it is called a *low oblique*. *Composite photos* taken with multilens cameras cover a larger area than those taken with single-lens cameras. *Mosaics* may be made of a large area by patching together individual verticals or obliques and may serve to some extent as regional maps. Verticals are more widely used than obliques in geologic field work, whereas obliques are particularly valuable for illustrative purposes; for certain types of problems they may also be useful in field studies.

Colored photographs will undoubtedly come into use for some types of geological field work but as yet barely a beginning has been made in their application to geological studies. During World War II the Navy made use of kodacolor reversal film in determining depths of water along coast lines. The same technique should be applicable to photogeologic mapping and study of the continental shelves in the exploration of these areas for possible oil structures. The United States Geological Survey has experimented in the use of color photos as a means of recognizing zones of oxidation in the Tintic, Utah, mining district.

Advantages and disadvantages of aerial photographs. The advantages of aerial photographs are manifold, but they do not eliminate the need for topographic maps or field work. They are a supplementary tool which can, when used effectively, reduce greatly the amount of field work and permit much more effective use of a topographic map.

FIG. 21.6. A mosaic of the area around Alcova, Wyoming. (Photo by Jack Ammann Photogrammetric Engineers.)

The biggest advantage of aerial photographs lies in the enormous amount of detail which they show. Many things are discernible on them which do not show on a topographic map and which may be overlooked or not visible to a person on the ground because of his limited perspective or inability to detect all the tonal differences in soil, vegetation, and other features. To a beginner, the maze of detail on aerial photographs may be confusing, but practice will enable him to make fullest use of it. Stereoscopic study of photos is usually necessary to obtain their maximum usefulness. Aerial photographs give a continuity of topographic relief and other features not obtainable on the ground. Aerial photographs are great timesavers in that they make possible a better planning of the field approach to a problem. Study of them makes possible the laying out of the best traverse routes and the location of the more likely places where significant geomorphic features and rock outcrops may be found. Except in heavily wooded areas, it is usually possible to locate points in the field exactly from an aerial photograph, since individual trees, gullies, and other markers are readily recognizable. One indirect advantage of aerial photographs is that with them, much more so than with topographic maps, attention will be given to ecological factors and thereby the user will be led to see how differences in vegetation and soils may give clues to the geology and topography.

Aerial photographs, however, have their disadvantages and limitations, and these should be recognized in order to avoid mistakes or undue optimism regarding their value. In spite of the great detail which they show, they are not absolutely accurate maps. A camera lens photographs objects in perspective and hence they are not shown in true positions with respect to a horizontal plane, as may be done on a map. Aside from inaccuracies which may have resulted from tilt of the plane while the photography was done, there is *topographic distortion* resulting from the fact that an aerial photograph is a radial rather than isometric projection. In a radial projection, the distance of any point with respect to the center of the picture is proportional to its elevation and thus high points are shown relatively farther from the center than lower points. Thus, accurate distances and directions cannot be measured except at the center of a photograph.

The third dimension may be obtained through use of a stereoscope, but quantitative values of elevation are not as readily obtainable as on a topographic map. Methods have been developed for making elevation determinations directly from photographs (Desjardins, 1950), but these procedures are too technical for the beginner. A person observing photos stereoscopically must be careful not to get an exaggerated idea of relief. Low hills may look like mountains under a magnifying stereoscope. The exaggeration of relief which comes from stereoscopic viewing may be used to an advantage in the search for features of such slight relief that they do not show on most topographic maps. Thus, in the study of what may be called the microgeomorphology of an area, an aerial photograph has decided advantages. Features as small as ant hills may stand out distinctly, either because of the stereoscopic exaggeration of relief or because of the tonal differences which they produce on the photo.

Photographs may have limited use because of mechanical deficiencies in photography or printing. Those made at the wrong time of the year may fail

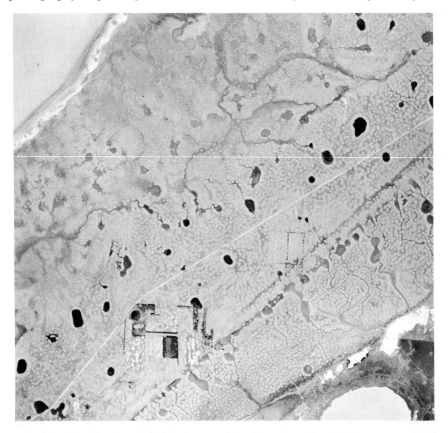

FIG. 21.7. Vertical photo of pimple mounds in Chambers County, Texas, an example of the detail which aerial photos may show. (Production and Marketing Administration photo.)

to have the tonal variations which are so frequently significant. Summer, or rather the dry season, in general, is the best time to take photographs, for then the variations in soil tones related to varying moisture content are most evident, although foliage may obscure details in forested areas. The time of day when the photography was done may affect the quality of a photograph since shadows influence the clarity with which features are brought out. Quality of print development may make the difference between photographs that show clearly a large amount of detail and those that show little. Thus, photographs may well lack the quality that characterizes most of the newer topographic maps. This, however, is not an inherent defect of an aerial photograph but rather a defect that is related to its production.

Aerial photographs may have limited geomorphic usefulness because of deep soil and vegetation cover. Heavy forests may obscure almost all topographic details. Thick glacial deposits or wind-blown silts and sands may cover the bedrock, but such deposits have their own distinctive topographic expression. Even though vegetation covers or obscures land forms and bedrock, it may provide a clue to the geology through distinctive variations.

Stereoscopic study of aerial photographs. A simple folding type of stereoscope is best for showing the details of aerial photos, but such a stereoscope limits the view to a much narrower strip of terrain than does a mirror stereoscope. To get the best effects the two photographs to be viewed stereoscopically should be of equal legibility and equally well-illuminated. Unless the observer's eyes are of equal power, or have been properly corrected by glasses, difficulty may be encountered in obtaining stereoscopic vision. Shadows on overlapping photographs should fall away from the source of light, or otherwise a *pseudoscopic view* may result which causes the relief to reverse itself with valleys standing up as ridges and hills and trees and other positive forms showing as depressions. The points or areas common to the two photographs to be viewed should be placed about $2\frac{1}{2}$ inches apart (this is known as the interpupillary distance and varies with individuals). One looks through the two lenses of a stereoscope as if each eye were looking into infinity and projecting each line of sight parallel to the other. Stereoscopic vision can be obtained with practice without the use of lenses by holding two prints at arms length from the eyes. It is usually helpful at first to hold a piece of cardboard midway between the two points to be viewed. This method of observation, however, does not give as sharp a picture as when the photos are viewed under a stereoscope because magnification is lacking. In viewing an area of slight relief, an observer may find it advantageous to use alternate photographs rather than immediately adjacent ones. The result is a greater exaggeration of relief which brings out lesser forms. This method, however, is applicable only to those areas that are shown on three adjacent photographs, and hence complete stereoscopic coverage cannot be obtained this way.

Clues to airphoto interpretation. Features that will be of major value in the interpretation of aerial photographs are: distinctive land forms, drainage patterns and texture, variations in color tones of the photographs, outcrop patterns and structural relationships, degree and type of erosion, soils, and types and distribution of vegetation.

It is perhaps not too much to say that a thorough understanding of geomorphic forms with all their implications is the most basic tool in the interpretation of aerial photographs. This implies a thorough understanding of the response of different types of rocks to the denudational processes under varying climatic conditions, how rock structure may be reflected topographically, and particularly how lesser topographic feature may give a clue to the underlying lithology and structure. All noticeable changes in form, slope,

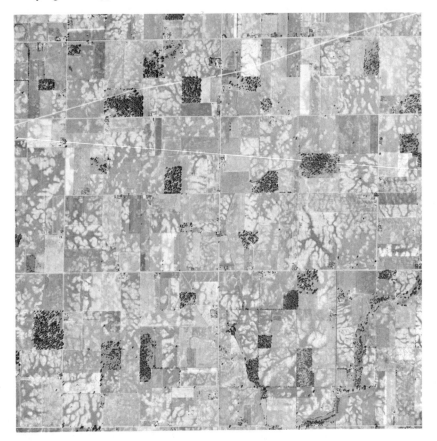

FIG. 21.8. Vertical photo of a portion of Wisconsin till plain in Shelby County, Indiana, showing the typical mottled airphoto pattern which characterizes till of the Tazewell substage. (Production and Marketing Administration photo.)

pattern, or trend of geomorphic features and any departure from normal topographic expression should be given consideration. DeBlieux (1949) has shown how on the deltaic plain of the lower Mississippi River, where there is a marked scarcity of geologic and topographic expression of subsurface structure, it may be possible to use such a slight detail as the widening of a natural levee as evidence for the existence of a salt dome beneath that portion of the natural levee that is broader than normal. At one place, a pronounced meander along an otherwise straight stretch in one of the relict distributaries of the Mississippi marked a salt dome.   One not only needs to know the normal topographic expressions for various types of lithology and structure but also one needs to be equally aware of the possible significance of what may be called *anomalous topographic features.*

FIG. 21.9.   Vertical photo of an area of Illinoian till plain in Scott County, Indiana. The lacy pattern along tributary stream is a diagnostic airphoto pattern of Illinoian till. (Production and Marketing Administration photo.)

What has been stated above regarding the significance of drainage patterns applies to interpretation of aerial photos as well as to topographic maps. Drainage texture reflects the relative permeability of the underlying bedrock, and differences in texture may differentiate areas of shale from those of sandstone, or areas of outwash sands and gravels from area of glacial till, as well as other rock variations.   Abandoned or even buried stream channels may be recognized on aerial photographs, and Jenkins (1935) has shown how aerial photographs were used to locate Tertiary river channels with gold placers in the Sierra Nevada region of California.

Practically all aerial photographs exhibit noticeable and commonly pronounced variations in color tones which result from the blending together of

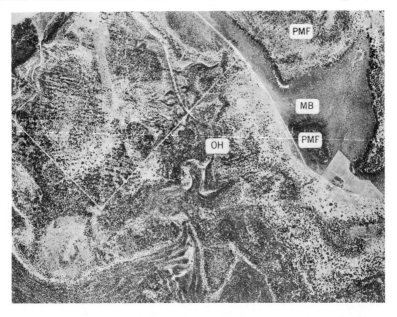

FIG. 21.10. Vertical photo of an area in San Saba County, Texas, which shows how outcrop belts of geologic formations are indicated by vegetation differences. PMF, Pennsylvanian Marble Falls limestone; MB, Mississippian Barnett shale; and OH, Ordovician Honeycut dolomite and limestone. Mesquite vegetation characterizes the areas of Barnett shale, whereas the areas of the Honeycut and Marble Falls formations have cedars, live oaks, and other trees. (Production and Marketing Administration photo.)

details usually too small to be recognized as individual features. These tonal variations result from differences in the reflective power of a land surface as influenced by such factors as the color of the soil or rock, moisture content, and amount of organic material in the soil. Such tonal variations commonly group themselves into definite patterns which are indicative of distinct types of rock, soil, land forms or vegetation and are usually shown on aerial photographs far more effectively than can be done on a map. This is particularly true in areas of low relief. It becomes a primary problem of the photo-interpreter to familiarize himself with the significance of these tonal variations. This may necessitate a field search for the causes, but, once they are recognized, they become helpful in mapping the features thus identified.

Where bare rock is exposed over a considerable area, as is commonly the case in arid regions, the outcrop pattern may be readily evident, and this, along with the associated tonal variations, may give a clear picture of geologic structure. Igneous intrusions, such as stocks and dikes, may be outlined distinctly in this way. Major structural features are often evident from the

FIG. 21.11. The Ganzo Azul oil dome near Pucallpa, Peru, a domal structure in a tropical rain-forest area where no bedrock is exposed, but which is outlined by tonal differences in vegetation related to varying lithology. (Courtesy American Geographical Society.)

pattern of outcrop and direction of dip of the rock strata. Secondary structural features, such as joints, stratification, or foliation, also may be apparent.

The degree and type of erosion shown on aerial photographs may be significant if the principles of geomorphic development are thoroughly understood. Steep and nearly vertical bluffs which characterize areas of loess, scarcity of gullies and drainage lines across permeable sands and gravels, lack of erosion on lacustrine plains, sinkholes and discontinuous stream courses of karst plains, and gullying of the badland type on clays and shales are but a few examples of how different types of geologic materials are suggested by particular geomorphic features.

Frequently, important clues to bedrock geology can be obtained from a careful study of the distribution of vegetation. Distinctive vegetation types or patterns commonly reflect the influence of soil differences related to differences in rock types or parent materials. Vegetation changes, however, rather than resulting from fundamental variations in geology, may be produced by difference in topographic exposure. Such contrasts are commonly noted, particularly in arid and semiarid regions, between the north and south exposures of east-west valleys. Even in regions where the vegetal cover is so thick as to prevent any actual exposures of soil or bedrock, differences in vegetation tones may reflect geologic structure. The significance of changes in vegetation distribution and types may not be apparent until field correlations have been made, but, once these relationships are understood, further work is greatly simplified.

Certain deceptive markings upon photographs may be misleading. In mountain areas, for example, patches may show where timber is either absent or very different from that of its surroundings. At first, these areas may be considered as having geologic significance when actually they may mark areas lacking trees or having second-growth timber, where slumping or landslides have removed the original forest. Abandoned roads and railways may show upon photos as distinct lines which might be thought to have geologic significance. Poor development or printing of photographs may produce tonal variations or patterns which may be misleading. Poor matchings of individual photographs or tonal variations in adjacent photographs used to make a mosaic may produce lines which reflect no change in topography or geology. Photographs of an area taken during different seasons may have marked differences in tonal qualities related to varying vegetation and soil moisture conditions.

Recognition of rock types. Subject to limitations set by lack of topographic relief, thickness of soil, or other type of cover and density of vegetation, fairly reliable inferences as to rock types can be drawn from aerial photographs. Areas of unconsolidated sediments, such as wind-blown sands, glacial till, and alluvium, may still retain some of the topographic features that originally distinguished them. Consolidated sediments are recognizable by banding in their outcrop pattern. If the beds are nearly horizontal, their outcrop pattern will follow the contours. Alternating strips of trees and grass may indicate the arrangement of beds, particularly if cliff-making formations are present to make certain tracts non-tillable. Thus alternating sandstones and shales are often reflected along hillsides by a sequence of woods and fields. If rock beds dip steeply, the outcrop pattern, or features which reflect it, are more likely to be linear or sinuous in trend. The outcrop belt will make a V pointed downstream where the dip is downstream and a V pointed upstream where the dip is upstream, unless the stream gradient is greater than the dip. Vertical beds pass across topographic features without lateral deflection of their outcrop belt, unless they are faulted. Joint patterns, if soil cover is not too great, may be outlined or suggested on a photograph by such features as aligned sink-holes, a grooved or striated appearance of the photo, rectangular or angulate drainage patterns, or alignment of vegetation. Faults are suggested by gaps, breaks or offsets in outcrop, and topographic or vegetal patterns. Rectilinear cliffs or escarpments and vegetation zones, especially if transverse to the drain-

FIG. 21.12. Vertical photo of a portion of Gellie County, Ohio, on which the contour pattern of the fields and vegetation indicate nearly horizontal sedimentary rock strata. (Production and Marketing Administration photo.)

age and general lineation of topography, should suggest field investigation for possible faulting. Folded structures may be quite evident or may be sug-gested by such features as parallel and converging ridges and valleys or zones of vegetation, looped or zigzag ridges and valleys and trellis, radial or centrip-etal drainage patterns. Where dips are not too steep the type of fold is usually evident.

Bodies of extrusive igneous rocks are usually readily recognizable from photos. Volcanic cones, craters, spatter cones, lava flows, volcanic necks, and other features associated with extrusive activity are usually distinctive in form and color. Small intrusive rock bodies, such as dikes and stocks, may be readily recognized. Dikes commonly give rise to distinctive ridges or occasionally to depressions, or, if not topographically conspicuous, show noticeable contrasts in color with the rocks that they intrude. The circular or subcircular outline of a stock is often apparent from the color contrast which it shows

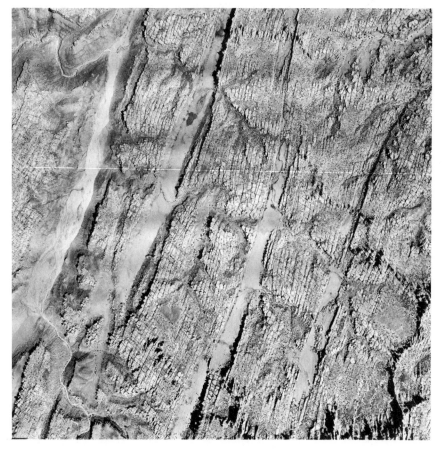

FIG. 21.13. Vertical photo of an area two miles south of the junction of the Green and Colorado rivers in southeastern Utah, which shows two sets of joint systems in the Cedar Mesa sandstone. (Photo by Jack Ammann Photogrammetric Engineers.)

with adjacent intruded rocks or from its topographic expression. A large intrusive body, such as a batholith, may be difficult to recognize because it covers such a large area that its contacts may not be shown. Recognition will then be more difficult, but its presence may be suggested by topography which reflects uniformity of rock material but lacks an outcrop pattern indicative of sedimentary rocks. Unless the rocks are notably fractured or faulted, the drainage pattern in a plutonic area is likely to be dendritic. Recognition from photos of such igneous masses as sills and laccoliths may be difficult, but there may well be suggestions of their existence which field checking can affirm or deny.

FIG. 21.14. Vertical photo of an area in Lawrence County, South Dakota, showing upturned sediments above a buried igneous intrusive mass. (Production and Marketing Administration photo.)

The delimiting of areas of metamorphic rocks from aerial photographs may be difficult unless the rocks are so highly metamorphosed that they have been crumpled and distorted, in which case their existence is usually quite evident. If, however, only a slight or moderate degree of metamorphism has taken place, it will be hard to distinguish them from the rock types from which they were derived. Banding and foliation may be recognizable, but it may be impossible to distinguish these from the banded patterns that characterize steeply dipping sedimentaries. In general, the effect of metamorphism has been to increase resistance of rocks to weathering and erosion and hence differential erosion and weathering effects are not likely to be as prominent on metamorphic rocks as on alternating weak and strong sedimentaries.

FIG. 21.15. Vertical photo showing early Pre-Cambrian sediments (dark gray) near Duncan Lake Northwest Territories, Canada, intruded by granite (light gray) bodies as much as one-half mile wide. Both of these are intruded by later Pre-Cambrian diabase dikes (dark gray). (Royal Canadian Air Force photo.)

SOIL MAPS: A MUCH-NEGLECTED TOOL

In Chapter 4, we pointed out that there are five major factors that influence the development of a soil profile: climate, soil biota, topography, parent material, and time. The last three are geologic in nature, and hence it should be evident that soil variations reflect to a considerable degree geologic variations of one type or another. Soil maps made prior to 1920 in general are not particularly useful in geologic field work, but today remarkably detailed and accurate soil maps are being made from field traverses and airphoto interpretation. Such soil maps, if properly interpreted, can furnish a great deal of geologic as well as geomorphic information.

An understanding of the significance of soil series and catenas (see p. 78) is essential to proper use of soil maps. All the members of a soil series were derived from the same type of parent material under essentially similar topographic conditions, and hence can be grouped together as designating a particular type of geologic material. Usually, however, in converting a soil map into a geologic map, it is more advantageous to work with soil catenas than with soil series. All the members of a soil catena (Bushnell, 1942) were derived from similar parent material, but they show notable differences in their profiles because of the varying types of topography under which each soil type formed. Soil catenas thus have both geologic and topographic implications. The first step in converting a soil map into a geologic map is to group together all members of each mapped catena. The next step is to determine the parent material from which each soil catena was derived. Until one has become familiar with the many soil catenas which have been recognized, this will usually entail field observations. When these two steps have been taken, it is a simple matter to make a geologic map from a soil map. Because each member of a soil catena evolved under slightly different topographic conditions, it will be possible to associate the individual soil types with such varying types of topography as upland flats, slight, moderate or steep slopes, depressions, or floodplains. Although a soil map does not give the quantitative information about topography that a contoured map does, it does give a general picture of the varying types of terrain present in the mapped area.

Soil maps do not show directly all topographic forms, but they almost always reflect them, and their topographic implications are readily grasped. They are particularly valuable in areas of continental glaciation, and in such regions one soon finds that the areas of kames and eskers, outwash plains and valley trains, ground and end moraines, lacustrine areas, and areas of wind-blown sand and loess are readily distinguishable by the different soils developed upon them. The geologic age of the glacial materials usually is indicated by the particular soil series present.

Soil maps also may be valuable in bedrock geology except in areas where there are many thin geologic formations. Under these conditions each geologic formation may not be reflected by a specific soil type. If, for example, several adjacent limestone formations are present, a soil map will show where the limestones are, but it will seldom be useful in distinguishing a particular limestone formation unless one possesses some marked mineralogic quality that gives rise to a distinctive soil type. Where geologic formations have an appreciable width of outcrop, however, as in a region of gently dipping formations of considerable thickness, a soil map, when properly interpreted, becomes a rather good geologic map. For many types of geologic work a soil map ranks with a topographic map or aerial photograph in usefulness in mapping areal geology.

CLIMATIC MAPS

Although it is generally recognized that climatic variations have an influence upon the operation of the geomorphic processes, surprisingly few attempts have been made to correlate the two. In many parts of the world this has been impossible because of scarcity or lack of detailed weather and climatic data, but for considerable areas data are available for long enough periods to give an adequate picture of weather and climatic conditions.

Many types of geomorphic problems can be studied better if detailed climatic information is brought to bear upon them. Information on precipitation, including not only the annual amount, but its seasonal distribution, variability, the percentages which fall as rain or snow, intensity of rainfall, particularly of the extremely heavy rains, and many other aspects of the precipitation may well contribute to a better understanding of problems involving types and rates of erosion. Visher (1937, 1945) has emphasized the importance of extreme rainfalls in the problem of erosion and has shown, for example, that the greater erosional rate in southern Indiana as compared with northern Indiana is not solely the result of the greater relief in southern Indiana but is related also in part to greater rainfall there, a greater percentage of rainfall during the winter months when the protective effects of vegetation are at a minimum, greater frequency of torrential rains, fewer days of frozen ground in winter, and other factors that contribute to maximum erosional effects.

Consideration should be given to such temperature factors as average annual temperature, annual range of temperature, daily range of temperature, number of days with temperatures above and below freezing, depth of frost penetration, frequency of days with freeze and thaw, and such other temperature factors as may influence the operation of geomorphic processes (Visher, 1945).

REFERENCES CITED IN TEXT

Barrell, Joseph (1920). The Piedmont terraces of the northern Appalachians, *Am. J. Sci., 199*, pp. 242–258.

Bates, R. E. (1939). Geomorphic history of the Kickapoo region, Wisconsin, *Geol. Soc. Am., Bull. 50*, pp. 814–880.

Baulig, Henri (1935). *The Changing Sea Level*, p. 43, George Philips and Son, Ltd., London.

Bushnell, T. M. (1942). Some aspects of the catena concept, *Soil Sci. Soc. Am., Proc., 7*, pp. 466–476.

Bushnell, T. M. (1944). The story of Indiana soils, *Purdue Agr. Expt. Sta., Spec. Cir. 1*, 52 pp.

DeBlieux, Charles (1949). Photogeology in Gulf Coast exploration, *Am. Assoc. Petroleum Geol., Bull. 33*, pp. 1251–1259.

Desjardins, Louis (1950). Techniques in photogeology, *Am. Assoc. Petroleum Geol., Bull. 34*, pp. 2284–2317.

Jenkins, O. P. (1935). New techniques applicable to the study of placers, *Calif. J. Mines Geol., 31*, pp. 143–200.

Lobeck, A. K. (1924). *Block Diagrams*, 206 pp., John Wiley and Sons, New York.

Raisz, Erwin (1948). *General Cartography*, 2nd ed., pp. 103–123 and 297–308, McGraw-Hill Book Co., New York.

Rea, H. C. (1941). Photogeology, *Am. Assoc. Petroleum Geol., Bull. 25*, pp. 1796–1799.

Rich, J. L. (1938). Recognition and significance of multiple erosion surfaces, *Geol. Soc. Am., Bull. 49*, p. 1699.

Smith, G. H. (1935). The relative relief of Ohio, *Geog. Rev., 25*, pp. 272–284.

Strahler, A. N. (1952). Hypsometric (area-altitude) analysis of erosional topography, *Geol. Soc. Am., Bull. 63*, pp. 1117–1142.

Thompson, H. D. (1936). Hudson gorge in the Highlands, *Geol. Soc. Am., Bull. 47*, pp. 1831–1848.

Thompson, H. D. (1941). Topographic analysis of the Monterey, Staunton, and Harrisonburg quadrangles, *J. Geol., 49*, pp. 521–549.

Visher, S. S. (1937). Regional contrasts in erosion in Indiana, with special attention to the climatic factor in causation, *Geol. Soc. Am., Bull. 48*, pp. 897–930.

Visher, S. S. (1945). Climatic maps of geologic interest, *Geol. Soc. Am., Bull. 56*, pp. 713–736.

White, G. W. (1939). End moraines of north-central Ohio, *J. Geol., 47*, pp. 277–289.

Additional References

Dietz, R. S. (1947). Aerial photographs in the geologic study of shore features and processes, *Photogram. Engineering, 13*, pp. 537–545.

English, W. A. (1930). Use of airplane photographs in geologic mapping, *Am. Assoc. Petroleum Geol., Bull. 14*, pp. 1049–1058.

Gwynne, C. S. (1942). Swell and swale pattern of the Mankato lobe of the Wisconsin drift plain in Iowa, *J. Geol., 50*, pp. 200–208.

Hanson-Lowe, J. (1935).   The clinographic curve, *Geol. Mag., 72,* pp. 180–184.

Jenkins, D. S., D. J. Belcher, L. E. Gregg, and K. B. Woods (1946).   The origin, distribution, and airphoto identification of United States soils, *Tech. Div. Rept. 52* and *Appendix B, U. S. Dept. Commerce,* Washington.

Lobeck, A. K. (1917).   The position of the New England peneplain in the White Mountain region, *Geog. Rev., 3,* pp. 53–60.

Lobeck, A. K., and W. Tellington (1944).   *Military Maps and Air Photographs,* 256 pp., McGraw-Hill Book Co., New York.

Loel, Wayne (1941).   Use of aerial photographs in geologic mapping, *Trans. Am. Inst. Mining Met. Engrs., 144,* pp. 356–409.

Melton, F. A. (1945).   Preliminary observations on geological use of aerial photographs, *Am. Assoc. Petroleum Geol., Bull. 29,* pp. 1756–1765.

Putnam, W. C. (1947).   Aerial photographs in geology, *Photogram. Engineering, 13,* pp. 557–564.

Rich, J. L. (1939).   A bird's-eye cross section of the central Appalachian Mountains and Plateaus: Washington to Cincinnati, *Geog. Rev., 29,* pp. 561–586.

Rooney, G. W., and W. S. Levings (1947).   Advances in the use of air survey by mining geologists, *Photogram. Engineering, 13,* pp. 570–584.

Smith, H. T. U. (1941).   Aerial photographs in geomorphic studies, *J. Geomorph., 4,* pp. 171–205.

Smith, H. T. U. (1943).   *Aerial Photographs and Their Applications,* 372 pp., D. Appleton-Century Co., New York.

Smith, H. T. U. (1947). Aerial photographs in geologic training, *Photogram. Engineering, 13,* pp. 615–619.

Smith, K. G. (1950).   Standards for grading texture of erosional topography, *Am. J. Sci., 248,* pp. 655–668.

van Nouhuys, J. J. (1937).   Geological interpretations of aerial photographs, *Trans. Am. Inst. Mining Met. Engrs., 126,* pp. 607–624.

Visher, S. S. (1941).   Climate and geomorphology: some comparisons between regions, *J. Geomorph., 5,* pp. 54–64.

Wooldridge, S. W., and R. S. Morgan (1937).   *The Physical Basis of Geography,* Chapter 16, Longmans, Green and Co., Ltd., London.

22 · Applied Geomorphology

INTRODUCTION

Land forms are the most common features encountered by anyone engaged in geologic work and various phases of engineering. If they can be properly interpreted, they have their meaning in terms of geologic history, structure, and lithology. Failure to see their application to a particular problem too often arises from unfamiliarity with pertinent geomorphic principles. No claim is made that even the most thorough knowledge of geomorphology equips one to become an economic geologist, petroleum geologist, engineering geologist, or mining geologist, but it is the author's belief that too often these "practical geologists" fail to make maximum possible use of basic geomorphic concepts. In attempting to discuss practical applications of geomorphology it is impossible to separate this branch of geology from other aspects of geology that affect the evolution of landscapes. The reader should realize by now, if he is ever to do so, that an understanding of land forms presumes an appreciation of the influence of such geologic controls as lithology and structure. In the following discussion we shall of necessity consider problems that have their structural, lithologic, and stratigraphic aspects, but we shall attempt to select those that involve an application of fundamental geomorphic principles.

APPLICATION OF GEOMORPHOLOGY TO HYDROLOGY

Hydrology of limestone terrains. Hydrologic problems in limestone terrains are best understood when the geomorphology of such areas is fully comprehended. No rock varies more in its ability to yield water than limestone. Permeability in limestones is in part primary and in part secondary or acquired. *Primary permeability* is dependent upon the presence of initial interconnecting voids in the calcareous sediments from which the rock was formed. *Secondary* or *acquired permeability* results from joints and fractures produced by diagenetic and diastrophic processes and from openings created by solution along joints and bedding planes. The presence or absence of large

553

solutional cavities depends largely upon whether the limestone has been so situated in the past as to allow joints and bedding planes to be actively enlarged. This in turn depends mainly upon whether the limestone is now capped by an erosion surface, or has been at some time in the past so situated with respect to the local water table so as to permit groundwater to move downward through the rock and carry on solution. Hamilton (1948) has shown that limestones in the Lexington, Kentucky, area differ in their yield of water largely according to the degree to which their joints and bedding planes have been solutionally enlarged. This secondary permeability varies most notably with respect to the topography of the region, being greatest beneath and adjacent to topographic lows or valleys. In this area, solutional openings are limited largely to depths of less than 80 to 100 feet below the surface.

Solutional openings with their increased permeability are of significance not only in present-day topography but also in buried karst landscapes. It has been shown that the large artesian yield in the Roswell, New Mexico, artesian basin (Fielder and Nye, 1933) is largely possible because a buried Permian karst terrain upon the Pichacho limestone was later buried under sediments of the upper Permian Pecos formation. Five areas within the artesian basin have notably high yields, and they all owe their existence to more highly cavernous limestone adjacent to ancient drainage lines. Thus the ability to recognize ancient karst topographies now buried becomes important in the evaluation of the water-yielding possibilities of some regions, for such terrains lowered below the water table under hydrostatic head may yield water in great quantities.

How far advanced a particular area is in the karst cycle will determine the best sources of water in a karst terrain. In early stages of karst evolution conditions are not greatly different from those of other types of landscapes with similar relief, for diversion of runoff to underground routes has not progressed far, but as the cycle proceeds an increasing percentage of runoff is diverted to solutionally opened passageways, and surface streams become few in number and often short-lived. Karst springs then become the main sources of water. Many of the great springs in the United States are of this type. Such springs may supply sufficient quantities of water to meet moderate demands, but the quality of the water usually is a matter that should be carefully considered. If their water becomes turbid after heavy rains, it has had no filtration after abandoning surface for underground routes. Bacterial contamination may run high under such conditions. It thus becomes important that the sources of the spring water be determined so that the possibilities of serious contamination can be ascertained. This involves determining which swallow holes and sinkholes feed water to the underground drainage systems which emerge as springs. This can be done by putting some coloring material, such as fluorescein, into the water entering near-by sink-holes and swallow holes and making tests at the various springs to determine which sinkholes and swallow holes feed them.   A knowledge of the structural geology of the region is pertinent here, for groundwater moves more readily down than up the regional dip.

Obtaining water from wells in a limestone terrain may be easy or difficult. If the limestone has great primary permeability, to which has been added secondary permeability in the form of solutional openings, there may be no difficulty in obtaining wells of large yields.   Such conditions exist in the Eocene Ocala limestone of Florida and Georgia.   Few wells in it are dry. If, however, the limestone is a dense, compact limestone with little mass permeability, movement of groundwater will be largely through secondary openings.   Under these conditions, the obtaining of a satisfactory well to a large degree depends upon chance intersection of one of the solutionways through which the groundwater moves.   Such wells, even if yielding satisfactory quantities of water, are subject to the possibility of contamination indicated above for karst springs.   Not all water obtained from limestones is subject to contamination.   If there is a suitable filter rock such as sandstone above the limestone, the quality of the water may be good even when it moves through secondary openings.   Karst plains obviously lack a filtering cover, and any sinkholes, swallow holes, or karst valleys within an area of clastic rocks should at least cast doubt upon the purity of the water of near-by springs.

The early history of the Bloomington, Indiana, water supply is a good example of how lack of the most elementary knowledge of geomorphic and geologic principles led to a long-continued series of water crises.   When the town had reached such a size that individual wells and cisterns were inadequate and need for a municipal water supply was imperative, the first municipal water supply was obtained by building a dam below a karst spring west of town and ponding its water.   The dam was built over limestone having many solution channels.   No water ever went over the dam's spillway; instead it went under the dam and emerged as seepages below the dam.   Water shortages soon developed, and it became necessary to enlarge the water system. Geologists strongly recommended that the karst area west of the city be abandoned as a source of water and a dam to pond surface runoff be built to the northeast of the city where the valley floors were cut below limestone into relatively impervious siltstone.   City officials ignored this advice and built another dam in the karst plain area which ponded the waters of two karst springs.   For a time the water supply was adequate despite leakage, but as the city grew water shortages became increasingly frequent.   Finally, the geologists' advice was followed, and a dam was built across a valley underlain by siltstone and the surface runoff from some 8 square miles was stored.   This supply proved adequate and dependable, and the reservoirs in the karst region were abandoned along with the taxpayer's money that had gone into building them.

Groundwater in glaciated areas.   Over much of North America groundwater for domestic and industrial purposes is obtained from glacio-fluviatile deposits.   Recognition of the possibilities of large supplies of groundwater from glaciated areas largely depends upon familiarity with the types of deposits from which large yields can be obtained, along with an understanding of the geomorphic history of the area during both preglacial and glacial times. Outwash plains, valley trains, and intertill gravels are particularly likely to yield large volumes of water.   Till is a poor aquifer because of the clay in it, but most tills contain local lenses of sand and gravel which may supply enough water to meet domestic needs.   Surface topography rarely gives a clue to the existence of such water-bearing lenses; they are likely to be more abundant in areas adjacent to lines of glacial drainage and usually are elongated in the direction of glacial movement.

It has been found that buried preglacial and interglacial valleys may be sources of large volumes of groundwater.   Recognition of their presence within an area depends upon a detailed study of the preglacial topography and geomorphic history of the area.   In Chapter 16, we discussed briefly one of the great preglacial drainage lines of North America, the so-called Teays Valley, which is now buried beneath glacial drift across Ohio, Indiana, and Illinois (Horberg, 1945, 1950).   It and its tributaries represent an enormous potential groundwater supply as yet hardly touched.   This buried valley crosses Indiana in a general westerly direction and is buried beneath as much as 400 feet of glacial drift, a considerable part of which is sand and gravel. The city of Peru, Indiana, is near this buried valley and gets its water supply from wells sunk into it.   Two ten-inch wells into this buried valley have a potential daily yield of 2,200,000 gallons of water.   Other examples of cities that get their water supplies from other buried valleys are Champaign-Urbana, Illinois, Canton, Ohio (Schaefer, White, and Van Tuyl, 1946), Schenectady, New York (Simpson, 1949), and Dayton, Ohio (Norris, 1948).   Buried valleys have proved important sources of water in such states as Nebraska (Lugn, 1935), where a subhumid climate puts a premium on water.   There is little question that buried valleys are destined to play an increasingly important role in water supply, and their full utilization awaits a determination of their distribution by geologists versed in the preglacial and glacial history of the regions in which they may be found.

Location of buried valleys is done by constructing bedrock topography maps of glaciated areas.   These are made by first determining the thickness of drift cover from well records or from geophysical data obtained by seismic

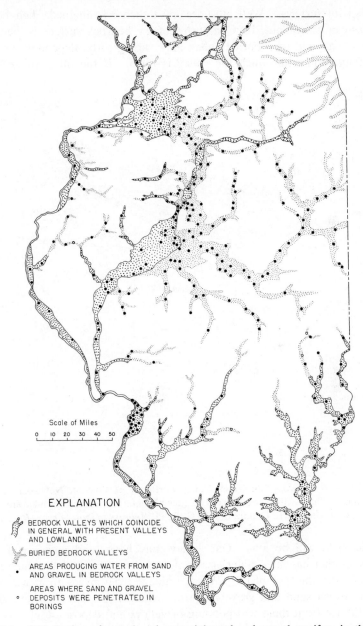

EXPLANATION

BEDROCK VALLEYS WHICH COINCIDE
IN GENERAL WITH PRESENT VALLEYS
AND LOWLANDS

BURIED BEDROCK VALLEYS

AREAS PRODUCING WATER FROM SAND
AND GRAVEL IN BEDROCK VALLEYS

AREAS WHERE SAND AND GRAVEL
DEPOSITS WERE PENETRATED IN
BORINGS

FIG. 22.1. Distribution of actual and potential sand and gravel aquifers in the major buried preglacial valleys of Illinois. (After Leland Horberg, *Illinois Geol. Survey, Bull. 73.*)

or resistivity surveys. Well altitudes must then be obtained; then bedrock altitudes can be computed. Contouring of the bedrock surface is then done.

The water yield of a buried valley may vary greatly, depending upon the type of material with which the valley is filled. If the fill is till, clays, or

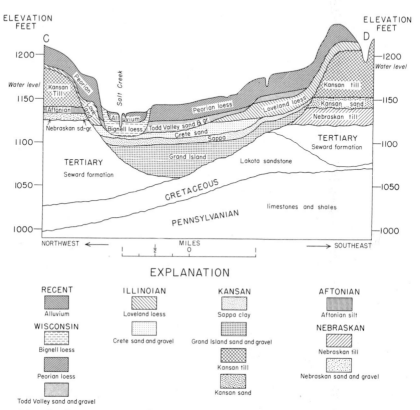

Fig. 22.2. Section across Salt Creek Valley northeast of Lincoln, Nebraska. A preglacial valley is filled with various types of glacial materials, some of which are potential aquifers. (After G. E. Condra and E. C. Reed, *Nebraska Geol. Survey, Bull. 15-A.*)

silts, the yield will be low. One valley may have sand and gravel in it, whereas another has finer materials, or even different parts of the same valley may vary as to the type of fill. When drainage from an ice front was unimpeded, outwash sand and gravel were deposited, but where the land sloped toward the ice front there was ponding of valleys and deposition of clays and silts in them. Thus a thorough knowledge of the directions of glacial movement and lines of glacial outwash is important in evaluating the groundwater potentialities of a buried valley. The same principles apply to the possibilities of tributary buried valleys. Not all buried valleys are of preglacial age. Interglacial valleys may be present which contain water-bearing sands and gravels. Recognition of them requires proper interpretation of intertill sands and gravels.

An interesting dual use of a buried preglacial valley at Louisville, Kentucky, has been described by Guyton (1946). The Ohio Valley, in which the city of Louisville is located, received glacial outwash during both Illinoian and Wisconsin times. The preglacial Ohio Valley here is filled with outwash as much

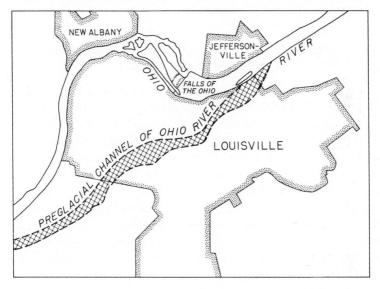

F<small>IG</small>. 22.3. Relation of preglacial Ohio River channel to present course at Louisville, Kentucky. (After W. F. Guyton.)

as 125 feet deep. The present course of the Ohio River at Louisville does not coincide with its preglacial valley trench but lies to the north of it. The municipal water supply of Louisville is obtained from the Ohio River, but there are a large number of wells in the south part of the city which obtain water from glacial outwash in the preglacial valley. In summer, groundwater is preferable to river water for air conditioning because it has a much lower temperature than that of river water. Groundwater in this locality has a fairly uniform temperature of about 57 degrees Fahrenheit, whereas the river water may become as warm as 80 or 85 degrees in summer. Wells had been drawing so heavily upon water in the glacial gravels that a serious shortage was imminent. It was decided not to pump the wells during winter months but to use river water which was cool enough at that season. During the winter months the glacial gravels in the buried valley were recharged with cool filtered river water from the municipal supply. It was feasible to do this because the water level in the buried valley had been drawn down so by pumpage that little of the city water put into it flowed out and hence most of it was recoverable. It was further possible to maintain a desired low temperature in the groundwater by recharging it during the winter months. Water drawn from the wells and used for cooling purposes could have been put back into the aquifer in the buried valley, but its temperature after use was high and its continued return would have raised the temperature of the groundwater above that suitable for air conditioning.

GEOMORPHOLOGY AS AN EXPLORATORY TOOL IN ECONOMIC GEOLOGY

That an understanding of the geomorphic features and history of a region may be profitably utilized in the exploration for some minerals is not too generally appreciated by economic geologists. McKinstry (1948) has written an excellent discussion of the possibilities of using geomorphic guides in mining geology, and the following discussion is patterned in part on his treatment of this topic. Geomorphic features may serve in three ways as guides in the search for mineral deposits: (1) some mineral deposits have a direct topographic expression; (2) the topography of an area may give a clue to the geologic structure which has favored accumulation of the minerals, and (3) an understanding of the geomorphic history of an area may make possible a better appreciation of the physical condition under which the minerals accumulated or were enriched.

Surface expression of ore bodies. Not every ore body has an obvious surface expression, but many do as topographic forms, as outcrops of ore, gossan, or residual minerals, or as such structural features as fractures, faults, and zones of breccia. Numerous examples of positive topographic features which mark the sites of lodes exist. The lead-zinc lode of Broken Hill, Australia, is marked by a conspicuous ridge. At Santa Barbara, Chihuahua, Mexico (Schmitt, 1939), massive quartz veins stand out conspicuously because they are much more resistant to erosion than the unsilicified country rock. Veins make ridges, whereas the intervein areas generally give rise to arroyos, although major drainage lines may cut across veins where they are narrow or have been weakened by cross faulting. In the Parral district in Chihuahua, Mexico, veins, because of the abundant silica in them, are more resistant to weathering and erosion and stand out as hills and ridges. The Codero vein outcrops in the San Juan district of Chihuahua, Mexico, are in most places capped by caliche several feet thick, and the plan of outcrop is well shown by bands of caliche bounded by limestone or rhyolite. Numerous other examples exist, such as in the Oatman district, Arizona, and the California Mother Lode, where quartz veins stand out in distinct topographic relief.

Not all ore outcrops are reflected in positive topographic forms. Some veins and mineralized areas may either lack conspicuous topographic expression or be reflected by depressions or subsidence features. Calcite veins may be inconspicuous or be reflected topographically by depressions, as at Oatman, Arizona, and Beaverdell, British Columbia. As Schmitt (1939) has stated, "the topographic expression, positive or negative, of an *unoxidized* ore shoot is, of course, dependent on the relative resistance to erosion by shoot and walls. A difference in resistance is the result of the many vagaries of mineralization in its effect on the veins and walls, but the type of effect may be constant for any given district thus providing a guide to the valuable mineralization." The problem commonly is to find what the specific geomorphic reflection of an ore body is.

Shrinkage of ore bodies during oxidation, giving rise to what has been called *mineralization slump* (Locke and Billingsley, 1930), has been observed at several localities, particularly at Sierra Mojada, Mexico, and Bisbee, Arizona. Referring to the Bisbee area, Locke and Billingsley noted that mineralization slump was most strikingly present in limestones where, as at Bisbee, the effect was much the same as that produced by stope mining. They concluded that, had this topographic effect been recognized early in the development of the Bisbee region, an area which eventually yielded three thousand million pounds of copper could have been marked out.

Wisser (1927), who made a careful study in the Bisbee area of the effects of what he termed *oxidation subsidence,* described its effects as follows:

These characteristics resemble those established in subsidence due to mining. In other words, the effects of oxidation shrinkage resemble those of stoping. As the ore, its walls, and its floor lose volume during oxidation, the unsupported roof fails, and, by the dropping out of blocks from its center, begins to take the form of an arch or dome. Successive doming cracks, one above the other, form curved shells above the ore, and, as subsidence proceeds, the cracks open wider and wider and the shells successively fail and break into blocks, which fall from the roof and make a jumble of fragments, large and small, like the filling of a caved stope····. It eventually results in a slight subsidence of the whole block overlying the ore and in the development around that block of steep fissures that separate it from the uncracked surrounding rock and outline on the surface the region of subsidence.

Application of this surface manifestation of ore bodies resulted in the mapping of nineteen areas where oxidation subsidence was definitely shown to be over known ore bodies, four areas where oxidation subsidence was mapped but was not over known ore bodies, and two areas where known ore bodies were not reflected at the surface by subsidence.

Malcolmson (1902) has described the effects of oxidation subsidence in the Sierra Mojada in the southern part of the State of Coahuila in the north of Mexico as follows:

A noticeable peculiarity in the eastern part of the camp, particularly in the Salvador mine, is the presence of a shattered and completely broken-up limestone

FIG. 22.4. Subsidence resulting from block caving at Ray, Arizona. This artificially induced subsidence simulates oxidation subsidence. (Photo by Paul Whitehead, Kennecott Copper Co.)

roof over the lead-carbonate ores. This shattering appears to have been due to the collapse of the enormous roof after reduction in bulk of the ore-body, due to oxidation. This disintegration is in places so complete that the line, over a considerable area, is made of small angular fragments, which sometimes run into the stopes like very coarse sand.

Topography may be helpful in searching for iron ore. Although no broad generalization can be made about the exact type of topography necessary for the accumulation of iron ore, for a particular deposit there is likely to be a certain definite topographic expression. The iron ore deposits of the Lake Superior region are so commonly associated with hills or ridges that the term iron range has become implanted in the literature dealing with them.

The almost fabulous find of high-grade iron ore at Cerro Bolivar, Venezuela (Lippert, 1950), having a proved reserve in 1949 of over 500,000,000 tons resulted from field investigation by geologists of two small mountains that were shown on aerial photographs. Residual iron deposits are the results of concentration of iron during long periods of weathering, and thus old erosion and weathering surfaces are favorable sites for their accumulation provided there are present beneath these surfaces rock types that make possible a concentration of iron oxides. The residual iron ores of the Mayari and Moa districts in the Province of Oriente, Cuba, are found upon a peneplain surface now 1500 to 2000 feet above sea level. They were probably produced (Leith and Mead, 1911) by alteration of serpentine rock which was itself probably derived from some such rock as peridotite. When the higher-grade ores of the middle latitudes have been exhausted, the lateritic iron deposits of tropical and subtropical regions may become important economic ores, and search for them will entail application of some knowledge of ancient geomorphology and climates. Bog iron deposits are usually of local and limited extent but they are found under definite topographic conditions that are easily recognizable.

Weathering residues. Several important economic minerals are essentially *weathering residues* of present or ancient geomorphic cycles, and in the search for these geomorphology can play an important role. In addition to certain types of iron deposits mentioned above, such materials as the clay minerals, caliche, bauxite, and some manganese and nickel ores are of this nature. An important concept in the search for economic minerals has come into use. It may be stated briefly as follows: weathering and erosion are constantly at work on the rocks of the earth's surface; the products of rock weathering may have economic value; these products in the late stages of a geomorphic cycle are likely to be left in situ as residues; in the earlier stages of a cycle they either do not form or they are removed and concentrated elsewhere by deposition. Residual weathering products may be found upon recent weathering surfaces or they may lie upon ancient weathering surfaces now buried or exhumed. The surfaces upon which they most commonly form are peneplain or near-peneplain surfaces. Unconformities in the geologic column often represent ancient peneplain surfaces and hence may be logical locations for certain residual minerals. More commonly such minerals are to be found upon remnants of Tertiary erosional surfaces above present base levels of erosion.

The above principles are well-illustrated in the formation of bauxite. There are two modes of origin of bauxite: (1) that in which the bauxite represents the residuum of an exceedingly small amount of insoluble aluminous material in limestones and dolomites and (2) that in which it is the direct product of the weathering of aluminous minerals. In either case, bauxite is a weathering residue. The aluminous laterites of the West Indies (Goldich and Bergquist, 1947, 1948) apparently are examples of the first class, for they seem to represent concentrations through weathering of small amounts of alumina in limestones and dolomites of probable upper Eocene age. The old weathering surface upon which these deposits accumulated is probably of Pliocene age. Sinkholes upon this surface provided the most favorable sites for accumulation of aluminous laterite.

The Arkansas bauxite deposits (Bramlette, 1936; Mead, 1915) were formed upon an Eocene erosion and weathering surface as a direct product of the weathering of nepheline syenite. Although the major deposits in Arkansas are residual in nature, some of the bauxite produced by weathering was removed by erosion and is found interbedded with other types of Tertiary deposits.

It is not entirely clear why the weathering of igneous rocks produces both hydrous oxides of aluminum, such as bauxite, and clay minerals or hydrous aluminum silicates. One explanation is that the difference in the final product is determined by the climatic conditions under which weathering takes place. In temperate climates the residual products from the weathering of igneous rocks are clay minerals. This type of weathering has been termed *kaolinization,* but it should be recognized that numerous minerals other than kaolin may form. On the other hand, under tropical climates final weathering products are hydrous oxides of such metals as aluminum, iron, and manganese. This type of weathering has been called *laterization.* If these conclusions are valid, we may infer that the climate in Arkansas at the time the bauxite was formed was tropical rather than temperate as it is now. Harrison (1910) has claimed, however, that both laterite and clays are formed under tropical weathering and that whether one or the other develops depends upon the composition of the parent rock. Laterites were thought by him to be derived from basic igneous rocks and clays from granitic rocks. Mohr (1944) reached a somewhat similar conclusion, arguing that calcic feldspars give rise to bauxite and related minerals and alkalic feldspars to clay minerals. Extensive bauxite deposits (Singewald, 1938) at Gant, Hungary, seem to have formed upon a weathering surface of probable lower Cretaceous age through concentration of the alumina in Triassic dolomites in a way similar to that postulated in the West Indies. Scattered bauxite deposits are found in some southern states (Adams, 1927) in pockets in Paleozoic limestones and dolomites which had been subjected to deep weathering during the geomorphic cycle which culminated in the formation of the Tertiary Highland Rim peneplain.

Woolnough (1930), as previously stated, thought that a definite relation existed between the nature of a duricrust and the character of the rock from which it was derived, aluminous laterite being formed from rocks such as granite which are low in iron; ferruginous laterite, from basic feldspathic rocks; opaline crusts, from argillaceous rocks; and travertine crusts (caliche), from sediments rich in lime.   There may be differences of opinion as to why

FIG. 22.5.   Vertical photo of part of the Parish of St. Ann, Jamaica, showing the karst topography on which bauxite is found.   Bauxite occupies the untimbered basins and sinkholes between residual limestone knobs.   (Photo courtesy Reynolds Jamaica Mines, Ltd.)

one type of residuum forms in one place and a different type in another, but there is general agreement that weathering residues are definitely to be associated with old weathering surfaces at or near the peneplain stage of a geomorphic cycle.   That this principle can be used in a practical way in the exploration for bauxite was shown when, in the search for bauxite during World War II, geologists of one of the aluminum companies (Schedeman, 1948) applied these principles and figured out where bauxite ought to be found and then sent field parties to Haiti in 1943 and discovered bauxite where they had predicted that it should be.

Manganese deposits associated with the Harrisburg peneplain in the Shenandoah Valley of Virginia have been described (Hewett et al., 1918). The deposits resulted from a concentration by groundwater of manganese which was sparingly present in Cambrian limestones. The deposits are not restricted to the level of the peneplain surface, being found as much as 400 to 500 feet above and 250 feet below it, but they are most abundant on the peneplain surface.

Pecora (1944) has described deposits of garnierite (nickel-silicate) and associated cobalt and manganese oxides near São José do Tocatíns, Brazil, which are on a topographic surface of low relief called the Goiáz Upland surface. They were derived from serpentinized rocks which have weathered to a yellowish-brown ferruginous-siliceous clay and jasperoidal chalcedony. Stages postulated in the development of the deposits were as follows: (1) formation of a thick mantle of weathered material upon the Goiáz surface; (2) regional uplift with resulting stream dissection of the Goiáz surface with probable continuation of deep weathering; and (3) further regional uplift accompanied by canyon cutting and secondary enrichment and alteration of the weathered material to form the nickel-silicate and manganese-oxide deposits. Similar nickel-silicate (Pecora and Hobbs, 1942) deposits in Douglas County, Oregon, form a blanket varying from a few feet to as much as 60 to 70 feet in thickness upon a later Tertiary erosion surface known as the Klamath Upland surface. This phase of geology which concerns itself with the study and recognition of ancient weathering surfaces and soils has come to be known as *paleopedology* (Nikiforoff, 1943); although yet in its infancy, it offers great possibilities in the search for economic minerals here designated as weathering residues.

Geomorphic principles applied to placer deposits. Geomorphic principles have been applied to the location of placer deposits more than to any other phase of economic geology. Placer concentrations of minerals result from definite geomorphic processes, are found in specific topographic positions, and may have a distinctive topographic expression. As many as nine types of placer deposits have been recognized. They are residual, colluvial, eolian, bajada, beach, glacial including those in end moraines and valley trains, and buried and ancient placers. Alluvial placers are most important.

Residual placers or "seam diggings" are residues from the weathering of quartz stringers or veins, are usually of limited amount, and grade down into lodes. Colluvial placers are produced by creep downslope of residual materials and are thus transitional between residual placers and alluvial placers. Gold placers of this type have been found in California, Australia, New Zealand, and elsewhere. Part of the tin placers of Malaya (Westerveld, 1937) are colluvial placers (the *koelits*) and part are alluvial placers (the *kaksas*). About one-third of the world's platinum is obtained from alluvial placers in Russia, Colombia, and elsewhere. Gold, tin, and diamonds are among the more important minerals obtained from alluvial placers. Diamonds in the Vaal and Orange river districts of South Africa, the Lichtenburg area of South Africa, the Belgian Congo, and Minas Geraes, Brazil, are obtained from alluvial placers. About 20 per cent of the world's diamonds comes from placer deposits. Eolian placers have yielded gold in Australia and Lower California, Mexico. Bajada placers form in the gravel mantle of a pediment and in the confluent alluvial fans of a bajada. They are more likely to be found nearer a mountain base than out on the more gentle slopes of a basin fill. Beach placers have yielded gold in California and Alaska, diamonds in the Namaqualand district of South Africa, zircon in India, Brazil, and Australia, and ilmenite and monazite from Travancore, India. Glacial placers are not too important, but gold has been found in the sand and gravels of valley trains and end moraines in the Tarryall area in Park County, Colorado (Singewald, 1942).

Placers may become buried under landslides, mudflows, lava flows, glacial drift, ash falls, marine sediments, marine waters, or later alluvium. Some buried placers are of Recent geologic age, but others belong to the class here designated as ancient placers.

Ancient placers of Cambrian age have been found in a conglomerate in the Black Hills region but the most important ancient placers are of Tertiary age. Placers submerged beneath oceanic waters as a result of late glacial and postglacial rise of sea level have been found at various places. The river systems of Malaysia, which contain tin placers, have been previously discussed (p. 465). Where depth of water is not too great they have been dredged, but undoubtedly great quantities of tin are submerged beneath oceanic waters.

Probably the most famous buried placers are those found in Tertiary valleys in California. Their description by Lindgren (1911) has become a classic and in more recent years Jenkins and his associates have published several papers on them (Jenkins, 1935; Jenkins and Wright, 1934). Since recovery of gold from these ancient placers is dependent to a large degree upon a full appreciation of the past and present geomorphology of the region, they will be discussed in detail as a type example of the application of geomorphic principles to exploitation of placer deposits.

Tertiary river channels containing gold placers have been found in California in the Sierra Nevada and the Klamath Mountains. Those of the Sierra Nevada have a volcanic cover, whereas those of the Klamath Mountains do not. It is particularly in the Sierra Nevada that a full appreciation of past and present geomorphology has been particularly useful in the search for placers. Quaternary and Recent alluvial gravels have produced about three times as much gold as the Tertiary gravels, but they are largely exhausted. Around $300,000,000 worth of gold has been obtained from Tertiary gravels, and it has been estimated that $600,000,000 worth remains.

In the Sierra Nevada region, there are three sets of stream systems which need to be distinguished in order to recognize the possibilities of buried placers. They are the *prevolcanic valleys* of early Tertiary or Eocene age, the *intervolcanic valley systems* of Miocene and Pliocene age, and the *Quarternary* or *postvolcanic valley systems*. Buried placers are chiefly in the early Tertiary or prevolcanic valley systems. Evidently the favorable conditions for the accumulation of gold placers which existed during the Eocene were never again quite duplicated. Early Tertiary streams were able to deposit important placers because they flowed over the bedrock which contained the gold, and apparently some stream courses were influenced by the same structural features (chiefly faults) which localized the gold lodes. Eocene climate apparently was tropical or subtropical, and this favored deep chemical weathering and release of large quantities of gold. Intervolcanic and postvolcanic streams locally obtained and reconcentrated gold from Eocene stream channels, but in few instances came in contact with the gold veins. The problem of locating buried placers is, in the first place, one of reconstructing the early Tertiary bedrock topography and determining the positions of buried bedrock valleys, but this in itself is not enough, for the age of the materials overlying the bedrock is significant as well. It is necessary to distinguish buried Eocene valleys from buried Miocene and Pliocene valleys. If Miocene and Pliocene valleys did not make contact with the rocks of the Jurassic batholith from which the gold came, they are not potential sources of gold, except where locally they may have intersected, reworked, and reconcentrated earlier Eocene placers.

The oldest volcanic deposit is a rhyolitic ash, and recognition of it aids in the identification of Eocene channel fillings. Gravels resting upon bedrock and covered with this rhyolitic ash can be regarded as of Eocene age. In places, the rhyolitic ash was deposited in lakes where it forms thinly laminated deposits, known to the miners as "pipe-clay." Following deposition of the rhyolitic ash, a mass of andesitic volcanic material, known to the miners as "cobble wash," covered the Eocene valleys, the rhyolitic ash, and even intervening ridges. The stream systems which developed across the cobble wash during Miocene and Pliocene times bore no necessary correlation with buried Eocene valleys and their gravel fills. Lava flows extended down many of the Miocene and Pliocene valleys and, at some places, as at Table Mountain, stand out as topographic ridges, representing most interesting examples of inversion of topography. The buried intervolcanic stream channels have often been confused with Eocene channels, and much money has been wasted in tunneling into them. Quaternary streams have in many places cut into Eocene channels and carried their gold down their valleys, but this reconcentrated gold has been largely recovered, so that most of the gold that remains as placer gold is to be found in Eocene placers.

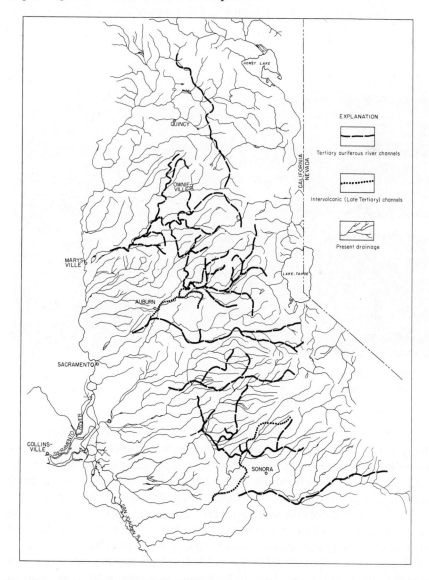

Fig. 22.6. Map showing the relationships between present drainage lines and the Tertiary intervolcanic and auriferous channels in the Sierra Nevada. (After Waldemar Lindgren, *U. S. Geol. Survey, Profess. Paper 73*.)

Geomorphic principles can be further applied to the search for gold after the buried channels have been properly identified. Gold placers are likely to be richest where there was slowing down of stream velocities. Thus an understanding of the effects of varying gradients along the buried channels, changes in shape and size of the channels, and irregularities on the floor of the

FIG. 22.7. Table Mountain near Sonora, California. The ridge marks a Tertiary lava flow down an Eocene valley and is a striking example of inversion of topography. (Photo by Spence Air Photos.)

channel which would encourage deposition of the metal becomes important. The type of rock forming the bedrock floor may influence the deposition of placers because some rocks erode more smoothly than others. Some of the richest placers have been found where highly irregular and pitted bedrock surfaces provided obstructions to cause settling of gold. Schists erode into small ridges and grooves which may serve as natural riffles to collect gold. On the other hand, where diorite forms the bedrock of a buried channel, gold is less likely to be found because diorite usually erodes to a smooth surface.

Location of placers may be aided by drilling and geophysical testing. A magnetic survey will usually be helpful because magnetite is likely to be associated with gold. If the bedrock is a basic type with a higher magnetic intensity than the placer gravels, areas of magnetic "lows" may reflect the positions of the filled channels.

Knowledge of the bedrock geology, application of geophysical surveying, test drilling, and aerial-photograph interpretation all contribute their parts to exploration for these buried placers, but most fundamental to this search is a thorough understanding of the geomorphic history of the region.

APPLICATION OF GEOMORPHOLOGY TO ENGINEERING PROJECTS

Most engineering projects involve evaluation of geologic factors of one type or another; terrain characteristics are among the most common factors. An appreciation of the geomorphic history of the area may greatly aid the proper evaluation of surficial materials and the configuration of the bedrock profile.

Highway construction. Topography obviously plays an important role in determining the most feasible highway route, although sometimes cultural features may necessitate routes not in best accord with topography. Different types of terrain impose varying problems in engineering. A route over a karst plain necessitates repeated cut and fill, otherwise the road will be flooded after heavy rains as sinkholes fill with surface runoff. Bridge abutments in a karst region should be so designed that they will not be weakened by enlarged solutional cavities which are likely to be present. Occasionally small caverns encountered in the process of grade construction create special problems.

Glacial terrains present many types of engineering problems. A flat till plain is topographically ideal for road construction, but in areas where end moraines, eskers, kames, or drumlins exist there is need for cut and fill to avoid circuitous routes. In an area of Wisconsin glaciation, there may be numerous lakes or former lake sites now filled with lacustrine materials which must be avoided. Muck areas, which mark sites of former lakes, are unsuited for roads which are to carry heavy traffic. If a road is built across them as they are, heavy traffic will cause the plastic materials beneath the lake floor to flow, and "sinks" in the road bed will result. To avoid this, the lacustrine fill may have to be excavated and replaced with materials that will not flow under heavy load.

Areas with the considerable relief which characterizes late youth and early maturity will necessitate much bridge construction and many cuts and fills. In such areas landslides, earthflows, and slumping become serious problems. Whenever a stabilized hillside slope or talus slope has a cut made through it, downslope movement of material by one process or another is invited. It may be impossible to plan a highway route so as to avoid completely the possibility of earth movement, but too many highway routes have been laid

FIG. 22.8.   Rockfall on State Route 7, Washington County, Ohio.   (Photo by Ohio Department of Highways.)

FIG. 22.9.   Slumping of a road built upon plastic lacustrine material, Bayfield County, Wisconsin.   (Photo by Wisconsin Geol. Survey.)

out without recognition of the effects that follow disturbance of stabilized slopes.

Landslides and other types of mass-wasting present problems not only in highway construction but in various other phases of engineering. Landslides and earthflows near Ventura, California, caused displacement of roads, breaks

Fig. 22.10. Ejection of a mud-water mixture (pumping) from a concrete pavement joint immediately after passage of a car. (Photo by Engineering Experiment Station, Purdue University.)

in pipe lines, tilting of oil derricks, and shearing of well casings (Putnam and Sharp, 1940). In the Dunedin district of New Zealand (Benson, 1946), rock slides, in an area of feldspathic-quartz-schists and argillaceous mudstones, resulted in the displacement of pipes feeding a hydroelectric system, roadways, railway fills, and buildings. Landslides engendered in the excavation of the Culebra section of the Gaillard cut in the Panama Canal delayed completion of the canal and greatly increased its cost. Total excavations in the Gaillard cut amounted to 168,265,924 cubic yards of material, of which it was estimated 73,062,600 cubic yards were necessitated by landslides.

In highway construction designed to carry heavy traffic, the nature of the soil beneath a road surface, or what is called the subgrade, has become increasingly significant because of its control over the drainage beneath a highway. The lifetime of a highway, under moderate loads, is determined largely by two factors: the quality of the aggregate used in the highway and the soil texture and drainage of its subgrade. Thus an appreciation of the relationships of soils to varying topographic conditions and type of parent material becomes essential in modern highway construction. A knowledge of soil profiles,

FIG. 22.11. Deterioration of a highway built upon a clay-shale subgrade. (Photo by Engineering Experiment Station, Purdue University.)

which to a large degree reflect the influence of geomorphic conditions and history, is basic. Pumping, a term applied to the expulsion of water from beneath road slabs through joints and cracks, is one of the most serious problems encountered by highway engineers. Pumping is largely a result of poor drainage in a subgrade. Displaced water carries up soil particles and thus creates voids beneath a pavement, which in turn result in cracking of the pavement. Poor highway performance characterizes silty-clay subgrades with a high water table, and best performance is found on granular materials with a low water table.

The accompanying summary shows the variation in number of cracks in relation to several types of subgrade materials found along a section of U. S. Highway 31 across a glaciated part of Indiana (Sweet and Woods, 1947).

Number of Cracks per 1000 Feet

Subgrade Material	Grades	Cuts	Fills
Wind-blown sand	9.7	9.9	9.8
Granular terrace	10.3	7.5	13.4
Wisconsin till	27.5	25.2	19.8
Illinoian till	23.9	23.9	25.2

It is obvious that pumping is notably greater over glacial till than over permeable materials such as wind-blown sand and outwash gravel.   A till

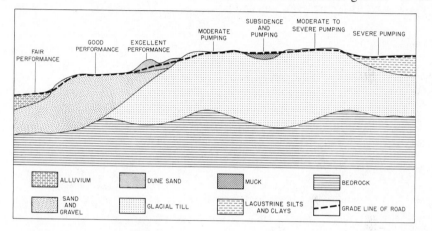

Fig. 22.12.   Diagram suggesting road performance to be expected over various types of glacial deposits.

plain, which may present ideal topographic conditions for construction of road beds with minimum cut and fill, may give poor performance unless granular materials are introduced into the subgrade or surface drainage conditions are improved.   The poor internal drainage of a till plain is in part related to the impermeability of most tills and in part to the formation by weathering of impermeable clay minerals in the B horizon of the soil profile.   A thorough understanding of the differences in drainage characteristics of different soils involves a study of the clay mineralogy of the soils.   As has been pointed out (Ekblaw and Grim, 1936), two clay minerals alike in chemical composition may have different moisture-absorptive qualities and behave quite differently under heavy loads.

On a glacial terrain (and many of our superhighways are being constructed on such terrains) we may find variations in topography and materials between till plains, end moraines, lacustrine plains, outwash plains, drumlins, kames, eskers, wind-blown sand, and loess.   Each of these presents a different type of subgrade material and topography for highway construction.   An appreciation of these differences will inevitably lead to better highway con-

struction.   Some states—Massachusetts, for example—are making seismic
tests of the materials upon which their highways are being built in order to
get a better idea of the engineering problems that they will present.   With
increasing use of granular materials in subgrades and fills, availability of such
materials becomes an item of increasing importance.   It is possible to map
all such materials along a proposed highway route with the aid of aerial photographs or soil maps.

Location of sand and gravel pits.   Sand and gravel have many engineering as well as commercial and industrial uses.   Selection of suitable sites for

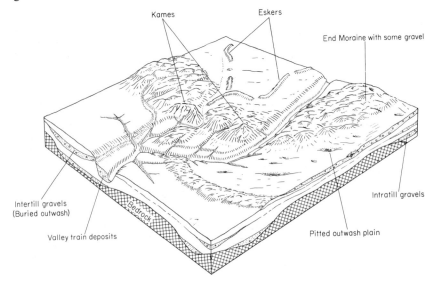

Kames

Eskers

End Moraine with some gravel

Intertill gravels
(Buried outwash)

Bedrock

Valley train deposits

Intratill gravels

Pitted outwash plain

FIG. 22.13.   Diagram showing modes of occurrence of various types of glacial gravels
which have economic significance.   (Drawing by William J. Wayne.)

sand and gravel pits will entail evaluation of such geologic factors as variation in grade sizes, lithologic composition, degree of weathering, amount of
overburden, and continuity of the deposits.   Sand and gravel may be found
as floodplain, river terrace, alluvial fan and cone, talus, wind-blown, residual,
and glacial deposits of various types.   All have distinctive topographic relationships and expressions and varying inherent qualities and possibilities of
development.   A recognition of the type of deposit is essential to proper evaluation of its potentialities.   Demand for gravel is generally greater than for
sand, although near cities the demand for sand for plaster and other uses may
be considerable, and a pit which is to produce both must be evaluated in
terms of the ratios of the two grade sizes.   This involves an appreciation of
the type of deposit and the distance that the material has been carried by
water.

Floodplain deposits are likely to contain high proportions of silt and sand and show many variable and heterogeneous lateral and vertical gradations. Alluvial fan and cone gravels are angular in shape as well as variable in size, especially near their apices. Talus materials, in addition to being angular, are too large to be suitable for most uses and are limited in extent. Wind-blown sands may be satisfactory sources of sand but, of course, have no gravel in them. Residual deposits lack assortment and are likely to contain pebbles

FIG. 22.14. Portion of a kame, showing the heterogeneous nature of kame deposits. (Photo by Conrad Gravenor, Geol. Survey of Canada.)

that are too deeply weathered to be suitable for use as aggregate in cement work. A high percentage of iron-coated chert when used as aggregate usually has deleterious effects. Residual deposits are furthermore likely to be limited in extent.

Terraced valley trains and outwash plains are usually favorable sites for pits, for they do not have a thick overburden and usually are extensive. Whether a terrace deposit is of Wisconsin age, Illinoian, or older age is significant. A Wisconsin gravel terrace has a relatively shallow cover of weathered material which would have to be removed, probably not over 5 or 6 feet, as compared with 10 to 15 feet for Illinoian gravels and more for older glacial outwash. Height of the terrace above the present valley floor will determine whether the pit will be operated as a wet or dry one. An esker may contain excellent materials for a gravel pit because of its good assortment, but the amount of gravel available may be limited because eskers usually are not large and typically are segmented rather than continuous for long distances. Even though an esker has sufficient continuity, its development as a pit is likely to involve buying a large tract of land or acquiring rights from several property owners. Kame deposits show a poor degree of assortment, and hence a large amount of the material handled will have to be discarded because it is too large or too fine. Furthermore, a kame deposit is likely to grade into till within a short distance and thus limit the possible extent of operation. Many gravel pits have been opened in intertill gravels without an appreciation on the part of the operator that overburden will rapidly become excessive as the pit is extended into a hillside.

Dam-site selection. Application of geology to dam-site selection involves a synthesis of knowledge concerning the geomorphology, lithology, and geologic structure of terrains. As Bryan (1929) has pointed out, there are at least five requirements of good reservoir sites that depend upon geologic conditions: (1) a water-tight basin of adequate size; (2) a narrow outlet of the basin with a foundation that will permit economical construction of a dam; (3) opportunity to build an adequate and safe spillway to carry surplus waters; (4) availability of materials needed for dam construction (this is particularly true of earthen dams); and (5) assurance that the life of the reservoir will not be too short as a result of excessive deposition of mud and silt.

We have previously pointed out some of the difficulties encountered in dam construction in a limestone terrain (see p. 555). The Hondo reservoir in southeastern New Mexico (Bryan, 1929) was built over limestone with a water table some 20 feet below the surface. Leakage was so rapid as to cause abandonment of the reservoir. Burwell and Moneymaker (1950) have described the difficulty that arose in a limestone area when a dam site was selected wholly on the assumption that a narrow stretch in a valley is a favorable location. Hales Bar Dam on the Tennessee River, about 12 miles west of Chattanooga, Tennessee, was built between 1905 and 1913 by private interests and was acquired in 1939 by the Tennessee Valley Authority. Regarding it, Moneymaker stated:

That the rock might be unsound or cavernous does not seem to have occurred to anyone before construction was actually undertaken. Hales Bar Dam was built without sufficient exploration, without sufficient foundation excavation, and without remedial treatment of a very cavernous foundation. Leakage through foundation rock was so great that completion was long delayed, and the cost was much greater than anticipated. The project, estimated at $3,000,000 and scheduled for completion in 2 years, cost $11,536,889 and required 8 years for completion.

Legget (1939) gave an example of how failure on the part of a contractor to realize how irregular a limestone bedrock surface may be because of differential solution led to grave financial consequences. Instead of having to excavate 450 cubic yards of rock as estimated, 4719 cubic yards had to be removed; instead of drilling 3300 feet of drill holes, 10,450 feet were drilled; instead of using 4200 bags of cement for grouting, 6368 bags were used, and the job instead of costing $82,992 as bid, cost $209,018.

A constriction in a valley is desirable from the standpoint of the size of the dam that will have to be built, but it may not always be a good dam site. In glaciated areas, where buried bedrock valleys containing sand and gravel fills are common, surficial topography may not give an adequate picture of subsurface conditions. The writer recalls an example, in Indiana, where the site selected for a proposed dam was at a constriction in a valley between a valley wall on one side and a spur end on the other. Study of the subsurface topography showed, however, that the spur to which one end of the dam was to be tied was underlain by a buried preglacial valley with sand and gravel in it. Despite warnings that the dam would leak, it was built here and has steadily lost water by leakage. Legget (1939) pointed out the difficulties encountered in the construction of a dam in Northern Ireland which arose in part from interpretation as bedrock of glacial boulders encountered in test holes. Instead of finding a bedrock foundation for the dam at a depth of 50 feet as anticipated, it was necessary to go to a depth of 180 feet.

Geomorphology applied to airfield site selection. Many of the factors that determine the choice of a site for an airfield are geomorphic or geologic in nature. Local climatic factors may be significant, but to a considerable degree even they are related to geomorphic conditions. Some of the major considerations that must be taken into account are: (1) possibilities of long runways from several directions; (2) drainage conditions; (3) amount of grading necessary; (4) danger of floods and fog; and (5) water supply, if municipal supply is not available.

We may see how varying geomorphic and geologic conditions affect development of an air site by considering the possibilities of several different types of terrains: a till plain, a floodplain, a karst plain, a lacustrine plain, an outwash plain, and a terrace underlain by granular materials. Each will be evaluated in terms of the five factors listed above.

A till plain

1. Long runways from several directions will be possible.

2. Drainage will present a problem, particularly if the plain is flat. Tiling will be necessary, and permeable subgrade materials will have to be introduced beneath runways.

3. Grading will be at a minimum on a flat till plain, but if such features as end moraines, drumlins or kames are present, there will be the problem of cut and fill. The probability is that cut and fill will about balance and will not be excessively expensive, because of the unconsolidated nature of the materials.

4. Floods and fogs should present no serious problem.

5. Wells probably can be obtained in local lenses of sand and gravel in the till, and unless the personnel is to be unusually large the water supply would be adequate.

A floodplain

1. Runway possibilities will depend upon the width of the floodplain.

2. Drainage will be a problem because of a high water table and flat topography. Tiling and ditching will be necessary to improve this, and permeable subgrade materials under the runways will be needed.

3. Grading will be at a minimum.

4. The danger of floods will be serious and will likely necessitate a system of artificial levees to prevent flooding. Fogs may at times present problems.

5. Water should be easily obtainable from the river or wells in alluvium.

A karst plain

1. Long runways from several directions will be possible.

2. Drainage will present a problem in that the many sinkholes will fill up with water after heavy rains, and those that are to be crossed by runways will have to be filled.

3. There will be much grading. Sinkholes will have to be filled and cuts made in bedrock. There may be some difficulty in holding fills in sinkholes unless all inlets into subterranean runways are closed.

4. Except for local and temporary flooding of sinkholes, there will be no danger from floods, and fog should be no problem.

5. Water supply may present a problem because of the uncertainty of getting wells in limestone terrains. Even if adequate water were found, it would require filtration and treatment to insure its purity.

A lacustrine plain

1. Runway possibilities will be excellent.

2. Natural drainage will be poor, and ditching and tiling will be necessary. Permeable subgrade materials beneath the runways will be required.

3. Grading will be at a minimum.

4. There will be no danger from floods, and fogginess should not be above normal unless the field is adjacent to a large lake.

5. Water supply may present a problem because lacustrine deposits consist largely of clays and silts which are not good aquifers. Local lenses of sandier materials may provide wells.

An outwash plain

1. Runway possibilities will be excellent.

2. Drainage will be excellent because of the permeable nature of the underlying materials.

3. Little grading will be necessary, unless the outwash plain is a pitted one, in which event kettles will have to be filled. This will not be too expensive because of the unconsolidated nature of the outwash deposits.

4. Floods and fogs should present no problems.

5. Water can be easily obtained from wells.

A terrace underlain by granular materials

1. Long runways from several directions will not be possible unless the terrace is unusually extensive.

2. Drainage will be excellent.

3. Grading will be at a minimum.

4. There will be no danger from floods, but fogs over the adjacent river may at times be troublesome.

5. Water supply can probably be easily obtained from wells in the granular deposits or from the adjacent stream.

Application of geomorphology to military geology. In contrast to German and Japanese military leaders, American and British have been slow to make the maximum use of geology in warfare. Some geologists were utilized during World War I but on a limited scale. The Military Geology Unit of the United States Geological Survey rendered great service to the Corps of Engineers during World War II, but the war was well-advanced before military authorities saw the needs for and possibilities of the use of geologic experts. It is natural for geologists to blame this situation upon the failure of the military to realize the usefulness of geologic information, but it is perhaps not entirely their fault, for geologists were slow to realize that to be useful in military operations geologic knowledge must be translated into terms that a non-geologic-minded person can understand.

World War I was largely stabilized trench warfare, and the information that was most useful was more geologic than geomorphic in nature (Brooks, 1921). Information about the kind of rock that would be encountered in digging trenches, in mining and countermining and the possibilities of water supply and supplies of other geologic materials was most utilized. Topography did play a role in maneuvering and planning routes of attack, but it can hardly be said that the Allies utilized basic geomorphic knowledge to any great extent.

With development of the blitzkrieg type of warfare during World War II, topography became more important, because the effectiveness of a blitz depends to a large degree upon the trafficability of the terrain. As a consequence, in more recent years *terrain appreciation* or *terrain analysis* have become semimagic words with the military.   A geomorphologist may lack knowledge as to how best to utilize terrains in military operations, but certainly his concepts of terrain conditions are far more adequate than those of the military specialist or other geologists for that matter.   He appreciates that land forms are the result of an interaction of geomorphic and geologic processes through time; that land forms are not groups of unrelated and haphazard individual forms but have systematic relationships that reveal their origins and tell much about the underlying bedrock geology and structure, as well as the soils and vegetation of a region.   As Erdmann (1943) put it, the geomorphologist has "an eye for the ground or an instinctive eye for configuration, the judgment of how distant ground, seen or unseen, is likely to lie when you come to it · · ·.   Terrain is the common denominator of geology and war." Whether it is in connection with the interpretation of topographic maps or aerial photographs, this basic appreciation of different types of terrain is fundamental to a proper planning of military campaigns.   If geologic maps are available, they aid greatly, but geologic and topographic maps for most of the world's area have not yet been made; deductions must be drawn from interpretations of aerial photographs, and these deductions are based chiefly upon the differences that photos show in topography, soils, and vegetation.   Regarding the use of aerial photographs in this connection, Hunt (1950) stated:

Even where geologic maps are lacking or are on such a small scale as to be practically useless for tactical intelligence, geologic principles can be applied with advantage to interpreting the terrain from aerial photographs.   Little training in reading vertical photographs is required to recognize mountains, hills, lakes, rivers, woods, plains, or some kinds of swamps.   But much more than that can and should be interpreted from the pictures for the purposes of acquiring complete terrain intelligence.   It is essential to know the kind of hill, the kind of plain, the kind of river or lake, and so on, because by knowing this it is frequently possible to reconstruct the geology.   The interpreter, with some confidence, can then make predictions as to water supply, the kind and depth of soil, trafficability, ground drainage and other construction problems, construction materials, movement and cover, and many of the other elements that are essential to an adequate estimate of the terrain situation.   In brief, therefore, aerial photographs are useful to the preparation of terrain intelligence insofar as they provide information on the geology of the area.   Identification of a hill or other terrain feature is but a small part of the story that can be read from a photograph; all important is the recognition of the significance of the particular land form, in terms of kind of ground and slope.

Application of geomorphology to oil exploration.   It is probably true that for much of the earth's surface most of the oil structures that are recognizable from their surface expression have been discovered, but there are considerable areas where surface exploration is still being carried on; most oil companies have parties doing surface geology. For this type of work an appreciation of the geomorphic expression of geologic structures may be helpful. In such oil fields as the Rangely, Colorado, Kettleman Hills, California, and Elk Basin, Wyoming-Montana fields, anticlinal structures are conspicuously reflected in the topography. Many of the Gulf coast salt dome

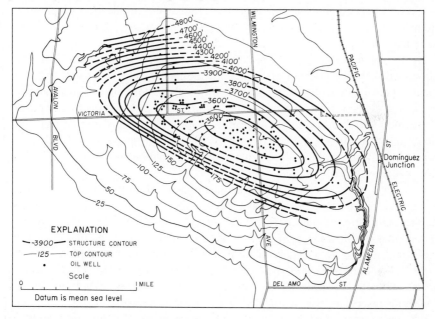

Fig. 22.16. Map showing the close correlation between present topography and geologic structure in the Dominguez Hills oil field, California. (After F. P. Vickery, *Am. Assoc. Petroleum Geol., Bull. 12.*)

structures are evident in the topography, at least when viewed upon aerial photographs. Salt domes are not always topographic highs but may form topographic basins because of solution of part of the salt plug.

In general, it is a fairly good working principle for the student of geomorphology to suspect that areas that are topographically high may also be structurally high, keeping in mind the possibilities of inversion of topography which may result where weak beds exist at the crest of a structural high. This principle is particularly applicable to areas of late Tertiary or later diastrophism. Vickery (1927) has pointed out that there is a striking correspondence between the location of the oil fields in the Los Angeles, California, coastal plain belt and its topography. Not all topographic highs have become oil fields, but in the Signal Hill, Dominguez, Baldwin Hills, Sante Fe Springs, Montebello, and other oil fields there is correspondence between topography and structure. This striking correlation results from the fact that deformation has taken place so recently that there has not been time for erosion and mass-wasting to destroy the effects of diastrophism. In regions of heavy tropical forest, where the topography cannot be seen through the forest, tonal differences in the vegetation may outline an anticlinal or domal structure. All abnormalities in drainage patterns and courses should be seriously considered as possibly reflecting structural controls. Increasing application of photogeology to oil exploration makes an understanding of land forms a basic requirement of the geologist using this technique.

According to Leverson (1934), many if not most oil and gas pools are associated with unconformities. Unconformities are ancient erosion surfaces; hence a petroleum geologist must deal with buried landscapes. Where ancient erosion surfaces truncate permeable beds and are later sealed over by deposits, the erosion surfaces become stratigraphic traps. Most stratigraphic traps are along unconformities. Three types of stratigraphic traps will be discussed— buried karst topographies, shoestring sands, and angular unconformities which have become sealed.

In several oil fields, such as the Lima-Indiana, Michigan Basin, and West Texas fields, carbonate rocks have been important oil and gas reservoirs. Permeability in carbonate rocks, as discussed above, may be either primary or secondary. Most large oil yields from limestones have been obtained from rocks that have a high degree of permeability produced by solution. Bybee (1938) has suggested that the West Texas reservoirs are largely cavernous limestones, and Howard (1928) estimated that 95 per cent of oil yields from limestone reservoirs was associated with erosion surfaces on limestones beneath which there had been solutional enlargement of joints and bedding planes. Limestone formations several hundred feet thick may yield oil in quantity only from the upper 25 to 50 feet, in which solutional openings are present. As noted above in the discussion of karst groundwater reservoirs, permeability is usually greatest adjacent to ancient drainage lines.

Shoestring sands are elongate buried sand bodies. They are particularly common in southeastern Kansas and northeastern Oklahoma (Rich, 1923; Bass, 1936). There is probably no phase of petroleum exploration which can use to better advantage a knowledge of the detailed characteristics of specific topographic features than that which deals with the exploitation of shoestring sands. Shoestring sands may be buried coastal beaches and bars, offshore bars, river channel fillings, delta distributary channel fillings, or tidal channel fills. It is important that the specific origin of a shoestring sand be recognized as soon as possible to properly plan drilling programs (Rich, 1938). An offshore bar differs greatly in cross section, pattern, and trend with respect to the ancient shore line from a stream channel filling. Offshore bars extend in smooth, crescentic curves parallel to a former shore line; they are likely to be discontinuous or segmented by former tidal inlets and taper toward the ends of the individual segments; bar segments may be productive of oil, whereas gaps between them will not; commonly the individual bar seg-

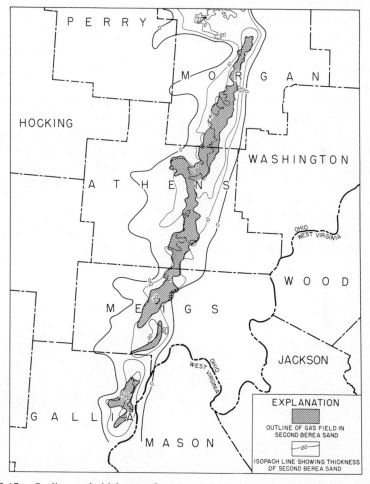

FIG. 22.17. Outline and thickness of gas pool in second Berea sand in southeastern Ohio. Buried topography is reflected here, for the gas field is outlined by an ancient offshore bar. (After J. F. Pepper et al., *U. S. Geol. Survey, Oil and Gas Investigations Preliminary Map 79.*)

ments are offset en echelon; points or protuberances representing spits or hooks may extend shoreward from the bars; their tops may be irregular because of dunes upon them.

River channel fillings have plans distinctly different from those of offshore bars. It may be difficult to anticipate the plan of a particular channel filling until several wells have been drilled. The sand body of a channel fill is continuous, instead of broken, and has nearly uniform thickness and width; it may curve first in one direction and then another as would be expected of a meandering stream course; its trend is roughly at right angles to the ancient shore line in contrast to that of an offshore bar. These differences in two types of shoestring sands should make it obvious that familiarity with the characteristics of the various types of topographic features which evolve into shoestring sands is necessary for most economical exploitation of their oil and gas possibilities.

The East Texas oil field has produced more oil than any other field in the United States. It is but one of many examples of a stratigraphic trap produced by the sealing of an old erosion surface. Oil is produced from the Cretaceous Woodbine sand. This formation is truncated by an erosion surface which has many irregularities upon it. Later this erosion surface was buried and sealed by deposition of the Eagle Ford clay and Austin chalk. The hills upon this buried landscape are now most productive of oil.

Other applications of geomorphology. The uses discussed above represent some of the more striking practical applications of geomorphology, but there are other fields in which geomorphology may be useful. Soil maps are

Fig. 22.18. Photo of a road cut in which the much greater erosibility of the plastic *B* horizon in a soil profile is shown by the change from sheet erosion to gullying. (Photo by Engineering Experiment Station, Purdue University.)

to a considerable degree topographic maps, and the differentiation of the various members of any soil series rests fundamentally upon the different topographic conditions under which each member of the soil series developed. Modern beach engineering (Mason, 1950; Krumbein, 1950), to be successful, must be based upon an appreciation of the processes of shore-line development. The problem of soil erosion (Brown, 1950; Peterson, 1950) is essentially a problem involving recognition and proper control of such geomorphic processes as sheetwash erosion, gulleying, mass-wasting, and stream erosion. Severity of erosion is not determined by the angle of slope alone. It may not be serious on steep slopes where those slopes are underlain by permeable materials, and it may be serious on slight slopes where they are on impermeable materials. The related problem of land classification also entails an appreciation of varying types of terrains and the best uses that may be made of them.

Other fields might be mentioned in which an appreciation of geomorphic features is important, but those discussed should be sufficient to show that a thorough understanding of geomorphic principles, along with a knowledge of the geomorphic history of specific areas, may contribute significantly to the solution of problems in applied geology.

REFERENCES CITED IN TEXT

Adams, G. I. (1927). Bauxite deposits of the southeastern states, *Econ. Geol., 22,* pp. 615–620.

Bass, N. W. (1936). Origin of the shoestring sands of Greenwood and Butler counties, Kansas, *Kansas Geol. Survey, Bull. 23,* pp. 69–126.

Benson, W. N. (1946). Landslides and their relation to engineering in the Dunedin district, New Zealand, *Econ. Geol., 41,* pp. 328–347.

Bramlette, M. N. (1936). Geology of the Arkansas bauxite region, *Arkansas Geol. Survey, Inf. Circ. 8,* pp. 4–31.

Brooks, A. H. (1921). The use of geology on the western front, *U. S. Geol. Survey, Bull. 128-D,* pp. 85–124.

Brown, C. B. (1950). Effects of soil conservation, pp. 380–406, in *Applied Sedimentation,* edited by Parker D. Trask, John Wiley and Sons, New York.

Bryan, Kirk (1929). Geology of reservoir and dam sites, *U. S. Geol. Survey, Water Supply Paper 597,* pp. 1–33.

Burwell, E. B., Jr., and B. C. Moneymaker (1950). Geology in dam construction, *Geol. Soc. Am., Berkey Vol.,* pp. 37–39.

Bybee, H. P. (1938). Possible nature of limestone reservoirs in the Permian basin, *Am. Assoc. Petroleum Geol., Bull. 22,* pp. 915–924.

Ekblaw, G. E., and R. E. Grim (1936). Some geological relations between the constitution of soil materials and highway construction, *Illinois Geol. Survey, Rept. Invest., 42,* 16 pp.

Erdmann, C. E. (1943). Application of geology to the principles of war, *Geol. Soc. Am., Bull. 54,* pp. 1169–1194.

Fielder, A. G., and S. S. Nye (1933). Geology and groundwater resources of the Roswell artesian basin, New Mexico, *U. S. Geol. Survey, Water Supply Paper 639,* pp. 179–189.

Goldich, S. S., and H. R. Bergquist (1947). Aluminous lateritic soil of the Sierra de Bahoruco area, Dominican Republic, W. I., *U. S. Geol. Survey, Bull. 953-C,* pp. 57–77.

Goldich, S. S., and H. R. Bergquist (1948). Aluminous lateritic soil of the Republic of Haiti, W. I., *U. S. Geol. Survey, Bull. 954-C,* pp. 99–109.

Guyton, W. F. (1946). Artificial recharge of glacial sand and gravel with filtered river water at Louisville, Kentucky, *Econ. Geol., 41,* pp. 644–658.

Hamilton, D. K. (1948). Some solutional features of the limestone near Lexington, Kentucky, *Econ. Geol., 43,* pp. 39–52.

Harrison, J. B. (1910). The residual earths of British Guiana commonly termed "laterite," *Geol. Mag., 7,* pp. 553–562.

Hewett, D. F., G. W. Stose, F. J. Katz, and H. D. Miser (1918). Possibilities for manganese ore on certain undeveloped tracts in the Shenandoah Valley, Virginia, *U. S. Geol. Survey, Bull. 660-J,* pp. 280–281.

Horberg, Leland (1945). A major buried valley in east-central Illinois and its regional significance, *J. Geol., 53,* pp. 349–359.

Horberg, Leland (1950). Bedrock topography of Illinois, *Illinois Geol. Survey, Bull. 73,* pp. 67–72.

Howard, W. V. (1928). A classification of limestone reservoirs, *Am. Assoc. Petroleum Geol., Bull. 12,* pp. 1153–1161.

Hunt, C. B. (1950). Military geology, *Geol. Soc. Am., Berkey Vol.,* pp. 295–327.

Jenkins, O. P. (1935). New technique applicable to the study of placers, *Calif. J. Mines Geol., 31,* pp. 143–200.

Jenkins, O. P., and W. Q. Wright (1934). California's gold-bearing Tertiary channels, *Eng. Mining J., 135,* pp. 497–502.

Krumbein, W. C. (1950). Geological aspects of beach engineering, *Geol. Soc. Am., Berkey Vol.,* pp. 195–223.

Legget, R. F. (1939). *Geology and Engineering,* pp. 304–352, McGraw-Hill Book Co., New York.

Leith, C. K., and W. J. Mead (1911). Origin of the iron-ores of central and northeastern Cuba, *Trans. Am. Inst. Mining Engrs., 42,* pp. 90–102.

Leverson, A. I. (1934). Relation of oil and gas pools to unconformities in the mid-continent region, pp. 761–784, in *Problems of Petroleum Geology,* American Association of Petroleum Geology.

Lindgren, Waldemar (1911). The Tertiary gravels of the Sierra Nevada, *U. S. Geol. Survey, Profess. Paper 73,* pp. 9–81.

Lippert, T. W. (1950). Cerro Bolivar—saga of an iron ore crisis averted, *J. Metals, 188,* pp. 222–236.

Locke, Augustus, and Paul Billingsley (1930). Trend of ore hunting in the United States, *Eng. Mining J., 130,* pp. 565–566 and 609–612.

Lugn, A. L. (1935). The Pleistocene geology of Nebraska, *Nebraska Geol. Survey, Bull. 10,* 2nd ser., pp. 35–37 and 201–209.

McKinstry, H. E. (1948). *Mining Geology,* pp. 219–232, Prentice-Hall, New York.

Malcolmson, J. W. (1902). The Sierra Mojada, Mexico, and its ore deposits, *Trans. Am. Inst. Mining Engrs., 32,* p. 130.

Mason, M. A. (1950). Geology in shore-line control problems, pp. 276–290, in *Applied Sedimentation,* edited by Parker D. Trask, John Wiley and Sons, New York.

REFERENCES

Mead, W. J. (1915). Occurrence and origin of the bauxite deposits of Arkansas, *Econ. Geol., 10,* pp. 28–54.

Mohr, E. C. J. (1944). *The Soils of Equatorial Regions with Special Reference to the Netherlands East Indies,* p. 96, J. W. Edwards, Ann Arbor.

Nikiforoff, C. C. (1943). Introduction to paleopedology, *Am. J. Sci., 241,* pp. 194–200.

Norris, S. E. (1948). The water resources of Montgomery County, Ohio, *Ohio Water Resources Board, Bull. 12,* 100 pp.

Pecora, W. T., and S. W. Hobbs (1942). Nickel deposits near Riddle, Douglas County, Arizona, *U. S. Geol. Survey, Bull. 931-I,* pp. 211–212.

Pecora, W. T. (1944). Nickel-silicate and associated nickel-cobalt-manganese-oxide deposits near São José do Tocantíns, Goiáz, Brazil, *U. S. Geol. Survey, Bull. 935-E,* pp. 261–283.

Peterson, H. V. (1950). The problem of gullying in western valleys, pp. 407–434, in *Applied Sedimentation,* edited by Parker D. Trask, John Wiley and Sons, New York.

Putnam, W. C., and R. P. Sharp (1940). Landslides and earthflows near Ventura, southern California, *Geog. Rev., 30,* pp. 591–600.

Rich, J. L. (1923). Shoestring sands of eastern Kansas, *Am. Assoc. Petroleum Geol., Bull. 7,* pp. 103–113.

Rich, J. L. (1938). Shorelines and lenticular sands as factors in oil accumulation, pp. 230–239, in *The Science of Petroleum,* Oxford University Press, London.

Schaefer, E. J., G. W. White, and D. W. Van Tuyl (1946). The ground-water resources of the glacial deposits in the vicinity of Canton, Ohio, *Ohio Water Resources Board, Bull. 3,* pp. 20–26.

Schmedeman, O. C. (1948). Caribbean aluminous ores, *Eng. Mining J., 149,* June number, pp. 78–82.

Schmitt, Harrison (1939). Outcrops of ore shoots, *Econ. Geol., 34,* pp. 654–673.

Simpson, E. S. (1949). Buried preglacial groundwater channels in the Albany-Schenectady area in New York, *Econ. Geol., 44,* pp. 713–720.

Singewald, Q. D. (1938). Bauxite deposits at Gant, Hungary, *Econ. Geol., 33,* pp. 730–736.

Singewald, Q. D. (1942). Stratigraphy, structure, and mineralization in the Beaver-Tarryall area, Park County, Colorado, *U. S. Geol. Survey, Bull. 928-A,* pp. 37–42.

Sweet, H. S., and K. B. Woods (1947). Mapcracking in concrete pavements as influenced by soil textures, *Purdue Eng. Expt. Sta., Highway Research Reprint 33,* pp. 286–301.

Vickery, F. P. (1927). The interpretation of the physiography of the Los Angeles coastal belt, *Am. Assoc. Petroleum Geol., Bull. 11,* pp. 417–424.

Westerveld, J. (1937). The tin ores of Banca, Billiton and Singkep, Malay archipelago —a discussion, *Econ. Geol., 32,* pp. 1019–1041.

Wisser, Edward (1927). Oxidation subsidence at Bisbee, Arizona, *Econ. Geol., 22,* pp. 761–790.

Woolnough, W. G. (1930). The influence of climate and topography in the formation and distribution of products of weathering, *Geol. Mag., 67,* pp. 123–132.

Additional References

Cross, W. P. (1949). The relation of geology to dry-weather stream flow in Ohio, *Trans. Am. Geophys. Union, 30,* pp. 563–566.

George, W. O. (1948). Development of limestone reservoirs in Comal County, Texas, *Trans. Am. Geophys. Union, 29,* pp. 503–510.

Gester, G. C., and John Galloway (1933). Geology of the Kettleman oil field, California, *Am. Assoc. Petroleum Geol., Bull. 17,* pp. 1161–1193.

Moneymaker, B. C. (1948). Some broad aspects of limestone solution in the Tennessee Valley, *Trans. Am. Geophys. Union, 29,* pp. 93–96.

Price, P. H., and H. P. Woodward (1942). Geology and war, *Am. Assoc. Petroleum Geol., Bull. 26,* pp. 1832–1838.

Stearns, H. T. (1942). Hydrology of volcanic terrains, *Physics of the Earth,* Pt. IX, Hydrology, pp. 678–702, McGraw-Hill Book Co., New York.

Author Index

591

Subject Index

Marine erosion, processes, 47, 432, 433
Marine plains, 178, 190, 191
Mass-wasting, 36, 44, 60
  factors favoring, 46, 60
  geomorphic significance, 84, 95, 103
  types, 45, 84
Mass-wasting as an engineering problem,
    571, 572, 573
Mathematical analysis of topographic
    maps, 529, 530, 531
Matterhorn, Switzerland, 373
Mature topography, arid cycle, 291
  characteristics, 137
  fluvial cycle, 131, 137, 138
  karst cycle, 351
  periglacial cycle, 413
  shore-line cycle, 454, 455, 457
Maturity, stage in geomorphic cycle, 20,
    137, 138
Mauna Loa, eruption of, 490
Meander, 130, 145
  entrenched, 145, 146, 147
  incised, 145, 146
  inclosed, 145, 146
  ingrown, 145, 147
  intrenched, 145
Meander bar, 166, 168
Meander belt deposits, 168
Meander core, 130
Meander cutoff, 130, 169
Meander neck, 130
Meander scar, 130, 131, 167
Meander scroll, 168
Meander sweep, 131, 147
Megathermal Phase, 363
Mendips, 134
Mesas, 136
Mesotil, 81, 422
Meteor Crater, Arizona, 516
Meteorite craters, 515
  diagnostic features, 517, 518
  examples, 516, 517
Meteorites, as geomorphic agents, 53
Miami soil series, 78
Microclimatology, relation to geomorphic
    processes, 58
Microgeomorphology, 537
Mid-Atlantic Ridge, 477, 478, 479
  origin, 478
  topography of, 477, 478

Middle-latitude deserts, 282
Military applications of geomorphology,
    581
Mindanao Trench, 476
Mineral deposits, topographic expression,
    560
Mineral stability, 41, 42
  relation to weathering, 41, 42
Mineralization slump, 561
Misfit rivers, 156
Mission fault, California, 263, 265
Mississippi River, 147, 148, 149
  delta, 149
  description of valley, 167
  history of lower, 147, 148, 149
  Pleistocene history, 147, 148, 149, 402
  pre-Pleistocene bedrock valley, 402
  valley terraces, 148, 149, 415
Missouri River, Pleistocene history, 398
Mitchell Plain, Indiana, 186, 318, 327
Mogotes, 334
Monadnocks, 181, 182
Monoclinal blocks, 259
Mono Craters, 499
Monocyclic landscape, 22, 23
Monterey submarine canyon, 470
Monument, 373
Moraine, 374
  ablation, 386
  dump, 389
  end, 374, 375
  ground, 374, 375, 376, 387, 388
  interlobate, 389
  kame, 379, 389
  lateral, 374, 375, 379, 381
  lodge, 389
  push, 389
  recessional, 374, 389
Morehouse Lowland, 403
Morphogenetic regions, 60
  definition, 63
  outline of, 64
Morphographic maps, 534
Morvan, 194
Mosaic, aerial, 535, 536
Mosores, 181
Moulin, 379
Mountain glaciation, 367, 380
  cycle of, 368, 372
  dating, 380